INSTRUMENTATION
DESIGN STUDIES

INSTRUMENTATION DESIGN STUDIES

Ernest O. Doebelin

CRC Press
Taylor & Francis Group
Boca Raton   London   New York

CRC Press is an imprint of the
Taylor & Francis Group, an **informa** business

CRC Press
Taylor & Francis Group
6000 Broken Sound Parkway NW, Suite 300
Boca Raton, FL 33487-2742

© 2010 by Taylor and Francis Group, LLC
CRC Press is an imprint of Taylor & Francis Group, an Informa business

No claim to original U.S. Government works

Printed in the United States of America on acid-free paper
10 9 8 7 6 5 4 3 2 1

International Standard Book Number: 978-1-4398-1948-7 (Hardback)

Library of Congress Cataloging-in-Publication Data

Doebelin, Ernest O.
  Instrumentation design studies / Ernest Doebelin.
    p. cm.
  Includes bibliographical references and index.
  ISBN 978-1-4398-1948-7
    1. Engineering instruments--Design and construction. 2. Systems engineering. 3. Automatic control. I. Title.

  TA165.D547 2010
  620'.0042--dc22
                                                      2009028747

Visit the Taylor & Francis Web site at
http://www.taylorandfrancis.com

and the CRC Press Web site at
http://www.crcpress.com

Contents

Preface

All my earlier books (*Dynamic Analysis and Feedback Control* [1962]; *Measurement Systems*, five editions [1966–2004]; *System Dynamics* [1972]; *System Modeling and Response* [1980]; *Control System Principles and Design* [1985]; *Engineering Experimentation* [1995]; and *System Dynamics: Modeling, Analysis, Simulation, Design* [1998]) were designed as engineering textbooks to be used as aids in teaching undergraduate and graduate courses in the areas of system dynamics, measurement, and control. They were thus organized to progress in a carefully designed sequence of chapters, which led the student from simple basic concepts toward progressively more comprehensive and practical views of the field under study. As is usual in textbooks, each chapter included homework problems designed to stimulate students' personal understanding of important concepts. While these books were originally intended for teaching purposes in engineering schools, their judicious blending of useful theory with practical hardware and design considerations made them appealing also to engineering practitioners who wanted to update their education in specific areas.

This book is still devoted to the same general areas (system dynamics, measurement, and control) but departs from the textbook format to address the needs of practicing engineers working in those fields, which are sometimes collected under the heading "instrumentation." As is common with this type of book, homework problems are not included. While all the chapters certainly have a common interest in the overall field, each is largely self-contained in addressing an important subarea of the subject. As such, they are readily accessible to readers with a specific interest in improving their expertise in the chosen topic. While the book is not designed for a specific academic course, it could be profitably used as additional enrichment reading for any number of specific courses, or possibly for a single seminar-type experience.

Central to all the chapter treatments is the close integration and widespread use of appropriate software, such as MATLAB®/Simulink® (dynamic system simulation), Minitab® (statistical tools), and Mathcad (general engineering computation). To facilitate readers' comprehension of software applications, detailed appendices in the form of sharply focused and user-friendly *mini-manuals* are provided for MATLAB/Simulink and Minitab. (Most Mathcad applications are sufficiently self-explanatory and user-friendly that additional explanation is not warranted.) While engineering software packages provide extensive printed manuals and/or online help, in my experience these aids are too voluminous and unfocused to allow efficient use for specific application areas, such as instrumentation. These appendix manuals are specifically addressed to the text's application areas, and thus can be used by the reader in an efficient and time-saving manner.

This new book is largely based on a series of homework projects that I developed over many years for an advanced measurement course/lab populated by a mix of engineering seniors and graduate students. This experience was valuable in showing me the best ways to present the material, which was continuously revised over the years. The homework project manual included extensive notes that led the student through the particular topic and required certain calculations and explanations at each of the steps in the development. In adapting this manual to the needs of this book, I replaced the homework sections with a complete presentation and explanation of the solutions required of the students. I also adapted the format to meet the needs of the new audience, and augmented the technical material with any new developments that I was familiar with. I hope this book will be a useful and interesting learning tool for engineers in the instrumentation field.

Ernest O. Doebelin

MATLAB® is a registered trademark of The MathWorks, Inc. For product information, please contact:

The MathWorks, Inc.
3 Apple Hill Drive
Natick, MA 01760-2098 USA
Tel: 508 647 7000
Fax: 508-647-7001
E-mail: info@mathworks.com
Web: www.mathworks.com

Author

Ernest O. Doebelin was born in Germany in 1930, but left for the United States in 1933. His elementary and secondary schooling were in the public schools of Cleveland, Ohio, and North Ridgeville, Ohio. He received his BSc in mechanical engineering (1952) from the Case Institute of Technology, Cleveland, Ohio, and his MSc and PhD (in 1954 and 1958, respectively) in mechanical engineering from the Ohio State University, Columbus. While working on his PhD, he was a full-time instructor, and under the guidance of the department chairman, S.M. Marco, taught many of the undergraduate mechanical engineering courses. This experience contributed to his lifelong interest in the entire mechanical engineering curriculum, and gave his subsequent teaching and writing in more restricted areas a "generalist" flavor.

As an assistant professor, he was assigned to develop the curricular area of *instrumentation and control*, which in those early years consisted of only a single course. Over the years, he developed and taught eight courses in the areas of *system dynamics, measurement, and control*, ranging from the sophomore to the PhD level. Seven of these courses had laboratories which he designed, developed, and taught. Textbooks for all these courses were written along with comprehensive lab manuals and software mini-manuals. In a career that was focused on teaching, Prof. Doebelin was fortunate to win many awards. These included several departmental, college, and alumni recognitions, and the university-wide distinguished teaching award (five selectees yearly from the entire university faculty). The American Society for Engineering Education (ASEE) also presented him with the Excellence in Laboratory Instruction Award. After retirement in 1990, he continued to teach in lectures and labs, but for one quarter a year. He also worked on a volunteer basis at Otterbein College, Westerville, Ohio, a local liberal arts school, developing and teaching a course on *understanding technology*, as an effort to address the nationwide problem of technology illiteracy within the general population. As a further hobby after retirement, he has become a politics/economics junkie, focusing particularly on alternative views of globalization.

1

Introduction to Statistical Design of Experiments: Experimental Modeling of a Cooling System for Electronic Equipment

1.1 Introduction

The *statistical design of experiments* (DOE) is the subject of entire large books and academic courses. Its various techniques are widely practiced in industry and have achieved many successful practical applications. Many engineers have little or no familiarity with this important approach and the purpose of this chapter is essentially to raise your consciousness of this topic. The development of true expertise must of course depend on further study and practical experience. Hopefully this introduction will at least make you aware of the general approach so that you will consider it when facing new experimental projects as they arise. Widely available statistical software (such as the Minitab whose use is explained in Appendix B) makes the application of the methods much easier and quicker than was the case in earlier years. Because I was convinced of the importance of making these methods accessible to all undergraduate mechanical engineers, I included two chapters on these topics in a textbook published in 1995.* Chapter 2 of that book introduces general basic concepts to readers with no background in statistics while chapter 4 develops the methods of DOE. My idea was that the existing books and courses required so much time and effort that most engineers and students would not make this investment, so I tried in these two chapters to simplify and streamline the material by extracting what I thought were the essential ideas and methods. If in the future you want to go beyond what is presented in this short chapter, you might start with these two chapters since they will "get you going" in the shortest time. Of course, if you find that you use these methods regularly, you might go to the more detailed texts (or short courses offered by many companies and software suppliers) for deeper background. Chapters 2 and 4 of my 1995 textbook provide links to such resources.

---

* E.O. Doebelin, *Engineering Experimentation: Planning, Execution, Reporting*, McGraw-Hill, New York, 1995.

1.2 Basic Concepts

We now present, in a severely condensed (but hopefully still useful) form, the basic concepts of DOE. The problems dealt with can be described as follows. We have, say, some manufacturing process that produces a product or material that has one or more attributes associated with quality and/or cost. This quality parameter depends on several process variables that we are able to set within a certain range of values. The process is sufficiently complex that physical/mathematical modeling to reveal the relation of the quality parameter to the set of process variables has proven not possible or insufficiently accurate. We therefore propose to *run an experiment* in which we "exercise" the process by setting the process variables at several combinations of values and then measure the value of the quality parameter that results. We then analyze this data to develop a *mathematical model*, which predicts the effects of the process variables on the quality parameter. Many times such modeling allows us to find which variables are the most significant, and also the *optimum combination* of process variable settings; that is, one that maximizes quality or minimizes cost.

While the study of manufacturing processes is perhaps the application of most economic significance, DOE methods are directly applicable to other situations. For example, the NASA Johnson Space Center (Houston, Texas) ran experiments on the Space Shuttle's life-support system, which removes water vapor and carbon dioxide from the cabin atmosphere. The rate of removal of carbon dioxide was the quality parameter and the process variables were: temperature of a bed of absorbent material, partial pressure of water vapor in the inlet stream, partial pressure of carbon dioxide in the input stream, and total gas-flow rate. Physical/mathematical modeling of this system had not provided a good understanding of process behavior or reliable predictions of the effects of the process variables on CO_2 removal rate, so an experimental approach was undertaken. (More details of this application, including a complete set of real-world data and its analysis to provide a useful model are given at the end of this chapter.)

Finally, DOE methods are used for *computer experiments*, where the data are generated, not by a physical experiment but rather by a computer simulation.* For example, a finite element analysis (FEA) study of a machine part might be interested in the effects of various dimensions and material properties on the stress, deflection, or natural frequency. One can, of course, run such a simulation over and over with various combinations of input parameters in an attempt to find parameter values which minimize stress or deflection, or maximize natural frequency. Since each such run may be quite expensive, and the search for the optimum lacks much guidance as to "which way

---

* A. Rizzo, Quality engineering with FEA and DOE, *Mechanical Engineering*, May 1994, pp. 76–78.

to go," this approach requires many runs and thus may be quite inefficient. DOE methods allow us to choose a relatively small number of parameter combinations to run, formulate a model relating our quality parameter to dimensions and material properties, and then use this model to predict the optimum combination.

1.3 Mathematical Formulation

With the above background, we can see that all these applications can be thought of mathematically as a problem of finding a functional relation between a set of process variables and some quality parameter. In DOE parlance, the process variables are called *factors* and the quality parameter is called the *response* (y). Mathematically

$$y = b_0 + b_1 f_1(x_1) + b_2 f_2(x_2) + b_3 f_3(x_3) + \cdots \tag{1.1}$$

Here the f_i's are functions which can involve any of the process variables in any way. If we call the process variables x_a, x_b, x_c, ... then, for example, f_1 might be $\sqrt{x_a(x_b/x_c)}$. The standard methods of DOE require that the functional relation in Equation 1.1 be linear in the coefficients b, but the $f(x)$ functions can take *any* form. While the restrictions put on Equation 1.1 prevent the use of certain kinds of functions, this form is sufficiently versatile to meet most, but not all practical needs. The advantage realized by the restrictions is that the solution for the unknown b values is readily accomplished by routine computerized methods of linear algebra. An experiment consists of choosing the functional forms of the $f(x)$'s, running the experiment to get a numerical value for y (the dependent variable) that results from each set of x's, and then analyzing these data to find the numerical values of the b's. We have to use at least as many sets of x's as there are b's in our model if we want to get a solution (n linear equations in n unknowns). Usually we use *more* sets of x's, which makes the equation set *overdetermined* and requires use of least-squares solution methods, but these fortunately are also part of standard linear algebra. Each set of x's and the associated response y constitutes one run of our experiment.

While Equation 1.1 allows a very large variety of functions to be used, many useful applications employ a much more restricted class of functions. A major class of such applications is the so-called *screening experiment*. Here, we have a situation where we have identified, by using our familiarity with the physical process, a number (sometimes as large as 10 or 15) of process variables (*factors*) which might influence the quality parameter of interest. We want to run a frugal experiment that will narrow this rather long list down to a few factors that really matter, which we will then study in more detail. Such experiments often use only two values of each process variable, a high value and a low value. Since we generally know the allowable ranges of the process variables, we can choose these high and low values numerically for each process variable. (An approach to *multivariable* experimentation much used in science and engineering is to hold all variables except one at *constant* values and then change this one variable over some range, thus isolating the effect of this variable. Doing this in turn for each of the variables, we hope to discover useful relations. While such an approach is common and can lead to useful results, the whole premise of DOE is that a more efficient method lets all the variables change *simultaneously*.) Thus the next step in the DOE procedure is to define the *combinations* of variable settings that will be used; each such combination is called a *run* of the experiment. For example, a run of a four-factor screening experiment might be to set all four factors to their individual high values. Another run might be to set factors 1 and 2 at their high values and factors 3 and 4 at their low values. While one might use "common sense" to define the set of runs, more systematic and efficient ways are available.

1.4 Full-Factorial and Fractional-Factorial Experiments

If there are k factors and each is to be restricted to two values, it becomes clear that to explore all possible combinations will require an experiment of 2^k runs. Such an experiment is called a *full-factorial* type. When k gets large, a full-factorial experiment can be prohibitively expensive in time and money, so we sometimes use *fractional-factorial* experiments. These use a carefully chosen subset of the runs of the full factorial, reducing the amount of information we can glean, but also cutting the costs. From Equation 1.1, however, it is clear that to find, say, four b values, we must have at least four runs (four equations in four unknowns). Fractional-factorial experiments usually define their runs using an orthogonality principle. Our abbreviated presentation will not attempt to explain this concept, and fortunately, standard statistics software (such as Minitab) provides the desired run definitions. The most common screening experiment attempts to find only the so-called *main effects* of the factors. Then Equation 1.1 takes the simple form:

$$y = b_0 + b_1 x_1 + b_2 x_2 + b_3 x_3 + \cdots \tag{1.2}$$

where the x's are now the factors (independent process variables) themselves. That is, we seek only the linear effects of the individual factors. This simple model has some theoretical foundation in that any smooth nonlinear function can be approximated (for small changes away from some chosen operating point) by linear terms. (For $y = f(x)$, the tangent line to the curve and for $z = f(x, y)$, the tangent plane to the surface give a geometrical interpretation.)

Sometimes, the model will benefit from the so-called *interaction terms* such as $b_i x_1 x_2$ (two-factor interaction) or $b_i x_1 x_2 x_3$ (three-factor interaction), with interactions higher than two-factor being very rarely used. If Equation 1.2 were to be augmented with higher powers of factors, such as $b x_1^2$, we would find that the analysis software would fail; to deal with such terms we would need a screening experiment which uses *three* settings (high, medium, low) for each factor, which expands the scope and cost of the experiment, but is sometimes necessary. The intuitive reason for this behavior is that *two points can only determine a straight line; it takes three to allow curvature.*

An important consideration in choosing between full-factorial and fractional-factorial experiment designs is the issue of *confounding*. In a full-factorial experiment we are able to distinguish both main effects and interactions. This capability is lost, to some extent, in fractional-factorial experiments; the main effects and some interactions are said to be *confounded*. The *degree* of confounding is given by the *resolution level* of the design; common designs being designated as Resolution III, Resolution IV, or Resolution V. Resolution III designs have the smallest number of runs, but can only isolate the *main effects* of the factors; interaction terms cannot be reliably identified. Resolution IV designs require more runs, but can find main effects and two-way interactions. Higher order (three-way, four-way, etc.) interactions are confounded and thus not identifiable. Resolution V designs can find main effects, two-way interactions and three-way interactions. Since three-way interactions are not common, most fractional-factorial designs use either Resolution 3 or 4. See the appendix material on Minitab for further details on this topic.

1.5 Run-Sequence Randomization

Another consideration is that of *run-sequence randomization*. If an experiment has, say, 8 runs, does it matter what *sequence* we use in actually performing these runs? There are a number of possible reasons for randomizing the sequence of the runs rather than blindly using a nominal sequence given by common sense, some book, or software. When we originally list the factors which we believe effect the process response variable, there are always some *other* factors that we consciously or unconsciously leave out of our list. Often these are subtle effects of the human operator who carries out the experiment. If operators are somewhat inexperienced, their task performance may improve, due to learning as they go from the first runs to the last; this can bias the results. On the other hand, operators may become fatigued as they work through the sequence causing poorer performance later in the sequence. If an apparatus is operated at several different power levels, as we change from one run to the next, we approach the new steady-state condition through some sort of transient. These transients usually approach the new steady state in an asymptotic fashion, thus if the run sequence goes monotonically from low power levels to high, we will always be somewhat below the intended steady state when we take our data, biasing the results in that direction. We need to be on the lookout for these and similar effects and adjust our run sequence to counteract the biases. Most software (such as Minitab) will give both a *nominal* run sequence and a randomized sequence. Unless there are good technical or economic reasons to the contrary, you should use the randomized sequence.

1.6 Validation Experiments

We will have a number of specific ways to judge the quality of the process model generated by the data analysis procedure, but one technique is highly recommended by expert practitioners: the *validation experiment*. When we set up our run sequence, if practical, we should split the total number of runs into two groups. The first, larger group, is used to find the process model (find the best functions and numerical b values). The smaller group (most experts recommend about 25% of the total runs) is not used to find the model. Rather, *after* we have found the model, we use these validation runs to see how well our model predicts the response variable for these process variable combinations *that were not used to define the model*. Such an approach seems intuitively desirable since we use our process models to predict behavior over the entire span of process variable values, not just those used in the model definition.

1.7 Example Experiment: Modeling an Electronics Cooling Process

Summarizing the procedure for developing a math model of some process, we note that there are basically two steps. We first have to choose the factors to be studied and define the set of experiment runs to be performed. We then take the raw data generated by our experiment and analyze it in some way to find the best values of the model coefficients and then judge the utility of the model. Rather than continuing with a general discussion of DOE methods, we now use a specific example to complete the introduction to these methods. This example treats an actual experimental apparatus that was designed and built to provide practical experience with a minimum of time and effort. While the electrical design of electronic equipment is of course the purview of electrical engineers, it is undertaken in parallel with cooling system design by mechanical engineers, since there is a significant interaction between

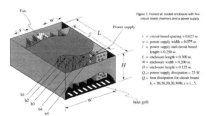

fan cooled enclosure analysis
using a first order method

Figure 1: Forced air cooled enclosure with five
circuit board channels and a power supply.

s = circuit board spacing = 0.025 m.
w = power supply width = 0.075 m.
l = power supply and circuit board length = 0.250 m.
L = enclosure length = 0.300 m.
W = enclosure width = 0.200 m.
H = enclosure height = 0.125 m.
Q_{ps} = power supply dissipation = 25 W.
Q_i = heat dissipation for circuit board b_i = 20,30,20,20,30W; i = 1...5.

Introduction

Most Electronics Packaging Engineers and Analysis Specialists have access to a wide range of high quality analysis tools. While a few computer programs use analytical methods, the majority are based on numerical analysis techniques. Although these programs are becoming easier to use, they typically require considerable skill, and they are certainly capable of producing results with a high order of accuracy. However, even with these elegant tools at his/her disposal, the Design Engineer should have a working knowledge of less sophisticated methods.[1] A first order analysis may, for example, predict temperatures sufficiently high so as to indicate that further design consideration is more necessary than in-depth thermal analysis. Additionally, many engineers work in small companies that cannot afford sophisticated analysis tools. Although these smaller companies have the option of hiring a consultant to perform the most complex calculations, it is in the interest of the responsible design engineer to perform a first order analysis in order to ascertain the magnitude of the thermal design problem.[2] An example of a first-order analysis of a fan cooled enclosure is the subject of the following paragraphs.

Gordon N. Ellison
Thermal Computations, Inc.

Gordon N. Ellison has more than twenty-five years experience in the field of thermal design and analysis of electronic equipment. He recently retired from Tektronix, Inc., where he was a Chief Scientist and Tektronix Fellow. He is currently the President of Thermal Computations, Inc., through which he provides thermal analysis/design consultation and custom software solutions. Ellison is the author of numerous publications including, "Thermal Computations for Electronic Equipment" Robert E. Krieger Publishing Co.

Footnote:
1. Application of first order methods is also recommended for problems such as conduction solutions using fixed temperature or convection/radiation boundary conditions. In addition, a first order method may often be used to determine a boundary condition value for input into a finite element or finite difference program.

2. The reader is cautioned that a first order analysis may not have sufficient resolution to permit detection of all problem areas.

FIGURE 1.1
Electronics Cooling magazine article title page. (From Ellison, G.N., *Electronics Cooling*, 1, 16, 1995. With permission.)

the two technologies. Figure 1.1 shows the beginning of a typical article in the trade journal called *Electronics Cooling*,* which is devoted entirely to this subject and runs useful articles on theoretical and experimental methods used in this area. This particular article discusses a relatively simple analytical technique useful for quick initial studies. Many other articles in this journal deal with various experimental methods. Our example will deal with a simple scheme for cooling electronics components inside a "box." Four variable-speed fans are used to inject cool air into the box at four locations. Heated air exits the box at three locations, two of which have adjustable vents which allow us to vary the flow area for the leaving air. Baffles located at several stations along the airflow path may be positioned at different angular attitudes, to direct the local air stream into desired directions.

While we would certainly be interested in *temperature* at various locations, taking temperature as the dependent variable is not desirable for a demonstration experiment because the transients in going from one run to the next would be too slow to carry out all the runs in a reasonable time. Instead, since we are here interested mainly in explaining DOE methods (rather than heat

---

* G.N. Ellison, Fan cooled enclosure analysis using a first order method, *Electronics Cooling*, 1(2), October 1995, 16–19. www.electronics-cooling.com.

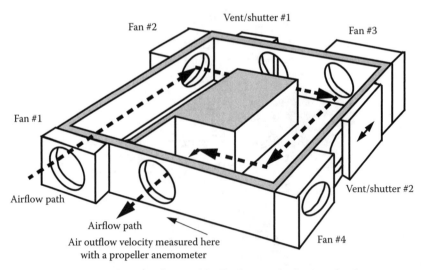

There are no heating elements. The top of the "box" is completely covered with a transparent plastic sheet. Rotatable baffles (not shown) can be located at various points along the airflow path and oriented at various angles with the airflow.

FIGURE 1.2
Apparatus for demonstration experiment on electronic cooling.

transfer concepts) I chose to make the *exit flow velocity* the dependent variable since its transients are only a few seconds long. This flow velocity is quickly and easily measured with a propeller type of anemometer which has a digital readout in convenient units, such as ft/min. The experiment factors will be the individual speeds of the four fans and the area settings of the two airflow exits, giving a total of six independent variables. (The baffles were not present in this experiment; I chose to not study their effect at this time.) The speed of the fans is adjusted with a potentiometer supplied with each fan while the speed measurement relies on a pulse generator built into each fan; it gives two pulses per fan revolution. Connecting the output of the pulse generator of each fan to a digital counter/timer gives convenient RPM readings. The two vent areas are measured by the location of a shutter which can be closed or opened manually. Figure 1.2 shows a simplified view of this apparatus.

1.8 Using Minitab® to Design the Experiment and Then Analyze the Results

I decided to use three levels (low, medium, and high) for each of the six factors (four fan speeds and two vent area settings). In a practical application, we would know what the allowable ranges of the factors were, so we could then choose numerical values for the three levels of each factor. If we wanted to run all possible combinations of six factors at three levels each (a full-factorial experiment), this would require $3^6 = 729$ runs, clearly impractical. In choosing fractional-factorial experiment designs, we can appeal to statistics textbooks and also statistics software. In the Minitab software available to us, of the various designs offered there, many engineers use the so-called Taguchi orthogonal arrays. Genichi Taguchi is a Japanese engineer who revolutionized Japanese (and later worldwide) manufacturing process experimentation. At this point you may want to take a look at Appendix B, where I have provided a condensed manual for Minitab which allows one to use this software for common engineering applications with a minimum of time and effort. In Minitab you can access Taguchi designs by using the following sequence:

STAT>DOE>TAGUCHI>CREATE TAGUCHI DESIGN   then

select "3-level design" and "6 factors"   then

click on "designs" (don't select "add a signal factor for dynamic characteristics")   then

click on "OK" and then "options"   then

select "show design in worksheet" then "OK" and again "OK"

A table will then appear on the worksheet which shows all the runs of the experiment as an array of rows (factor values) and columns (one column for each factor); the six factors are called A, B, C, D, E, and F. This experiment has 27 runs and is shown in Table 1.1. It uses –1, 0, 1 to denote the low, medium, and high values of a factor. This is the way most texts "code" the factor values. Each of the 27 rows represents one run of the experiment. In Minitab, the default coding method uses 1 for low, 2 for medium, and 3 for high, as shown in Table 1.2. Tables 1.1 and 1.2 contain identical experiment plans but use two different coding schemes; I show both tables so that you become familiar with the two coding methods in common use. My own preference is not to use coding at all; we just enter the *actual* numerical values of the variables. This retains more of a physical feel for the data, saves the time required for coding, and may reduce errors by eliminating one step in the procedure.

We earlier mentioned the concept of confounding and resolution level as it related to our choice of the "size" of the experiment. Minitab does not give the resolution level of Taguchi designs, but such information is available from a 1988 text by Ross.[*] For the experiment given above (6 factors, 3 levels, 27 runs) the resolution level is such that this experiment is intended to reliably find only the main effects, not any interactions. In fact, the text states on page 83, "Taguchi does not place much emphasis on the confounding that exists in a low-resolution experiment …." "Taguchi views interactions

---

[*] P.J. Ross, *Taguchi Techniques for Quality Engineering*, McGraw-Hill, New York, 1988, p. 233.

TABLE 1.1

Taguchi Experiment with 27 Runs, Coded –1 0 1

Run	A	B	C	D	E	F
1	–1	–1	–1	–1	–1	–1
2	–1	–1	–1	–1	0	0
3	–1	–1	–1	–1	1	1
4	–1	0	0	0	–1	–1
5	–1	0	0	0	0	0
6	–1	0	0	0	1	1
7	–1	1	1	1	–1	–1
8	–1	1	1	1	0	0
9	–1	1	1	1	1	1
10	0	–1	0	1	–1	0
11	0	–1	0	1	0	1
12	0	–1	0	1	1	–1
13	0	0	1	–1	–1	0
14	0	0	1	–1	0	1
15	0	0	1	–1	1	–1
16	0	1	–1	0	–1	0
17	0	1	–1	0	0	1
18	0	1	–1	0	1	–1
19	1	–1	1	0	–1	1
20	1	–1	1	0	0	–1
21	1	–1	1	0	1	0
22	1	0	–1	1	–1	1
23	1	0	–1	1	0	–1
24	1	0	–1	1	1	0
25	1	1	0	–1	–1	1
26	1	1	0	–1	0	–1
27	1	1	0	–1	1	0

as being unimportant ...." If we wanted to first run a more frugal experiment, we might opt for two levels rather than three, since the main reason for using 3 is the capability for detecting and modeling *nonlinear* effects. For initial experimentation, linear effects are often preferred. In this case we might use a *standard* fractional-factorial (rather than Taguchi) design, as shown in Figure B.30 of the Minitab manual in Appendix B. There we see that for 6 factors and 2 levels, we can get a resolution IV design with only 16 runs, and this resolution level allows identification of both main effects and two-way interactions. Table 1.3 shows the layout of this experiment.

A Taguchi design for six factors and two levels of each factor is easily found from Minitab as we did above for the three-level experiment. We find that we have a choice among four possible designs, ranging from 8 to 32 runs. Of these four, we will display the smallest and the largest. Table 1.4 shows

TABLE 1.2

Taguchi Experiment with 27 Runs, Coded 1 2 3

Run	A	B	C	D	E	F
1	1	1	1	1	1	1
2	1	1	1	1	2	2
3	1	1	1	1	3	3
4	1	2	2	2	1	1
5	1	2	2	2	2	2
6	1	2	2	2	3	3
7	1	3	3	3	1	1
8	1	3	3	3	2	2
9	1	3	3	3	3	3
10	2	1	2	3	1	2
11	2	1	2	3	2	3
12	2	1	2	3	3	1
13	2	2	3	1	1	2
14	2	2	3	1	2	3
15	2	2	3	1	3	1
16	2	3	1	2	1	2
17	2	3	1	2	2	3
18	2	3	1	2	3	1
19	3	1	3	2	1	3
20	3	1	3	2	2	1
21	3	1	3	2	3	2
22	3	2	1	3	1	3
23	3	2	1	3	2	1
24	3	2	1	3	3	2
25	3	3	2	1	1	3
26	3	3	2	1	2	1
27	3	3	2	1	3	2

the most frugal Taguchi experiment supplied by Minitab (8 runs). This is close to the *absolute* most frugal experiment possible when there are six factors. That is, it takes at least seven equations (runs) to solve for six factor coefficients and the one constant in the process model. Table 1.5 shows the 32-run Taguchi model. Careful examination shows that this model *replicates* the 8-run experiment four times. That is, it runs that experiment four times. Ross (1988) tells us that this experiment has resolution such that both main effects and two-way interactions can be identified.

When we *repeat* (replicate) an experiment in this way, we get information on the effects of process variables that we have *ignored* in our model, since these effects have not been controlled in any way and thus are free to vary as they wish and thus affect the response variable (*y*). Also, even if there were

TABLE 1.3

Fractional-Factorial Design with Resolution
Level IV

Run	A	B	C	D	E	F
1	1	1	1	−1	1	−1
2	−1	−1	1	1	1	−1
3	1	1	−1	1	−1	−1
4	−1	1	1	−1	−1	−1
5	−1	−1	−1	−1	−1	−1
6	−1	1	1	1	−1	1
7	−1	1	−1	−1	1	1
8	1	1	1	1	1	1
9	1	1	−1	−1	−1	1
10	−1	−1	1	−1	1	1
11	1	−1	−1	−1	1	−1
12	1	−1	1	1	−1	−1
13	−1	−1	−1	1	−1	1
14	1	−1	1	−1	−1	1
15	−1	1	−1	1	1	−1
16	1	−1	−1	1	1	1

TABLE 1.4

Taguchi Experiment Design for 6 Factors,
8 Runs, and 2 Levels

Run	A	B	C	D	E	F
1	1	1	1	1	1	1
2	1	1	1	2	2	2
3	1	2	2	1	1	2
4	1	2	2	2	2	1
5	2	1	2	1	2	1
6	2	1	2	2	1	2
7	2	2	1	1	2	2
8	2	2	1	2	1	1

no effects from such un-modeled factors, replication will reveal *how well* we
are actually able to repeat all the settings of the modeled factors, and how
accurately we are able to *measure* all the factors and the response variable. All
these extraneous effects are called *noise* and the analysis software provides
results which indicate the severity of this noise, but such information is *only*
available if we replicate the experiment at least once. The use of replication
is an important nuance of DOE, but because we are limiting our discussion
to the basic essentials, we will only make you aware of it, not pursue the
details.

TABLE 1.5

Taguchi 32-Run Design with Quadruple
Replication

Run	A	B	C	D	E	F
1	1	1	1	1	1	1
2	1	1	1	1	1	1
3	1	1	1	1	1	1
4	1	1	1	1	1	1
5	1	1	1	2	2	2
6	1	1	1	2	2	2
7	1	1	1	2	2	2
8	1	1	1	2	2	2
9	1	2	2	1	1	2
10	1	2	2	1	1	2
11	1	2	2	1	1	2
12	1	2	2	1	1	2
13	1	2	2	2	2	1
14	1	2	2	2	2	1
15	1	2	2	2	2	1
16	1	2	2	2	2	1
17	2	1	2	1	2	1
18	2	1	2	1	2	1
19	2	1	2	1	2	1
20	2	1	2	1	2	1
21	2	1	2	2	1	2
22	2	1	2	2	1	2
23	2	1	2	2	1	2
24	2	1	2	2	1	2
25	2	2	1	1	2	2
26	2	2	1	1	2	2
27	2	2	1	1	2	2
28	2	2	1	1	2	2
29	2	2	1	2	1	1
30	2	2	1	2	1	1
31	2	2	1	2	1	1
32	2	2	1	2	1	1

Returning now to our example, I chose the three levels of fan speed to be 1200, 2400, and 3600 rpm. These speeds correspond to pulse rates (as measured by the counter/timer) of 40, 80, and 120 pulses per second. We could use either the RPM values or the pulse rates as the numbers entered into our experiment matrix; I chose the pulse rates. When we shortly show the actual numbers they will not be exactly 40, 80, and 120 since in real-world experimentation, we *try* to set certain values but the apparatus and measurement method are never perfect. The three levels of vent area correspond to the vent being entirely closed (0 area), 1/2 open (0.5 area), and wide open (1.0 area). We could have measured the *actual* areas and used these numbers, but will instead use 0, 0.5, and 1.0, since these are proportional to the actual area. This kind of coding is widely used and perfectly legitimate. In fact, many engineers will always code such data as –1, 0, and +1, as seen in most of the statistics software when they display a certain type of experiment design. To see the correctness of this, note that for a three-level design, we always have a low value, a high value, and a medium value which is exactly half way between the high and the low. For example, if the high value were 58 and the low value 20, we must choose the medium value to be 39. We could then transform the actual physical variable x into a coded variable w by defining the coded variable to be $w = (x - 39)/19$. The analysis could then be carried out in terms of w and the results converted back to x values if we so wished. Such transformations could be done for all the factors.

Up to this time we have implicitly assumed that all our variables (dependent and independent) are available as *numerical* values. While the response variable must always be measurable in numerical terms, the factors need not be, and this is sometimes the case. Doebelin (1995)[*] treats a study of surface finish in machined parts, as determined by four factors: brand of cutting fluid (A, B, C), tool material (D, E, F), part material vendor (G, H, I), and brand of lathe used (J, K, L). That is, three brands or grades of cutting fluid could be used, three types of cutting tool would be suitable, the part material is nominally some aluminum alloy, but we can purchase it from three different mills, and finally we have lathes in our shop that come from three different manufacturers. The reference shows several ways of dealing with such situations, but we again, in the interest of brevity, only make you aware that such problems can be easily dealt with, but do not take the time to show you how.

Another technique which we will not take time to explain but want to at least make you aware of is the so-called response-surface method.[†] When a process exhibits a peak (local maximum) or valley (local minimum) for the quality variable (response), it may be desirable to find the combination of factor settings that produces this optimum response (maximum of quality or minimum of cost). Furthermore, if we can find factor combinations which make this peak or valley *broad* rather than *narrow*, such process settings produce what is called a *robust design*. That is, for a broad peak, if our process settings unavoidably wander somewhat from our desired values, the

[*] E.O. Doebelin, *Engineering Experimentation, Planning, Execution, Reporting*, McGraw-Hill, New York, 1995, pp. 239–254.

[†] E.O. Doebelin, *Engineering Experimentation, Planning, Execution, Reporting*, McGraw-Hill, New York, 1995, pp. 273–284.

performance will still be close to optimum. Methods to find such optima are called response-surface methods and entire books* have been written explaining the procedure. Doebelin (1995) gives a condensed treatment which explains the essence of the approach.

1.9 Multiple Regression: A General Tool for Analyzing Experiment Data and Formulating Models

Texts on DOE derive various types of experiment designs and explore concepts such as confounding and resolution levels. "Catalogs" of experiment designs are provided and one must choose a design from these lists. They then use the technique called analysis of variance (ANOVA) to analyze the data and reach certain kinds of conclusions. Even though the calculations provide the basic parameters that would be needed to express it, the mathematical model relating the response variable to the factors is often not much discussed or used. Instead, the conclusions offered are statements that a certain factor is *statistically significant* or insignificant.

The mathematical tool called *multiple regression* was developed without regard to the *design* of experiments; it was just a general tool which finds the "best" set of coefficients in a given linear math model, based on some measurements of the independent and dependent variables. One can use any number of runs and any number of levels, and the levels can be chosen *anywhere* within the physically possible range of the factors. As we indicated earlier, we *are* required to use only models which are *linear* in the *b* coefficients of the model, to keep the mathematics within the requirements of linear algebra tools, and there must be at least as many runs as there are unknown coefficients. In choosing the various runs of the experiment, we can use common sense concepts such as choosing factor combinations that "exercise" the entire space defined by the ranges of all the factors. What we mean by this is best seen by a simple example with only three factors. Each of these factors can be assigned an axis of an *xyz* coordinate system, and any run of the experiment defines a single point in this space. The ranges of each factor can also be plotted on this coordinate system, thus defining the "volume" in which any run-defining point must fall. Common sense then tells us that we should choose our runs so as to explore this volume as completely as is possible in the chosen number of runs. When there are more than three factors, we no longer have this convenient geometrical interpretation, but the idea of a "space" within which any run must lie is still useful.

---

* G.E.P. Box, *Empirical Model Building and Response Surfaces*, Wiley, New York, 1987.

TABLE 1.6

Minitab Worksheet with Electronics Cooling Data

	C1	C2	C3	C4	C5	C6	C7
	fan1	fan2	fan3	fan4	vent1	vent2	V
1	38.4	38.7	38.7	38.7	0.0	0.0	380
2	38.4	38.7	38.7	38.7	0.5	0.5	291
3	38.4	38.7	38.7	38.7	1.0	1.0	240
4	35.8	80.5	78.9	79.5	0.0	0.0	697
5	35.8	80.5	78.9	79.5	0.5	0.5	571
6	35.8	80.5	78.9	79.5	1.0	1.0	545
7	35.8	117.0	118.0	120.0	0.0	0.0	1011
8	35.8	117.0	118.0	120.0	0.5	0.5	825
9	35.8	117.0	118.0	120.0	1.0	1.0	819
10	81.0	38.9	79.7	124.0	0.0	0.5	693
11	81.0	38.9	79.7	124.0	0.5	1.0	618
12	81.0	38.9	79.7	124.0	1.0	0.0	695
13	80.7	81.6	124.0	39.5	0.0	0.5	768
14	80.7	81.6	124.0	39.5	0.5	1.0	770
15	80.7	81.6	124.0	39.5	1.0	0.0	778
16	80.4	120.0	40.0	81.7	0.0	0.5	811
17	80.4	120.0	40.0	81.7	0.5	1.0	663
18	80.4	120.0	40.0	81.7	1.0	0.0	770
19	117.0	38.6	121.0	80.0	0.0	1.0	774
20	117.0	38.6	121.0	80.0	0.5	0.0	845
21	117.0	38.6	121.0	80.0	1.0	0.5	784
22	121.0	82.5	40.5	120.0	0.0	1.0	740
23	121.0	82.5	40.5	120.0	0.5	0.0	782
24	121.0	82.5	40.5	120.0	1.0	0.5	646
25	118.0	124.0	80.5	40.3	0.0	1.0	862
26	118.0	124.0	80.5	40.3	0.5	0.0	858
27	118.0	124.0	80.5	40.3	1.0	0.5	789

Table 1.6 shows a Minitab worksheet with the actual measured data for our electronic cooling experiment. Note that we can give descriptive names to the factors (rather than the generic A, B, C, etc.) and that the three factor levels may vary a little from run to run, even though we intend for them to be identical. The next step is somehow to analyze this data to get a mathematical model relating the outlet airflow velocity (response) to the six experiment factors (four fan speeds and two vent area settings). In the DOE literature, two major methods are available for this task: ANOVA and multiple regression. In my 1995 text I tried to make the case* for concentrating

---

* E.O. Doebelin, *Engineering Experimentation, Planning, Execution, Reporting,* McGraw-Hill, New York, 1995, pp. 239–254.

on the regression approach rather than ANOVA. This is definitely an unconventional position, and would be questioned by many statisticians, but I still feel that it is valid for most engineers, who want some practical working tools with a minimum of learning time and effort. I feel that the regression approach is much easier to understand and handles most practical applications. The simplest version of regression is that which you have all used to fit the best straight line to a set of scattered data, using so-called least-squares methods. Here you are fitting a linear model to a set of measured data in such a way as to minimize the sum of the squares of the deviations of the fitted line from the actual data. Multiple regression is an extension of these least-squares methods to more complex models with more than one independent variable. Software such as Minitab performs these calculations with a minimum of effort on your part. While we do not need to have any statistical background to use multiple regression, when there are more than three factors, the choice of a *good* set of runs for our experiment using only common sense becomes much more difficult. Our approach is thus to use available statistical software or textbooks to choose the experiment design, and then use multiple regression to analyze the data.

1.10 Stepwise Regression

While we have already decided that we will use six factors in our model, it is often not clear at the beginning of a modeling study whether all the initially considered factors are really necessary. That is, we want the *simplest* models that *adequately* explain the data. *Stepwise regression* is a technique often used at these early stages of model-building to explore models of various degrees of complexity, so that we can choose the simplest models that meet our needs. With Table 1.6 entered into the Minitab worksheet, to access the stepwise regression routine, we enter

STAT>REGRESSION>REGRESSION>BEST SUBSETS

now enter for "response"   v
for "predictors":   fan1 fan2 fan3 fan4 vent1 vent2
for "predictors in all models"   leave blank
now click on "options"
now enter for "free predictors in all models"   Min1   Max   leave blank
now enter for "models of each size to print"   1
select "fit intercept"   (it is already selected as the default)

These entries request that Minitab find those predictor (factor) combinations that give the best 1-variable, 2-variable, 3-variable, 4-variable, 5-variable, and 6-variable models. That is, if you wanted, say, the best 4-variable model, *which* 4 of the 6 available variables should you use? When you run this procedure, Minitab displays a table showing the six models, together with some statistical "figures of merit" for each model.

Best Subsets Regression: V versus fan1, fan2, fan3, fan4, vent1, vent2

```
Response is V

                                          v  v
                             f  f  f  f   e  e
                             a  a  a  a   n  n
                             n  n  n  n   t  t
Vars  R-Sq  R-Sq(adj)   Cp       S   1  2  3  4   1  2
   1  30.4       27.6 362.5 148.61   X
   2  61.5       58.3 192.1 112.78   X  X
   3  79.0       76.2  97.4 85.129   X  X  X
   4  88.9       86.9  44.3 63.180   X  X  X  X
   5  93.2       91.6  22.4 50.513   X  X  X  X      X
   6  96.4       95.3   7.0 37.845   X  X  X  X   X  X
```

We see that the best one-variable model is that which uses fan2 as the only factor, however, as we might guess, this model would not do a very good job of fitting the data. Of the four "fit quality" parameters listed (R-sq, R-sq adj, C_p, S), the most useful is R-sq ("R square") the *coefficient of determination*.[*] For a model which perfectly fits the data, R^2 would equal 1.0, the highest possible value; values near 1.0 denote good models. We see that the 5-variable and 6-variable models have quite good R^2 values, so we might want to study them further. Note that the Best Subsets routine does *not* provide the actual model equations; we need to run separate regressions to get that information. To save time in this presentation, we will study just the 6-variable model. (The procedure is the same for any size model, so you will see how to get this more detailed information for the 5-variable model if you wanted to.)

We mentioned earlier the desirability of *validation experiments*. This can be implemented in our present situation by deleting from Table 1.6 about 25% of the runs (called *validation runs*), and then doing the model-fitting with the remaining runs. We then check the model obtained, by entering the validation runs into the model and comparing the model-predicted values with the actual measured values of response v. Let us decide to

---

[*] E.O. Doebelin, *Engineering Experimentation: Planning, Execution, Reporting*, McGraw-Hill, New York, 1995, pp. 201, 210–213.

use 7 runs as our validation experiment. The question is, *which* 7 runs? There is no routine procedure for picking these runs, but a choice must somehow be made. One possible criterion would be to *scatter* the 7 runs more or less evenly over the total space of the factors. If there were only three factors, the *xyz* geometrical view would facilitate this. For six factors it is not nearly so obvious, so we have to somehow choose one method from several which might seem equally reasonable. I decided to choose 7 runs which had *v* values somehow scattered over the full range of the measured *v* values (240–1011 ft/min). Of course there is still more than one way to do this, so my choice is not unique nor necessarily the best. The statistics texts which I have read do not give any advice on this question. Once we have picked these 7 runs, we need to move their data into columns 8 through 13 on the worksheet, putting the columns in the same left-to-right sequence as used for the original data set. (These new columns will be used by Minitab to compute the model predictions for these runs.) Table 1.7 shows this new Minitab worksheet.

To run any multiple regression in Minitab, use the *sequence* shown below, but of course enter parameters and values that relate to your specific application.

STAT>REGRESSION>REGRESSION

for "response" enter   v

for "predictors" enter the 6 factors

now click on "graph", then select   "normal plot of residuals"

"residuals versus fits"

then click on OK

now click on "options"   leave "weights" blank, select "fit intercept" (this is the constant in the model)

for "prediction intervals for new observation" enter c8 c9 c10 c11 c12 c13 (this asks for the 7 predicted v values)

for "confidence level" enter 95 (this is the default)

for "storage" select "fits" and " prediction limits"   then click on OK

now click on "results"   select "regression equation", "table of coefficients", and "R^2"   then click on OK

now click on "storage"   select "residuals", "coefficients", and "fits" click on OK, then OK again to start the calculations.

TABLE 1.7

Minitab Worksheet for Experiment with Validation Run

Run	fan1	fan2	fan3	fan4	vent1	vent2	v	fan1	fan2	fan3	fan4	vent1	vent2	v
1	38.4	38.7	38.7	38.7	0.5	0.5	291	38.4	38.7	38.7	38.7	0.0	0.0	380
2	38.4	38.7	38.7	38.7	1.0	1.0	240	35.8	80.5	78.9	79.5	0.0	0.0	697
3	35.8	80.5	78.9	79.5	0.5	0.5	571	35.8	117.0	118.0	120.0	0.5	0.5	825
4	35.8	80.5	78.9	79.5	1.0	1.0	545	81.0	38.9	79.7	124.0	1.0	0.0	695
5	35.8	117.0	118.0	120.0	0.0	0.0	1011	80.4	120.0	40.0	81.7	1.0	0.0	770
6	35.8	117.0	118.0	120.0	1.0	1.0	819	117.0	38.6	121.0	80.0	0.0	1.0	774
7	81.0	38.9	79.7	124.0	0.5	1.0	618	118.0	124.0	80.5	40.3	0.5	0.0	858
8	81.0	38.9	79.7	124.0	0.0	0.5	693							
9	80.7	81.6	124.0	39.5	0.0	0.5	768							
10	80.7	81.6	124.0	39.5	0.5	1.0	770							
11	80.7	81.6	124.0	39.5	1.0	0.0	778							
12	80.4	120.0	40.0	81.7	0.5	1.0	663							
13	80.4	120.0	40.0	81.7	0.0	0.5	811							
14	117.0	38.6	121.0	80.0	0.5	0.0	845							
15	117.0	38.6	121.0	80.0	1.0	0.5	784							
16	121.0	82.5	40.5	120.0	0.0	1.0	740							
17	121.0	82.5	40.5	120.0	0.5	0.0	782							
18	121.0	82.5	40.5	120.0	1.0	0.5	646							
19	118.0	124.0	80.5	40.3	0.0	1.0	862							
20	118.0	124.0	80.5	40.3	1.0	0.5	789							

The results of the calculation appear in the *session window*. For our example, they appear as follows:

Regression Analysis: v versus fan1, fan2, fan3, fan4, vent1, vent2

```
The regression equation is
v = 8.8 + 2.05 fan1 + 2.94 fan2 + 3.15 fan3 + 1.62 fan4
    - 91.2 vent1 - 75.0 vent2
```

Predictor	Coef	SE Coef	T	P
Constant	8.83	49.18	0.18	0.860
fan1	2.0506	0.2437	8.42	0.000
fan2	2.9376	0.2530	11.61	0.000
fan3	3.1511	0.2431	12.96	0.000
fan4	1.6191	0.2374	6.82	0.000
vent1	-91.18	20.19	-4.52	0.001
vent2	-74.95	22.36	-3.35	0.005

```
S = 35.1287 R-Sq = 97.5% R-Sq(adj) = 96.3%
```

Analysis of Variance

Source	DF	SS	MS	F	P
Regression	6	623910	103985	84.26	0.000
Residual Error	13	16042	1234		
Total	19	639952			

Predicted Values for New Observations

Obs	Fit	SE Fit	95% CI	95% PI
1	385.86	31.47	(317.88, 453.85)	(283.98, 487.75)
2	696.06	23.00	(646.36, 745.75)	(605.34, 786.77)
3	908.99	21.74	(862.03, 955.95)	(819.74, 998.23)
4	649.93	22.39	(601.57, 698.29)	(559.94, 739.92)
5	693.35	26.37	(636.38, 750.32)	(598.45, 788.24)
6	798.00	26.81	(740.09, 855.91)	(702.54, 893.46)
7	888.38	23.06	(838.56, 938.20)	(797.60, 979.16)

In this tabulation, there are several items that are of little use, and we simply ignore them: SE Coef, T, p, S, R-sq adj, Analysis of Variance table, SE Fit, 95% CI, and 95% PI. The *main* result here is of course the *regression equation*, which is the math model that relates the response to the various factors. The quality of the fit of the model to the measured data is indicated by the value of R^2, which in this case is 0.975 (97.5%), quite close to the *perfect model* value of 1.0. Perhaps a better model quality indicator is a *graph* which plots the model-predicted values of v against the measured values of v. Such a graph would be a perfect straight line at 45° for a perfect model. If we plot v versus v (the perfect model) on this same graph, it is easy to see where the predictions differ from the perfect model. We will also plot on this graph the predicted v values for the 7 validation runs which were *not* used to formulate the model.

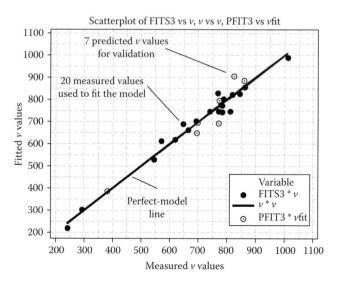

FIGURE 1.3
Graphical check of model quality using 20 runs.

Figure 1.3 is such a graph, which can be produced using the Minitab general graphing procedures explained in the appendix Minitab manual. We see there that the model-predicted values, for both the 20 runs used to formulate the model and also the 7 runs used as the validation experiment are quite close to the perfect-model line.

(I need at this point to add some caution to our recommendations on the use of R^2 and the graph of predicted versus measured response values. These *are* the best tools for evaluating the quality of a model, but just as with any tools, there are some caveats on using them *blindly*. I refer here to the fact that for *any* model, if you make only as many runs as there are unknown coefficients in the model, you will always get a *perfect* fit, an R^2 value of 1.0, and a graph which is a perfect straight line, yet this model will generally *not* be very good at predicting behavior for factor combinations *not* used to formulate the model.)

Now that we see that our simple model form (only linear terms in the six factors) seems to work well, we might rerun the regression using this same form but now use *all 27 runs* for the model formulation. That is, we are *wasting* quite a bit of actual data when we exclude 7 runs for our validation experiment. The regression equation is now given as

Regression Analysis: V versus fan1, fan2, fan3, fan4, vent1, vent2

```
The regression equation is
V = 25.6 + 2.12 fan1 + 2.87 fan2 + 2.91 fan3 + 1.62 fan4 - 74.4
    vent1 - 87.2 vent2
```

```
Predictor         Coef
Constant         25.56
fan1            2.1163
fan2            2.8652
fan3            2.9137
fan4            1.6177
vent1           -74.44
vent2           -87.22

    R-Sq = 96.4%
```

Note that in the above display I have deleted the extraneous results mentioned earlier and displayed only results that are truly useful. Figure 1.4 shows that this model also is a good fit to the measured data. From the numerical values of the coefficients we see that the effect of fan speed is a positive one for each fan; increasing the speed increases the outlet velocity, which seems reasonable. Fan 2 is the most effective and fan 4 the least; something that would not be obvious by inspection of Figure 1.2. The effect of the two vents is negative; increasing the vent area results in a decrease in outlet velocity. This seems reasonable since the vents are really *leaks* that allow air to escape before it reaches the outlet. Vent 2 is more effective than vent 1, which also seems reasonable since vent 2 is closer to the outlet. Note that the maximum effect of the vents on response v is −74.4 ft/min for vent1 and −87.2 ft/min for vent 2. When v is a large value, like 1011 ft/min, we see that the contribution of the vent area to this value is quite small. For lower values of

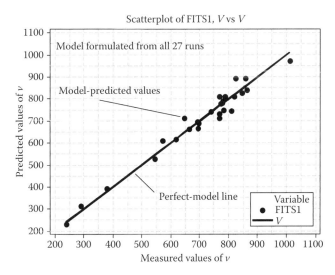

FIGURE 1.4
Graphical check of model quality using 27 runs.

v, the effect of the vents is more pronounced. The model thus gives us a good insight into the effects of the various factors, something that would be difficult to do reliably from a theoretical model study because such studies always require *simplifying assumptions* that deviate from the actual behavior.

While the regression results which we have emphasized are always of interest, one other graphical result that is available without any additional effort is sometimes useful. Minitab routinely creates four graphs that relate to the *residuals*. A residual is nothing but the difference (or *error*) between the measured value of the response variable and the model-predicted value, so there are as many residual values as there are runs in the experiment. The graphs referred to explore some of the statistical characteristics of the residuals from a given experiment. Figure 1.5 shows these graphs for our experiment which used all 27 runs to formulate the model. The two graphs in the left column examine the probability distribution of the residuals as compared with the *Gaussian* (also called *normal*) distribution widely used as a model in much statistical work. A residual distribution that is close to Gaussian is a desirable situation in any multiple regression study. The upper plot is the most useful; if the data points fall close to the plotted straight line, then our residual distribution is close to Gaussian. For our example, the result is very favorable. The lower plot is a histogram that would, for an *infinite* number of residuals, approach the well-known *bell-shaped curve* of the Gaussian distribution. Usually this histogram is too *coarse* to be very useful and we rely more on the upper plot. Of the two plots in the right column, the upper one is the more useful. In this graph we are looking for any *pattern* in the displayed points. What we *want* is *no*

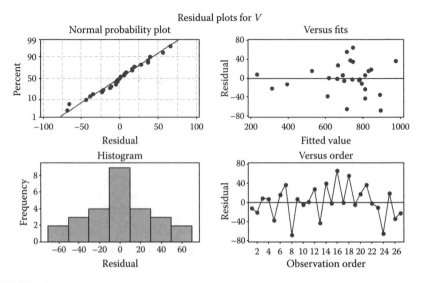

FIGURE 1.5
Minitab graphs for evaluating residuals.

pattern; just a random scatter. Again, our example gives a good result; the residuals seem to be randomly scattered. Sometimes, when a linear model is used but the real process actually has, say, a square term in its behavior, the residuals will show a roughly parabolic pattern.* When we observe this, it is evidence that we should try adding some sort of nonlinear term to our model. Unfortunately, random effects of un-modeled variables may obscure such a trend, so this graph has limited utility in this regard.

We conclude this chapter with a brief description of a model study that used our methods for a more practical purpose than the demonstration experiment which was the main subject of this chapter. I came across this material in *NASA Tech Briefs*, a monthly journal devoted to transferring engineering technology from the U.S. space program to other industries. This publication includes short articles describing some engineering work and then offers, upon request, more detailed Technical Support Packages; the one I used was TSP MSC-21996. One of the documents included was a conference paper[†] coauthored by Dr. Chin H. Lin, Chief of Systems Design and Analysis Branch, Lyndon B. Johnson Space Center, 2101 NASA Road 1, Houston, Texas, 77058-3696. Dr. Lin was most helpful with several phone calls and letters when I inquired about details not obvious from the documents.

I found out from Dr. Lin that this study was carried out without the benefit of formal DOE methods such as factorial and fractional-factorial designs and ANOVA, the standard tools described in textbooks for many years. The approach was strictly one of defining a number of experiment runs based on *common sense* and then a routine application of multiple regression to find an acceptable model. I make a point of this because the *standard methods* are often promoted as being the *only* proper way to do this kind of work. Dr. Lin told me that some time after this particular study, his group did begin to use the available information on DOE to choose experiment designs that allowed adequate modeling using experiments with fewer runs. This fits in with the approach I suggest in this chapter; use DOE methods to define the experiment and then use multiple regression to fit the model.

Part of the Space Shuttle orbiter life support system has the task of removing water vapor and carbon dioxide from the cabin atmosphere. For short missions (less than 8 days) the system uses lithium hydroxide cartridges for this purpose. For missions exceeding 8 days, this design has a severe weight and volume penalty and an alternative method was desired. A solid amine regenerative system was proposed using as adsorbent material

---

* E.O. Doebelin, *Engineering Experimentation: Planning, Execution, Reporting*, McGraw-Hill, New York, 1995, pp. 210–212.
† F.F. Jeng, R.G. Williamson, F.A. Quellette, M.A. Edeen, and C.H. Lin, Adsorbent testing and mathematical modeling of a solid amine regenerative CO_2 and H_2O removal system, 21st International Conference on Environmental Systems, San Francisco, CA, July 15–18, 1991.

polyethylenimine (PEI)-coated acrylic ester containing 20% by weight of PEI. The solid amine sorbent material adsorbs carbon dioxide and water vapor when the cabin atmosphere gas is passed through it. When the adsorbent gets saturated it is exposed to space vacuum to desorb the CO_2 and H_2O, whereupon it can be reused in a regenerative cycle. Experiments were to be run on a scale model of the proposed new system to verify its performance and derive by multiple regression, a math model relating performance to design parameters.

Figure 1.6 shows a schematic diagram for the scale model of the process. The adsorbent bed is contained within a metal tube of length 3 in. and inside diameter 7/8 in. The test procedure is as follows. Compressed dry air with the desired CO_2 partial pressure is drawn from the bottles and enters the test system at the desired operating pressure and flow rate after passing through a regulator and metering valves. The dry gas then flows through a bubble-bath type of humidifier to obtain the desired dew point, simulating the cabin atmosphere of interest. The gas stream then flows through ducting which is wrapped with heating tape in order to obtain the desired inlet gas temperature. The gas stream is vented (diverted from the adsorbent bed) until a desired inlet temperature is reached. The sealed adsorption bed is submerged in a constant-temperature water bath to maintain the desired adsorption temperature. Once the desired gas temperature and adsorption bed temperature are reached, the test is started by opening the valve to allow flow through the bed.

The inlet CO_2 and H_2O vapor are partially adsorbed on the solid amine. Inlet airflow rates were monitored using Hastings mass flow meter model NALL 5XPX. The CO_2 concentration in the inlet and outlet gas streams was continuously measured using MSA Infrared Gas Analyzers model 3000. Dew points in the inlet and outlet streams were measured with General Eastern Dew Point Monitors model HYGRO-M1. The temperature and pressure of the solid amine bed, and the inlet and outlet gas streams were monitored continuously. The bed vacuum pressures during desorption were measured using a pressure transducer manufactured by MKS Instruments, model 222BA and a Hastings vacuum gage. Once started, the adsorption process was allowed to continue for 13 min. In the modeling studies, the dependent variable (response) will be the rate at which CO_2 is removed from the gas stream flowing through the adsorbent bed. This rate is computed by using the CO_2 measurement, the total mass flow measurements at inlet and outlet, and the CO_2 concentration measurement. While continued use of the bed will eventually lead to saturation and loss of CO_2 removal capability, the measured data showed that, during the 13 min tests, no evidence of saturation appeared; the CO_2 removal rate was very constant.

Table 1.8 shows the 44 runs that were completed over the time interval from November 13, 1990 to April 23, 1991. All these runs were used to formulate the model; none were set aside for a validation experiment. The results are displayed as follows:

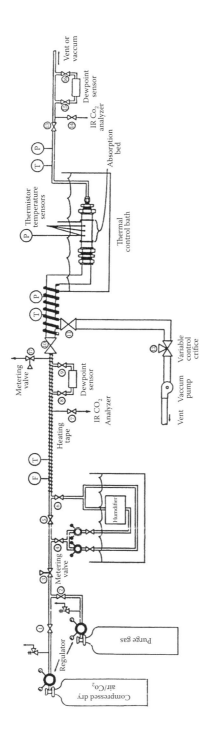

FIGURE 1.6

Scale model of Space Shuttle cabin atmosphere regeneration system.

TABLE 1.8

Data for NASA Cabin Atmosphere Experiment

Run	tbed	flow	pco2	ph2o	rco2
1	85	290	1.0	6.05	33.5
2	85	290	1.0	9.21	38.0
3	85	290	1.0	13.73	39.1
4	85	390	1.0	6.05	49.8
5	85	390	1.0	9.21	51.8
6	85	390	1.0	13.73	46.9
7	85	290	3.0	6.05	114.2
8	85	290	3.0	9.21	131.2
9	85	290	3.0	13.73	125.6
10	85	390	3.0	6.05	154.8
11	85	390	3.0	9.21	159.2
12	85	390	3.0	13.73	149.2
13	85	290	5.0	13.73	211.1
14	85	390	5.0	6.05	246.1
15	85	390	5.0	9.21	260.5
16	85	390	5.0	13.73	264.5
17	85	484	5.0	9.21	295.6
18	85	484	5.0	13.73	296.7
19	85	290	5.0	6.05	188.8
20	85	290	5.0	9.21	199.4
21	95	290	3.0	6.05	112.3
22	95	290	3.0	9.21	118.4
23	95	290	3.0	13.73	124.7
24	95	390	3.0	6.05	163.9
25	95	390	3.0	9.21	166.3
26	95	390	3.0	13.73	156.5
27	95	290	5.0	6.05	205.8
28	95	290	5.0	9.21	195.9
29	95	290	5.0	13.73	192.7
30	95	390	5.0	6.05	256.6
31	95	390	5.0	9.21	252.0
32	95	390	5.0	13.73	252.0
33	95	484	5.0	9.21	286.6
34	95	484	5.0	13.73	258.9
35	95	290	7.6	6.05	263.1
36	95	290	7.6	13.73	271.2
37	95	390	7.6	6.05	352.9
38	95	390	7.6	13.73	348.9
39	95	484	7.6	13.73	403.9
40	110	390	7.6	6.05	353.3

TABLE 1.8 (continued)

Data for NASA Cabin Atmosphere Experiment

Run	tbed	flow	pco2	ph2o	rco2
41	110	390	7.6	9.21	328.2
42	110	390	14.8	6.05	561.5
43	110	390	14.8	9.21	575.7
44	120	390	14.8	9.21	578.6

Regression Analysis: rco2 versus tbed, flow, pco2, ph2o

```
The regression equation is
rco2 = - 93.7 - 0.517 tbed + 0.473 flow + 37.7 pco2 + 0.564
       ph2o


Predictor       Coef
Constant      -93.67
tbed         -0.5175
flow         0.47319
pco2          37.711
ph2o          0.5642


     R-Sq = 98.2%
```

The good fit of the model is indicated by the high value of R^2 and confirmed by the graph of Figure 1.7. It is quite likely that a good model could

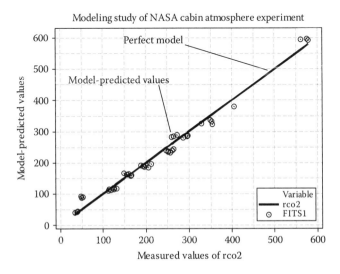

FIGURE 1.7
Graph to check accuracy of the fitted model.

be achieved in fewer runs (and thus lower cost and time delay) using DOE guidelines for designing the experiment. We see from Table 1.8 that up to five levels of a factor were used, whereas two are often sufficient. Since we got a good fit with strictly linear terms in the regression equation, there may be no need to use more than two levels. Because there are only four factors, a *full-*factorial design would require only 16 runs, many fewer than NASA used. We should note however that the 44 runs used more thoroughly explore the range of the process than would a 16-run full factorial and may thus give a more reliable model. Unfortunately, the 44 runs of Table 1.8 do not include the factor combinations and levels needed in 16-run full factorial, so we cannot make a proper comparison. If one uses DOE methods regularly, one can explore such comparisons and, over time, get some insight into making these choices more rational. After these particular experiments, Dr. Lin's group *did* adopt some DOE methods, so they may have studied this process beyond what we have seen here, however I am not aware of any publications that addressed these issues.

2

Vibration Isolation for Sensitive Instruments and Machines

2.1 Introduction

Measurement and/or control systems whose performance would be adversely affected by vibration may require a mounting system that "filters out" the existing vibrational force or motion of the structure to which the system is attached. Optical systems are a good example, as testified to by the inclusion of a section on anti-vibration mounts in almost every optical company's catalog. Some laser and holography measurement techniques cannot tolerate base motions more than a small fraction of the wavelength of the light being used, but the existing and unavoidable floor vibrations greatly exceed these levels. This means that the optical table must include some kind of vibration-isolating system. Similar requirements exist in some manufacturing processes, particularly those used for producing microelectronic circuits or microelectromechanical systems (MEMS). Here, the exceedingly small dimensions and tolerances of the parts being produced require close control of spurious vibrations.

Over the years, engineers have invented and developed a number of technologies for implementing vibration isolation. In this chapter, we will explore six of these approaches:

1. Simple passive spring/mass systems using metal or elastomer springs
2. Passive air-spring systems
3. Active air-spring systems
4. Low-frequency isolation using negative-spring-constant devices
5. Active electromechanical systems
6. Tuned vibration absorbers and input-shaping methods

Our approach will be conventional but you should be aware that unconventional methods* are worthy of consideration. When discussing the conventional methods, Rivin (2006) states: "Unfortunately, such an approach is considered 'politically correct' in vibration-isolation applications and is universally used."

2.2 Passive Spring/Mass Isolators

The classical vibration-isolation approach is to mount the mass to be isolated on properly designed springs so as to create a vibrating system whose natural frequency is much lower than the lowest frequency to be suppressed. The springs will of course have some inherent damping, and in some cases additional designed-in damping may be added. Many readers will already be familiar with this approach since it is covered in introductory system dynamics and vibration courses of mechanical engineering curricula around the world. We now quickly review the behavior of this type of system. Figure 2.1 shows a schematic diagram representing this class of

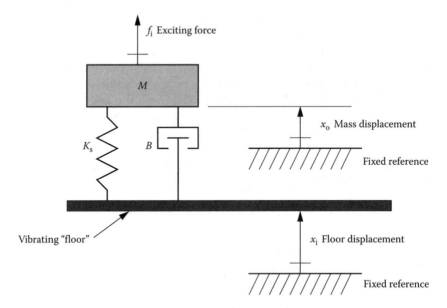

FIGURE 2.1
Analysis model for vibration isolation.

---

* E.I. Rivin, Vibration isolation of precision objects, *Sound and Vibration*, July 2006, pp. 12–20, downloadable at www.sandv.com.

isolation devices. While we mentioned earlier the use of such systems to attenuate floor *motions*, they are also useful for reducing motion due to fluctuating *forces* acting on the mounted mass. These might come from rotating or reciprocating "machinery" mounted on the mass. For example, many optical setups use piezoelectric motion control systems (called *stages*) to position lenses, mirrors, etc. The moving masses in such systems create vibration-exciting forces. For the most critical situations, *acoustic* forces originating in the surrounding air may need to be isolated. For these reasons, we include both input motions and input forces in our analysis model of Figure 2.1. We should note that a *general* study of vibration isolation would consider all six degrees of motion freedom possible for a single rigid mass. Such a study[*] becomes quite complicated because the equations of motion are *coupled* and their solution and interpretation for design purposes is difficult, so we restrict ourselves here to a single axis of motion. Many practical problems are adequately handled by this approach since the mentioned coupling may be nonexistent or weak, allowing sufficient accuracy with this simplified method. Modeling the *payload* (instrument, machine, etc.) as a single rigid mass is also an approximation, but is often adequate because the isolator guarantees that the motion will be *low frequency*. At low frequencies, all parts of our payload tend to move together, as if they were one mass.

Applying Newton's law to a free-body diagram of the mass, we can get the system differential equation:

$$f_i - K_s(x_o - x_i) - B(\dot{x}_o - \dot{x}_i) = M\ddot{x}_o \tag{2.1}$$

Using the *D*-operator notation ($D \underset{\Delta}{=} d/dt$) this equation can be rewritten as

$$(MD^2 + BD + K_s)x_o = f_i + (BD + K_s)x_i \tag{2.2}$$

We now define the standard parameters and use the superposition theorem of linear systems to break the equation up into two operational transfer functions[†] relating the response x_o to the two inputs f_i and x_i:

$$\frac{x_o}{f_i}(D) = \frac{K}{\dfrac{D}{\omega_n^2} + \dfrac{2\zeta D}{\omega_n} + 1} \tag{2.3}$$

[*] H. Himelblau and S. Rubin, Vibration of a resiliently supported rigid body, *Shock and Vibration Handbook*, Vol. 1, C.M. Harris and C.E. Crede (Eds.), McGraw-Hill, New York, 1961, pp. 3-1–3-52.

[†] E.O. Doebelin, *System Dynamics: Modeling, Analysis, Simulation, Design*, Marcel Dekker, New York, 1998, p. 56.

$$\frac{x_o}{x_i}(D) = \frac{\dfrac{2\zeta}{\omega_n}D+1}{\dfrac{D^2}{\omega_n^2}+\dfrac{2\zeta D}{\omega_n}+1} \tag{2.4}$$

where

$$\omega_n \triangleq \sqrt{\frac{K_s}{M}}, \quad \zeta \triangleq \frac{B}{2\sqrt{K_s M}}, \quad K \triangleq \frac{I}{K_s} \tag{2.5}$$

The symbol $\triangleq$ means *equal by definition*, ω_n is the *undamped natural frequency*, ζ is the *damping ratio*, and K is the *steady-state gain* for the transfer function relating x_o to f_i. The steady-state gain of transfer function 2.4 is 1.0; therefore, it does not appear explicitly. For those readers who prefer Laplace transform methods, the D operators can be replaced by the Laplace complex variable s everywhere.

For the situation where we wish to isolate the mass from *floor vibrations*, Equation 2.4 is the transfer function of interest. Since all time-varying vibration signals (periodic, transient, random, etc.) can be expressed in terms of their *frequency content*,[*] we usually want the *frequency response* of our systems, which is easily obtained from the operational (or Laplace) transfer functions by converting them to the *sinusoidal transfer function*. (By frequency response we mean the sinusoidal-steady-state response of the system to a perfect sine wave input of a given amplitude and frequency. Sinusoidal steady state means that we wait for all transients to die out before we "measure" the output. This output will always be a perfect sine wave of the same frequency as the input, but its amplitude and phase will differ from that of the input.) To convert to the sinusoidal transfer function, we substitute the term $i\omega$ everywhere we see a D (or an s), which makes the right-hand side of the transfer function a complex number, which can be numerically calculated (for numerical values of the physical parameters) for any frequency of interest. This complex number has a magnitude (which is the output/input amplitude ratio) and a phase angle that will vary with frequency, and we plot two curves (called the *frequency response curves*) for frequencies ranging from 0 to some highest chosen value. MATLAB® has many useful routines that make all these calculations easy and quick, and the appendix material on MATLAB/Simulink® shows you how to use these convenient tools.

$$\frac{x_o}{x_i}(i\omega) = \frac{\dfrac{2\zeta}{\omega_n}i\omega+1}{\dfrac{(i\omega)^2}{\omega_n^2}+\dfrac{2\zeta i\omega}{\omega_n}+1} \tag{2.6}$$

---

[*] E.O. Doebelin, *Measurement Systems: Application and Design*, 5th edn., McGraw-Hill, New York, 2003, pp. 149–200.

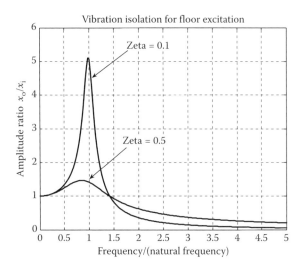

FIGURE 2.2
Effect of damping on isolation.

Figure 2.2 is obtained by using these tools and choosing some numerical values (0.1 and 0.5) for the damping ratio zeta. We need not choose specific values for the natural frequency because in Equation 2.6, wherever ω_n appears, it appears as the ratio ω/ω_n; thus we plot our curve against this ratio, making the display much more useful since it can now be used for any natural frequency we wish. We see from this curve that if we want the motion of mass M to be *less* than the input (*floor*) motion, the natural frequency must be designed to be well below the lowest frequency present in the input motion. The curves for the two values of damping cross at a frequency ratio of $\sqrt{2}$; this is true for *all* values of damping. Below this value the input motion is *amplified*, above that we get some degree of isolation. Also, we see that at the *high* frequencies where vibration isolation is effective, we get a better isolation with *less* damping. Here we need to be careful because low damping is seen to give excessive motion *amplification* rather than isolation in the neighborhood of the natural frequency. This is a practical concern because real-world output motions will never be perfect sine waves already in steady state. Rather, real systems will need to be *started up and stopped* from time to time, and these transients can cause large transient motions of the mass if the damping is too low. Also, in addition to start-up and stopping transients, the input motion waveform will include some transient effects which again can cause momentary large motions. We see thus that choice of damping will require some compromises between steady-state isolation and transient amplification.

We can use Simulink simulation (see Appendix A if you are not familiar with this software) to clearly explain some of these effects. In Figure 2.3 I have simulated two systems as in Figure 2.1, with no force input, but a floor

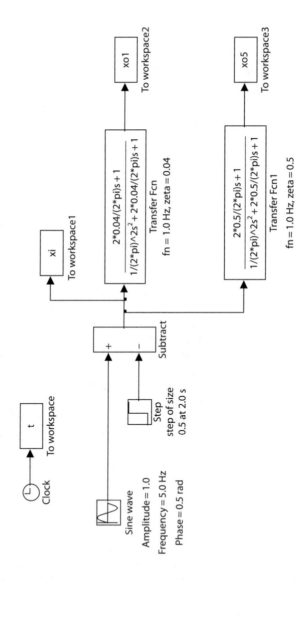

FIGURE 2.3
Simulink simulation of effect of damping.

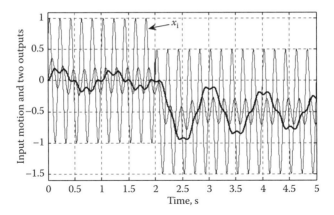

FIGURE 2.4
Light damping causes *ringing* at natural frequency.

motion input. Both systems have a natural frequency of 1.0 Hz, but one has a damping of 0.5 while the other has a damping of 0.04. The input motion is a combination of a sine wave with a phase angle of 0.5 rad (this makes it *jump* at time 0) and a step input of size 0.5 that occurs at $t = 2.0$ s. Figure 2.4 shows the results of this simulation. We see that with damping 0.5, the system does not show much response at the natural frequency and quickly goes into sinusoidal steady state after a disturbance. However, it allows quite a bit of the input motion to appear at the output (poor isolation). With a damping of 0.04, any disturbance is seen to cause a lot of motion at the natural frequency (this is called *ringing*), and it takes quite a long time for the system to *settle* into sinusoidal steady state after a disturbance. In fact this steady state has not been achieved at the end of this graph (5 s). If we run the simulation longer, we will see (Figure 2.5) that it does finally settle to a small value of output motion, as we would expect from Figure 2.2.

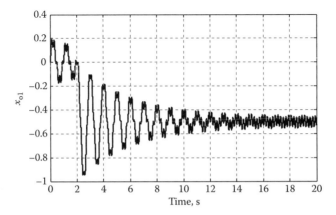

FIGURE 2.5
Light damping gives good steady-state isolation but is sensitive to transient inputs.

While the response to sine waves (Figure 2.2) gives us a good understanding of the principle of isolation and is the basis for the design of isolation, isolation is also effective for transient and random inputs, as we study in the simulation of Figure 2.6. A short-duration (relative to system natural period) triangular pulse of floor displacement is formed in a lookup table and applied to our isolator system with the good results shown in Figure 2.7. If the pulse were of *longer* duration, the isolation would not be very good. A random input of floor displacement, which has a frequency content from about 5–100 Hz, is seen in Figure 2.8 to be very well attenuated. If our random signal had frequency content down to 1.0 Hz or less, then the isolation would not be so good.

When our interest is in isolating the mass from input *forces* f_i, we want the frequency response curve for transfer function 2.3, as shown in Figure 2.9. Here we can again nondimensionalize the horizontal axis so that the graph applies to any natural frequency. On the vertical axis, it is now convenient to plot $x_o/(f_i/K_s)$, which is also nondimensional, and makes the curve applicable to any K_s value. We see that good isolation requires the natural frequency to be well below the lowest forcing frequency, and that in the high-frequency range where isolation is good, damping has *no* effect; we get the same quality of isolation for low or high damping. The *physical* reason for this is that at high frequency, the mass tends to stand still, thus neither the spring or the damper has much effect; the good isolation is strictly the result of the *inertia* of the mass. Very light damping, however, is still to be avoided because of possible transient *ringing*.

For the isolation of both input forces and motions, we have found that the main requirement is a low natural frequency (relative to the exciting frequencies) and a judicious choice of damping. Since the payload mass is usually fixed by the application and thus not available as a design variable, getting a sufficiently low natural frequency usually comes down to choosing a suitably soft spring (low K_s). When the exciting frequencies are below about 10 Hz, the required springs are so soft that they may cause unacceptable static deflections, or buckling problems. Since the static deflection is given by $Mg/K_s = g/\omega_n^2$, we see that low natural frequencies are bound to give large static deflections. In choosing a specific spring for a given application we can use one of the many *general* types of springs (coil, leaf, etc.) or resort to one of the many forms of vibration and shock *mounts* available from a wide range of manufacturers. Metal, elastomer, or metal/elastomer combinations are available, with those using elastomers providing damping in addition to the spring function. While our Figure 2.1 shows a separate *damper*, it is much more common to choose a form of spring which includes some kind of built-in damping effect. Such damping effects are usually not *viscous*, so the damping term in our differential equation is an approximation relating to some kind of *equivalent* viscous damping.

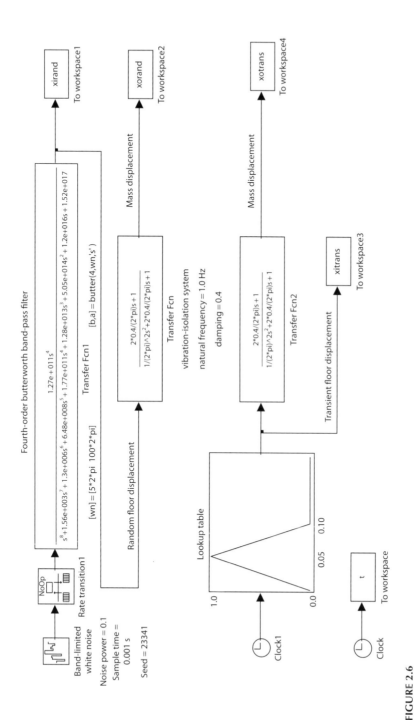

FIGURE 2.6
Transient and random input simulation.

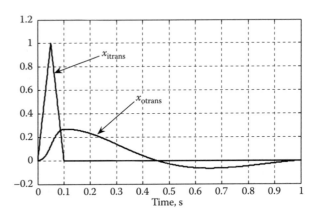

FIGURE 2.7
Pulse input of floor motion is well attenuated.

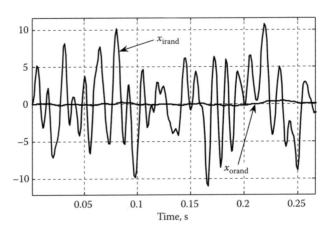

FIGURE 2.8
Random input of floor motion is well attenuated.

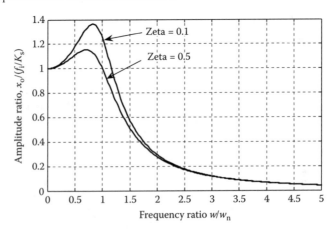

FIGURE 2.9
Isolation of input forces.

2.3 Passive Air-Spring Systems

For some applications, *air springs* offer certain advantages, so we consider them next. Air springs are of course used for purposes other than vibration isolation, just as are other forms of springs. They consist essentially of a sealed elastic "container" of roughly cylindrical shape, which can be pressurized with a gas. The container is often made of some type of elastomer, which may be reinforced with fabric or metal in various ways. Whereas an ordinary metal spring, say a coil type, must deflect more and more as the payload gets larger, causing stability (buckling) problems when low spring constants are needed, an air spring can support a larger load at the *same* deflection if we simply increase its internal pressure. This higher pressure *will* affect the spring stiffness adversely from a vibration-isolation standpoint (it stiffens the spring, and raises the natural frequency), but a net advantage is often possible.

We saw in Figure 2.2 that a high damping ratio was desirable to limit amplitude at resonance but made isolation worse at higher frequencies, where isolation was most necessary. A different configuration of springs and dampers than that in Figure 2.1 allows us to resolve this conflict and get *both* a low resonant peak and isolation similar to the ideal *undamped* system at high frequencies. This configuration (Figure 2.10) is what is used in most air-spring isolators, but we will first study it as if we were using ordinary metal

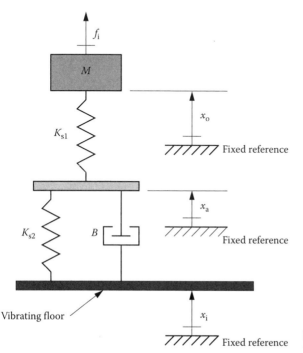

FIGURE 2.10
System for improved isolation.

springs. At high frequencies, the damper becomes "stiff" and input motion x_i is effectively applied directly to the bottom of spring K_{s1}. This creates an *undamped* vibrating system consisting of mass M and spring K_{s1}. Such a system has, for this type of isolator, the *best possible* isolation at high frequency. At lower frequencies, near resonance, the damper is still active, allowing us to choose numerical values so as to get an acceptable resonant peak.

One way to get the differential equations of this system is to place a fictitious mass m at the coordinate called x_a. We can then write two Newton's law equations, one for m and one for M. Having written the equations, we then recall that $m = 0$ and delete that term. This is not the only way to deal with coordinates that have no mass associated with them, but I have found it useful in avoiding mistakes such as wrong algebraic signs. Following this procedure we get

$$K_{s1}(x_o - x_a) + K_{s2}(x_i - x_a) + B(\dot{x}_i - \dot{x}_a) = m\frac{d^2 x_a}{dt^2} = 0 \qquad (2.7)$$

$$f_i + K_{s1}(x_a - x_o) = M\ddot{x}_o \qquad (2.8)$$

These two simultaneous equations can be manipulated to eliminate the coordinate x_a (which is of no interest to us) and get a single equation relating x_o to x_i, and thus the transfer function:

$$\frac{x_o}{x_i}(D) = \frac{\dfrac{B}{K_{s2}}D + 1}{\dfrac{MB}{K_{s1}K_{s2}}D^3 + \left(\dfrac{1}{K_{s1}} + \dfrac{1}{K_{s2}}\right)MD^2 + \dfrac{B}{K_{s2}}D + 1}$$

$$= \frac{\tau D + 1}{\dfrac{MB}{K_{s1}K_{s2}}D^3 + \left(\dfrac{1}{K_{s1}} + \dfrac{1}{K_{s2}}\right)MD^2 + \dfrac{B}{K_{s2}}D + 1} \qquad (2.9)$$

(If the manipulation mentioned above is unfamiliar, see Appendix A) An equation similar to Equation 2.9 could be obtained relating x_o to f_i but here we concentrate on isolating floor vibrations.

Whenever we analyze a new dynamic system, we try to put it into *standard form*, using standard terms such as first-order systems, second-order systems, etc., since this relates the new system to the more familiar *basic building blocks*. We have done this in the numerator of Equation 2.9 by defining the *time constant* τ as B/K_{s2}. Unfortunately the denominator must be left *as is* because the cubic form cannot easily be factored in letter form. (There *are* formulas for getting the roots of a cubic polynomial in letter form, but they are so complex that nobody uses them.) We do know, however, that the factored form must be either 3 first-order terms or 1 first-order and an underdamped second-order.

From this, we can show the correctness of our earlier *intuitive* statement about the improved high-frequency isolation of this design. In Equation 2.4, since the numerator is first-order and the denominator is second-order, at high frequency the *net* effect is a first-order term in the denominator, which gives an attenuation of 20 dB/decade. In Equation 2.9, the net effect at high frequency is a *second-order* term in the denominator, giving a 40 dB/decade attenuation. Recall that 20 dB/decade means that the amplitude ratio drops by a factor of 10 when the frequency changes by a factor of 10; 40 dB/decade gives a change of 100-to-1, 10 times better.

If the mass M is considered given, we have three numerical values (two springs and a damper) to choose in designing a specific system. Considering the system differential equation, the characteristic equation will have three roots, and for the damping values that give good isolation, these roots will consist of one real root and two complex roots. The two complex roots will define a damped natural frequency and this frequency can be estimated by letting B equal zero, giving the result

$$\omega_n = \sqrt{\frac{\left(\dfrac{K_{s1}K_{s2}}{K_{s1} + K_{s2}}\right)}{M}} \tag{2.10}$$

The simplest design situation would be one where the floor displacement is measured (or estimated) to be a single sine wave of known frequency and amplitude, say 0.01 in. at 10 Hz. We next need to know what vibration level our payload can tolerate, say 0.0005 in. Sometimes instrument manufacturers will supply such data.* Our design problem in this case is to provide 20-to-1 isolation at 10 Hz. To estimate the needed damping and spring coefficients, we will assume that, at 10 Hz, we are in the range where the system can be approximated as the simpler form (Figure 2.1) with the active spring being K_{s1}. In this range, we can also neglect the effect of damping; the damper is stiff. We see that the amplitude ratio at 10 Hz should be 0.05, and we can estimate the needed natural frequency from

$$\left| \frac{1}{\dfrac{(i\omega)^2}{\omega_n^2} + \dfrac{2\zeta i\omega}{\omega_n} + 1} \right| = 0.05 \tag{2.11}$$

---

* H. Amick and M. Stead, Vibration sensitivity of laboratory bench microscopes, *Sound and Vibration*, February 2007, pp. 10–17; W.R. Thornton et al., Vibration isolation of a medical facility, *Sound and Vibration*, December 2006, pp. 10–18; E.E. Ungar et al., Predicting football-induced vibrations of floors, *Sound and Vibration*, November 2004, pp. 16–22 (above 3 references are downloadable at www.sandv.com); E.E. Ungar and J.A. Zapfe, Well-designed buildings minimize vibration, *Laser Focus World*, September 2004, pp. 113–117.

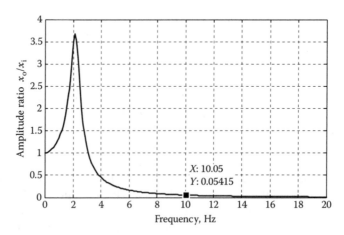

FIGURE 2.11
Results of trial design.

Neglecting the damping, this is easily solved for the needed natural frequency as 2.29 Hz, which in turn makes K_{s1} = 53.5 lb$_f$/in. If we choose a damping ratio of, say, 0.4 to prevent excessive transient *ringing*, we can estimate B from Equation 2.5 as 11.5 lb$_f$/(in./s). Finally, we need to choose the other spring constant, but there is really no guidance available for this based on our current information. Since *all* of our parameter values so far are *estimates* which will be adjusted shortly, let us consider the two spring constants to be equal.

We now insert all our *trial values* into Equation 2.9 and compute the sinusoidal transfer function for this relation. Using the available MATLAB tools, we get the curve of Figure 2.11. We see that the attenuation (0.0542) is quite close to our desired specification and the resonant peak has an amplification of about 3.7, quite a bit larger than one expects for a damping ratio of 0.4. The explanation of course is that this is a *third* order system, not the second order of our approximate calculations, and as such, it *does not have* a damping ratio! We can, however, now factor the cubic polynomial into one real root and a pair of complex conjugate roots (we could not do that when the parameters were *letters*). The complex root pair represents a second-order term, which *does* have a damping ratio. The MATLAB *roots* and *damp* routines make these calculations easy. We have to first define a vector of the coefficients in the denominator of Equation 2.9; I called it *a*, and then enter the commands:

```
roots(a)

ans =
    -2.0583e+000 + 13.4501e+000i
    -2.0583e+000 - 13.4501e+000i
    -5.1907e+000

>> damp(a)
```

Eigenvalue	Damping	Freq. (rad/s)
-2.06e+000 + 1.35e+001i	1.51e-001	1.36e+001
-2.06e+000 - 1.35e+001i	1.51e-001	1.36e+001
-5.19e+000	1.00e+000	5.19e+000

We see that there is a first-order factor with a time constant of $1/5.19 = 0.193\,s$ and a second-order factor with a damping ratio of 0.151 and a natural frequency of $13.6/6.28 = 2.17\,Hz$. This low damping ratio (we tried for 0.4 but got 0.15) explains the large peak in the frequency response. The natural frequency is quite close to what we wanted, and in fact we did get very close to the attenuation desired, which is of main interest.

Whether this tentative design requires some fine-tuning depends on what kind of transient effects might occur in the specific application. If they are minor, we might be satisfied with this first trial. If not, we could increase the B value, but we then should really run a *simulation* of our redesigned system, which includes a simulation of the transient inputs that we are concerned about. (A larger B value would certainly reduce the resonant peak in the frequency response, but this *does not* tell you what the transient response of x_o would be. We will shortly do such a simulation related to a different design problem.)

The new design problem stipulates a more realistic description of the floor input displacement than the rather idealized single sine wave that we just studied. We take the floor vibration displacement as a Gaussian random variable with an RMS value of about 0.01 in. and a bandwidth of 10–100 Hz. (Our design procedure could actually work with a *measured* record of the *actual* floor vibration if such a record were available from a data acquisition system; MATLAB readily accepts such imported file data making the simulation closer to reality.) Suppose also that our payload can only tolerate an RMS displacement of 0.0005 in.; thus we require an isolation of about 20-to-1. Our design approach is again to use any available approximate relations to rough out a design, and then use simulation to fine-tune it. Since the lowest frequency is about 10 Hz and the 20-to-1 attenuation specification is the same as before, we can use the design we just completed as the *first trial* of the new design. For the frequencies above 10 Hz, the attenuation will be better than this, so our trial design is conservative in this respect. Of course, if our new isolator design had *different* numbers, we would have to rough out a *new* trial design using the same methods explained above.

The MATLAB material on random signals provided in the appendix is used to set up the simulation of the floor displacement signal using Simulink's *band-limited white noise* module. The output of this module is sent to a Butterworth band-pass filter, set to the 10–100 Hz bandwidth (see Appendix A.1.12). While the white noise module can be adjusted to give a desired RMS value for *its* output signal, our requirement is on the band-pass filtered signal, so we will have to actually *measure* its RMS value and make adjustments on a trial-and-error basis to get the 0.01 in. RMS value stipulated. Figure 2.12

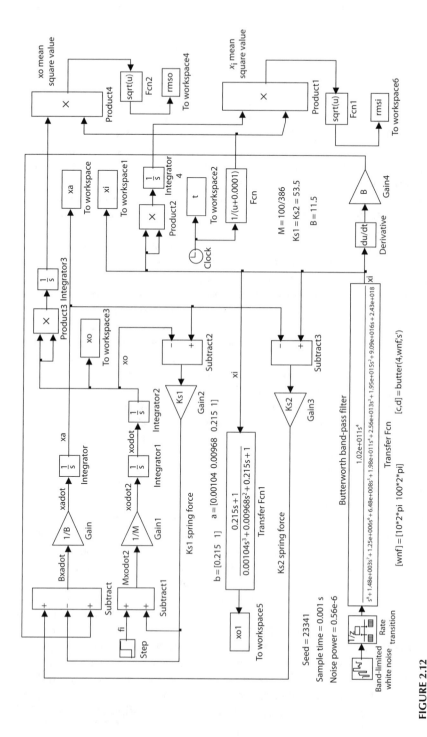

FIGURE 2.12

Simulation of system of Figure 2.10.

shows this simulation. At the bottom of this diagram we see the simulation of the random input. The *seed* and *sample time* of the noise generator can be set directly, without any trial and error. In most simulations, we can just accept the default value of the seed (23341). The sample time determines the frequency content of the random square-wave signal. The appendix discussion of this tells us to use a sample time whose reciprocal is 10 times the highest frequency we want; in our case we want content up to about 100 Hz, so the sample time is taken as 0.001 s. The noise power setting determines the "size" of the random signal and in this case has to be set by trial and error, by running the simulation with a guessed value and then adjusting this until we get an RMS value of about 0.01 in. for the x_i signal.

The standard way of computing the mean-square value of any signal (including random ones) is to square that signal, then integrate it with respect to time, and divide that integral by the running value of time. Since time starts out at a value of 0.0 (which we dare not divide by), we add a small number to t before the division. To get the RMS value, we just take the square root of the mean-square signal. You can see all these operations at the right of the simulation diagram, for both the x_i and x_o signals. Once we have a proper x_i signal, we can send it into a transfer function block to produce x_o; this is shown near the middle of the diagram. We now have all that is really needed in this simulation, but I wanted to show an *alternative* simulation of the isolation system itself, because this alternative method has some features that we sometimes need. When we analyze *any* dynamic system we usually begin with a set of simultaneous differential equations derived from the physics of the system. The *preferred* simulation approach in all cases is to draw the simulation diagram directly from these equations, *without* any preliminary manipulation, such as the derivation of transfer functions. This method is preferred because it gives a simulation diagram that explicitly shows all the *internal* events, whereas a transfer function *hides* these and displays *only* two *external* signals, the system input and output. The upper left part of our present diagram is devoted to this more detailed simulation, which is easily obtained using routine methods shown in the appendix. Note that in this diagram all the system parameters appear explicitly in *gain blocks,* whereas the transfer function shows only the coefficients of the numerator and denominator, which are *combinations* of the basic parameters. Also, we see all the forces, velocities, and accelerations, which again are *hidden* in the transfer function. One advantage of the transfer function method is *compactness*; it takes up much less space on our diagram, which might be useful if our overall system were more complicated. In the present diagram, in addition to making you aware of the two schemes, using both allows us to *check* results since xo and xo1 are of course the same signal and if they do not agree, we look for errors in our simulation.

In Figure 2.13, we see that when the input RMS value is 0.01, 20 times the output RMS value is *less* than this, which means that our system is *overdesigned;*

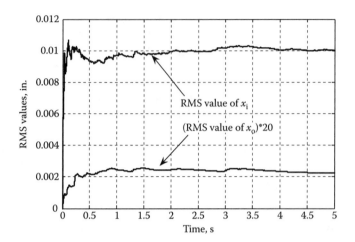

FIGURE 2.13
Root-mean-square (RMS) values of x_i and x_o.

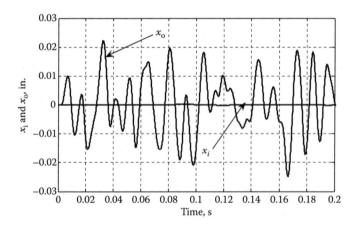

FIGURE 2.14
Time histories of x_i and x_o.

it is better than we need. (The input RMS value was set at the desired value of 0.01 in. by trial and error, using the noise generator parameter *noise power* to do the adjusting.) Figure 2.14 shows a portion of the time histories of input and output. Note there that the *peak values* of x_i (a random signal does not have a *single* peak value, the way a sine wave does) are the same order of magnitude as the RMS value (0.01); this is a general fact that is useful in our case for initially adjusting the noise power. The fact that our system works better should not be too surprising. We set the isolation at 20-to-1 *at* 10 Hz whereas most of our signal is *above* 10 Hz, where the isolation gets much better. We probably should revise our design since we can now use stiffer springs, which is preferred when possible. Before this can be done, we need to work a little more on damping. From Figure 2.10 we see that if B is made very small we lose damping, but this also happens when we make B very large, because when B is stiff, the system *ignores* both B and K_{s2}. If damping *goes to zero* for *both* large and small B, then there must be an *optimum B* value which gives the best damping. We could search for this optimum B value in various ways, but a "brute force" approach through simulation is perhaps the quickest.

To implement this search for the optimum B, it is useful to embed a frequency-response calculation *within a MATLAB program* (M-file). Since this approach is useful in many practical problems, I want to present it here for this example. (As usual, more details can be found in the appendix.)

```
% crc_book_chap2_optz.m
% finds the optimum damping ratio for a vibration isolator
% set parameter numerical values
M=100/386
Ks1=53.5
Ks2=53.5
% set up some empty vectors to hold values of B and peak of
   curve
BB=zeros(1,50);
peak=zeros(1,50);
% start a for loop
for n=1:50
    B=n*0.4;
    b=[B/Ks2 1];
    a=[M*B/(Ks1*Ks2) ((1/Ks1)+(1/Ks2))*M B/Ks2 1];
    f=linspace(0,10,50);
    w=f*2*pi;
    [mag,phase]=bode(b,a,w);
    peak(n)=max(mag);
    BB(n)=B;
end
plot(BB,peak)
```

Figure 2.15 shows that the optimum value does exist and that it is quite *broad*; a large range of B values gives peaks close to the optimum.

Until now, we have taken the two spring constants to be equal, without any real justification for this choice. To explore this question, we now let them be unequal and see the effect. Since there are an infinite number of combinations that can be tried, we need to somehow narrow this down. One requirement that seems reasonable is to insist that the *series combination* of the two springs, which would have the spring constant $K_{s1}K_{s2}/(K_{s1}+K_{s2})$, have the same value (26.75) as for equal springs. This would keep the static deflection the same for all the combinations. If we do this and use a range of K_{s1} values ranging from 30 to 100, we can compute the amplitude ratio for each combination, holding

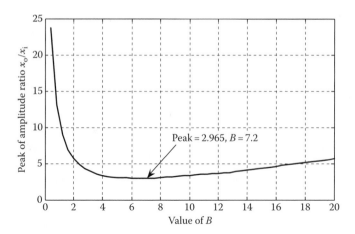

FIGURE 2.15
Optimum damping study.

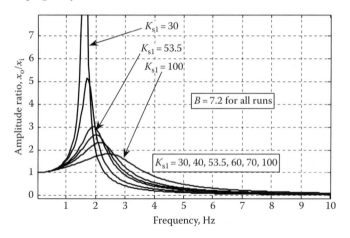

FIGURE 2.16
Study of unequal spring constants.

damping constant at the optimum value (7.2) that we found for equal springs. Actually, the optimum damping *changes* for each combination, but we saw earlier that this seems to be a broad optimum, so using 7.2 for all the runs is not unreasonable. Figure 2.16 shows our choice of equal springs may well be a good one. Some of the other choices show a smaller peak value, but their isolation at high frequency is worse, as we see in Figure 2.17.

With our added knowledge, we might now go back and revise our original design, which gave better isolation than our specification required. We can use stiffer springs than we did before, which reduces the chance of buckling and instability. Stiffer springs raise the natural frequency and make the isolation worse, which is what we want. Again, we might adjust the

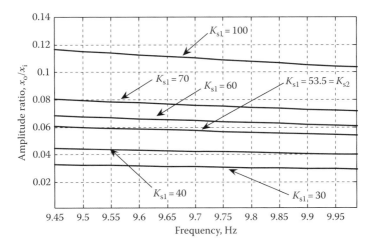

FIGURE 2.17
Combinations that have good peaks have bad high-frequency isolation.

damping for a new optimum, but this will not have a big affect, so we stay with $B = 7.2$. We now need to rerun the simulation of Figure 2.10 with various values of K_{s1}, taking K_{s2} equal, until we get (20)(xorms) to be about the same as xirms. We find that setting the two springs at $250\,lb_f/in.$ gives the desired result, as seen in Figure 2.18. Figure 2.19 shows the time histories for our final design.

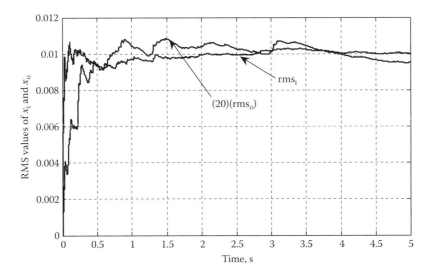

FIGURE 2.18
Spring stiffness adjusted to meet the isolation specification.

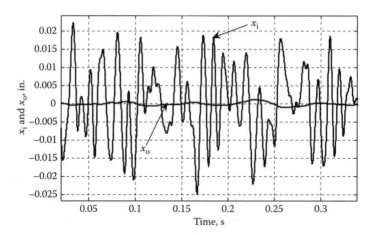

FIGURE 2.19
Isolation results for final design.

Recall that all the above work for the system of Figure 2.10 was preliminary to applying these results to *air springs,* and we now want to return to that topic. Actual air springs usually employ elastomer (*rubberlike*) diaphragms to contain the air, since these can be made to have very little friction. Low friction is important if we want to isolate really small floor displacements. Optical tables, for instance, must isolate floor motions much less than 0.001 in. Coulomb (*dry*) friction requires a threshold force to break it loose, allowing small floor displacements to be transmitted straight through to the payload, defeating the isolation. In Figure 2.20, we model the elastomer diaphragm as an equivalent piston/cylinder for ease of analysis. The piston/cylinder contact is assumed to be frictionless. Unless you have some previous experience in analyzing systems of this type, a one-to-one correspondence between elements in Figures 2.10 and 2.20 will probably not be obvious. The compressibility of the air in the two chambers will certainly give spring effects, and the orifice (a flow resistance) is known to be an energy dissipating device, however we will have to rely on analysis to convince us that the equations of the two systems have the same form.

In analyzing systems with fluid elements, several approaches are possible, each with its own advantages and drawbacks. We will here use a relatively simple scheme which, however, has been found to predict the actual behavior with useful accuracy. The compressibility of the gas (usually air or nitrogen) is modeled by the use of the fluid property called the *bulk modulus E*. It is defined by the relation

$$E \underline{\Delta} \text{ bulk modulus } \underline{\Delta} - \frac{dp/dV}{V} \qquad (2.12)$$

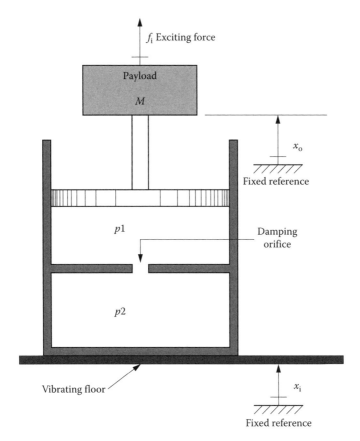

FIGURE 2.20
Air-spring analysis model.

and relates pressure changes dp to volume changes dV in a volume V of gas. Actually there are two common forms of bulk modulus, the *adiabatic* and the *isothermal*, which have different numerical values and each one appropriate for certain situations. When the changes in the system are relatively slow, heat transfer within the gas and its surrounding container keeps the temperature nearly constant (often the *room temperature*) and therefore we use the isothermal value of bulk modulus. (Because the temperature to be used *must* be an absolute temperature, processes where the temperature actually changes by, say, 20°F are regularly treated as isothermal because the change is *not* 20°F out of 70° (a 29% change) but rather 20°F out of 530°R, which is a 4% change.)

When changes take place *rapidly*, there is not enough time for much heat transfer to take place, so we here use the adiabatic value of bulk modulus. For example, in acoustic studies, the air pressure changes of interest are usually sound pressure fluctuations of frequency 20–20,000 Hz, much too fast for significant heat flow to occur. Air-spring vibration isolators also deal with relatively rapid vibrations, so we will use the adiabatic bulk modulus in our present analysis. From texts on fluid mechanics or thermodynamics, we find that the adiabatic bulk modulus for gases is kp, where k is the ratio of specific heats (1.4 for air) and p is the gas absolute pressure. The fact that this property varies with pressure makes the behavior *nonlinear*, however for the small pressure changes caused by vibration in air springs, it is quite accurate (as a linearizing approximation) to treat the bulk modulus as a fixed number, with a value corresponding to the *average* pressure.

Figure 2.20 shows a system with three unknowns: the displacement x_o and the two pressures p_1 and p_2. Therefore, we need to develop three differential equations to describe the system and solve for the unknowns. As usual for linearized analysis of nonlinear systems, we assume small changes about a specified *operating point* for all the variables. We choose this operating point to coincide with the initial *steady state* of the system at time zero, when the weight of the payload is supported motionless by the air pressure force of the piston. Thus the operating-point values for the two pressures are both given by W/A_p, and x_o, x_i, and f_i are all defined as zero. At time equal to zero we allow the inputs x_i and f_i to vary in any way we wish, but always with *small* changes. Our first equation is a simple Newton's law for the mass M:

$$f_i + p_1 A_p = M \frac{d^2 x_o}{dt^2} \tag{2.13}$$

where A_p is the piston area. Note that the gravity force on M (its weight) does *not* appear in this equation because we choose the origin of the coordinate x_o to be at the static equilibrium position, where this force is just balanced by the steady-state pressure force. (This is routinely done in *any* vertical vibration problem.) All pressure symbols in our equations represent not the actual pressures but the small perturbation away from the initial operating-point value. (Actual air springs may require an *experimental test* to get a good number for the effective area. We would need to apply various pressures and measure the force produced, giving a nearly straight-line graph of force versus pressure. The slope of the best-fit line would give us the number A_p.)

The volume V_1, where p_1 exists, will be written as $L_1 A_p$, where L_1 is the length of this chamber. In an actual air spring, the shape may not be a simple cylinder, so we would then just measure the actual volume and not try to relate it to any length. Also, we will treat the volumes as fixed constants, even though they actually vary slightly. (If we do not make these kinds of assumptions, the differential equations will be *nonlinear* and thus analytically unsolvable. We could of course use *simulation* to study the nonlinear equations, but this requires numerical values (not just letters) for all the parameters.) At the design stage, we *never know* these numerical values since we are trying to choose them to meet system performance specifications. Our linearized analysis will usually give us some useful *relations* (formulas) between parameters and performance, and these aid us in preliminary design (simulation *never* supplies such formulas). This gives us trial numerical values of parameters and we can *then* use simulation to refine the initial design. This design sequence (linearized analysis, *then* simulation) is commonly used for all kinds of dynamic systems, not just our example. During a short time interval dt, the pressures will each vary by a small amount, and we need to relate the pressure changes to motions and flow rates. In volume 1 the pressure changes for two reasons. First, the relative motion $d(x_i - x_o)$ causes a volume change $A_p d(x_i - x_o)$. This volume change causes a pressure change which we can relate to the bulk modulus.

Over the same time interval, there will be a flow rate through the damping orifice, adding or removing some volume from chamber 1 and again causing pressure change. We now must model the relation between this flow rate and the pressure difference which causes it. We again assume linear behavior and define a fixed *flow resistance* R_f, so that the instantaneous mass flow rate is given by $(p_1 - p_2)/R_f$. The flow resistance can be estimated from fluid mechanics theory as soon as the detailed shape of the flow passage ("orifice") is known. Some isolator manufacturers use a porous plate as the flow resistance, giving a multitude of small diameter flow paths, which encourages laminar flow. I could not find any published results which establish whether such a flow resistance is really superior to a simple single orifice. In any case, (pressure-drop)/(flow rate) *experiments* are necessary to get accurate values of flow resistance for whatever form of flow passage is used.

The change in pressure p_1 during the time interval dt is attributed to two effects—that due to volume change caused by piston motion, and that due to volume change caused by inflow or outflow at the orifice:

$$dp_1 = \frac{dV_1 E}{V_1} = \frac{d(x_i - x_o)EA_p}{V_1} - \frac{(p_1 - p_2)dt}{\rho R_f}\frac{E}{V_1} \qquad (2.14)$$

$$\frac{dp_1}{dt} = \left(\frac{dx_i}{dt} - \frac{dx_o}{dt}\right)\frac{A_p E}{V_1} - \frac{p_1}{\tau_1} + \frac{p_2}{\tau_1} \quad \text{where} \quad \tau_1 \triangleq \frac{\rho R_f V_1}{E} \quad \tau_2 \triangleq \frac{\rho R_f V_2}{E} \qquad (2.15)$$

Here ρ is the density (mass/volume) of the gas, and is assumed constant at a value corresponding to *room temperature* and the initial steady pressure. A similar analysis for volume 2 yields our third and last equation:

$$\frac{dp_2}{dt} = -\frac{p_2}{\tau_2} + \frac{p_1}{\tau_2} \qquad (2.16)$$

Having obtained three equations in three unknowns, the physical analysis is now complete. If we were going to simulate this system, it would be best to do it directly from the three simultaneous equations (assuming we had the needed numerical values). If, however, we want to show that this system is analogous to the system of Figure 2.10 and Equation 2.9, we need to eliminate the two pressures and get a *single* equation for just x_o. There are various ways to combine n linear equations in n unknowns into one equation for just one of the unknowns, but the most systematic one uses determinants, as explained in the appendix. The equations must be written in D-operator (or Laplace s) form and arranged into rows and columns, with each row being one of the equations and each column being devoted to one of the n unknowns. The system inputs (in our case x_i and f_i) are moved to the right of the equal sign, forming the right-most column. We can then solve for any one of the unknowns, but only *one at a time*. For our example, we have

$$(MD^2)x_o + (-A_p)p_1 + (0)p_2 = f_i \tag{2.17}$$

$$(0)x_o + \left(-\frac{1}{\tau_2}\right)p_1 + \left(D + \frac{1}{\tau_2}\right)p_2 = 0 \tag{2.18}$$

$$\left(\frac{A_pE}{V_1}D\right)x_o + \left(D + \frac{1}{\tau_1}\right)p_1 + \left(\frac{-1}{\tau_1}\right)p_2 = \frac{A_pED}{V_1}x_i \tag{2.19}$$

To solve for any one of our three unknowns, we first *replace* the coefficients in the column of that unknown with the column of inputs on the right. We then form two determinants whose ratio is equal to the unknown and thus will give us the single equation we desire. The *numerator determinant* is formed from the coefficients of the terms on the left. The *denominator determinant* is formed from these coefficients *as they stand* in Equations 2.17 through 2.19; that is, *before* we replaced the desired output column with the input column. Note that we only have to compute the denominator determinant once, since it will be the same no matter which output we seek. The numerator determinant will have to be recomputed for each desired output, since it *changes*.

For simple determinants like our "3-by-3's," manual calculation is not too tedious. For larger determinants (more equations in more unknowns) computer assistance is welcomed, and is available if you have *symbolic processor* software such as MAPLE. I used Mathcad, which includes a subset of MAPLE to do our example. Its output requires only a little manual interpretation to get the results we want. By gathering the terms in the Mathcad printout and simplifying we get

PROGRAM IS CALLED air_spring_determinants.xmcd

numerator determinant $\qquad\qquad$ denominator determinant

$$\begin{pmatrix} \text{fi} & -Ap & 0 \\ 0 & \dfrac{-1}{\text{tau2}} & D+\dfrac{1}{\text{tau2}} \\ \dfrac{\text{ApEDxi}}{\text{V1}} & D+\dfrac{1}{\text{tau1}} & \dfrac{-1}{\text{tau1}} \end{pmatrix} \quad \begin{pmatrix} MD^2 & -Ap & 0 \\ 0 & \dfrac{-1}{\text{tau2}} & D+\dfrac{1}{\text{tau2}} \\ \dfrac{\text{ApED}}{\text{V1}} & D+\dfrac{1}{\text{tau1}} & \dfrac{-1}{\text{tau1}} \end{pmatrix}$$

numerator determinant

$$\frac{-(\text{fi}\cdot D^2\cdot \text{V1}\cdot \text{tau2}\cdot \text{tau1}+\text{fi}\cdot D\cdot \text{V1}\cdot \text{tau2}+\text{fi}\cdot D\cdot \text{V1}\cdot \text{tau1} + \text{ApEDxi}\cdot \text{Ap}\cdot \text{tau1}\cdot D\cdot \text{tau2}+\text{ApEDxi}\cdot \text{Ap}\cdot \text{tau1})}{\text{tau2}\cdot \text{tau1}\cdot \text{V1}}$$

denominator determinant

$$\frac{-(MD^2\cdot D^2\cdot \text{V1}\cdot \text{tau2}\cdot \text{tau1}+MD^2\cdot D\cdot \text{V1}\cdot \text{tau2}+MD^2\cdot D\cdot \text{V1}\cdot \text{tau1} + \text{ApED}\cdot \text{Ap}\cdot \text{tau1}\cdot D\cdot \text{tau2}+\text{ApED}\cdot \text{Ap}\cdot \text{tau1})}{\text{tau2}\cdot \text{tau1}\cdot \text{V1}}$$

$$\left[\frac{ML_1\tau_2}{EA_p}D^3 + \frac{ML_1}{EA_p\tau_1}(\tau_1+\tau_2)D^2 + \tau_2 D + 1\right]x_o$$

$$= (\tau_2 D+1)x_i + \frac{L_1\left(1+\dfrac{\tau_2}{\tau_1}\right)}{EA_p}\left(\frac{\tau_1\tau_2}{\tau_1+\tau_2}D+1\right)f_i \qquad (2.20)$$

Comparing the x_o/x_i relation here with that of Equation 2.9, we see that they have the same form, so we can associate coefficients in one equation with those in the other, and use numerical values from our earlier study to help in choosing numerical values for the air spring. Before doing this, we can get some useful *general* results from Equation 2.20 by looking at the steady-state behavior, which is obtained by setting $D = 0$. We see that for any constant x_i, x_o is equal to x_i. For any *steady* force input, x_o will be that force times the steady-state gain of the x_o/f_i transfer function, as given by Equation 2.20. The ratio force/deflection is of course the *spring constant* of the air spring:

$$\text{spring constant of air spring } \Delta K_{as} \triangleq \frac{EA_p}{L_1+L_2} \qquad (2.21)$$

Equation 2.20 provides another useful result if we remove the damping effect by setting the fluid resistance R_f equal to zero (recall that we did this also for our earlier *metal spring* system). The D^3 and D terms then disappear, giving an undamped second-order system with natural frequency given by

$$\omega_n = \sqrt{\frac{EA_p}{(L_1 + L_2)M}} = \sqrt{\frac{EA_p^2}{(V_1 + V_2)M}} \tag{2.22}$$

Note that this also verifies the spring constant definition of Equation 2.21. The version of Equation 2.22 that uses the volumes rather than the lengths is useful for real air springs, where the shape is not exactly a cylinder. For an air spring, the stiffness depends on the volume of air under compression, and when we set R_f equal to zero, it means that the *hole* between the two air chambers is *wide open* and the two *separate* volumes are now effectively one volume equal to the sum of the two. Be sure to remember that when the damping orifice *is* present, the system is third order, with one first-order factor, and one underdamped second-order whose natural frequency will be close to but *different* from the value in Equation 2.22.

We can now design our air spring to have the same good isolation properties that we designed into the earlier *metal spring* system by requiring that the coefficients of the terms be numerically equal. Because the air spring has a number of parameters that can be adjusted, there is not one unique set of values that gives this correspondence. We thus need to choose some of the parameters based on *other* considerations and then adjust the remaining parameters. Depending on the application, this could be done in various ways. Let us suppose that the space limitations dictate that the piston diameter be 3 in. or less (so we decide to use 3 in.), and that the two chamber lengths be equal (because we can see no advantage to making them unequal, and our earlier *metal spring* study showed no advantage for unequal springs).

Since our payload weighs $100\,\mathrm{lb_f}$, and $A_p = 0.785D^2 = 7.07\,\mathrm{in.^2}$, the steady-state pressure must be set at 14.15 psig (28.85 psia). This makes the adiabatic bulk modulus E equal to 40.39 psi. Comparing Equations 2.9 and 2.20 we see that $\tau_2 = \tau = 7.2/250 = 0.0288\,\mathrm{s}$. Since the chamber lengths have been assumed equal, τ_1 will also be 0.0288. The D^2 coefficient in Equation 2.9 is $0.008(100/386) = 0.00207$, so $2ML_1/EA_p$ must equal this value, allowing us to solve for L_1, which turns out to be 1.15 in. If you check the values of the D^3 coefficient in each equation you will see that they are equal. The only parameter not yet determined is the flow resistance, which can be found from the known values of the time constants. To find the density of the air in the system at steady state, we use the known absolute pressure (28.85 psia) and assume the temperature to be 70°F (530°R):

$$\text{air-specific weight} = \frac{p}{RT} = \frac{28.85 \cdot 144}{53.3 \cdot 530} = 0.147\,\frac{\mathrm{lb_f}}{\mathrm{ft^3}}$$

$$\text{density} = \frac{0.147}{386 \cdot 1728} = 2.20\mathrm{e}-7\,\frac{\mathrm{lb_f\text{-}s^2}}{\mathrm{in.^4}} \tag{2.23}$$

We can now calculate the needed flow resistance:

$$R_f = \frac{\tau_1 E}{\rho V_1} = \frac{0.0288 \cdot 40.39}{2.20 \cdot 10^{-7} \cdot 8.13} = 6.50 \cdot 10^5 \frac{\text{psi}}{\frac{\text{lb}_f}{\text{s}} / 386} \tag{2.24}$$

Accurate values of mass-flow resistance R_f usually require experimental testing, as we mentioned earlier. Proper sizing of the orifice (or other flow passage) is relatively easy to do experimentally by trial and error. We simply operate the system with different size orifices until we get the performance we want. To determine some trial values we can use some theoretical formulas available in the literature* for laminar flow in a sharp-edged orifice or a capillary tube:

$$\text{sharp-edged orifice } R_f = \frac{50.4\mu}{\pi\rho D^3} \quad \text{capillary tube } R_f = \frac{128\mu L}{\pi\rho D^4} \tag{2.25}$$

where
  D is the inside diameter of the orifice or tube
  μ is the fluid viscosity
  L is the length of the tube
  ρ is the density of the gas

Air viscosity varies with temperature but is almost independent of pressure; for a temperature of about 70°F it is about 2.72e-9 lb$_f$-s/in.^{2}. Using the density value computed above, and trying an L value of 1.0 in., we find the diameter of the orifice to be 0.00677 in. and the capillary tube D is 0.0296 in.

While we have computed the air-spring dimensions to meet a specific requirement, commercial isolators of this type are more likely to be offered as a range of *pre-designed* systems, from which the customer will choose an appropriate one. This is more practical than building each unit to a unique set of specifications. For the *passive* air-spring isolator we have discussed, the air pressure must be manually adjusted to support whatever payload weight is present. Since it is common for the equipment on, say, an optical table to be changed and/or moved around from time to time, the height changes and tilting that occur make the needed manual adjustments tedious and possibly inaccurate. For this reason, almost all such isolators add an automatic height control system (*servosystem*) to ease this chore. Whether the height is manually or automatically adjusted, the needed pressure changes cause the air-spring *stiffness* to also change whenever the payload mass changes, because bulk modulus E is proportional to pressure. To support any payload weight W, the air-spring *gage* pressure must be W/A_p psig, but the bulk modulus depends

---

* E.O. Doebelin, *System Dynamics: Modeling, Analysis, Simulation, Design*, Marcel Dekker, New York, 1998, pp. 223, 231.

on the *absolute* pressure, according to $E = 1.4\, p_{abs}$. For many applications, the gage pressure is quite large (often up to about 100 psig) compared with the atmospheric pressure (14.7 psia), so we can approximate the absolute pressure as the gage pressure. This makes the spring constant directly proportional to payload weight, which in turn *makes the system natural frequency independent of the payload,* a feature usually thought of as desirable.

2.4 Active Air-Spring Systems

When some kind of self-leveling system is added, we will call that an *active* air-spring system. In addition to self-leveling (which is a nearly static operation), some systems use electromechanical leveling valves which are fast enough to allow some control of dynamic operation. The simplest self-leveling systems use valves which are entirely mechanical. When payload weight increases for some reason, the payload will move downward slightly, causing the valve to admit air under the diaphragm, increasing the pressure and moving the payload upward. As the payload rises toward the desired height (set manually by an adjustment screw), the valve gradually closes as the payload returns to the proper position. If payload weight should *decrease*, the reverse sequence would again return the payload to the proper position. This simple arrangement is a *feedback system* and thus must be properly adjusted to prevent instability (continual up and down cycling). The manufacturer makes this setting and the user rarely needs to change it. Figure 2.21 (from Barry Controls[*]) shows a possible arrangement. Another manufacturer[†] offers standard mechanical valves which position to about 0.1 in., and precision valves which give about 0.01 in. The standard valves are tightly sealed for motions less than 0.1 in. and thus conserve gas in systems that use a pressurized bottle as the gas supply. (The precision valves leak a very small amount of air.) Technical Manufacturing Corporation (TMC) also offers systems with electromechanical valves and eddy-current displacement sensors; these have a position stability of about 0.0001 in.

While we had earlier stated that we would not go into the dynamic *analysis* of multiaxis isolation systems, we do want to show some clever mechanical/pneumatic isolators which provide isolation in the vertical and all horizontal directions. When combined into sets of four or more such isolators, they also provide angular isolation. The devices we will show are from TMC and are called *gimbal piston* isolators. Vertical isolation is provided by a self-leveling air spring, and horizontal by a sort of inverted pendulum arrangement. These isolators *must* be used in sets of at least three

---

[*] Barry Controls (Burbank, CA, www.barrycontols.com) no longer offers such pneumatic isolators.

[†] Technical Manufacturing Corporation, Peabody, MA, www.techmfg.com

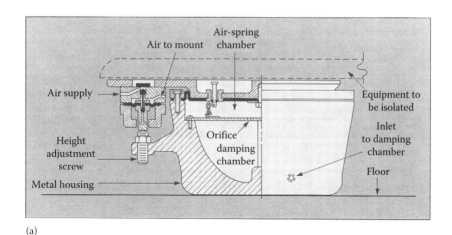

(a)

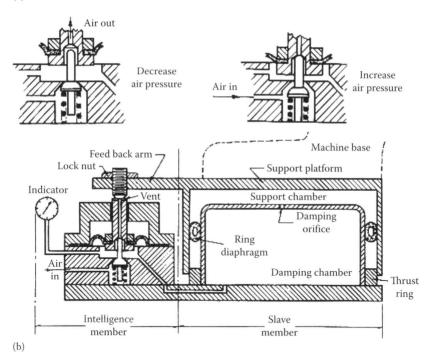

(b)

FIGURE 2.21
Mechanical leveling valve arrangement. (a) Overall system. (b) Valve details.

(four are preferred). (Most payloads are more-or-less *rectangular*, so four isolators, one at each *corner* is a common arrangement.) Figure 2.22 shows a simplified diagram of a single isolator. A design detail emphasized in the TMC catalog is the thin-wall rolling diaphragm used to provide friction-free motion even for vibration inputs of only a few microinches. Such small

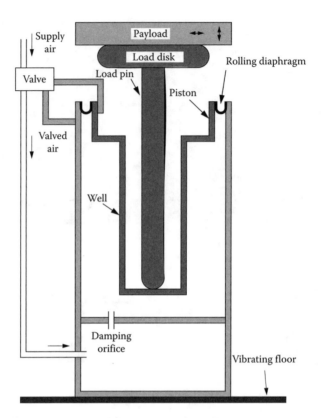

FIGURE 2.22
Gimbal piston isolator of TMC.

vibratory inputs (often from the factory floor) must be isolated in many production machines used in the semiconductor and MEMS industries. TMC
makes a big issue of how they and their competitors formulate performance
specifications, claiming that some vendors base isolation performance on
large vibratory inputs, which are much easier to isolate than the small micro-
inch motions. TMC prefers to send their engineers to the customer's factory
to carefully measure the *actual* vibration environment, so that a suitable
isolation system can be reliably designed. Manufacturers of semiconductor
production machinery usually provide data on how much *floor* vibration
is allowable for proper operation of their equipment. Knowing the *input*
(measured floor vibration) and the *output* (allowable machine vibration) one
hopefully can design an isolation system to meet these needs.

Let us now discuss the detailed operation of the gimbal piston device. The
payload (the machine to be isolated) is fastened to the load disk at the top of
the isolator. This load is transmitted by the load pin (a simple vertical shaft)
to the bottom of the piston well. This well and the piston act vertically as one
rigid piece which feels the pressure difference across the rolling diaphragm.

The pressure self-adjusts (as discussed earlier) so as to always balance the payload's weight and maintain the payload at a fixed vertical position, even when the mass of the payload might change. (The leveling valve is shown very schematically; Figure 2.21 showed details.) The spherical bottom end of the load pin is hardened and rests on a hard flat surface at the bottom of the piston well. As horizontal motion of the isolator base occurs, the load pin will *rock* back and forth on this spherical contact with essentially frictionless motion. Since we might be isolating only a few microinches of floor motion, this rocking motion will of course also be very small. Notice that if we place a mass load on the isolator load disk, the piston well can *tip* since the flexible diaphragm allows a *gimbal-type* rotation of the well. Such tipping would proceed until moving parts of the isolator would bump into the stationary members.

This action is clearly not allowable and is the reason that one must use at least *three* such isolators to support a given payload mass. Figure 2.23 shows how this is done; the payload is rigidly bolted to each of the three gimbal piston isolators. Since the payload is essentially a rigid body, tipping of any of the isolators is now *not possible* since the three load disks must stay in the same horizontal plane. Thus the payload can move horizontally but cannot tip. (This assumes that the three self-leveling air springs maintain the same level.) When the *payload* is a semiconductor production machine of some sort, such machines often have internal parts which move in several axes, giving

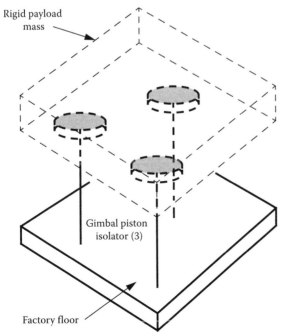

Rigid payload mass

Gimbal piston isolator (3)

Factory floor

FIGURE 2.23
Isolators must be used in groups of at least three.

rise to inertial forces on the payload. These forces are a source of undesired vibration, just like floor motions. The isolation system must counteract both of these effects, so as to keep the payload as steady as necessary. The TMC catalog states that at least three isolators must be used but that the most common situation uses four. When four are used, the air-spring leveling systems must use only three leveling valves, the fourth air spring must be *slaved* to one of the other three, since *three points determine a plane*. Similar considerations apply to systems with more than four isolators and the catalog explains this in some detail.

Having established that a system of three or more isolators allows only horizontal and vertical payload motion, but *no* tipping, we can now try to understand how an individual isolator suppresses horizontal payload motions, and get a formula for the natural frequency of the horizontal vibration of the isolated system. To do this it is helpful to use some simplified diagrams which strip away confusing details but retain the essence of the isolator behavior for horizontal motion. Figure 2.24 shows the essential features of the horizontal action. The (fictitious) ball joint allows the two-dimensional tipping (*gimbal motion*) of the piston well in the same way as the actual rolling diaphragm. Note that the payload and isolator load pin are essentially a single rigid mass which can translate horizontally but *cannot* tip. The piston *well*, however *can* tip (rotate on the ball joint) producing the angular motion θ.

In Figure 2.24a, think of grasping the payload and then moving it horizontally, slightly to the left, causing the piston well to tip through the angle θ,

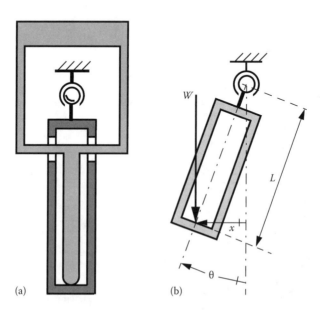

FIGURE 2.24
Analysis model of gimbal piston isolator. (a) Schematic diagram. (b) Analysis sketch.

as shown in Figure 2.24b. The load pin, which is rigid with the payload, also translates to the left, but stays vertical, because the payload cannot tip. (Of course, in Figure 2.24b the load pin does not "go outside" the well as shown there, since the angle is very small. The angle is shown large here just to give a clearer picture.) If you were holding the payload in this deflected position, your hand would feel a horizontal force. This is the horizontal force that the payload feels when it is vibrating in the horizontal (x) direction, and is the only force that would enter into a Newton's law equation ($F = MA$) for the horizontal motion. The payload weight W exerts a moment about the ball-joint pivot point and you can use this fact to easily find the horizontal force on the payload in terms of the x displacement. Remembering that the angles will be very small, we are able to derive a formula for the horizontal natural frequency:

$$\text{horizontal force} = F_x = W(\sin\theta)(\cos\theta) \approx W\theta \quad \text{for small angles} \quad (2.26)$$

$$\sum F_x = -W\theta = -W\frac{x}{L} = M\ddot{x} \quad \frac{\ddot{x}}{g} + \frac{x}{L} = 0 \quad \omega_n = \sqrt{\frac{g}{L}} \quad (2.27)$$

This is of course the well-known formula for the frequency of a simple pendulum, for small motions. We see that for horizontal isolation, the natural frequency depends only on the length L, *and is independent of payload weight.* We found this independence also for the vertical isolation when the air spring was self-leveling.

When designing this system for horizontal isolation, we are severely restricted by the availability of only one adjustable parameter, the length L. Some horizontal isolation applications require a natural frequency of about 0.3 Hz, which would lead to $L = 109$ in., clearly impractical. To meet this need, TMC invented the compact subhertz pendulum (CSP) shown in Figure 2.25. Here the air-spring housing is suspended from three angled cables, again forming a pendulous structure. The cables' inclination with the vertical results in a natural frequency that depends on both L and the angle θ, resulting in the ability to get low natural frequencies without excessive L values. In the analysis model of Figure 2.26 we analyze a cable arrangement that would provide for only one axis (called x) of motion, while the actual device, just as in the gimbal piston isolator, provides for horizontal inputs in *any* direction in the horizontal plane. The two cables shown lie in the same vertical plane, and the horizontal motion also takes place in this plane. This simple model is easier to visualize than the actual system and we will later be able to use its results for the real system by means of a simple geometrical interpretation.

Recall that such isolators cannot be used *singly*; they must be used as a set of at least three, all bolted rigidly to the payload, so that no tipping, only horizontal motion is possible. When the system is in its equilibrium position, the cables each make an angle θ_0 with the vertical. If we then displace the

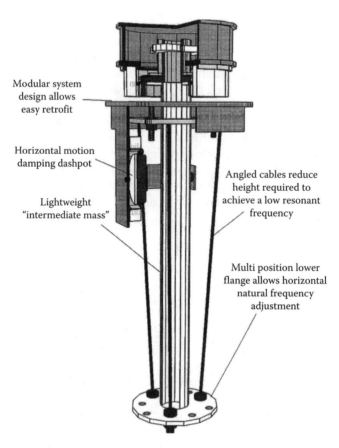

FIGURE 2.25
CSP of TMC.

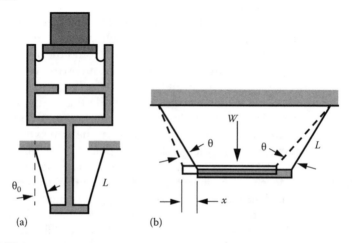

FIGURE 2.26
Analysis model for CSP. (a) Schematic diagram. (b) Analysis sketch.

payload horizontally by an amount x (there is also a much smaller y motion which we neglect), the cable angles will each change by a small amount θ. The motions are so small that the *left and right* angular changes can be assumed to be equal. Since we are interested in the horizontal natural frequency of the system we need to find the horizontal force F_x that the payload and attached isolator mass will feel, in terms of the x displacement. As in the gimbal piston study, we can assume that the moving mass cannot tip because we use at least three isolators rigidly bolted to the payload. Assuming small angles and using some common identities from trigonometry, we get

$$F_x = \frac{W}{2\cos(\theta_0 + \theta)} - \frac{W}{2\cos(\theta_0 - \theta)} \qquad x \approx L\theta \tag{2.28}$$

$$\cos(\theta_0 + \theta) = \cos\theta_0 \cos\theta - \sin\theta_0 \sin\theta \approx 1 - \sin\theta_0\theta \tag{2.29}$$

$$\cos(\theta_0 - \theta) = \cos\theta_0 \cos\theta + \sin\theta_0 \sin\theta \approx 1 + \sin\theta_0\theta \tag{2.30}$$

$$F_x \approx \frac{W}{2}\left[\frac{1}{1-(\sin\theta_0)\theta} - \frac{1}{1+(\sin\theta_0)\theta}\right] = \frac{(2\sin\theta_0)\theta W}{2} = \frac{2\sin\theta_0 x W}{2L} \tag{2.31}$$

$$\frac{Mg}{2}\left(\frac{-2\sin\theta_0}{L}\right)x = M\ddot{x} \qquad \omega_n = \sqrt{\frac{g\sin\theta_0}{L}} \tag{2.32}$$

It remains now to relate this *one-dimensional* result to the true *three-dimensional* geometry of the actual cable suspension. In Figure 2.27 we assume for convenience that the motion of interest (which *could* be chosen in any direction in the horizontal plane) is in the *left/right* direction. Then the cable location at "a" would contribute *no* horizontal force, while those at "b" and "c" would

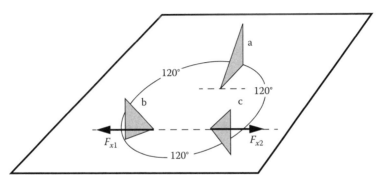

FIGURE 2.27
Relating the simplified model to the real system.

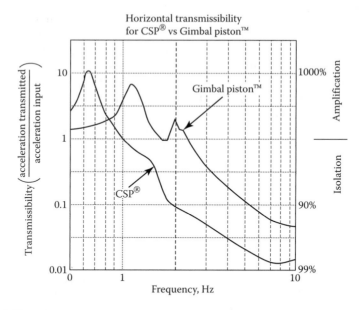

FIGURE 2.28
Comparison of gimbal piston and CSP isolators.

contribute components equal to those we used in Figure 2.26 multiplied by the cosine of 30°. This would give our final corrected natural frequency as

$$\omega_n = \sqrt{\frac{\cos(30°)(g\sin(\theta_0))}{L}} = 0.93\sqrt{\frac{g\sin\theta_0}{L}} \tag{2.33}$$

The TMC catalog quotes a CSP isolator with a length of 16 in. as having a natural frequency of 0.3 Hz. Using our results, this would require an angle of about 9.75°. Figure 2.28 (from the TMC catalog) shows the low-frequency improvement obtained with this CSP isolator.

We complete this section on active air-spring isolators by showing another TMC design in Figure 2.29. This figure is taken from U.S. patent 5,918,862 and is for a unit with improved damping, attained by using silicone oil (about 100,000 centistoke viscosity) in an added damping mechanism. The pneumatic (orifice) damping that is a standard feature of gimbal piston isolators is usually sufficient, particularly because *too much* damping, as we learned earlier, gives poorer isolation at high frequency. Higher damping is helpful in applications such as semiconductor manufacture where wafer-handling robots attached to the isolated mechanism create disturbing forces with their rapid "pick and place" motions. Here the desired damping is often close to *critical damping* (damping ratio of 1.0), which is unattainable with pneumatic orifice damping but often feasible with the added liquid damping. The TMC Web site and printed catalog list a number of patents related to their products. Some of these documents go into great detail and are thus a useful source of

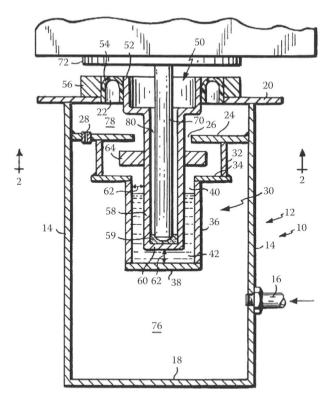

FIGURE 2.29
Gimbal piston isolator with added liquid damping. (From U.S. Patent 5,918,862, sheet 1 of 3, July 6, 1999.)

information. On the Web site there are links that take you to the U.S. patent office site, but I have found this difficult to use. I much prefer Google's patent site which is very easy to use and provides PDF downloads of the patents.

2.5 Low-Frequency Isolation Using Negative-Spring-Constant Devices

As we have discovered earlier in this chapter, when the vibration-exciting effects have low-frequency content, good isolation requires the isolator to have a low natural frequency, which usually means that rather soft springs must be used. Since the static deflection is g/ω_n^2, a natural frequency of 1 rad/s results in a static deflection of 386 in., which is thoroughly impractical. A clever mechanical device can produce *negative* spring constants, which when combined with *ordinary* springs, allows us to get very low spring constants without the buckling and large space requirements associated with conventional spring isolators.

As many readers will already know, a simple spring/mass/damper system using a spring with a negative spring constant is dynamically *unstable* and thus useless for any practical work. In fact, when we are writing Newton's law equations for any vibrating system, if we see *negative signs* on *any* terms in the system's characteristic equation, we start looking for *errors* in our analysis. That is, the physical parameters of mass, viscous damping, and elasticity are assumed to all carry *positive* numerical values. As a simple example, consider a spring/mass/damper system with the equation

$$5 - 1\dot{x} + 1x = 1\ddot{x} \tag{2.34}$$

which has a negative spring constant $-1.0\,lb_f$/in. and has an applied force of $5.0\,lb_f$. If we assume that the initial displacement and velocity are both zero when the 5 lb step input force is applied, we can easily get an analytical solution for the displacement x:

$$x = 1.94e^{-2.74t} + 3.06e^{1.74t} - 5.0 \tag{2.35}$$

The term $3.06e^{1.74t}$ clearly goes toward infinity as time goes by and this behavior exists now matter what the applied force is, or if there is no applied force at all! (If there is no applied force, the *slightest* nonzero initial displacement or velocity results in the same instability.) This example makes it clear that if we are to use springs with negative stiffness, we need to be very careful how we employ them!

Dr. David L. Platus (use Google Patents and his name, to see some of his inventions) is the inventor of a practical isolator concept using negative spring constants and is the founder of a company[*,†,‡] which manufactures and markets these totally mechanical devices (they use no electrical or pneumatic power sources). Figure 2.30 shows a unit which combines a tilt isolator, a vertical isolator, and a horizontal isolator in one compact assembly. To reveal the basic concept of a negative spring constant without the confusion of various practical details, we will study the simple toggle mechanism of Figure 2.31. The two horizontal forces FH are assumed constant; there are various ways this may be implemented. For static equilibrium, we can write

$$\text{force in the diagonal links of length } L \triangleq F_L = \frac{FH}{\cos\theta}$$

$$\text{vertical force } F = 2F_L \sin\theta = 2FH\tan\theta = 2FH\frac{x}{\sqrt{L^2 - x^2}} \tag{2.36}$$

* Minus K Technology, Inglewood, CA, david@minusk.com, www.minusk.com
† D.L. Platus, Smoothing out bad vibes, *Machine Design*, February 26, 1993, pp. 123–130.
‡ D.L. Platus and D.K Ferry, Negative-stiffness vibration isolation improves reliability of nano-instrumentation, *Laser Focus World*, October 2007, pp. 107–109.

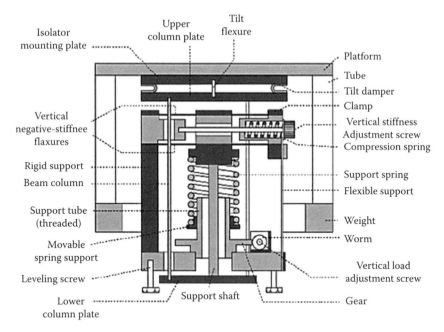

FIGURE 2.30
Negative spring constant isolator from Minus K technology.

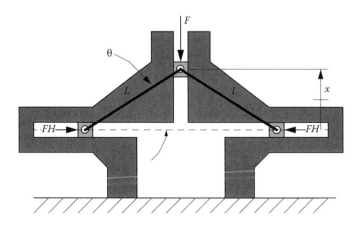

FIGURE 2.31
Analysis model of toggle-type negative spring constant mechanism.

Any spring is defined by the relation between the applied force and the resulting deflection; in our case the relation between F and x, which is clearly nonlinear is shown in Figure 2.32 for the numerical values $L = 5.0$ in. and $FH = 250\,\text{lb}_f$. Near the origin, this nonlinear spring is nearly linear, with spring constant $-2FH/L$ ($-100\,\text{lb}_f/\text{in}.$ in our case). To stabilize this unstable mechanism and also provide support for a payload of weight W, we add an

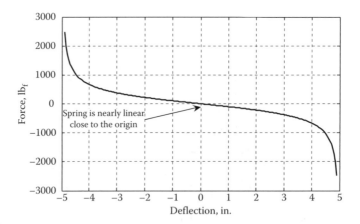

FIGURE 2.32
Toggle mechanism has a negative (and nonlinear) spring constant.

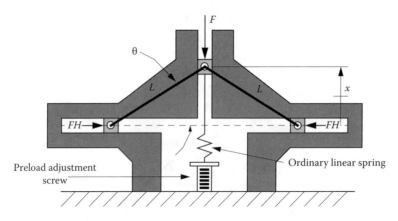

FIGURE 2.33
Practical isolator uses both positive and negative springs.

ordinary linear spring as in Figure 2.33. This spring must have a *free length* such that when the payload is present, it deflects to the $x = 0$ position. If we made this spring's (positive) stiffness exactly equal to the negative stiffness of the toggle mechanism at $x = 0$, the *net* spring constant would be *zero*, as would the *natural frequency* of the mass/spring system. Since zero is the lowest possible natural frequency, such an isolator would have the best possible isolation. This result is of course "too good to be true"; the slightest reduction (due to temperature, aging, etc.) in the ordinary spring's stiffness would make the total system unstable. We thus design the ordinary spring's stiffness to be slightly *greater* than this critical value to allow a margin of safety, and provide an adjustment screw to raise or lower this spring to suit a given payload weight. In Figure 2.34, we have set the numbers such that a 100 lb payload settles into equilibrium at the $x = 0$ position.

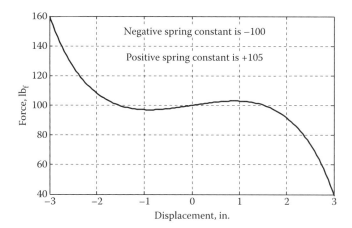

FIGURE 2.34
Isolator force/deflection curve with ordinary spring adjusted to support a 100 lb payload.

While the analysis model of Figure 2.33 is correct in principle and easy to analyze, the actual hardware used by Minus K Technology often takes a different form. In particular, to realize an adjustable means of providing the negative spring constant, they employ *beam/columns*. In Figure 2.35 we recall the accepted engineering meanings of the terms *beam, column,* and *beam/column*. The formulas for computing column and beam/column behavior are different for different *end conditions*. For given end conditions, one can calculate the *critical buckling load* for the column. For a beam/column, the horizontal spring constant (F/x in Figure 2.35) will vary, depending on the vertical load that is simultaneously applied. As the vertical load (called P_{design} in Figure 2.35) approaches the critical buckling load, the horizontal spring constant approaches zero. Here we do not really want a *negative* spring constant since we saw earlier (Figure 2.34) that the *net* spring constant must

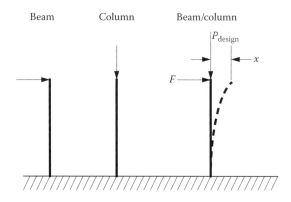

FIGURE 2.35
Use of beam/column as a low-stiffness spring.

be small but positive. This can be achieved in the beam/column by providing an adjustable vertical force. Some of these features are recognized in the practical isolator of Figure 2.30.

To design a proper beam column we need theoretical formulas for the horizontal deflection caused by a horizontal force, in the presence of a given vertical force which is somewhat less than the critical buckling load. For the simple configuration of Figure 2.35 (which is *not* the same as the *both ends built-in* conditions in the practical isolator), I found the needed formulas.* The simple Mathcad program below shows some results.

program name is beam_column_negative_springs.xmcd

computes data for beam/columns as negative springs (one end fixed, one free)

first set column width b, thickness h, length L, and modulus E all dimensions in inches and pounds force

$b := 1.0 \qquad h := 0.05 \qquad E := 3.10^7 \qquad L := 5.0$

$I := \dfrac{b \cdot h^3}{12} \qquad I = 1.042 \cdot 10^{-5}$

$Pbuck := \dfrac{\pi^2 \cdot E \cdot I}{4 \cdot L^2} \qquad Pbuck = 30.843 \quad n := 0.099, 0.199 \ldots 0.999 \quad Pdesign(n) := n \cdot Pbuck$

$M(n) := \sqrt{\dfrac{E \cdot I}{Pdesign(n)}}$

$Kspring(n) := \dfrac{Pdesign(n)}{M(n) \cdot \tan\left(\dfrac{L}{M(n)}\right) - L}$

$M(n) =$
10.117
7.135
5.821
5.039
4.506
4.113
3.807
3.561
3.357
3.185

n =	Kspring (n) =	Pdesign (n) =
0.099	6.766	3.053
0.199	6.023	6.138
0.299	5.279	9.222
0.399	4.532	12.306
0.499	3.783	15.39
0.599	3.033	18.475
0.699	2.28	21.559
0.799	1.525	24.643
0.899	0.767	27.727
0.999	$7.61 \cdot 10^{-3}$	30.812

---

* W.C. Young and R.G. Budynas, *Roark's Formulas for Stress and Strain*, 7th edn., McGraw-Hill, New York, 2002, p. 229.

The desired horizontal spring constant is called K_{spring}, and we compute it for a range of P_{design} values from 0.099 to 0.999 times the critical buckling load P_{buck}. We see that we can get, by adjusting the vertical force, horizontal spring constants ranging from 6.77 to 0.00761 lb_f/in. We of course need to *back off* from the critical buckling load by a sufficient amount to give stable operation, but very low spring constants are practically obtainable, giving the low horizontal natural frequencies we desire.

Knowing the payload mass and the horizontal spring constant, we can estimate the natural frequency, treating the system as linear for the small motions usually encountered. To get a little practice in handling nonlinear systems, let us use Simulink to simulate the simple nonlinear model of Figure 2.33. The system parameters are as follows: payload weight is 100 lb_f, linear spring constant is 105 lb_f/in., $FH = 250 \, lb_f$, $L = 5.0$ in., and we assume a small amount of viscous friction, say, 0.2 lb_f/(in./s). In Figure 2.33 we assume that the payload is attached to the vertical sliding member, whose displacement is now called x, and that the assumed viscous damping effect acts on the velocity dx/dt. The *foundation* in this figure now becomes the vibrating floor, with displacement called x_i, and the motion called x in that figure is now the relative displacement $(x - x_i)$ in our new terminology.

Applying Newton's law to the payload mass M, we get the following non-linear differential equation:

$$F_i - K_s(x - x_i) - B\dot{x} + \frac{(2)(250) \cdot (x - x_i)}{\sqrt{25 - (x - x_i)^2}} = M\ddot{x} \tag{2.37}$$

which leads to the simulation diagram of Figure 2.36. The disturbing inputs F_i and x_i could be any time functions associated with a specific practical application; in this simulation I provided for step and sinusoidal variations of force, and sinusoidal and random variations of floor displacement.

The linearized natural frequency is obtained from the payload mass (100/386) and the net spring stiffness of $105 - 100 = 5 \, lb_f$/in.; it is numerically 0.70 Hz. We can check our simulation by applying a small input force step, and *measuring* the frequency of the damped sine wave of x that results. Figure 2.37 shows a measured value of 0.69 Hz for the *damped* natural frequency, a little less than the theoretically predicted *undamped* value, as expected. In Figure 2.38, the input force is sinusoidal with amplitude 0.001 lb and frequency 2.0 Hz, while the floor vibration is sinusoidal with amplitude 0.001 in. and frequency 10 Hz. The displacement x of the isolated mass is graphed directly, but the input force and floor vibration are scaled and shifted to give a clearer graph. We see that due to the low frequency of the force, this input is not well isolated, while the higher frequency floor vibration is well attenuated. When we input a random floor vibration with frequency content from 0 to about 100 Hz, Figure 2.39a shows the high frequencies well isolated, but a visually apparent low-frequency response at about 0.7 Hz, the system's natural frequency. (Figure 2.39b is a zoomed version which shows the input waveform more clearly.)

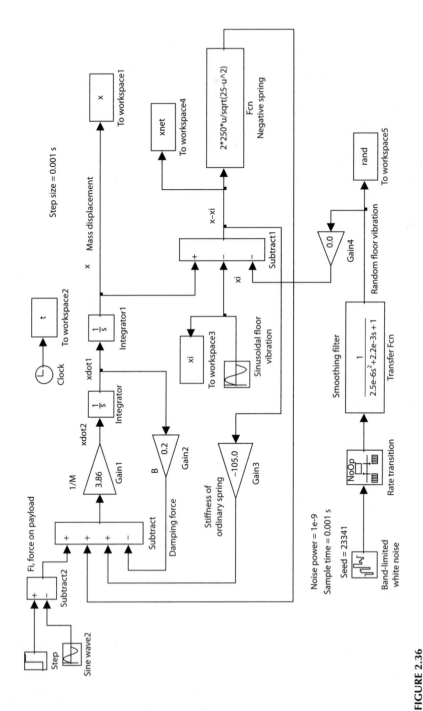

FIGURE 2.36
Simulink simulation of negative spring vibration isolator.

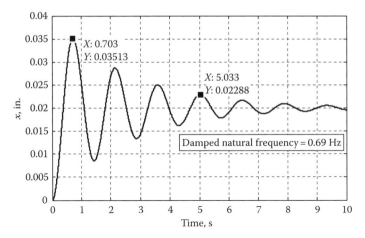

FIGURE 2.37
Simulation check of predicted natural frequency.

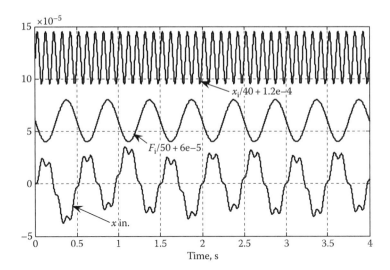

FIGURE 2.38
Isolator response to simultaneous inputs of force and floor vibration.

When discussing Figure 2.35 and the calculations based on that type of beam/column, we mentioned that this was *not* the proper configuration to represent the practical isolator of Figure 2.30, for which we show a simplified analysis model in Figure 2.40. There we can make use of some symmetry to show that the formulas we gave for the simpler situation can be adapted to the more complicated one. When there are *two* columns as in Figure 2.40, with F_v centered between them, the buckling value of F_v will be twice the

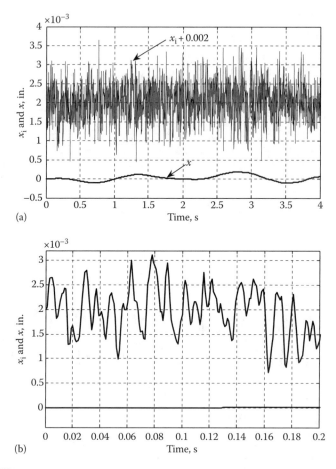

FIGURE 2.39
Isolator response to random floor vibration. (a) Graph for long times. (b) Expanded time scale.

force which causes the individual column to buckle, since each column feels only half of F_v, and the horizontal force at the column midpoint will also be half of the applied force F_h. Looking at the midpoint of the left column, we see that this is an *inflection* point, where the local curvature changes from concave upward to concave downward, so the local curvature is zero, which means that there is *no* bending moment at that point. Thus the lower half (and also the upper half) of this column are in the same loading situation as in Figure 2.35 and we can use the same formulas (properly applied) to get the horizontal spring constant F_h/x of the entire mechanism, since it is twice (two columns) the value of the spring constant $(F_h/2)/(x/2)$ at the midpoint. The value of $F_v/2$ that causes buckling is given by our earlier formula, with the column length taken as $L/2$; that is, $\pi^2 EI/[(4)(L/2)^2]$. Then the total load F_v which causes buckling is $2\pi^2 EI/L^2$, which is twice what we got earlier for the

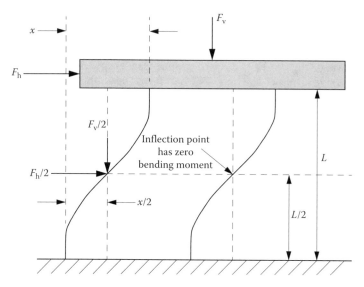

FIGURE 2.40
Analysis of double beam/column.

simpler column (taking length there as $L/2$). Again, a Mathcad program is used to get some numerical results. Here La is used for L in Figure 2.40.

$$La := 5.0 \quad Fvbuck := \frac{2 \cdot \pi^2 \cdot E \cdot I}{La^2} \quad pdesigna(n) := \frac{n.Fvbuck}{2} \qquad \begin{aligned} b &= 1 \\ h &= 0.05 \\ E &= 3 \cdot 10^7 \end{aligned}$$

$$Ma(n) := \sqrt{\frac{E \cdot I}{Pdesigna(n)}} \qquad Kspringa(n) := \frac{2 \cdot Pdesigna(n)}{Ma(n) \cdot \tan\left(\dfrac{0.5 \cdot La}{Ma(n)}\right) - \dfrac{La}{2}}$$

n =	Kspringa (n) =	Pdesigna (n) =
0.099	108.261	12.214
0.199	96.375	24.551
0.299	84.459	36.888
0.399	72.513	49.225
0.499	60.534	61.562
0.599	48.523	73.899
0.699	36.477	86.236
0.799	24.396	98.573
0.899	12.278	110.91
0.999	0.122	123.247

Fvbuck = 246.74

As far as I know, Minus K Technologies is the only maker of *negative spring constant* vibration isolators, probably due to patent protection. They offer a variety of models for different applications. While we, in our simulation, used viscous damping to model frictional effects and thus prevent unending natural vibrations, Platus (1993) explains how the *hysteretic* damping inherent in any stressed material may in some cases be sufficient. In fact it is argued that the low spring constants achieved in these designs actually *magnify* these damping effects in a useful way.

2.6 Active Electromechanical Vibration Isolation

The air-spring systems discussed earlier are certainly active systems since they are feedback control systems and use an external power source (the compressed air supply for the leveling servosystems) to achieve their desired performance. Pneumatic servosystems of this type are, however, usually too slow to be effective in counteracting the floor vibration and/or exciting force disturbances which cause payload vibration. They are only useful in maintaining the payload level and keeping the natural frequency constant for payloads of different mass. If we want to use feedback systems to *dynamically* fight against vibration, we must use hardware that is capable of much faster response. Various forms of electromechanical vibration-isolation systems have been developed for this purpose. Some are *one-of-a-kind* schemes intended for a specific purpose. Many of the space experiments (such as the Hubble telescope) carry very precise instruments of one kind or another that will tolerate only miniscule vibration, and part of the experiment design is to develop vibration isolation that meets these needs. However, a number of commercial manufacturers offer *general purpose* isolation systems that need only some specific *tuning* to meet the customers' isolation needs. It is also common to *combine* active and passive isolation concepts when a single method is unable to meet specifications.

We will concentrate on the general-purpose type of system because the special-purpose applications have problems and solutions that are to some extent unique to each application, and also tend to be quite complex and thus unsuitable for our introductory treatment. (If you should at some point need to study some of these more exotic systems, since many of them are part of government-sponsored space projects, you will find quite detailed discussions in papers published in the open literature.) A basic form of an active isolation system is shown schematically in Figure 2.41 and in conventional block-diagram form (as used in feedback system studies) in Figure 2.42. There we use two motion sensors (accelerometer and geophone) to measure payload absolute acceleration and velocity and an electromechanical

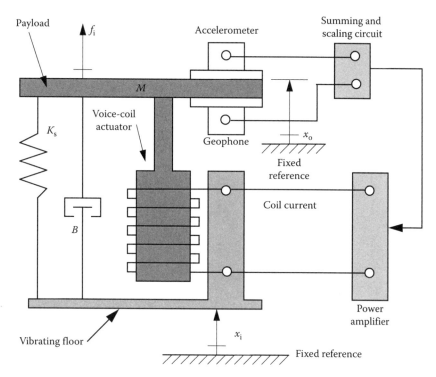

FIGURE 2.41
Basic active vibration-isolation system.

force producer called a *voice-coil actuator*.* The voice-coil actuator is a simple device identical in principle with the loudspeaker, but designed for precision force and motion control rather than acoustic sound production. It consists of a permanent magnet and a wound coil so arranged that a force proportional to the coil current is produced. The *power amplifier* shown is of the type called *transconductance*, which means that a voltage applied to the input terminals produces a proportional output current *without regard to the circuit being driven* (within limits, of course). This is a desirable feature since it suppresses some unwanted delays in the conversion from input voltage to output force that would be present if we used a conventional *voltage amplifier*. Thus we can model the amplifier and voice-coil actuator with the following equation:

$$\text{Magnetic force} = K_{mf}\, e_{ampi}$$

where e_{ampi} is the input voltage of the amplifier.

---

* http://www.bei-tech.com/products/actuators/voice_coil.htm

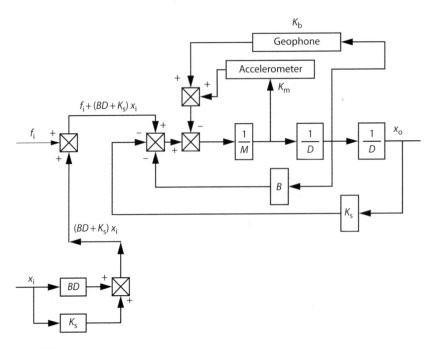

FIGURE 2.42
Feedback system diagram for active isolator.

The accelerometer and geophone sensors have their own dynamics but
we will initially assume them to be perfect sensors whose output voltages
are instantly proportional to payload acceleration and velocity. These are
often good assumptions because sensors are available in a wide range of
dynamic behavior and we may be able to choose ones whose dynamics
really are negligible. If not, we can always factor in the neglected dynamics *later*, after some preliminary calculations, which we try to keep simple.
The *geophone* is an absolute velocity sensor[*][†] originally marketed for seismic
well-drilling exploration but now useful in more general applications. It uses
the *moving-coil* velocity transducer and a mechanical spring/mass/damping
system to measure the absolute velocity of its case, which we fasten to our
payload. *Accelerometers* come in various forms[‡] most of which have simple
second-order dynamic behavior, but we again ignore these dynamics in our
preliminary studies. The output voltages from these two sensors are sent to
a scaling and summing circuit, so that we can multiply each voltage by an
adjustable factor and then sum these voltages to give a *single* input voltage
for the power amplifier. Our isolator will probably need some kind of *leveling*

---
[*] E.O. Doebelin, *Measurement Systems*, 5th edn., McGraw-Hill, New York, 2003, pp. 347, 356–357.
[†] http://www.geospacelp.com/industry2.shtml
[‡] E.O. Doebelin, *Measurement Systems*, 5th edn., McGraw-Hill, New York, 2003, pp. 357–372.

system, but this can be dealt with separately and does not interact significantly with the dynamic isolation, so we will not include it.

The payload mass is acted upon by only three forces, the spring force, the damper force, and the magnetic force (we can neglect gravity, as usual in vertical vibration problems), and we consider both input force f_i and floor vibration x_i.

$$f_i + K_s(x_i - x_o) + B(\dot{x}_i - \dot{x}_o) - K_m\ddot{x}_o - K_b\dot{x}_o = M\ddot{x}_o \qquad (2.38)$$

The magnetic force has two components, one proportional to acceleration and the other proportional to velocity; the coefficients K_m and K_b are products of the sensor sensitivity, the scaling factor applied in the scaling circuit, and K_{mf}. The magnetic force is really applied *between* the payload and the vibrating floor, so it acts (in the reverse direction) on the floor, not just on M. This does *not* affect our equations because we consider the motion x_i to be truly a motion *input*, unaffected by *any* forces. (In actuality, the moving floor *does* feel this force, but in most cases, the floor motion with the isolator present is essentially the same as with the isolator absent.) Note that the algebraic signs of K_m and K_b could be either positive or negative, because the sign of a voltage is easily reversed electronically and the voice-coil actuator can produce either upward or downward forces. Rewriting this equation,

$$(M + K_m)\ddot{x}_o + (B + K_b)\dot{x}_o + K_sx_o = B\dot{x}_i + K_sx_i + f_i \qquad (2.39)$$

Now we see that the constant K_m must have the same dimensions as M (they are added) and must have the same *effect* as M, so we have created a sort of "electronic mass," which can be either positive or negative! This is a well-known trick of feedback system design, and is usually used to *decrease* the inertia effect in motion-control systems which are too slow,[*] by making K_m a negative number.

In vibration isolation, we want *more* inertia because it gives lower natural frequency, but we often do not want to add more physical mass. Making K_m positive increases the *effective* inertia without changing M itself. A further improvement contributed by this feedback scheme is found in the damping term. We saw in earlier examples that damping was useful in limiting the resonant peak, but made transmissibility *worse* at higher frequencies. In our feedback system, we can now make B as small as possible (removing the *viscous coupling* between x_i and x_o that degrades isolation) but maintain a well-controlled resonant peak by providing a sufficiently large value of K_b. Another way of interpreting this is that K_b can be thought of as a *virtual*

---

[*] E.O. Doebelin, *Control System Principles and Design,* Wiley, New York, 1985, p. 365.

damper connected between M and the "*ground*," so it provides damping but does *not* transmit any undesired forces from x_i to M. In the vibration-isolation business, this is called *inertial damping*. We see here why feedback is used so often in designing high-performance dynamic systems of all kinds. Without feedback we have only two design parameters to adjust: the spring constant and the physical damping. This limits the system's performance. With feedback we have *two more* design parameters, allowing much more design freedom and thus improved performance. Of course, the system is more complicated and expensive, and the design procedure must assure *stability*,[*] since instability is possible in any feedback system. That is, in every feedback system, when we make adjustments to improve speed and accuracy, if we go too far, the system *will* become unstable.

The transfer function relating x_o to x_i is easily obtained from Equation 2.39:

$$\frac{x_o}{x_i}(D) = \frac{\dfrac{B}{K_s}D+1}{\dfrac{M+K_m}{K_s}D^2 + \dfrac{B+K_b}{K_s}D+1} = \frac{\tau D+1}{\dfrac{D^2}{\omega_n^2} + \dfrac{2\zeta D}{\omega_n}+1} \tag{2.40}$$

Considering the frequency-response version of this equation, we see that making B negligibly small will keep the numerator nearly constant at 1.0 as frequency increases, whereas with B present, the numerator (and thus the transmissibility) will get worse (larger) at high frequency. In some systems we considered earlier, we were *not allowed* to make B small since that gave too high a resonant peak at low frequencies. With feedback we can "have our cake and eat it too" since we can use K_b to get a desirable damping ratio even if B is zero. We can also make K_m large enough to achieve very low natural frequencies and thus good isolation at low frequencies. A simulation allows us to easily study these possibilities. I first set B and K_b equal to 0.0 to get an undamped system (see Figure 2.43). I then calculated the value of K_m needed to get a natural frequency of 0.5 Hz for a payload weight of 100 lb$_f$ and a spring constant of 100 lb$_f$/in. I then applied a step input of force of 0.001 lb$_f$, with x_i set equal to 0.0. This gave the undamped sine wave of 0.5 Hz frequency seen in the figure and verifies the theoretical calculation. I then calculated the value of K_b that would be needed (B is still 0.0) to get a damping ratio of 0.7. With K_b at that value (28.4) I again applied the step force, getting the well-damped response shown. Applying the same random floor motion used in Figure 2.36, we get the good isolation results of Figure 2.44.

Since all these good results assumed *perfect* sensors for velocity and acceleration, we should now model these instruments more realistically. Accelerometers come in many forms for different applications and we need to

---

[*] E.O. Doebelin, *Control System Principles and Design*, Wiley, New York, 1985, pp. 185–204.

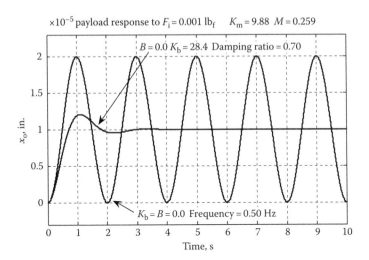

FIGURE 2.43
Feedback isolator for ideal conditions.

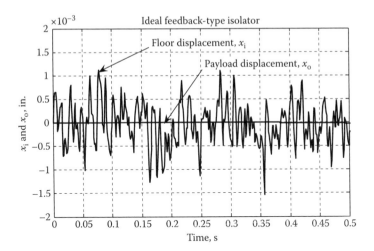

FIGURE 2.44
Ideal feedback-type isolator for random floor motion.

choose one suitable for ours. We *do not* want the piezoelectric type since they do not accurately measure low frequencies; most other types are possible candidates. Either a *deflection-type* or *servo-type* accelerometer with suitable specifications could serve our needs. Both of them have simple second-order dynamics and we need to choose a proper natural frequency and damping ratio. Many isolation applications involve frequencies up to about 30 Hz, so an accelerometer with 100 Hz natural frequency and 0.7 damping is a reasonable first choice. In the simulation diagram of Figure 2.45, I provide for

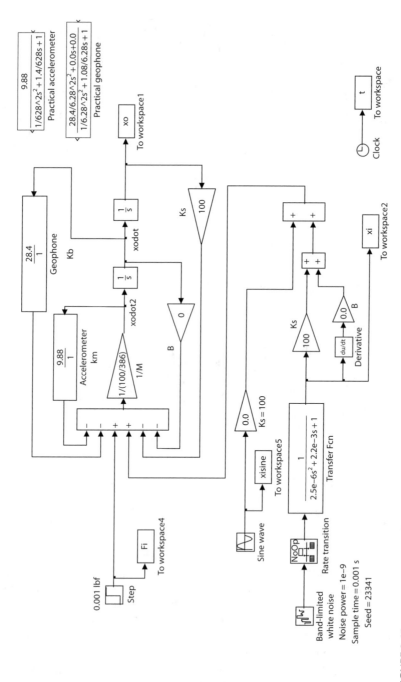

FIGURE 2.45
Simulation of feedback-type isolator.

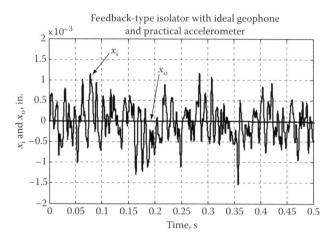

FIGURE 2.46
Accelerometer dynamics do not degrade performance.

both the ideal sensors and the practical versions. I next inserted the practical accelerometer but left the ideal geophone in place. We see in Figure 2.46 that a *real* accelerometer seems to work well here. Practical geophones, like all seismic absolute velocity pickups, have a transfer function that gives no response at zero frequency (a steady velocity) and poor response at frequencies below the instrument's natural frequency:

$$\frac{e_o}{v}(D) = \frac{K\dfrac{D^2}{\omega_n^2}}{\dfrac{D^2}{\omega_n^2} + \dfrac{2\zeta D}{\omega_n} + 1} \tag{2.41}$$

The lowest natural frequency geophone available from Geospace (see footnote[†] on page 82) has a 1.0 Hz natural frequency, a damping ratio of 0.54, and a sensitivity K of 3 to 15 V/(in./s). We did not choose a specific voice-coil actuator mainly because we did not know how much force might be needed, but there should be no difficulty in combining the geophone sensitivity with the actuator force constant and the power amplifier gain (A/V) to get the K_b value of 28.4 that we have assumed. Using Equation 2.41 we can compute the geophone amplitude ratio, as shown in Figure 2.47. We can now insert the practical geophone into our simulation (the practical accelerometer is already there) with the disastrous consequences shown in Figure 2.48.

This unstable system is of course not acceptable and we need to try and fix it. Note that until now, we have taken the physical damping B as zero, which is not possible in the real world; there *must* be some kind of frictional effects. Let us try to add some B to see if stability can be achieved,

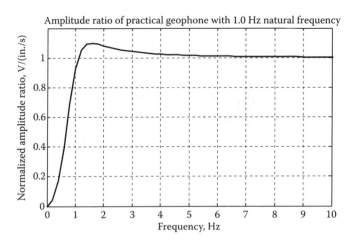

FIGURE 2.47
Frequency response of practical geophone.

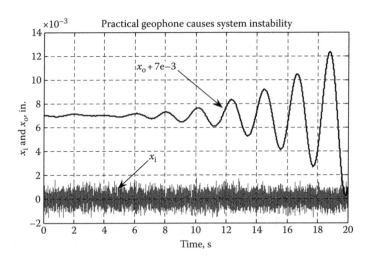

FIGURE 2.48
Geophone dynamics causes stability problem.

knowing that this will to some extent impair the excellent high-frequency isolation associated with $B = 0.0$. After numerous trials I was unable to find any satisfactory combination of settings. At that point, another well-known approach was considered. Why not use *only* the accelerometer and send its signal to an *integrating device*, to generate our velocity signal? While integration of time-varying signals is mostly a benign operation (it tends to smooth out most electronic noise effects), it does have a potential problem. That is, if the *input* to the integrator contains even a small steady bias signal (a real accelerometer with zero input acceleration will *not* have exactly zero at its output), then the integrator will gradually saturate and be disabled. A potential solution to this problem is to use an *approximate* integrator, which does not have this saturation problem. The simplest version of this is a first-order system $(K/(\tau D + 1))$ which approaches an ideal integrator as the frequency gets higher. At low frequencies it is *not* an integrator, but it does not go to *zero* at zero frequency, which was part of the problem with the geophone. We can easily try out this idea with a simulation modified as in Figure 2.49. There are only two parameters that need to be chosen in designing the approximate integrator, and suitable values are easily found by trial and error on the simulation and are shown in Figure 2.49. Figure 2.50 shows some sinusoidal tests with good results; isolation begins at about 0.5 Hz and improves as expected at higher frequencies. With the random input, Figure 2.51b shows some remaining small payload motion, mainly at low frequency. Our random input theoretically has frequency content down to 0 Hz, which of course is not easily isolated. If the *actual* random input did not have such low-frequency content (which is quite possible), then the isolator would look even better. To simulate such a random input, we take our present signal and *high-pass filter* it with the filter low-frequency cutoff chosen to match the real-world excitation.

Next we look at a very different approach to isolation, exemplified by a commercial product called STACIS, from the TMC.* We again need suitable sensors and actuators, but now the geophone *does* give good results, and the actuators are *piezoelectric* rather than the electromagnetic voice coil we used earlier. Piezoelectric actuators are *very* different from voice coils. If you take the two parts (permanent magnet and coil) of a voice coil in your hands, you can easily move one relative to the other when the coil is not energized. A piezo actuator does not have any *sliding parts*; it is essentially a *very* stiff *spring* made of piezoelectric ceramics. This spring can be made to expand and contract by applying a high voltage to the actuator terminals. The resulting motion is limited to the order of 0.001 in.; voice coils can provide much larger motions. For the STACIS actuators, the spring stiffness is about 1.9 million lb_f/in., and the voltage-to-motion relation is about 1 μin./V. Because these actuators do not work well in tension, a constant bias voltage of 500 V is applied to put the actuator into compression. The input voltage then is applied in the range ±500 V, to give a motion range of ±0.0005 in. without the actuator going into tension. (There may also be some mechanical preloading.) These isolators therefore are usable only for equally small-floor vibrations, but this range turns out to be practical for many ultraprecise applications.

Figure 2.52 is used to explain and analyze the STACIS concept. The commercial product provides three-dimensional vibration isolation and has a number of detailed mechanical and electrical features needed to

---

* www/techmfg.com

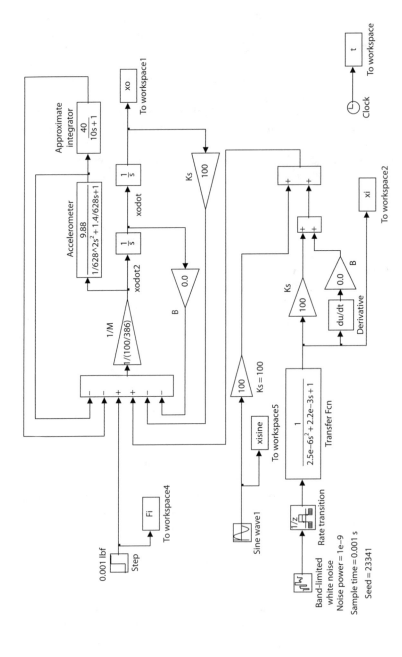

FIGURE 2.49
Feedback isolator using approximate integrator.

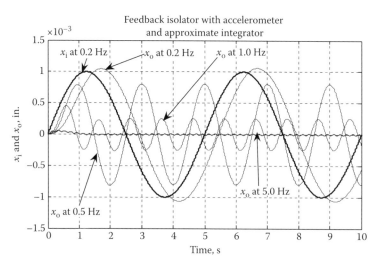

FIGURE 2.50
Frequency response of modified isolator.

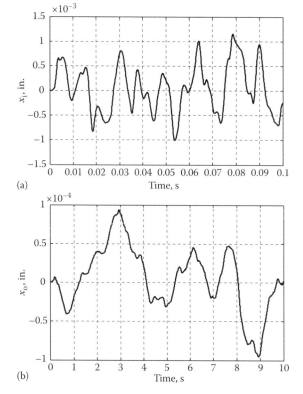

FIGURE 2.51
Isolator response with random floor vibration. (a) Input motion. (b) Output motion.

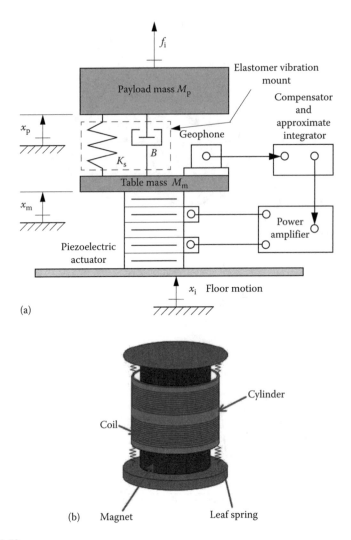

(a)

(b)   Magnet   Leaf spring

FIGURE 2.52
Simplified single-axis of STACIS active vibration isolator of TMC. (a) Overall system.
(b) Geophone details.

successfully implement the concept; we will consider only the basic single-axis operation, as depicted in Figure 2.52. The company's catalog has a
tutorial section which explains the operation of their various product lines,
including STACIS, but I got most of my information from one of their U.S.
patents, #5,660,255, August 26, 1997.* This patent is unusually detailed with
many drawings and numerical values which I studied to understand how
the system works. The numerical values I use are mostly from the patent, so

they may not correspond to current production models of the equipment. However, in some cases I had to figure out as best I could the details of some of the component dynamics and numerical values. *Also, various changes and improvements may have been made over the time period from 1997 to the present, so my discussion should not be taken as a precise description of the current product, but rather as a general explanation of the fundamental concepts.* The basic idea is to mount the payload on a low-mass "table" which is fastened to one end of the piezo actuator, the other end of which is attached to the vibrating floor. We measure the absolute displacement x_m of the table and use this as the feedback signal in a closed-loop system whose goal is to keep x_m as close to zero as possible. If this is successful, the bottom of the payload is then effectively attached, not to the vibrating floor, but to a nonmoving member, thus shielding it from the floor vibrations. By controlling the piezo actuator properly, every motion of the floor is *immediately* canceled out by an opposing motion of the actuator. This opposing motion is produced by comparing the x_m measurement to 0.0, amplifying this *error* signal, and then applying it to the actuator, which responds almost instantly by increasing or decreasing its length as needed. This is sometimes described as the "move out of the way" principle of vibration isolation.

It was found that this method worked well up to about 200 Hz, but floor vibrations above that frequency were too fast for the feedback system even when pushed to its practical limits. They then added, in "piggyback" fashion, an ordinary elastomer vibration mount (the patent mentions Barry Controls Model UC-4300) designed to give the payload mass and elastomer mount a natural frequency near 20 Hz. At 200 Hz and above this gives good isolation in the frequency range where the servosystem is inadequate. A rule of thumb used to choose the table mass M_m is that it should be about 1/100 of the payload mass M_p. In our example, the masses weigh 1000 and 10 lb$_f$. While we really want a displacement sensor, a geophone was used to measure table velocity and this signal was then approximately integrated to get the needed feedback signal. The geophone mentioned in the patent has a natural frequency of 4.5 Hz and damping ratio of about 1.0. We are now able to write some system equations:

$$f_i + K_{el}(x_m - x_p) + B_{el}(\dot{x}_m - \dot{x}_p) = M_p \ddot{x}_p \qquad (2.42)$$

$$K_{el}(x_p - x_m) + B_{el}(\dot{x}_p - \dot{x}_m) + K_{piezos}(x_i - x_m) - K_{piezox} x_{ma} = M_m \ddot{x}_m \qquad (2.43)$$

where
  K_{piezos} is the spring stiffness of the piezo actuator, in lb$_f$/in.
  K_{piezox} is a combination of the x_{ma} displacement transducer's sensitivity (V/in), the power amplifier gain (V/V), and the actuator voltage-to-displacement constant (in./V)
  x_{ma} is the approximate value of x_m "measured" by the compensated geophone (see Figure 2.53)

The last term on the left-hand side of Equation 2.43 models the actuator displacement response to input voltage as *instantaneous*, as given in the patent discussion. This is not exactly true, but piezo actuators often are so fast that this model is a reasonable approximation.

The geophone described in the patent has a natural frequency of about 4.5 Hz. I tried to use this with an approximate integrator to generate a usable x_m signal, but found it impossible; this geophone just does not work well enough at low frequencies. Geophones with lower natural frequencies might solve this problem, but they get quite large, heavy, and expensive. The patent hints at, but does not describe in detail, some op-amp circuitry used with the geophone. This reminded me of a velocity transducer,* recently introduced by Bently Nevada, which works on the same principle as a geophone, but performs well down to about 0.5 Hz. Bently Nevada (part of General Electric) would not reveal any details other than the fact that it was electronic compensation (rather than a low-frequency mechanical suspension) that gave this performance. After a little thought I was able to come up with some dynamic compensation that gave usable results. This turned out to be a good example of the utility, for dynamic system designers, to remain competent in *manual sketching*[†] of frequency-response curves, rather than relying only on computer-plotted curves, so I want to show some details here.

The manual sketches I refer to are the *Bode plots* of amplitude ratio using the straight-line approximations. All transfer function numerators and denominators can be factored into a limited number of *basic building blocks*: steady-state gains, integrators, differentiators, first-order terms, second-order terms, and dead times. Each of these terms has a simple straight-line asymptote for its amplitude ratio curve in dB, versus frequency on a log scale. By memorizing these simple basic shapes, and recalling that they can be *combined* by simple graphical addition, one can visually synthesize many practical dynamic systems to meet the desired performance. Having roughed out a design by manual sketching, we can then use, say, MATLAB computation and graphing to get accurate curves. Figure 2.53 shows the simulation in final form; Figure 2.54 shows the final accurate Bode plots. We can use Figure 2.54 to explain how the proper compensation was designed graphically, starting with the geophone itself. At low frequency it is a double differentiator; a straight line at +40 dB/decade. At high frequency (above its natural frequency of 4.5 Hz), it becomes a horizontal line at 0 dB (amplitude ratio of 1.0, since we ignore the geophone sensitivity (V/(in./s)) because we later combine all the gains in this path into an overall gain). The compensation needs to extend the flat 0 dB section down to lower frequencies, say to 0.5 Hz.

The obvious solution is to *cancel* the upward slope of 40 dB/decade with an equal downward slope using a *double integrator*. Since *perfect* integrators have the drift problem we encountered earlier, we must use two *approximate*

---

* Bently Nevada Model 330505.

† E.O. Doebelin, *System Dynamics: Modeling, Analysis, Simulation, Design*, Marcel Dekker, New York, 1998, pp. 631–637.

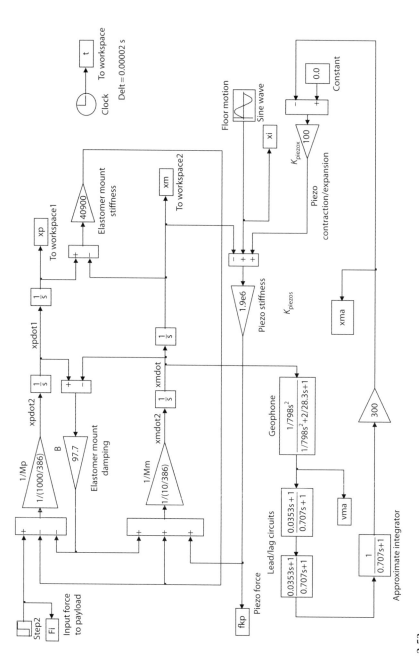

FIGURE 2.53
Simulation of simplified STACIS isolator.

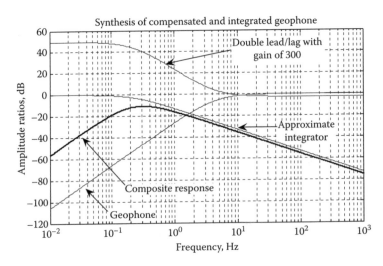

FIGURE 2.54
Explanation of geophone compensation and integration.

integrators, but some simple Bode sketching shows that this messes up the *high frequency* behavior (at high frequency the integrators pull the curve *downward* rather than keeping it flat). This problem is fixed by including two suitably chosen approximate *differentiators*. The net result is two *lead/lag* circuits as shown at the lower left of Figure 2.53. The *breakpoint* (1/(time constant)) of the lead/lag denominator is initially chosen at the desired low-frequency value of 0.5 Hz: 1/((0.5)(6.28)) = 0.318 s. When the accurate curves were plotted, this was changed to the 0.707 value shown, since the composite curve was *drooping down* to the left at 0.5 Hz rather than lying on the integrator line which slopes downward to the right. The lead/lag numerator time constant (0.0353 s) was chosen to put the breakpoint at the geophone natural frequency of 4.5 Hz (28.3 rad/s); 1/28.3 = 0.0353. This value did not have to be later adjusted. Finally, having *flattened* the geophone velocity response to about 0.5 Hz, we can now add the approximate integrator with breakpoint the same as the lead/lag denominator. The overall gain of about 300 was chosen by running the simulation with a sinusoidal floor motion input, comparing the perfect x_m signal with the geophone-based signal called x_{ma}, and adjusting this gain until the two curves were nearly the same. The K_{piezox} gain value of 100 in Figure 2.53 was established by trial and error but needed to be near 100 if an x_m (and thus x_p) motion was to be about 100 times smaller than x_i. Larger values would seem to make isolation even better, but as we know for every feedback system, excessive loop gain *will* ultimately cause instability. The composite response curve of Figure 2.54 shows that we have synthesized an approximate integrator for the geophone velocity signal.

Using this simulation, one can investigate many aspects of system performance and also try out any ideas for further improvement. Here we display some of the most obvious results, starting with Figure 2.55, which shows the

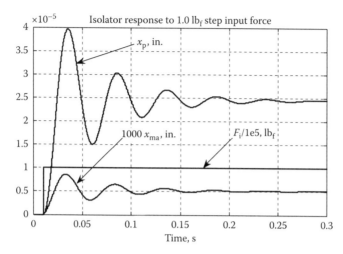

FIGURE 2.55
Response to step input of force.

response of the payload to a step force as being mainly that of the elastomer mount with a natural frequency of about 20 Hz and a damping ratio of about 0.15. The *table* motion is of course much smaller since its support "spring" (the piezo actuator) is *very* stiff. (The patent notes that another version of the system adds a velocity or relative displacement sensor to the payload, which adds to system complexity but provides another design parameter that might be used to reduce the response to payload force. I did not pursue this possibility.) Figure 2.56 shows that the isolator is working well even at the low

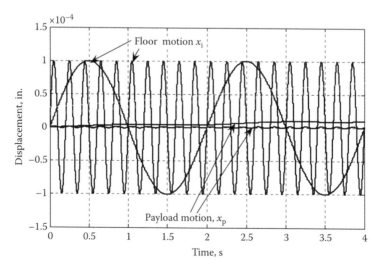

FIGURE 2.56
Isolator response to sinusoidal floor motion.

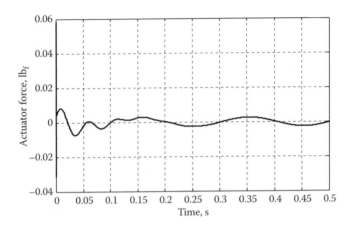

FIGURE 2.57
Actuator force response to 5.0 Hz floor input.

frequency of 0.5 Hz, and of course improves at higher frequencies as shown by the curve for 5.0 Hz. Since MATLAB graphs are auto-scaled, Figure 2.57 (which shows the actuator force) must be interpreted with some caution. It seems at first glance that the auto-scaling is not working, since the visible curve does not "fill the paper." Actually, because of the piezo's fast response (relative to a 5 Hz sine wave), part of the curve is squeezed onto the y-axis and is not visible. By zooming the graph to show early times, we produce Figure 2.58, and find that our integration time step (0.00002 s) that was used for this simulation does not give an accurate picture for time just after $t = 0.0$. This should cause some concern that perhaps *all* these simulation results might be in error, and should be checked. This was done and no errors were

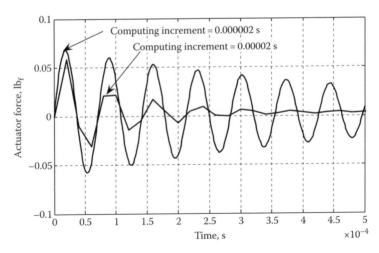

FIGURE 2.58
Accurate simulation for early times requires small computing increment.

found for the runs of Figures 2.55 and 2.56. This is fortunate since these runs cover *long* time intervals and would have been computed much more slowly at the 2 μs time step. Nevertheless, this experience should remind us to be wary of simulation results until proper checks have been carried out.

This completes our study of "move out of the way" active vibration isolation. More details are contained in the STACIS patent mentioned earlier, but as far as I know, a complete and detailed dynamic model of the system has not been published in the open literature. This is not unusual in the practical world of competitive business; hard won engineering lore is rarely revealed to potential competitors. Our simplified analysis hopefully serves to at least explain the basic concept.

2.7 Tuned Vibration Absorbers and Input-Shaping Methods

Tuned vibration absorbers are usually used to reduce vibratory motion that occurs *at a single frequency*, so they are more specialized than the methods we have earlier discussed, however, for completeness, we give a brief description. Figure 2.59 shows a typical arrangement; a *small* spring/mass/

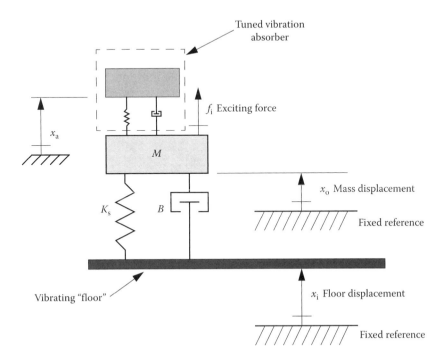

FIGURE 2.59
System with tuned vibration absorber.

damper system is attached to the main vibrating mass. If the exciting force or motion occurs at a single known frequency, under ideal conditions, *the motion of the main mass can be reduced to zero at that one frequency.* The absorber mass, damper, and spring are called M_a, B_a, and K_{sa}, respectively. Ideal conditions here means that B_a is zero and the natural frequency of the absorber spring/mass system is exactly equal to the frequency of the sinusoidal exciting force F_i or motion x_i. The need for such a damper is most obvious if the natural frequency of the *main* spring/mass system is the same as the exciting frequency, since then the main mass vibration would be at a maximum. The mass of the absorber is generally smaller than the main mass, ratios the order of 1/5 or 1/10 being typical. (Choice of a specific mass is often dictated by the application. In helicopters, which have many vibration problems, the electrical battery [which must be present anyway and whose location can be chosen] is sometimes used as the absorber mass, since for any *given* mass, we can always find a spring that gives the desired natural frequency.) At the designed frequency, the main mass stands still, but the absorber mass vibrates in such a way that its spring force exactly cancels the exciting force, giving zero net force (and thus zero motion) for the main mass.

Using Newton's law for each of the two masses we get

$$K_{sa}(x_o - x_a) + B_a(\dot{x}_o - \dot{x}_a) = M_a\ddot{x}_a \tag{2.44}$$

$$f_i + K_{sa}(x_a - x_o) + B_a(\dot{x}_a - \dot{x}_o) + K_s(x_i - x_o) + B(\dot{x}_i - \dot{x}_o) = M\ddot{x}_o \tag{2.45}$$

Working directly from these two equations we can set up the simulation diagram of Figure 2.60. The input motion and input force are generated at the lower left of this diagram. For the input force we provide both a single-frequency sine wave and a *chirp* (sine sweep), with gains set at 0.0 or 1.0 to select which we want. For the input motion, we provide only a single-frequency sine wave. While the ideal situation requires $B_a = 0.0$, we provide for nonzero damping to study real-world applications. The main mass weighs $20\,\mathrm{lb_f}$ while the absorber mass weighs $2\,\mathrm{lb_f}$. The excitation frequency is taken as $20\,\mathrm{Hz}$, which requires that both the main and absorber spring/mass systems have this same natural frequency; this leads to spring constants of 81.81 and $818.2\,\mathrm{lb_f/in.}$ for the two springs. The main system is given a (small) damping ratio of 0.05 so that the basic system has a large resonant peak, leading to the need for an absorber.

Our first study applies only a sinusoidal input force at $20\,\mathrm{Hz}$, for an absorber with no damping. The results are given in Figures 2.61 and 2.62. In Figure 2.61, the main mass displacement (with the absorber present) is rapidly approaching zero, compared with the large motion of the main mass when no absorber is attached. The *physical* explanation is clear in Figure 2.62. After a short transient startup period, the absorber spring force exactly cancels the applied excitation force, resulting in zero net force and thus a stationary main mass. We now remove the input force and apply an input displacement

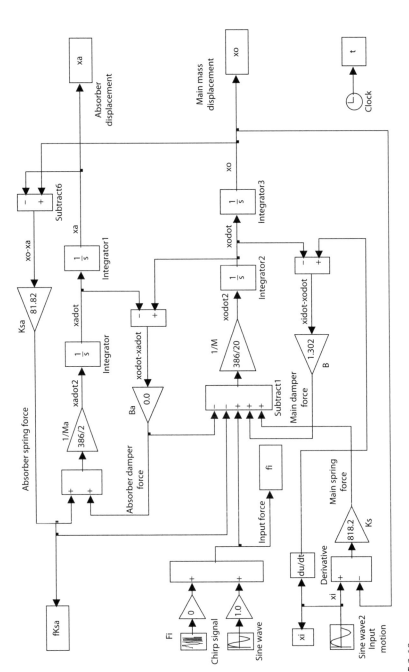

FIGURE 2.60
Simulation of tuned vibration absorber.

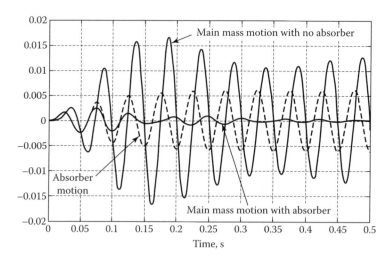

FIGURE 2.61
Mass motions for the startup period.

with the results of Figure 2.63. If we apply both the input force and the input motion, the absorber still reduces the main mass motion to zero. While these results confirm the vibration reduction claimed, they do not make apparent some important negative features of such absorbers. Since the overall system has two degrees of freedom, it will have *two* natural frequencies, and if lightly damped, there will now be *two* resonant peaks in the frequency response. That is, we have reduced the motion at the forcing frequency to zero, but now there are two *other* frequencies where excessive vibration might occur! These and other significant features will be made apparent by deriving the sinusoidal transfer functions relating output motion to input force and motion. Using our usual determinant method, we use Equations 2.44 and 2.45 to get

$$\frac{x_o}{f_i}(D) = \frac{M_a D^2 + B_a D + K_{sa}}{\begin{aligned}M_a M D^4 + (MB_a + M_a B_a + M_a B)D^3 + (M_a K_{sa} + M_a K_s + MK_{sa} + BB_a)D^2 \\ + (K_{sa}B + K_s B_a)D + K_{sa}K_s\end{aligned}}$$

(2.46)

$$\frac{x_o}{x_i}(D) = \frac{(M_a D^2 + B_a D + K_{sa})(BD + K_s)}{\begin{aligned}M_a M D^4 + (MB_a + M_a B_a + M_a B)D^3 + (M_a K_{sa} + M_a K_s + MK_{sa} + BB_a)D^2 \\ + (K_{sa}B + K_s B_a)D + K_{sa}K_s\end{aligned}}$$

(2.47)

If we set B_a to zero and $(K_{sa}/M_a)^{0.5}$ to be equal to the exciting frequency (*ideal conditions*), and then examine the numerators of the sinusoidal transfer functions, we can see mathematically why the main mass motion is exactly zero

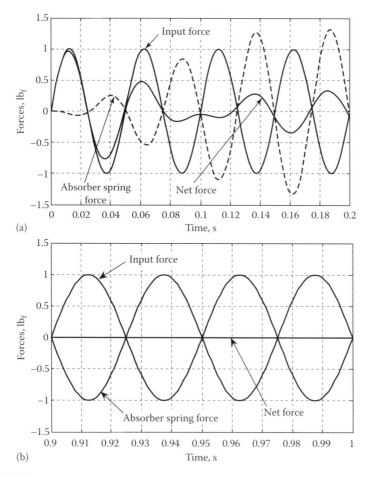

FIGURE 2.62
Forces (a) for startup and (b) in sinusoidal steady state.

at the exciting frequency; the numerators are exactly zero at that frequency while the denominators are not equal to zero. Using Equations 2.46 and 2.47, it is easy to insert numerical values and use our usual MATLAB routines to compute and plot the frequency-response curves, shown in Figure 2.64. We see that the response is zero at 20 Hz, but if the exciting frequency should *wander* away from this value, there are *two* frequencies at which large vibrations would occur. If the machine producing the vibrations is started up from rest, the excitation frequency must pass through the lower natural frequency peak before arriving at the null point. If this acceleration is fast enough, the resonance at that lower frequency may not have enough time to build up to dangerous levels. Once the machine starts working at the operating speed, the stability of speed becomes a concern. Some drives such as synchronous electric motors or systems with accurate feedback control of speed may have

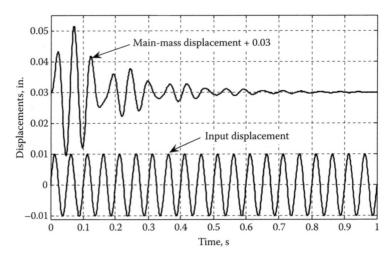

FIGURE 2.63
Absorber is effective for input displacement.

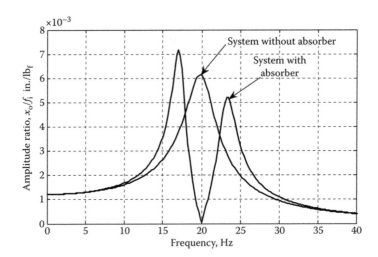

FIGURE 2.64
Absorber creates two new resonant peaks.

sufficient speed stability to make the absorber a practical solution. If this is not the case, one can try a *damped absorber*, which does not have a null point where the main mass motion is zero, but which does give some control over the height of the two resonant peaks. Damped absorbers are discussed in some detail in Den Hartog (1956)* and we will draw on this reference for some useful design formulas.

---

* J.P. Den Hartog, *Mechanical Vibrations*, 4th edn., McGraw-Hill, New York, 1956, pp. 87–105.

In order to derive useful analytical results, Den Hartog takes the case where the main mass is undamped. He shows that when damping is added to the absorber (B_a not zero), then it is advantageous to set the absorber natural frequency *away* from the exciting frequency by a certain amount. That is, the absorber natural frequency should be set at the value

$$\omega_{na} = \left[\frac{1}{1 + \dfrac{m_a}{m}} \right] \omega_{ex} \tag{2.48}$$

where ω_{ex} is the exciting frequency. For example, if the mass ratio is 0.25, then the absorber natural frequency should be set at 16 Hz if the exciting frequency is 20 Hz. He also suggests that the damping ratio ζ of the absorber be set as follows:

$$\zeta = \sqrt{\frac{3\mu}{8(1+\mu)^3}} \qquad \mu = \frac{m_a}{m} \tag{2.49}$$

For an undamped main system with $m = 20/386$, $K_s = 818.2 \, lb_f/in.$, a mass ratio of 0.25, and excitation frequency of 20 Hz, these design rules give the performance shown in Figure 2.65 (Figure 2.66 shows the response more clearly). The vibration reduction shown there is often sufficient to justify the use of the absorber. It is also necessary to determine the *relative* displacement of the

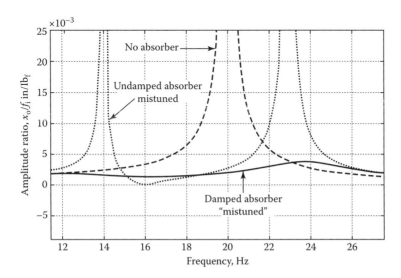

FIGURE 2.65
Use of intentional mistuning and absorber damping for an undamped basic system.

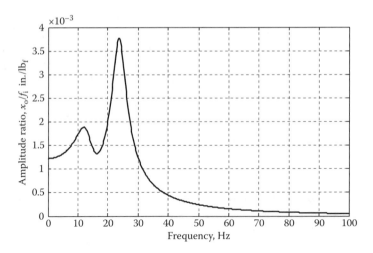

FIGURE 2.66
Damped, mistuned absorber.

two masses, since it determines the force and stress in the absorber spring. Den Hartog pursues these important design questions but we will leave these considerations to the reference.

We conclude our treatment of vibration control with a brief discussion of the technique called *input shaping*. Here we are concerned with suppressing the vibration exciting forces *at their source*, rather than mitigating their effect once they have occurred. In particular, we consider the vibration effects of certain motion-control systems on auxiliary equipment located nearby. The basic principles of input shaping have been known for some time but the practical application in the form of a commercial product is relatively recent. Based on his PhD dissertation at MIT, Dr. Neil Singer formed a small company to license the developed technology to a variety of industrial and government customers. His Web site[*] and paper[†] include much useful information, including some very convincing videos of various motion control applications. These videos focus on improving the behavior of the motion control system itself, but if such motion control systems are mounted on *platforms* that include vibration-sensitive instruments or machines, then reducing the vibratory motions of the motion-controlled object will also reduce the vibratory *reaction forces* felt by the platform and are passed on to the sensitive instruments. A more recent publication[‡] shows how to combine the tuned vibration absorbers we just discussed with the input-shaping concept.

---

[*] www.convolve.com
[†] N.C. Singer and W.P. Seering, Preshaping command inputs to reduce system vibration, *J. Dyn. Syst., Meas. Control*, 12, March 1990, 76–82.
[‡] J. Fortgang and W. Singhose, Concurrent design of vibration absorbers and input shapers, *J. Dyn. Syst., Meas. Control*, 127, 2004, 329–335.

Piezoelectric nanomotion controllers[*][†] have used input shaping (licensed from Dr. Singer's company) to greatly speed up scientific and manufacturing processes.[‡]

We now explain the basic concept of input shaping using a simple example. Suppose we have a motion-control system where the transfer function relating the load motion to the commanded motion is a lightly damped second-order system. Often we wish to move to a new position as rapidly as possible, by applying a step-input command. This will drive the mass quickly toward the desired position, but will cause large oscillations, and we will have to wait for these to decay sufficiently before we can continue with our processing operation (such as laser-drilling a hole at that location). The idea behind input shaping is that, instead of applying a step command, we find just the right shape of command that will get us to the new position quickly, *but without significant vibration.* Suppose the motion-control system transfer function has a natural frequency of 300 Hz and a damping ratio of 0.01. In defining our *shaped* input, we must be realistic in not exceeding the speed capabilities of the basic system. A 300 Hz system completes a single cycle of motion in about 0.0033 s, so if we ask for a non-oscillatory response which settles in about 0.01 s, that would not be unreasonable. We can choose a variety of *shapes* for our desired response; a simple choice would be the step response of a *critically damped* second-order system. Such a time response would have the Laplace transform $1/(s(\tau s + 1)^2)$. For any linear system, using the transfer function concept, there are always three entities: the input, the system, and the output. If any two of these are known, we can calculate the third. In our case, we know the system transfer function and the desired output, so we can easily compute the input, which will be our *shaped* input.

$$\text{input} = \frac{\text{output}}{\text{transfer function}} = \frac{\dfrac{s^2}{\omega_n^2} + \dfrac{2\zeta s}{\omega_n} + 1}{s(\tau s + 1)^2} \tag{2.50}$$

We need our input signal as a *time function*, not an *s* function, so we go to a Laplace transform table to find the *t* function that *goes with* the *s* function of Equation 2.50. This is found to be

$$f(t) = \frac{1}{\tau^2 \omega_n^2} \left[\tau^2 \omega_n^2 + \left[\frac{\dfrac{2\zeta\omega_n}{\tau} - \omega_n^2 - \dfrac{1}{\tau^2}}{\dfrac{1}{\tau}} t + \frac{\dfrac{1}{\tau^2} - \omega_n^2}{\dfrac{1}{\tau^2}} \right] e^{-\frac{t}{\tau}} \right] \tag{2.51}$$

---

[*] www.polytecpi.com
[†] http://www.physikinstrumente.com/en/products/prdetail.php?sortnr=400705
[‡] S. Jordan, Eliminating vibration in the nano-world, *Photonics Spectra*, July 2002, pp. 60–62.

For our example, we need to choose a value for τ that is as fast as possible, without exceeding the speed capability of the second-order dynamics. There is no "magic number" here but a value of 0.001 s is not unreasonable. With these values, Equation 2.51 becomes

$$f(t) = 1.00 - (1271t + 0.7186)e^{-1000t} \tag{2.52}$$

We can now use this input in a simulation of the input shaping method, as shown in Figure 2.67a, where we compare the response for the shaped input to that for a step input. I used a time delay of 0.001 s to show the signal behavior clearly near $t = 0.0$. The shaped input initially jumps up suddenly (just like a step input), and thus induces the beginning of an oscillation. However, this oscillation is immediately quenched by the input *reversing* its trend in

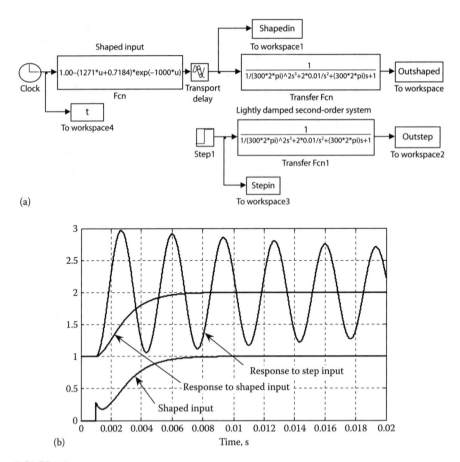

(a)

(b)

FIGURE 2.67
Shaped input suppresses vibration. (a) Simulation diagram. (b) Simulation results.

just the right way, giving a critically damped approach to the final value with no oscillation at all.

One can of course reduce such oscillations by using various *smooth* input commands rather than the step input, and this is sometimes done, but these methods may not give the response speed provided by the shaped input approach. Some dynamic systems have *several* natural modes of vibration (with associated frequencies) and it is possible to devise input shapes that will effectively cancel all these undesired vibrations. Dr. Singer's company (Convolve) has provided such solutions to many companies that have licensed his technology to improve the performance of their products.

3

Design of a Vibrating-Cylinder, High-Accuracy Pressure Transducer

3.1 Introduction

Most transducers have an *analog* output signal, usually in the form of a voltage proportional to the measured quantity. To obtain a digital version of such a signal, which can be accessed by a digital computer, we generally send it into an electronic analog-to-digital converter of some type. An alternative approach, used in many instruments, designs the transducer so that its output voltage is an oscillating signal whose *frequency* is proportional to the measured quantity. Note that this is still really an *analog* signal since the frequency changes smoothly with changes in the measured quantity. However, once our information is in the form of a frequency, we can use *digital* counter/timers like the general-purpose counter/timers regularly used in labs to measure frequency, time interval, etc., to get a digital display of our information. There are several advantages to this approach.

First, if we need to *transmit* our transducer signal over wires or by radio to a remote location, any electrical noise that might be present on the transmission channel will be largely ignored since, at the receiving end, we measure frequency, not amplitude. Second, the counters used will count the number of cycles over a chosen time period. Thus, if our signal is at, say, 1000 Hz, if we count over a 1 s period, we will get 1000 counts, with a resolution of 1 count, which is 0.1%. If our data is slowly changing, we might count over a 10 s period, getting 10,000 counts and an improved resolution of 0.01%. Of course, whatever counting period is used, the reading that we achieve is the *average* value of our measured quantity over the time interval corresponding to the counting period. While this averaging prevents us from getting an "instantaneous" reading of fast-changing data, it may actually be desirable in those cases where we are more interested in a short-term average value of a signal with small percentage fluctuations. Finally, we sometimes wish to get, in addition to the measured quantity itself, the *time integral* of that quantity. An obvious example is that of fluid flowmeters, where it is quite common to require both the instantaneous flow rate and the accumulated volume or mass that has passed through the meter over a specified time interval. This integration is accurately produced, for a varying flow rate, by setting the counter to *accumulate* the total counts over the specified time interval. Of course, this will only be accurate if the flow rate does not vary too rapidly.

In this chapter, we discuss how such transducers can be designed, using as example a high-accuracy pressure transducer. Such transducers can be used directly to measure pressure, but have also served as *standards*, used to periodically calibrate less stable (but much cheaper) pressure transducers in data acquisition systems. For example, one such standard can be used to calibrate many MEMS-type piezoresistive pressure sensors, whose short-term accuracy is very good, but which exhibit unacceptable temperature drifts and random zero-shift errors over longer time periods.[*] The *vibrating-cylinder* pressure transducer dates back to the 1950s, when it was introduced in Sweden by Svenska Flygmotor, an aircraft engine company, which later became part of the Volvo conglomerate. Later, this technology was transferred to England by the Solartron group.[†] Most recently (2008), the transducers are manufactured by Weston Aerospace.[‡]

3.2 Basic Concept

To explain the basic principles of the vibrating-cylinder transducer, we start by noting that all mechanical structures exhibit both elastic and inertial properties and are thus capable of free vibration at various natural frequencies. In most structures, however, these natural frequencies are *very little* affected by the magnitude of the force, torque, pressure, etc., being exerted on the structure, at least for the stress levels that structures usually experience without mechanical failure. For example, the first four natural frequencies of the familiar cantilever beam are given by

$$\omega_n = C_i \sqrt{\frac{EI}{\left(\frac{M}{L}\right)L^3}} \quad \frac{M}{L} \triangleq \text{mass per unit length}$$

(3.1)

$$C_1 = 3.52 \quad C_2 = 22.4 \quad C_3 = 61.7 \quad C_4 = 121.1$$

Note that the natural frequencies of the various modes of vibration depend on the dimensions and on the inertial and elastic properties of the beam

[*] D.B. Juanarena and C. Gross, Digitally compensated pressure sensors for flight measurement, *Sensors*, July 1985, pp. 10–20.

[†] SOL/AERO/84/P187, Solartron 3087 Series Digital Air Pressure Transducers, Farnborough, Hampshire, England, 1986.

[‡] www.westonaero.com

FIGURE 3.1
Cantilever beam with end force.

material, but *not* on the applied force (see Figure 3.1). This is typical of many structures and structural elements in general, including rods in torsion or tension/compression, plates in bending, etc.

To find structural elements whose natural frequencies *are* strongly dependent on applied forces, pressures, etc., we must examine strings and membranes under tension. Ideally, strings and membranes have *no* bending stiffness and exhibit "elastic" resistance to transverse forces or pressures only when externally tensioned. Real-world strings and membranes may be made of metals and thus exhibit *some* bending stiffness; however, this gets more and more negligible (relative to tension effects) as the diameter of a string (wire) or the thickness of a membrane is reduced. For a simple tensioned wire, completely neglecting bending stiffness, the lowest natural frequency is given by

$$\omega_1 = \frac{1}{2L}\sqrt{\frac{F}{M_L}} \tag{3.2}$$

where
$F \triangleq$ tensioning force
$M_L \triangleq$ wire, mass-per-unit-length

Figure 3.2 shows how this principle can be applied to pressure measurement. The vibrating-wire technology is also used in various other transducers.[*][†] In the vibrating-cylinder pressure transducer of our present study, the natural frequencies of pressurized thin cylindrical shells are of interest. This is an area of general interest in theoretical mechanics, not just with regard to pressure transducer design, and one can find numerous useful references in the book and paper literature of mechanical vibration. The reference I found most useful is a NASA report[‡] directed toward vibration problems in pressurized tanks of liquid-fueled rockets, and includes the effects of internal liquids. Fortunately, the equations include as a simpler special case, a tank with *no* internal liquid, the situation in our gas-pressure transducers.

---

[*] www.geokon.com
[†] E.O. Doebelin, *Measurement Systems*, 5th edn., Mc-Graw Hill, New York, 2003, pp. 474–476.
[‡] J.S. Mixon and R.W. Herr, An investigation of the vibration characteristics of pressurized thin-walled circular cylinders partly filled with liquid, NASA TR R-145, 1962.

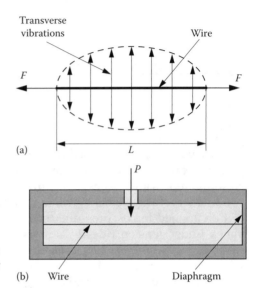

FIGURE 3.2
Vibrating-wire pressure transducer.
(a) Transverse vibrations. (b) Transducer
operating principle.

While all real-world mechanical structures behave as predicted by distributed-parameter (partial differential equation) models, and thus exhibit an infinite number of natural frequencies, cylindrical shells show some particularly complex and unusual vibration phenomena. Recall that for each natural frequency of any structure there is a corresponding *mode shape*, which is the dynamic deflection surface when the structure is vibrating in sinusoidal steady state at that natural frequency. For most structures, the mode shapes corresponding to the numerically higher natural frequencies display progressively more complex shapes (more nodes and antinodes), the cantilever beam again being a typical, though simple example (see Figure 3.3).

For thin cylindrical shells, the natural modes of vibration are described in terms of circumferential waveforms ($n = 2, 3, 4, \ldots$) and longitudinal waveforms ($m = 1, 2, 3, \ldots$), as shown in Figure 3.4 taken from Mixon and Herr (1962). For each combination of m (the number of longitudinal half waves) and n (the number of circumferential waves) there exist three distinct natural frequencies and associated mode shapes. Of these three, the lowest natural frequency is always associated with primarily *radial* motion, while the other two modes (of *much* higher frequency) consist of mainly tangential (either axial or circumferential) motion. Experimental testing in Mixon and Herr (1962) was able to excite only the radial (lowest frequency) mode in every case tried; the two associated tangential modes never appeared, thus *theory* also concentrated on predicting these radial modes only. Since these modes are readily excited, they are also the ones used in commercial vibrating-cylinder pressure transducers. To do preliminary design work on these transducers we need a theoretical equation relating natural frequencies to

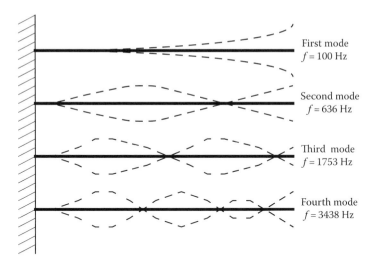

FIGURE 3.3
First four mode shapes of cantilever beam.

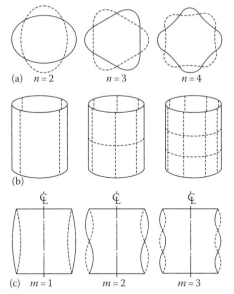

FIGURE 3.4
Radial mode shapes of thin-walled cylinders.
(a) Cross sections depicting circumferential wave fronts. (b) Node-line patterns. (c) Cross sections depicting longitudinal wave fronts.

dimensions, material properties, and the pressure (which we wish to measure). For closed-end cylinders (the configuration used in pressure transducers), Mixon and Herr (1962) provide their equation 48.

$$\omega^2_{m,n} = \left(\frac{Eg}{(1-v^2)\cdot\gamma r^2}\right)\left[\frac{(1-v^2)\lambda^4}{(n^2+\lambda^2)^2} + \frac{t^2}{12r^2}\cdot(n^2+\lambda^2)^2 + \frac{r(1-v^2)}{Et}\cdot\left(n^2+\frac{\lambda^2}{2}\right)p\right]$$

$$(3.3)$$

where
   L $\underline{\Delta}$ cylinder length
   r $\underline{\Delta}$ cylinder mean radius
   t $\underline{\Delta}$ wall thickness
   p $\underline{\Delta}$ pressure
   m $\underline{\Delta}$ 1, 2, 3, …
   E $\underline{\Delta}$ modulus of elasticity
   g $\underline{\Delta}$ acceleration of gravity
   n $\underline{\Delta}$ 2, 3, 4, …
   $\omega_{m,n}$ $\underline{\Delta}$ natural frequency, rad/s
   γ $\underline{\Delta}$ material specific weight
   ν $\underline{\Delta}$ Poisson's ratio
   λ $\underline{\Delta}$ $[\pi r(m+0.3)]/L$

Each combination of m and n defines one of the natural frequencies and associated mode shapes. One of the peculiarities of cylindrical shell vibrations is that some of the more complex mode shapes (higher values of m and/or n) have *lower* natural frequencies than the simpler shapes (lower m and/or n values). Since the lower natural frequencies are easier to excite, commercial pressure transducers often use them.

Inside the brackets of Equation 3.3 we see three terms, each of which makes a contribution to the numerical value of the frequency, however only the third (rightmost) term contains the pressure p. If we take $p = 0$ and let the wall thickness t approach zero, only the first term remains; it represents the contribution of elastic *stretching* of the shell midline. As t becomes larger, the second term, which represents a *bending* effect, becomes significant. Since the third (pressure) term has t in its denominator, it is clear that a strong effect of pressure on frequency can be obtained by making the shell sufficiently thin, which in fact is what is done in commercial pressure transducers. The material used for these transducers is usually a nickel–chromium–iron alloy called NI-SPAN-C. This material is used for the elastic elements of many mechanical transducers because its modulus of elasticity E has been designed to be nearly independent of temperature, thus eliminating or reducing the need for some kind of temperature *compensation*. In vibrating-cylinder pressure transducers of the class discussed here (Figure 3.5), the specifications are so stringent that, even with the use of NI-SPAN-C, there is still need for the inclusion of a built-in temperature sensor and use of its reading in the final calibration formula. Less precise transducers might meet their specifications simply by using this material, without any other corrections.

A very complete discussion of NI-SPAN-C properties and applications is available from the manufacturer.* One interesting feature brought out in this publication is that the *thermoelastic coefficient* (effect of temperature on E)

---

* www.specialmetals.com, Publication Number SMC-086.

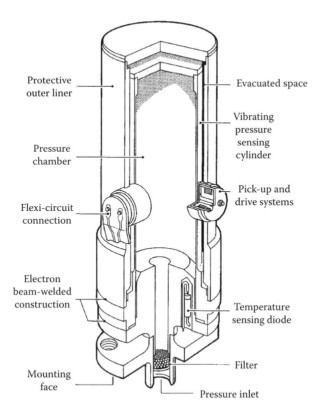

Protective outer liner

Evacuated space

Vibrating pressure sensing cylinder

Pressure chamber

Pick-up and drive systems

Flexi-circuit connection

Electron beam-welded construction

Temperature sensing diode

Filter

Mounting face

Pressure inlet

FIGURE 3.5
Construction of vibrating-cylinder pressure transducer.

has a frequency sensitivity that was only discovered recently. That is, if this parameter is carefully adjusted (say, by adjusting the percentage of nickel) to be near zero for a *static* application (like a pressure gage), if this material is then used at high frequency (as a vibrating-cylinder device would), the coefficient would be far from zero, defeating the purpose of using this material. Fortunately, when we know that an application will be at high frequency, the alloy formulation, heat treatment, and cold working can be adjusted so that we get near-zero temperature effect at operating frequencies. In our application, the frequency varies as the pressure varies, specifically from about 4000–5000 Hz for the 0–20 psia transducer we use as our example. The Web site, www.specialmetals.com, shows that the frequency dependence occurs only for frequencies between about 10 and 800 Hz, thus for our operating range, if we get an alloy that has near-zero temperature coefficient at 4000 Hz, that value will be maintained much the same up to 5000 Hz.

The elastic modulus E, specific weight, and Poisson's ratio take on a range of values, depending on the alloy formulation, heat treatment, and cold working. For our example, we will take E = 25.5 × 10⁶ psi, specific weight of 0.30 lb$_f$/in.³, and Poisson's ratio of 0.33. The frequency range given above (4000–5000 Hz) gives a change of about 1000 Hz when the pressure changes from 0 to 20 psia. This frequency change is selected to give good resolution; if we count for 1.0 s, we get about 1000 counts for the full-scale pressure change. Since the counters have a resolution of 1 count, this is 0.1%. As mentioned earlier, if we are satisfied with *average* values of pressure (often the case for nearly steady pressures), we can improve resolution with longer counting times. While the frequency changes just noted give good instrument performance, we need to now see whether they can actually be achieved with practical dimensions using Equation 3.3.

3.3 Cylinder Natural Frequency Calculations

If we were the original inventors of this transducer concept, a first task might be to decide on an appropriate *size* for the device, being guided by the desire in general to make transducers relatively small. Since today the transducer is a commercial product, we can avoid a study of size and simply start out with overall dimensions comparable to the existing transducers. A cylinder length of about 2.0 in. and radius of about 0.40 in. is typical. Having these numbers available, we can concentrate on the choice of modes (m and n) and cylinder thickness t, using Equation 3.3 and a Mathcad program:

$$t: 0.002 \qquad m := 1, 2..3 \qquad n := 2, 3..4 \qquad p := 0.0 \qquad ORIGIN := 1$$

$$L := 2.0 \qquad E := 25.5 \cdot 10^6 \qquad gam := 0.30 \qquad nu := 0.33 \qquad R := 0.40 \qquad g := 386$$

$$a1_{m,n} := \frac{E \cdot g}{(1 - nu^2) \cdot gam \cdot R^2} \qquad lam_{m,n} := \frac{\pi \cdot R \cdot (m + 0.3)}{L} \qquad a2_{m,n} := \left[\frac{(1 - nu^2) \cdot (lam_{m,n})^4}{\left[n^2 + (lam_{m,n})^2 \right]^2} \right]$$

$$a3_{m,n} := \frac{t^2}{12 \cdot R^2} \cdot \left[n^2 + (lam_{m,n})^2 \right]^2 \qquad a4_{m,n} := \frac{R \cdot (1 - nu^2)}{E \cdot t} \cdot \left[n^2 + \frac{(lam_{m,n})^2}{2} \right] \cdot p$$

$$fnp_{m,n} := \frac{\sqrt{a1_{m,n} \cdot (a2_{m,n} + a3_{m,n} + a4_{m,n})}}{2 \cdot \pi}$$

$$\text{fnp} = \begin{pmatrix} 0 & 1.032 \cdot 10^4 & 5.087 \cdot 10^3 & 3.42 \cdot 10^3 \\ 0 & 2.473 \cdot 10^4 & 1.363 \cdot 10^4 & 8.556 \cdot 10^3 \\ 0 & 3.735 \cdot 10^4 & 2.334 \cdot 10^4 & 1.543 \cdot 10^4 \end{pmatrix} \quad \begin{matrix} m = 1, n = 2,3,4 \\ m = 2, n = 2,3,4 \\ m = 3, n = 2,3,4 \end{matrix}$$

We see that $m=1$, $n=4$, and $t=0.002$ in. gives a frequency of 3420 Hz when the pressure is zero. This is lower than the 4000 Hz desired, so we try a larger thickness, $t=0.003$.

$$t := 0.003 \qquad m := 1,2..3 \qquad n := 2,3..4 \qquad p := 0.0 \qquad \text{ORIGIN} := 1$$

$$L := 2.0 \quad E := 25.5 \cdot 10^6 \quad \text{gam} := 0.30 \quad \text{nu} := 0.33 \quad R := 0.40 \quad g := 386$$

$$a1_{m,n} := \frac{E \cdot g}{(1 - nu^2) \cdot gam \cdot R^2} \qquad lam_{m,n} := \frac{\pi \cdot R \cdot (m + 0.3)}{L} \qquad a2_{m,n} := \left[\frac{(1 - nu^2) \cdot (lam_{m,n})^4}{\left[n^2 + (lam_{m,n})^2\right]^2} \right]$$

$$a3_{m,n} := \frac{t^2}{12 \cdot R^2} \cdot \left[n^2 + (lam_{m,n})^2 \right]^2 \qquad a4_{m,n} := \frac{R \cdot (1 - nu^2)}{E \cdot t} \cdot \left[n^2 + \frac{(lam_{m,n})^2}{2} \right] \cdot p$$

$$\text{fnp}_{m,n} := \frac{\sqrt{a1_{m,n} \cdot (a2_{m,n} + a3_{m,n} + a4_{m,n})}}{2 \cdot \pi}$$

$$\text{fnp} = \begin{pmatrix} 0 & 1.033 \cdot 10^4 & 5.224 \cdot 10^3 & 3.989 \cdot 10^3 \\ 0 & 2.474 \cdot 10^4 & 1.37 \cdot 10^4 & 8.842 \cdot 10^3 \\ 0 & 3.736 \cdot 10^4 & 2.34 \cdot 10^4 & 1.563 \cdot 10^4 \end{pmatrix} \quad \begin{matrix} m = 1, n = 2,3,4 \\ m = 2, n = 2,3,4 \\ m = 3, n = 2,3,4 \end{matrix}$$

Now, $m = 1$ and $n = 4$ gives 3989 Hz, very close to what we want. Next we need to raise the pressure to 20 psia, the full-scale value of the desired pressure range. We see below that $m=1$, $n=4$, and $t=0.003$ gives a frequency (at 20 psia pressure) of 4979, very close to our desired value of 5000 Hz.

$$\text{fnp} = \begin{pmatrix} 0 & 1.044 \cdot 10^4 & 5.689 \cdot 10^3 & 4.979 \cdot 10^3 \\ 0 & 2.48 \cdot 10^4 & 1.389 \cdot 10^4 & 9.351 \cdot 10^3 \\ 0 & 3.74 \cdot 10^4 & 2.353 \cdot 10^4 & 1.594 \cdot 10^4 \end{pmatrix} \quad \begin{matrix} m = 1, n = 2,3,4 \\ m = 2, n = 2,3,4 \\ m = 3, n = 2,3,4 \end{matrix}$$

We could of course adjust t to get *exactly* 5000 Hz but we realize that all theoretical formulas (except for very simple systems) may be somewhat in error, and we would be misleading ourselves to think that our predictions are within only a few percent of reality. In developing a new transducer, we of course expect to do experimental development before the design is finalized; the final t value would be based on those results, not our initial theoretical predictions. (The 4000–5000 Hz frequency range for a 0–20 psia transducer was the case when Solartron was the manufacturer. In 2008, when Weston Aerospace had taken over the device, it appears that the frequency was now in the 8000 Hz range. I tried to contact Weston with this question but was unable to get a response.)

3.4 Use of an Unstable Feedback System to Maintain Continuous Oscillation

Having established that pressure causes a usable change in cylinder natural frequency, it remains to invent some method of actually producing and measuring this frequency. Since various forms of mechanical vibrating elements have been successfully used in other applications for many years, an original invention is not really needed; we can simply adapt a known and proven principle to the current problem. The use of feedback methods in transducers is well known, with two examples being servo accelerometers[*] and hotwire anemometers.[†] In the usual situation, great pains are taken to ensure that the feedback system is *stable*. Now, however, we intentionally design for an *unstable* system. Most feedback analysis and design works with *linear* models for all the components found in the feedback loop. Such models predict that instability occurs when any of the roots of the closed-loop system *characteristic equation* have positive real parts, leading to terms in the solution of the form Ce^{at} or $Ce^{at} \sin (bt + \phi)$ where a is a positive number. Such terms go to infinity as time t progresses. This instability is initiated by the *slightest* disturbance of any kind. In a real physical system, rather than our idealized linear models, as the system variables grow due to the instability, various nonlinearities (neglected in the linear model) become important and usually result in the oscillations leveling off rather than continuing to increase in amplitude. In our vibrating-cylinder pressure transducer, we must arrange things so that the linear model is unstable, and that the amplitude-limiting nonlinear effect be *designed in* and under our control, rather than being a parasitic effect in one or more of the system components.

Most feedback system models must have a characteristic equation that is at least third order if instability is to be predicted analytically, so we must choose the components and their dynamic effects to conform to this.

---

[*] E.O. Doebelin, *Measurement Systems*, 5th edn., McGraw-Hill, New York, 2003, pp. 369–372.
[†] E.O. Doebelin, *Measurement Systems*, 5th edn., McGraw-Hill, New York, 2003, pp. 596–610.

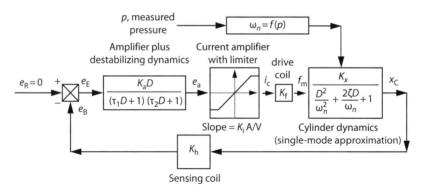

FIGURE 3.6
Unstable feedback system creates continuous oscillation.

The simplest amplitude-limiting nonlinearity which we can easily implement is that of *saturation*. Most components such as sensors, amplifiers,
motors, etc., will themselves exhibit saturation as their input signals get
larger, but we prefer to build in an intentional saturation nonlinearity
whose behavior we can specify. This is most easily done in an electronic
amplifier, if such is already needed as a basic system function in the linear
model. Figure 3.6 shows the type of feedback loop that is employed in
many transducers based on the frequency variation of some elastic element. While the cylinder itself has *many* natural frequencies, when it is
driven by oscillating forces of frequency very close to our chosen design
frequency, its dynamics can be simplified to a single second-order system,
as shown at the right of the figure; the other modes of vibration contribute
little to the motion. In choosing the *application point* of the driving force,
we of course avoid any *nodes* of our chosen mode shape and try to stay
close to *antinodes*, where the structure's response to external forcing is a
maximum. Electromagnetic forcing is convenient since the material of the
cylinder is NI-SPAN-C, which is magnetic and exhibits the other desirable properties discussed earlier. The driving force is created by a small
electromagnet (drive coil) which exerts an oscillating force on the cylinder wall when the coil carries an oscillating current. The small oscillating
radial motion of the cylinder wall is measured near another antinodal
point, using a non-contacting inductive displacement sensor (sensing coil)
to produce the feedback voltage e_B.

In a conventional stable feedback controller for which the motion x_C would
be the controlled variable, e_B would be compared with a reference input
voltage e_R, where e_R would act as a command to x_C. Since the mean value
desired for our oscillatory motion is zero, we can set e_R identically equal to
zero, making error voltage $e_E = -e_B$. Voltage e_E is amplified, and dynamically
modified (to ensure instability) with an op-amp circuit, creating voltage e_a.
A *transconductance amplifier* with gain K_i A/V is given a current-limiting

feature (saturation) and provides current i_c to the drive coil, which in turn produces magnetic force f_m, according to $f_m = K_f i_c$. The linear model (which neglects the saturation nonlinearity) of this closed-loop system is easily obtained directly from the block diagram in the usual way:

$$(e_R - K_h x_C)\left[\frac{K_a K_i K_f K_x D}{(\tau_1 D + 1)\cdot(\tau_2 D + 1)\cdot\left(\dfrac{D^2}{\omega_n^2} + \dfrac{2\zeta D}{\omega_n} + 1\right)}\right] = x_C \qquad (3.4)$$

To get the system differential equation, we can easily manipulate (cross multiply) Equation 3.4 into the form

$$(A_4 D^4 + A_3 D^3 + A_2 D^2 + A_1 D + A_0)x_C = B_1 D e_R \qquad (3.5)$$

where the A's and B_1 are easily found combinations of the K's, τ's, ζ, and ω_n. Since this system is of fourth order, it should be capable of instability. We will now use two of the common analysis tools of linear feedback system design to explain the choice of the destabilizing dynamics; the *root-locus*[*] method and the *frequency-response*[†] method, with the frequency-response method being augmented by a useful nonlinear tool called *describing function*.[‡]

In the root-locus method, the transfer functions in Figure 3.6 are treated as *Laplace transform transfer functions* rather than the *D*-operator form, but this is an easy conversion; we simply replace every *D* with an *s*, where *s* is a complex number $\sigma + i\omega$. Whereas *D* is an *operator* which cannot be assigned numerical values in a transfer function, s *can* take on numerical values. For those unfamiliar with the terms *poles* and *zeros*, a pole is a numerical value of *s* that makes a transfer function take on an *infinite* value, whereas a zero is an *s* value that makes the transfer function equal to *zero*. The root-locus method starts by plotting in the real/imaginary plane, the poles and zeros of the *open-loop transfer function*, defined as the transfer function relating B to E in a *generic* block diagram, which is e_B to e_E in our example. Thus in Figure 3.6 there is one zero at $s = 0$, two real poles at $s = -1/\tau_1$ and $s = -1/\tau_2$, and a pair of complex poles at $s = -\zeta\omega_n \pm i\omega_n\sqrt{1-\zeta^2}$, giving the graph of Figure 3.7. The steady-state gain of the open-loop transfer function is called the *loop gain*, and is a very important number in any feedback system; in our example the loop gain is $K_a K_i K_f K_x K_h$. In most feedback systems, raising the loop gain improves accuracy and speed, but will always cause instability if carried too far.

---

[*] E.O. Doebelin, *Control System Principles and Design*, Wiley, New York, 1985, pp. 305–323.
[†] E.O. Doebelin, *Control System Principles and Design*, Wiley, New York, 1985, pp. 191–197.
[‡] E.O. Doebelin, *Control System Principles and Design*, Wiley, New York, 1985, pp. 221–255.

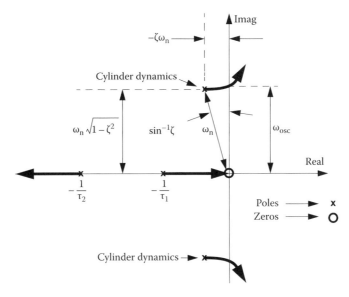

FIGURE 3.7
Root-locus for the vibrating-cylinder pressure transducer.

In derivations of the root-locus method, it is shown that if we plot the open-loop poles and zeros (as in Figure 3.7), when we set the loop gain to zero, the roots of the *closed-loop* system characteristic equation will coincide with these pole locations. As loop gain is increased, the closed-loop roots will move away from the open-loop poles along specific paths (the *root loci*), so for any specific value of loop gain there will be specific values for the closed-loop roots. Some of these paths (those which start at a pole and end at a zero) are "complete," while others start at a pole and move off toward infinity. In Figure 3.7 we have four open-loop poles, so there are four closed-loop roots, and their four paths are shown as the heavy lines, with arrowheads showing the direction the roots move as loop gain is increased. Since we have three more poles than zeros, three of the root loci head off toward infinity. Feedback system designers learn a set of simple rules which allow quick *sketching* of approximate root loci as an aid to system preliminary design. When tentative values of parameters have been selected, feedback system design/analysis software (such as MATLAB®'s control system toolbox) compute and plot *accurate* curves.

In Figure 3.7, we see that the two loci which start at the poles associated with the cylinder natural frequency and damping ratio do in fact move into the right half of the complex plane, indicating that the loop gain has been made too high, resulting in instability. If we set the loop gain at exactly the correct value, we could place these two roots *exactly* on the imaginary axis, giving two pure imaginary roots, which indicates an oscillation at frequency ω_{osc} that neither builds up or dies down, but stays at a fixed amplitude, and has a frequency *close* to the cylinder natural frequency. In our pressure transducer this is what we want, but unfortunately this theoretical prediction cannot be realized in practice. In a real-world system (rather than a mathematical model) all the system parameters are subject to small random fluctuations, meaning that the roots which we must place *exactly* on the imaginary axis will in reality be at least a little to the left or right, and therefore the oscillation will either build up or die down. To get a "stable" oscillation, we must have some nonlinear effect which somehow exerts a *restoring force* when the oscillations tend to either build up or die down. The *saturation nonlinearity* provides exactly this effect.

Most nonlinear differential equations that correspond to practical engineering devices are not solvable analytically. However, certain approximate methods have been developed which allow useful predictions. Perhaps the most common of these is the *describing function* method. This method extends the linear frequency-response stability technique called the *Nyquist diagram* to certain common nonlinearities. Before pursuing this discussion, I want to use simulation to illustrate the effect of saturation on sinusoidal signals. I will show that whereas the gain of linear components is not affected by the *size* (amplitude) of the input sine wave, the saturation nonlinearity has a "gain" that becomes *smaller* as the size of the input sine wave increases. This is exactly what we want in our transducer system. That is, in Figure 3.7, if for some reason the two critical roots should momentarily wander into the right half plane, the amplitude would begin to grow, but this growth results in a *lowering* of the loop gain, which restores stability, causing the oscillation to start to die down. This dying down of course *increases* the gain of the saturation, which tends to cause instability, building up the amplitude. We see that there is now a *restoring effect* which will keep the amplitude close to some fixed value.

A simple simulation demonstrates this *variable gain* effect of the saturation nonlinearity in Figure 3.8. For input sine waves of amplitude less than the saturation limits of +1 and −1, there of course is no saturation and the *gain* (ratio of output amplitude over input amplitude) is 1.0. For an input amplitude of 2.0 we get the results of Figure 3.9. The output of the saturation is now not a sine wave, but it *is* periodic and one could compute a Fourier series for it. The describing function method does exactly that, and approximates the series using only the first harmonic term. In our simulation, we use a low-pass filter to extract the first harmonic without the need for a Fourier series calculation. We see that the ratio of output amplitude and input amplitude is about $1.2/2.0 = 0.6$, which shows the reduction in gain that we anticipated. (Note that using a low-pass filter to extract the first harmonic also shows a *phase shift* which of course would not be present in the Fourier series calculation of this waveform, so we ignore it.) The justification given for ignoring the higher harmonics of the Fourier series is that the dynamics of many feedback systems' open-loop transfer functions do in fact include low-pass filtering effects. That is, as the output signal of the nonlinearity

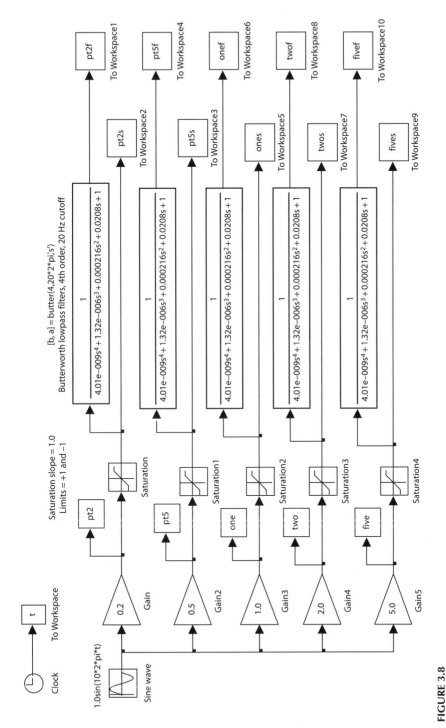

FIGURE 3.8
Demonstration of saturation nonlinearity as an adjustable gain.

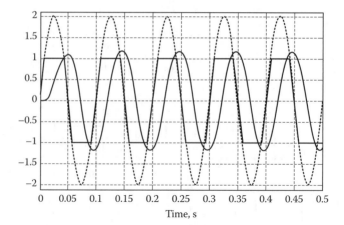

FIGURE 3.9
Saturation input (dotted), output (solid), and Fourier series first harmonic (solid).

travels downstream, the corners of the waveform will be rounded off. Figure
3.10a shows the simulation results for a larger input signal, which gives a
lower gain value (1.2/5.0 = 0.24), as expected. A formula* for the saturation
describing function (obtained using Fourier series) is shown in Figure 3.10b,
together with a Mathcad numerical calculation and graph. Our simulation
results correspond to $n = 2$ and 5 in the table, where we see good agreement of the approximate simulation method with the exact Fourier series
calculation.

We now want to show the Nyquist and root-locus methods applied to our
system, but this requires that all the numerical values be known. The Nyquist
method of stability determination uses a polar plot of the open-loop sinusoidal transfer function, in our case $(e_B/e_E)(i\omega)$. In Figure 3.6 this is shown in
factored form, so its general shape can be seen without any numerical calculation. At low frequency, because of the D in the numerator, the amplitude
ratio approaches zero and the phase angle +90°. Since all the other terms are
in the denominator, the phase angle will gradually get more lagging, going
from +90 to 0, to −90, to −180, and ending at −270 (the two first-order terms
each contribute −90, and the second-order −180). The amplitude ratio is not
easily visualized without having some numerical values, but the polar graph
clearly must cut across the −180 line at some point. This fact is useful since
the Nyquist method says that if this graph cuts the −180 line *to the left* of the
point −1, the closed-loop system will be unstable, while stable systems cut
this line *to the right* of the −1 point. A system whose graph goes *exactly through*
the −1 point is called *marginally stable*, that is, it oscillates with a fixed amplitude. This corresponds in the root-locus method to closed-loop poles that fall
exactly on the imaginary axis.

* E.O. Doebelin, *Control System Principles and Design*, Wiley, New York, 1985, p. 241.

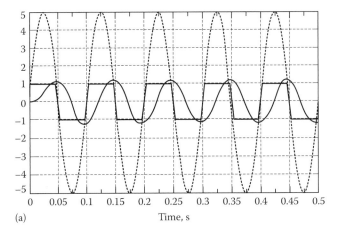

(a)   Time, s

computes describing function for saturation nonlinearity

n:= 1, 2.. 10

R is the ratio: saturation level/saturation input amplitude

$R(n) := \dfrac{1}{n}$

$c1(n) := asin\ R(n)$ $c2(n) := R(n)\left(1 - R(n)^2\right)^{0.5}$

$Gd(n) := \dfrac{2}{\pi} \cdot (c1(n) + c2(n))$ Gd is the saturation describing function.

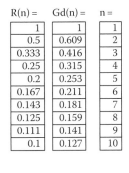

R(n) =	Gd(n) =	n =
1	1	1
0.5	0.609	2
0.333	0.416	3
0.25	0.315	4
0.2	0.253	5
0.167	0.211	6
0.143	0.181	7
0.125	0.159	8
0.111	0.141	9
0.1	0.127	10

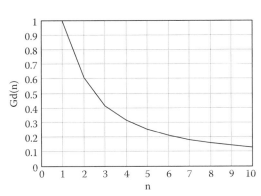

(b)

FIGURE 3.10
(a) Saturation input (dotted), output (solid), and Fourier series first harmonic (solid) and (b) describing function calculation for saturation nonlinearity.

To force the closed-loop system to oscillate at a frequency close to the cylinder natural frequency, we must arrange that the open-loop transfer function cuts through the −180° line near this frequency. This dictates our choice of numerical values for the two time constants. The lightly damped cylinder will have a −90° phase angle exactly at its natural frequency, so we try to choose the time constants to contribute about −180 at this frequency. First-order systems only *approach* −90°, so we can only design for, say, −87.5° at

4000 Hz. There seems to be no good reason to make the two time constants different, so we make them the same. Then, to get $-175°$ from both together, the tau (τ) value would need to be

$$(4000) \cdot (2\pi)\tau = \tan 87.5° = 22.903 \quad \tau = 0.0009118 \ s \qquad (3.6)$$

At 5000 Hz, the total phase angle would be about $-176°$, so the frequency would only have to shift away from 5000 slightly to shift the phase by about $4°$. For a lightly damped second-order system, the phase angle changes sharply near the natural frequency, so the oscillating frequency would not have to be much different from the cylinder natural frequency. Remember that all measurement systems need to be *calibrated* once they are built, so the oscillating frequency can deviate somewhat from the natural frequency and still give a useful instrument.

We have now justified the various terms of the open-loop transfer function given in Figure 3.6, except for the D in the numerator. (There are of course *other* open-loop transfer functions that could be made to give a self-oscillating system; our choice is just the simplest which meets the requirements.) The D in the numerator is also not essential, but might be a good choice, since it allows the amplifier to be an *AC* type, which has some advantages over the *DC* type in terms of cost and lack of drift. The choice of numerical value for the various component gains is somewhat arbitrary so long as the total gain exceeds that which causes instability. If we make the total gain much higher than needed for instability, then the gain of the saturation would have to adjust to a smaller value, which requires that the input to the saturation would need to get larger, possibly overloading some circuits. These and similar questions are usually resolved by the *experimental development* which normally follows the theoretical design stage. I thus picked for further study some component gains which seemed reasonable, but could of course be adjusted as needed when actual hardware is chosen. These values are

$$K_a = 3.0 \ V/V \quad K_i = 0.01 \ amp/V \quad K_f = 0.001 \ lb_f/amp \quad K_x = 1.0 \ in./lb_f$$

$$K_h = 1000 \ V/in. \quad \zeta = 0.10 \quad loop \ gain = 0.03 \ V/V$$

The loop gain value is somewhat larger than the minimum to cause instability. The damping ratio is probably higher than the actual value, which is impossible to predict from theory. We will shortly do a simulation of the entire system, where it is found that very small damping, while giving the desired oscillations, leads to long computing times because of the need to wait for the steady state. The real system, however, settles down in less than a second, so small damping does not create any practical problems.

3.5 Nyquist and Root-Locus Studies of System Operation

Now that we have numerical values of all system parameters, we can compute and plot these useful graphs. A MATLAB program follows:

```
01%   crc_book_chap3_rloc_nyq  Computes and plots Nyquist and
      root
02%   locus diagrams for vibrating-cylinder pressure transducer
03%   define parameter values
04    tau1=0.0009118
05    tau2=0.0009118
06    zeta=0.10
07    wn=4000*2*pi
08%   wn=5000*2*pi
09    Ka=3.0
10    Ki=0.01
11    Kf=0.001
12    Kx=1.0
13    Kh=1000.
14    K=Ka*Ki*Kf*Kx*Kh
15%   define open-loop transfer function using s=tf('s')
      statement
16    s=tf('s');sys=K*s/((s+1/tau1)*(s+1/tau2)*(s^2+2*zeta*wn*s
      +wn^2))
17%   extract numerator and denominator using tfdata statement
18    [num,den]=tfdata(sys,'v');
19    num1=num
20%   force denominator to have trailing term equal to 1.0
21    den1=den/den(5)
22    h=tf(num1,den1)
23%   plot Nyquist polar plot
24    f= 0:5:10000;
25    w=2*pi*f;
26    nyquist(h,w);hold on
27%   compute and plot Nyquist for positive frequencies only
28    [ha]=freqs(num1,den1,w);
29    R=real(ha);
30    I=imag(ha);
31    plot(R,I,'k','linewidth',2);pause
32%   plot root locus
33    hold off
34    rlocus(h)
```

The MATLAB Nyquist command (line 26) produces a *polar* graph of the open-loop sinusoidal transfer function; that is, a graph showing the amplitude ratio as a vector length, at an angle given by the phase angle, for the range of frequencies specified in the w statement (line 25). The most general

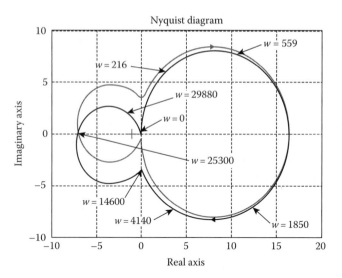

FIGURE 3.11

Nyquist plot for vibrating-cylinder open-loop transfer function (loop gain = 0.03).

form of the Nyquist stability criterion* requires that this plot include *negative* values of frequency. It turns out that this part of the plot is simply obtained by *reflecting* the curve for positive frequencies about the horizontal axis. The MATLAB Nyquist command automatically plots the graph for *both* positive and negative frequencies, giving a graph like Figure 3.11. Using the *data cursor* that is part of MATLAB graphing, one can click on any point on this graph and display the graph coordinates and frequency. For our purposes, I wanted to concentrate on the positive-frequency portion of this graph, so I used lines 28–31 to superimpose a heavy-line curve of just this portion on top of the original curve. According to the simplified Nyquist criterion applicable to this system, closed-loop instability is predicted if the positive-frequency curve cuts the –180° line to the left of the –1 point. Our example curve cuts this line at about –7, so instability is predicted. If we reduced the open-loop gain by about 7 times, the curve would go right through the –1 point, indicating *marginal* stability. The frequency associated with that point would be the predicted frequency of steady oscillations, in our example, about 25,300 rad/s (4027 Hz). (The scale of the graph, and the 5 Hz frequency resolution of program line 24, makes the location of the data cursor somewhat uncertain, so the 4027 frequency is not precise, but one can always get an accurate value by listing numerical values, rather than relying on the graph.) We used an open-loop gain value (0.03) that guarantees instability; it seems that we could use any value between 0.03 and 0.03/7. The actual value used would be determined by experimental development, once a prototype

---

* E.O. Doebelin, *Control System Principles and Design*, Wiley, New York, 1985, p. 192.

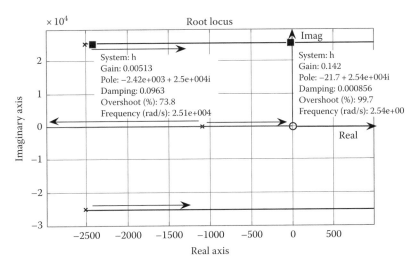

FIGURE 3.12
Root-locus plot for vibrating-cylinder system.

has been constructed. Whatever value is used, the saturation nonlinearity will force the *effective* total gain to be such as to put the curve right through the −1 point.

We used the "generic" root-locus graph of Figure 3.7 to explain the operating principle of the transducer; having numerical values, we can now plot an accurate graph, using program line 34. The graph which MATLAB displays is scaled to show the "whole picture," but we want to focus on certain critical regions. By zooming the graph, we get Figure 3.12, which again uses the data cursor to access some useful numerical values. For a (low) gain value that puts the closed-loop root near the open-loop cylinder-dynamics pole, we see a *system* natural frequency of 25,100 rad/s. Raising the gain to (0.142)(0.03) puts this closed-loop root right on the imaginary axis, giving marginal stability. Note that 0.142 is close to 1/7, which we found by Nyquist was required to put the Nyquist graph right through the −1 point, so the two analysis methods seem to agree, as they should.

3.6 Simulation of the Complete System

Having demonstrated the use of conventional feedback system design tools to understand the operating principle of our transducer and establish some numerical values, it is now appropriate to check out this design using simulation. That is, both the Nyquist and root-locus methods are *linear* design tools, whereas the saturation required in our design is a nonlinearity. The describing function method deals with nonlinearity, but in an approximate fashion, while simulation deals *exactly* with this nonlinearity. In the simulation diagram of Figure 3.13 we use a lookup table to program a pressure variation that exercises the transducer over its full range, starting with a dwell at 0 psia, a ramp rise to 10 psia with a dwell, and finally a ramp rise to 20 psia and a dwell there (see Figure 3.14). Because of the relatively high frequency of oscillation, the transducer responds to pressure changes quite quickly, so the duration of the dwells can be short and we will still see the system settle into its steady-state vibration. This short timescale also allows us to resolve individual cycles of vibration, but requires a very short integration time step of 10^{-7} s. While we usually use a transfer function block to model second-order systems, the time-varying natural frequency of this application is handled by setting this up as a differential equation, as seen at the bottom of Figure 3.13.

All the signals within the feedback loop oscillate at the same frequency, so we could observe that frequency in any of them; Figure 3.15 shows the cylinder wall displacement x_c. We see that the oscillation rapidly builds up to its steady state after we start the system with a rectangular pulse input at e_r with amplitude 0.01 and duration 0.0001 s. (*Any* disturbance will start the oscillation, which will be maintained after the disturbance is removed. In the real-world system, the act of switching on the power would be sufficient.) In Figure 3.15, it is difficult to see the change in frequency as the pressure changes; we could *zoom* the graph in the dwell regions to make this more clear. In the real system, the frequency would be measured with a digital counter/timer, using perhaps a sampling interval of 1 s. This is not advisable in our simulation because of the long computing time associated with a time-step of 10^{-7} s and a 1 s sampling interval. Instead of counting and timing cycles, one can measure the *period* of each cycle and take its reciprocal to get frequency, with the advantage of updating the frequency every cycle, rather than every second. Many commercial counter/timers use this approach to measure *low* frequencies, and automatically switch to counting cycles for high frequencies, in order to provide a large number of significant digits in the display whether the frequency is high or low.

One method to measure the period of each cycle of oscillation uses the *hit crossing* module of Simulink. This module monitors the signal of interest and outputs a 1 for each zero-crossing and 0 for all other times. Sending this signal to a *zero-order hold* module and then multiplying this signal by the current time t produces the signal *hit* of Figure 3.16. To get the period and frequency associated with each cycle, we run a small MATLAB program which uses the *find* command to extract the nonzero values of the signal *hit*.

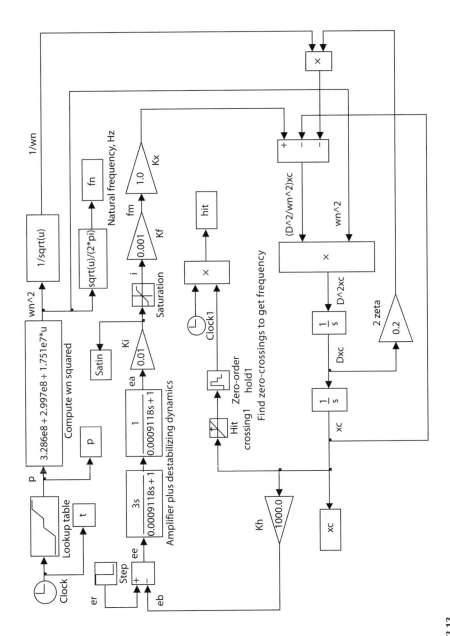

FIGURE 3.13
Simulation diagram for vibrating-cylinder pressure transducer.

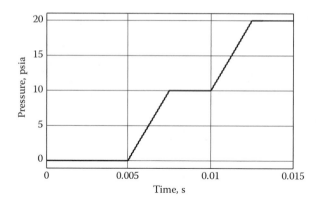

FIGURE 3.14
Pressure variation for transducer.

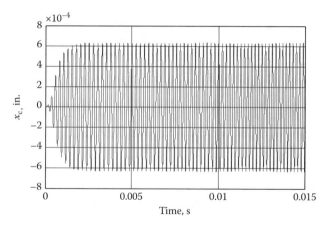

FIGURE 3.15
Cylinder wall displacement oscillations.

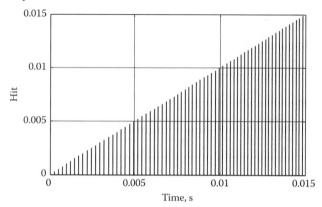

FIGURE 3.16
Zero-crossing times found by use of *hit-crossing* module.

```
>> [i,j,s]=find(hit);
>> whos
   Name              Size                Bytes    Class

   ans               10x1                   80    double array
   fn            150001x1              1200008    double array
   hit           150001x1              1200008    double array
   i                 67x1                  536    double array
   j                 67x1                  536    double array
   p             150001x1              1200008    double array
   s                 67x1                  536    double array
   satin         150001x1              1200008    double array
   t             150001x1              1200008    double array
   tout            1000x1                 8000    double array
   xc            150001x1              1200008    double array

Grand total is 901217 elements using 7209736 bytes

>> for n=3:67
fr(n)=1/(s(n)-s(n-1));
end
>> plot(i(3:67)*1e-7,fr(3:67))
>> hold on
>> plot(t,fn)
>>
```

In this program, the *s* values are the actual times of successive zero-crossings from negative to positive values. By subtracting the $n-1$ value from the n value we get the time for each cycle; its reciprocal is the frequency. The first few cycles do not give useful values, so we start with $n = 3$. The *whos* statement (*not* part of the program) tells us how many (67) zero-crossing values there are. The final plot statement superimposes a graph of the cylinder natural frequency; these values are available from the simulation diagram as the signal *fn*. In Figure 3.17, we see that during the dwell periods the oscillating

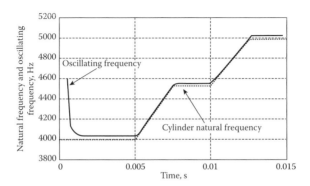

FIGURE 3.17
Comparison of cylinder natural frequency and system oscillating frequency.

frequency is slightly above the cylinder natural frequency. This agrees with the predictions of the root-locus and Nyquist analyses; to move the roots to the imaginary axis, or to shift the phase angle to −180°, the operating frequency has to increase. It appears that during the ramp changes the two frequencies are very close, perhaps because there is not enough time for these shifts to occur (note that *all* the time intervals in Figure 3.17 are small fractions of 1 s).

3.7 Ultraprecision Calibration/Measurement Using a 15-Term Calibration Equation, Built-In Temperature Compensation, and Microprocessor Data Reduction

We have mentioned the possible use of digital counter/timers with this transducer; however, a system using the Solartron transducers and discussed in Juanarena and Gross (1985) uses a variation of this technique because of a desire to achieve higher resolution without long counting times. (Any reference to *vibrating-cylinder transducers* in this chapter relates to the version produced and marketed by *Solartron Transducers, Farnborough, UK*, who I had communications with around 1986. I was unable to get any details from the current (2008) maker (*Weston Aerospace*), but from their Web site it appears that this technology has not changed in any significant way.)

In the system[*] described in Juanarena and Gross (1985), the near-sinusoidal voltage from the pickup coil is converted to a square wave by sending it to a simple comparator circuit since the zero-crossings of square waves may be more accurately sensed. An electronic counter (10 kHz full scale) counts 256 cycles of this square wave and produces a sharp pulse at the beginning and end of this time period. These timing pulses are used to start and stop the counting of a precision 10 MHz clock, giving, for example at 5000 Hz, a total count of 512,000 ± 1 count, a resolution of about 0.0002%. Taking into account all other sources of error in this measurement, the total uncertainty in the cylinder frequency measurement is quoted as less than 0.001% of full scale. An interesting sidelight on this frequency measurement is that the precision clock is also a feedback-type device which uses a piezoelectric crystal's natural vibration frequency to establish the 10 MHz timing reference. To keep the crystal's natural frequency constant within 5 ppm/100°, it is housed in a "can" whose temperature is closely controlled by a feedback system.

While the resolution of this type of frequency measurement can be made very fine by simply counting more cycles of cylinder vibration, this does *not* guarantee equal accuracy in pressure measurement. Perfect accuracy requires that a certain pressure always produces the same frequency.

[*] www.pressuresystems.com

Long-term tests of actual transducers at constant temperature show that the vibrating-cylinder technology *is* capable of 0.01% FS accuracy over a 12 month period. To achieve this desired level of accuracy for varying temperatures, cylinder temperature must be measured and its effects included in the calibration procedure. To provide temperature data, a temperature-sensitive silicon diode, which provides a voltage related to temperature, is built into the transducer housing. To measure an unknown pressure, one must thus measure the frequency and the transducer temperature (diode voltage) and then enter these numbers into a calibration equation.

We generally wish to keep calibration equations as simple as possible, however when the relations between dependent and independent variables are nonlinear, and the required accuracy is very high, calibration equations may need to include a large number of terms. The specific *form* of calibration equations may be suggested by theoretical results such as Equation 3.3 relating frequency and pressure, or we may just use *standard* forms, such as polynomials, adding more terms until the required accuracy is achieved. The Solartron engineers could not provide details on *how* they decided on the form of the equation, but they have published the form actually used for these transducers:

$$P = K_0 + K_1 F^2 + K_2 F^4 \tag{3.7}$$

$$K_0 \triangleq B_{00} + B_{01}V + B_{02}V^2 + B_{03}V^3 + B_{04}V^4 \tag{3.8}$$

$$K_1 \triangleq B_{10} + B_{11}V + B_{12}V^2 + B_{13}V^3 + B_{14}V^4 \tag{3.9}$$

$$K_2 \triangleq B_{20} + B_{21}V + B_{22}V^2 + B_{23}V^3 + B_{24}V^4 \tag{3.10}$$

where
   F is the oscillation frequency
   V is the voltage from the temperature-sensing diode
   the B's are constants to be chosen

Equation 3.7 shows the dependence of pressure on frequency, assuming *no* affect from temperature. Equation 3.3, solved for p, shows that p does depend on the square of frequency; perhaps the term in F^4 was included to allow for some deviation from the theory's ideal model. Equations 3.8 through 3.10 show the use of a fourth-degree polynomial to model the effects of temperature, since V is proportional to temperature. Combining Equations 3.7 through 3.10 gives a single equation that states the overall model, which can be submitted to multiple regression to find the best set of fitting coefficients:

$$P = C_1 S + C_2 SX + C_3 SX^2 + C_4 SX^3 + C_5 SX^4 + C_6 SY + C_7 SXY + C_8 SX^2 Y + C_9 SX^3 Y$$
$$+ C_{10} SX^4 Y + C_{11} SY^2 + C_{12} SXY^{2+C^2} + C_{13} SX^2 Y^2 + C_{14} SX^3 Y^2 + C_{15} SX^4 Y^2 \tag{3.11}$$

where
    $V \triangleq$ voltage of temperature-sensing diode, V
    $TP \triangleq$ cylinder vibration period, μs
    $P \triangleq$ absolute pressure, kN/m^2
    $S \triangleq 262.0 \times 10^{-6}$
    $X \triangleq 5.0(V - 0.60)$
    $Y \triangleq 5.0[(164.0/TP)^2 - 1]$

The numerical values of the 15 calibration coefficients C_1 to C_{15} are slightly different for each individual transducer and are found by static calibration against a very accurate pressure *standard*,* using 11 different pressures at each of 7 temperatures, giving a total of 77 sets of P, V, and TP values. The problem of finding the *best* C values based on 77 sets of data points is solved using the same *multiple regression* methods and software that we applied to the experiment design problem of Chapter 1.

The Solartron engineers graciously supplied me with one complete set of calibration data from an actual transducer, so we can use our Minitab software to process this data. Table 3.1 shows this data as 7 groups of 11 pressures each, ranging from about 0.5–38 psia. Note that the pressure values are *not* equally spaced; the spacing is closer at low pressures than at high. Perhaps this was done intentionally to give a greater weight (in the regression calculation) to low pressures, where the error, as a percentage of *reading* tends to be greater. (We see also that this transducer had a higher range than our example unit, which was 0–20 psia.) These numbers also allow us to conjecture to some extent about how the form of equation was chosen. Some of the choices appear to be mainly matters of *scaling* to improve the accuracy of the matrix calculations used in the multiple regression software. The definition of the parameter X has to do with the voltage from the temperature-sensing diode. We see from the data that this voltage ranged from about 0.4 to 0.8 V, so X appears to have been scaled so that its values are distributed more or less equally on each side of zero. The Y parameter deals with the measured frequency, but in the form of the *period*. If one solves Equation 3.3 for pressure p, the term involving the period appears squared in the denominator, so the term $(164/TP)^2$ seems reasonable. Again, the 164 is about the midpoint of the range of TP values, so the Y values will be distributed on either side of zero. If one computes the numerical values of the *variable terms* (like SX^2, SXY, etc.) from the given data (these calculations actually *have* to be done when we do the multiple regression), we see that most of these calculated values lie in the range 0.0001–0.00001. That is, all these numbers are fairly close together, again facilitating the matrix calculations, which get *ill-conditioned* when there is too wide a range of numerical values.

In trying to further "guess" why Equation 3.11 was given its actual form, we can see that the four terms which involve *only* the temperature, form a

---

* E.O.Doebelin, *Measurement Systems*, 5th edn., McGraw-Hill, New York, 2003, pp. 482–486.

TABLE 3.1

Calibration Data for Transducer, 11 Pressures at 7 Different Temperatures

TP	V	P	TP	V	P	TP	V	P	TP	V	P
180.6236	0.779680	3.4516	180.6306	0.726590	3.4514	180.6179	0.662160	3.4517	180.5856	0.588720	3.4517
178.4199	0.779680	17.2480	178.4193	0.726590	17.2487	178.4010	0.662160	17.2480	178.3647	0.588720	17.2485
175.7787	0.779680	34.4934	175.7689	0.726590	34.4941	175.7438	0.662160	34.4943	175.7027	0.588720	34.4942
173.2558	0.779680	51.7379	173.2370	0.726590	51.7390	173.2054	0.662160	51.7404	173.1596	0.588720	51.7383
170.8424	0.779680	68.9839	170.8150	0.726590	68.9851	170.7770	0.662160	68.9852	170.7264	0.588720	68.9853
168.5310	0.779680	86.2280	168.4950	0.726590	86.2301	168.4512	0.662160	86.2296	168.3961	0.588720	86.2303
166.3148	0.779680	103.4739	166.2710	0.726590	103.4757	166.2207	0.662160	103.4756	166.1612	0.588720	103.4751
162.9514	0.779680	131.0657	162.8952	0.726590	131.0679	162.8353	0.662160	131.0671	162.7687	0.588720	131.0670
158.2842	0.779680	172.4536	158.2094	0.726590	172.4575	158.1364	0.662160	172.4558	158.0596	0.588720	172.4563
154.0140	0.779680	213.8412	153.9224	0.726590	213.8441	153.8357	0.662160	213.8429	153.7495	0.588720	213.8436
149.4642	0.779680	262.1265	149.3525	0.726590	262.1300	149.2523	0.662160	262.1296	149.1551	0.588720	262.1306

TP	V	P	TP	V	P	TP	V	P
180.5393	0.519060	3.4515	180.4919	0.462260	3.4516	180.4473	0.417050	3.4518
178.3159	0.519060	17.2500	178.2668	0.462260	17.2491	178.2217	0.417050	17.2486
175.6508	0.519060	34.4953	175.5999	0.462260	34.4954	175.5538	0.417050	34.4943
173.1044	0.519060	51.7396	173.0516	0.462260	51.7412	173.0052	0.417050	51.7372
170.6683	0.519060	68.9864	170.6139	0.462260	68.9859	170.5665	0.417050	68.9835
168.3348	0.519060	86.2311	168.2787	0.462260	86.2303	168.2305	0.417050	86.2276
166.0968	0.519060	103.4773	166.0391	0.462260	103.4762	165.9899	0.417050	103.4724
162.6980	0.519060	131.0689	162.6390	0.462260	131.0677	162.5884	0.417050	131.0632
157.9834	0.519060	172.4591	157.9188	0.462260	172.4575	157.8660	0.417050	172.4509
153.6665	0.519060	213.8464	153.5979	0.462260	213.8445	153.5428	0.417050	213.8368
149.0643	0.519060	262.1332	148.9912	0.462260	262.1292	148.9334	0.417050	262.1217

quite conventional polynomial out to the fourth power. All the other terms are *interactions* of temperature and frequency (period), except for SY, which involves the *main effect* of period. Reasoning from the physics of Equation 3.3 to search for interaction effects, one can see temperature affecting dimensions, such as r, L, and t, and material properties such as E, γ, and v, but we are hard pressed to explain the particular interaction terms that were chosen. Perhaps the Solartron engineers simply tried various combinations and finally chose those given in Equation 3.11. If we simply accept their choice, then it is a routine calculation to find the *best* regression equation and the quality of that model. Using Minitab as explained in Chapter 1 and Appendix B, we get the following results.

The regression equation for the data of Table 3.1

$$P = 121 + 5622\ SX - 674\ SX^2 + 518\ SX^3 + 376\ SX^4 + 513748\ SY + 4211\ SXY + 957\ SX^2Y + 1136\ SX^3Y + 866\ SX^4Y + 2528\ SY^2 + 697\ SXY^2 + 180\ SX^2Y^2 + 264\ SX^3Y^2 + 210\ SX^4Y^2$$

Predictor	Coef
Constant	120.931
SX	5621.54
SX^2	−673.58
SX^3	518.39
SX^4	375.62
SY	513748.
SXY	4211.46
SX^2Y	957.17
SX^3Y	1135.91
SX^4Y	865.53
SY^2	2528.39
SXY^2	696.64
SX^2Y^2	179.89
SX^3Y^2	263.57
SX^4Y^2	209.58

R-Sq = 100.0% R-Sq(adj) = 100.0%

In the regression equation just given, the coefficients are given in rounded format; one must replace these by the accurate values shown in the list of coefficients when using the equation to predict pressures from measured vibration periods and diode voltages. Also, Minitab recognized S as a constant and eliminated it from the regression, replacing this effect with its equivalent, the constant term 120.931 (the *intercept*). Table 3.2 shows the error at each pressure and its percentage of that pressure. The largest percentage errors are at the low pressures, as expected, but typical errors are about 0.003%.

It may be of interest to briefly consider a regression which simply tries polynomials for the period and temperature effects, with perhaps one simple

TABLE 3.2

Error and Percent Error Using the 15-Term Calibration Equation

P	Error	%Error	P	Error	%Error	P	Error	%Error	P	Error	%Error
3.460	-0.008	-0.235	3.455	-0.004	-0.102	3.459	-0.008	-0.220	3.456	-0.004	-0.113
17.248	0.000	-0.003	17.246	0.003	0.017	17.250	-0.002	-0.014	17.247	0.002	0.009
34.491	0.002	0.007	34.490	0.004	0.012	34.495	0.000	-0.001	34.491	0.003	0.010
51.734	0.004	0.008	51.735	0.004	0.008	51.739	0.002	0.003	51.734	0.004	0.007
68.979	0.005	0.007	68.981	0.004	0.005	68.984	0.001	0.001	68.981	0.004	0.006
86.226	0.002	0.002	86.230	0.000	0.000	86.229	0.000	0.000	86.226	0.004	0.005
103.473	0.001	0.001	103.475	0.001	0.001	103.476	-0.001	-0.001	103.473	0.003	0.002
131.068	-0.002	-0.002	131.068	0.000	0.000	131.069	-0.002	-0.002	131.067	0.000	0.000
172.459	-0.005	-0.003	172.463	-0.006	-0.003	172.458	-0.002	-0.001	172.458	-0.001	-0.001
213.845	-0.004	-0.002	213.846	-0.002	-0.001	213.846	-0.003	-0.001	213.845	-0.001	-0.001
262.120	0.007	0.003	262.133	-0.003	-0.001	262.121	0.008	0.003	262.126	0.005	0.002

P	Error	%Error	P	Error	%Error
3.458	-0.006	-0.175	3.455	-0.006	-0.098
17.249	0.001	0.008	17.248	0.001	0.005
34.492	0.003	0.008	34.493	0.001	0.003
51.738	0.002	0.004	51.736	0.002	0.003
68.983	0.003	0.005	68.981	0.003	0.004
86.229	0.002	0.002	86.226	0.002	0.002
103.476	0.001	0.001	103.472	0.000	0.000
131.084	-0.015	-0.012	131.066	-0.003	-0.002
172.464	-0.005	-0.003	172.455	-0.004	-0.002
213.853	-0.006	-0.003	213.841	-0.004	-0.002
262.138	-0.005	-0.002	262.121	0.001	0.000

interaction term such as CXY. That is the sort of model one might apply as a "first try" for the calibration equation if there were no theoretical reasons for choosing other functions:

$$P = A_0 + A_1X + A_2X^2 + A_3X^3 + A_4X^4 + A_5Y + A_6Y^2 + A_7Y^3 + A_8Y^4 + A_9XY$$

$$(3.12)$$

The regression equation for this model is

$$P = 121 + 1.57\,X + 135\,Y - 0.190\,X^2 + 0.117\,X^3 + 0.086\,X^4 + 0.718\,Y^2 + 0.0108\,Y^3$$
$$- 0.037\,Y^4 + 1.30\,XY$$

Predictor	Coef
Constant	120.939
X	1.56674
Y	134.753
X^2	−0.1898
X^3	0.11676
X^4	0.0862
Y^2	0.7180
Y^3	0.01084
Y^4	−0.0370
XY	1.30311 0

$R\text{-Sq} = 100.0\%$ $R\text{-Sq(adj)} = 100.0\%$

The errors here are considerably larger (−0.2% to +0.45% of reading) than those for the more complex model (−0.015% to +0.008% of reading), as one might expect. Figure 3.18 compares the two sets of errors graphically; we see that the errors just quoted would be somewhat reduced if we set aside one *outlier* for each data set. The percent errors for the simpler model would actually be quite acceptable for many ordinary transducers, but not for an instrument which may be used as a *secondary standard*, as discussed in Juanarena and Gross (1985).

Pressure scanning systems, such as discussed in Juanarena and Gross (1985), have undergone some design changes in response to developments in piezoresistive (MEMS) pressure transducers over the years. In Juanarena and Gross (1985), these transducers were at an early stage of improvement where an *overall* recalibration (against the vibrating-cylinder standard) was required periodically, and involved three calibration points: zero, mid-range, and full scale. Currently (2008) this company's pressure scanners no longer require a built-in standard, although they still offer the vibrating-cylinder transducer (manufactured by Weston Aerospace) as a separate item. (Pressure Systems and Weston Aerospace are both part of the larger group of companies called Esterline Technologies.*) Pressure Systems also

---

* www.esterline.com

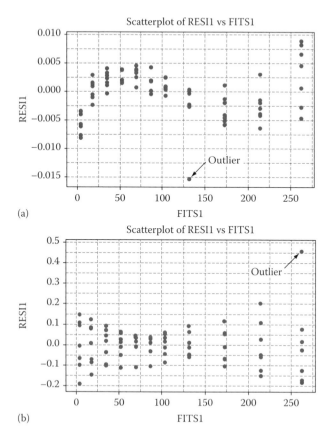

FIGURE 3.18
Residual graphs for (a) complex and (b) simple regression models.

manufactures a pressure standard which uses the frequency-modulated principle of the vibrating-cylinder device, but the elastic element is now a small,
thin quartz beam, to which a pressure-proportional axial force is applied to
change the frequency. An unstable feedback system is again used but the
force-applying and motion-sensing functions are performed by piezoelectric means using vibrating quartz tuning forks. The pressure-related force
is developed using a thin-wall metal bellows which has sensed-pressure
inside and a vacuum outside, to achieve an absolute pressure transducer. To
maintain the initial vacuum sufficiently close to zero psia, a *getter* material
is included. The entire transducer (except for the bellows) is micromachined
by special methods from a single block of quartz. Figure 3.19 shows some of
these details.

The current multichannel pressure scanning systems do not require a
pressure standard transducer such as the vibrating-cylinder or vibrating-beam
type, but still use a configuration as shown in Figure 3.20. Modules come
as blocks of 32 miniature piezoresistive differential pressure sensors with

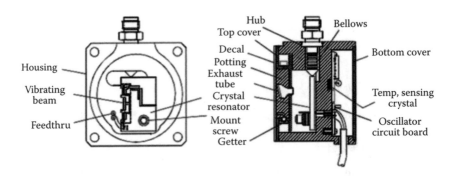

FIGURE 3.19
Quartzonix™ vibrating-beam pressure transducer.

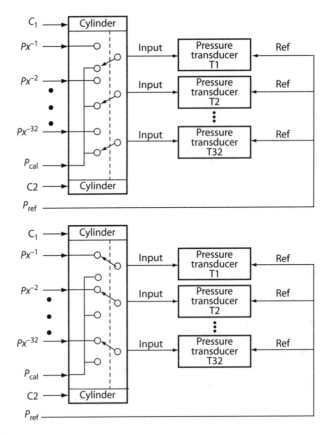

FIGURE 3.20
Calibration scheme for multichannel pressure scanner.

an integral pneumatically actuated multi-port valve in the form of a piston and cylinder. The sensors are differential-pressure types (psid) rather than gage (psig) or absolute (psia), but are not really used for the usual differential-pressure applications such as sensing the pressure drop across a flow-measuring orifice plate. Rather, one of the two pressure taps is connected to a pressure sensing location in an experimental setup and the second tap of each of the 32 sensors is connected to a constant (but selectable) common reference pressure. The *active* taps of all 32 sensors can be switched to a common calibration pressure by a single motion of the multi-port valve, by application of a control air pressure. In the earlier versions of the system, the calibration pressure was switched sequentially to one of three calibration pressures (zero, mid-range, full scale), the calibration pressure being accurately measured by a single *standard* transducer such as the vibrating cylinder. In the current systems, accurate temperature compensation of each piezoresistive sensor using digital methods and a single temperature sensor per module allows sufficient accuracy with only a periodic *re-zeroing* at one calibration pressure, usually ambient pressure, which is also the reference pressure, subjecting each sensor to the same pressure on each side (0 psid). No standard transducer is needed.

4

A Fast ("Cold-Wire") Resistance Thermometer for Temperature Measurements in Fluids

4.1 Introduction

The same apparatus used for constant-current hot-wire anemometry* can also be used, with proper adjustments, for measuring rapidly fluctuating fluid temperatures. When we analyzed (Doebelin, 2003, p. 599) the hot-wire anemometer as a fluid-velocity sensing device, we assumed for simplicity that the fluid temperature was steady. In some hot-wire applications, the fluid temperature *does* change significantly, causing the instrument output signal to be related to both temperature and velocity, and thus complicating the measurement of velocity. When this is the case, a separate measurement of fluid temperature is needed to correct the hot-wire reading so that it represents only velocity. If the temperature fluctuations are large but slow, this correction can be accomplished with relatively simple (slow) temperature sensors. When the fluid temperature variations are fast, then a fast thermometer is required. The instrument usually used here is called a *cold-wire*, and uses the same basic sensing element (a very fine wire) as does the velocity-measuring instrument, but operates it at such a small temperature difference (relative to the fluid temperature) that its sensitivity to flow velocity is almost zero, while its sensitivity to fluid temperature is good. Such cold-wire thermometers are also useful, not just for correcting hot-wire velocity measurements, but for general-purpose fast temperature sensing in fluids.

---

* E.O. Doebelin, *Measurement Systems*, 5th edn., McGraw-Hill, New York, 2003, p. 599.

4.2 Circuitry and Wire Details

The circuitry (see Figure 4.1) used is very simple and identical to the constant-current hot-wire circuit. To make the sensing wire essentially insensitive to velocity changes, it must be operated at much lower current values than are routinely used for hot-wires, which operate at a few hundred milliamps. Consulting the cold-wire literature, we find that a current of 0.5 mA is typical. Also, to get a fast thermal time constant for the wire, a very fine wire of a platinum alloy (90% platinum, 10% rhodium) must be used. Typical dimensions are a diameter of 1.0 μm and a length of 500 μm. Note that this gives a length/diameter ratio of 500, which (just as in hot-wire applications) makes conduction heat transfer to the supporting prongs negligible relative to convection at the wire surface.

We now want to say a little about how such *very* fine wires can actually be produced and applied. The method of manufacturing such fine wires was invented in the 1800s by a British scientist named Wollaston, and the wires are still marketed using his name. "Ordinary" wire is usually produced by a method called *wire drawing*, where a "fat" rod or wire is forced through a die of smaller diameter, thus reducing the cross section. Several stages of drawing may be necessary to produce fine wire. This basic method is, however, *not* capable of producing the very fine wires we need, since the force needed to pull the wire through the die would break the wire. Wollaston's invention was to take a wire larger than we need and encase it in a tube of silver. This tube/wire combination is strong enough to be drawn through a conventional die without breaking, and results in a very fine wire encased in a silver sleeve. For cold-wire applications, the wire is made of the platinum/rhodium alloy mentioned above. The finest Wollaston wires commercially available* have a diameter of 0.6 μm. The silver-encased wire is soft-soldered to the probe prongs and then the silver is etched away with, for example, nitric acid, leaving a central section of very fine platinum/rhodium wire as the sensitive part.

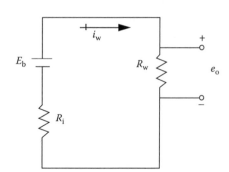

FIGURE 4.1
Circuit for cold-wire resistance thermometer.

---

* www.sigmundcohn.com

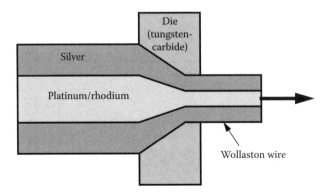

FIGURE 4.2
Manufacture of Wollaston wire.

The soldering process is quite difficult and is done under a microscope. Figure 4.2 shows the wire-drawing process in a simplified manner.

We first need to calculate the resistance of our wires from material properties and dimensions. The resistivity, ρ_w, of our alloy (at 20°C) is about 18.8 $\mu\Omega$ cm (pure platinum has 10.58), and the formula for the resistance of wire with diameter d and length L is

$$R_w = \frac{\rho_w L}{A} \tag{4.1}$$

where A is the cross-sectional area. While we will be using a specific d and L, it is of interest to see how the resistance varies with diameter, for $L = 500\,\mu$m (see Table 4.1). A resistance thermometer would always be designed for a specific range of temperatures. Let us take the range for our example to be from 20°C (293 K) to 220°C (493 K). To compute resistance as a function of temperature we need the material's temperature coefficient of resistance. For pure platinum (used in many "slow" precision thermometers), this is 0.00392 $\Omega/(\Omega$-K) while for the platinum/rhodium alloy, it is 0.0017. We see that platinum is more sensitive than the alloy, but for the fine wires we are using, the alloy tensile strength is about four times higher, so it is usually used, to reduce breakage problems during installation and use. Using $d = 1\,\mu$m and $L = 500\,\mu$m we can now make a table of resistance versus temperature for the sensing wire (see Table 4.2).

The concept of the thermometer circuit of Figure 4.1 is simply that the wire will come to a temperature close to that of the surrounding fluid, its resistance will change,

TABLE 4.1

Resistance versus Diameter for $L = 500\,\mu$m at 20°C

d, cm	R_w, Ω
0.0001	119.684517
0.0002	29.921129
0.0003	13.29828
0.0004	7.480282
0.0005	4.787381
0.0006	3.32457
0.0007	2.442541
0.0008	1.870071
0.0009	1.477587
0.001	1.196845

TABLE 4.2

Resistance versus Temperature for Wire
Element

$T := 293, 303, \ldots, 493$

$R_w(T) = 119.685 + (T - 293)(0.00170)(119.685)$

T	$R_w(T)$
293	119.685
303	121.719645
313	123.75429
323	125.788935
333	127.82358
343	129.858225
353	131.89287
363	133.927515
373	135.96216
383	137.996805
393	140.03145
403	142.066095
413	144.10074
423	146.135385
433	148.17003
443	150.204675
453	152.23932
463	154.273965
473	155.30861
483	158.343255
493	160.3779

and this will in turn cause a change in the output voltage, e_o. (Since the wire carries a nonzero current, its internal heat generation requires that it be somewhat hotter than the surrounding fluid; this is the so-called *self-heating error* present in *any* resistance thermometer or thermistor. The small current (0.0005 A) is chosen to make this error acceptable. If the system is calibrated against a standard and then used under the same conditions as at calibration, even this small error can, in principle, be removed.) While one can purchase commercial *constant-current sources*, in our case a simpler approach to constant-current operation is possible. That is, using readily available constant-voltage sources, we can choose the resistor, R_i, such that the current is nearly constant for the range of wire temperatures (and thus resistances) that we design for. To design the constant-current circuit (choose R_i and E_b), we state that we want the current to be near 0.5 mA, and that the current should stay nearly constant over the entire range of temperature. This requires that R_i be much larger than the wire resistance. As a trial design, we might take R_i to be

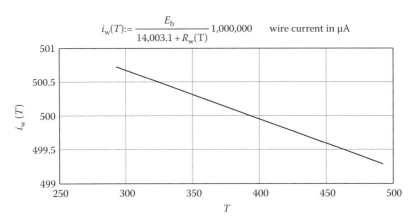

$$i_w(T):= \frac{E_b}{14{,}003.1 + R_w(T)} \cdot 1{,}000{,}000 \qquad \text{wire current in } \mu A$$

FIGURE 4.3
Check for constancy of current source.

100 times the mid-range value (140 Ω) of the wire resistance and find the E_b that is required to give 0.0005 A current. This simple calculation gives $R_i = 14{,}000\ \Omega$ and $E_b = 7.07$ V. We can now let the wire resistance vary over the design range (about 120–160 Ω) to see whether the current is really constant. Figure 4.3 shows that our initial design does give a current that is sufficiently constant. We are also interested in the range of output voltages that will be produced for the full range of temperature values. This is easily found to be 20 mV (60–80 mV). If the output voltage is read with a voltmeter of resolution 1 μV, we can resolve about 0.01 K temperature changes.

In considering limits on resolution it is necessary to also consider any *noise** sources in the circuit. If the voltage source is a battery, its noise contribution would probably be negligible, but if we use an electronic power supply, then, because such supplies rectify and filter the AC line power to create the DC, there is always some remaining *ripple*. We should consult the power supply specifications to find out how large this ripple might be, and what is its waveform. This ripple in supply voltage will cause a ripple in the current, and thus in our output voltage. Typical supply voltage ripple values might be 1 mV RMS and 3 mV peak. Using the peak value of 0.003 V, this would cause an output voltage change of about 29.7 μV when the wire resistance was at the mid-range value of 140 Ω. We see that this is much larger than the 1 μV voltmeter resolution, so we would *not* be able to resolve temperatures to 0.01 K as we first calculated, so a battery power supply might be required. Some low-noise instruments use rechargeable batteries with a built-in charger, which automatically keeps the battery charged. The other main source of noise in our circuit is the *Johnson noise* (also called thermal noise) generated in every resistor whose temperature is above absolute zero. This is usually

---

* E.O. Doebelin, *Measurement Systems*, 5th edn., McGraw-Hill, New York, 2003, pp. 88–90.

modeled as *white noise*, a random voltage with flat frequency content for all frequencies. To get numerical values, one must specify the *bandwidth* of the measurement, as determined by any filters that might be used in the system. If we want to measure fast-changing temperatures up to, say, 1000 Hz, then we could use a low-pass filter set to cut off at 1000 Hz, making the bandwidth also 1000 Hz. With the system bandwidth now defined, we can compute the Johnson noise as follows[*]:

$$e_{noise} = \sqrt{4kTR\Delta f} \text{ V, RMS} \tag{4.2}$$

where
 k = Boltzmann's constant = 1.38e−23 J/K
 T is the resistor's temperature (K)
 R is the resistance (Ω)
 Δf is the bandwidth (Hz)

Our circuit has two resistances; each one generates Johnson noise. For the wire resistance (taken as 140 Ω, for example), the noise appears directly in the circuit output voltage. For a bandwidth of 1000 Hz, the noise RMS value is about 3e−15 V, which would be negligible relative to the meter resolution of 1 μV. Our other resistor's Johnson noise is about 3e−13 V, which is treated as if it were the supply voltage, so its effect on the output voltage is even smaller, and thus negligible.

4.3 Estimating the Self-Heating Error

We chose the wire current to be very small so as minimize the temperature rise of the wire above the fluid temperature. Now we must check this assumption, which requires knowledge of the heat transfer between the fluid and the wire. We will shortly be needing some properties of air, which is the fluid in which measurements are to be made. Table 4.3 is based on data in a heat transfer text.[†] To access the above tabular data most efficiently in our upcoming calculations, we use Minitab multiple regression to fit cubic polynomial curves to represent the three air properties as functions of temperature. Because this data is quite smooth and nearly linear, low-degree polynomials, such as cubics give a very good fit:

---

[*] E.O. Doebelin, *Measurement Systems*, 5th edn., McGraw-Hill, New York, 2003, p. 251.
[†] A.F. Mills, *Basic Heat and Mass Transfer*, Irwin, Chicago, IL, 1995, p. 842

TABLE 4.3

Properties of Air at Atmospheric Pressure

Temperature, K	Density, kg/m³	Viscosity, kg/(m-s)	Conductivity, W/(m-K)
290	1.220	18.02e−6	0.0261
320	1.106	19.29e−6	0.0281
350	1.012	20.54e−6	0.0300
380	0.931	21.75e−6	0.0319
400	0.883	22.52e−6	0.0331
500	0.706	26.33e−6	0.0389
600	0.589	29.74e−6	0.0447
700	0.507	33.03e−6	0.0503

$$\text{den:} = 3.23155 - 0.0106688 \cdot T_f + 1.50741 \times 10^{-5} \cdot T_f^2 - 7.70731 \times 10^{-9} \cdot T_f^3$$

$$\text{vis:} = 3.42892 \cdot 10^{-6} + 5.81618 \cdot 10^{-8} \cdot T_f - 3.03056 \cdot 10^{-11} \cdot T_f^2$$
$$+ 1.08778 \cdot 10^{-14} \cdot T_f^3$$

$$\text{cond:} = 3.24662 \cdot 10^{-3} + 9.41753 \cdot 10^{-5} \cdot T_f - 6.28511 \cdot 10^{-8} \cdot T_f^2$$
$$+ 3.48159 \cdot 10^{-11} \cdot T_f^3$$

For steady-state conditions, we assume that the internal heat generation in the wire is exactly balanced by the convective heat transfer from the wire at temperature T_w to the surrounding fluid at temperature T_f:

$$i^2 R_w = i^2[119.685 + (T_w - 293)(0.00170)(119.685)] = hA_w(T_w - T_f) \tag{4.3}$$

where
  h is the convective heat transfer coefficient
  A_w is the surface area of the wire

Equation 4.3 must use SI units (electrical heating and heat transfer rates must both be given in W) or else a conversion factor must be included to reconcile watts on the left side and, say, Btu/s on the right side. The self-heating error is defined as the difference between wire temperature and fluid temperature:

$$\text{self-heating error } \underline{\Delta} T_w - T_f = \frac{i^2 R_w}{hA_w} \tag{4.4}$$

where we have chosen, for the time being, to not express R_w in terms of T_w. This form makes clear the need to use small currents to reduce the self-heating error.

At this point, we need to get information on the relation of h to fluid properties, temperature, and flow conditions, as given in Tavoularis (2005)*:

$$h = \frac{k}{D}\left(0.24 + 0.56\left(\frac{\rho DV}{\mu}\right)^{0.45}\left(1 + \frac{T_w - T_f}{2T_f}\right)^{0.17}\right) \quad 44 < Nr < 140 \quad (4.5)$$

where
   Nr is the Reynolds number $\rho DV/\mu$
   k is the air thermal conductivity
   D is the wire diameter
   μ is the air viscosity
   V is the fluid velocity
   T_w is the wire temperature
   T_f is the air temperature

Researchers found this equation for h by curve fitting experimental data for the flow regime called *vortex shedding*. That is, for this range of Reynold's number, the cylindrical wire sheds periodic vortices, whereas for lower Reynold's numbers the flow remains attached to the wire. A different formula is available for this lower range of velocities:

$$h = \frac{k}{D}\left(0.48\left(\frac{\rho DV}{\mu}\right)^{0.51}\left(1 + \frac{T_w - T_f}{2T_f}\right)^{0.17}\right) \quad 0.02 < Nr < 44 \quad (4.6)$$

When computing h values, one must thus be careful to use the correct formula, depending on the value of the Reynold's number (for given fluid properties and wire size, the velocity). At this point, we see that in addition to specifying the range of temperatures that we wish to measure, we also need to give the range of air velocities to be considered. Let us take this as the range 1–20 m/s. We can then compute the Reynolds number to see whether Equation 4.5 or 4.6 should be used for our application. Since air density and viscosity depend on temperature, we need to do this calculation for both the lowest (293 K) and highest (493 K) temperatures. Table 4.4 shows that for our conditions we need to use Equation 4.6.

To find the self-heating error, we need to know the value of h from Equation 4.6, but this equation includes the term $\left(1 + \dfrac{T_w - T_f}{2T_f}\right)^{0.17}$, which requires knowledge of T_w, which we cannot know until we use h to get the self-heating error. To avoid a tedious iterative calculation, we want to show that for small self-heating error, this term can be taken as 1.0 without serious

---

* S. Tavoularis, *Measurement in Fluid Mechanics*, Cambridge University Press, New York, 2005, p. 252.

TABLE 4.4

Reynold's Number versus Flow Velocity

$V := 1, 3, \ldots, 21 \quad Nr(V) := (\text{den})(d_{\text{wire}})(V) / \text{vis}$

V	$Nr(V), T_f = 293\,K$	$Nr(V), T_f = 493\,K$
1	0.066	0.027
3	0.199	0.082
5	0.332	0.137
7	0.465	0 191
9	0.598	0.246
11	0.731	0.301
13	0.864	0.355
15	0.997	0.41
17	1.13	0.465
19	1.263	0.52
21	1.396	0.574

error. Let us assume that the self-heating error never exceeds 10 K and we take T_f as its smallest value (293 K) to give the worst case. The above term then becomes 1.0029, which is close enough to 1.0 for our purposes, so we replace it by 1.0 in further calculations. We can now compute h for the given range of velocities if we pick a given air temperature. There is no need to do this for many temperatures; it should suffice to do it for the lowest and highest air temperatures, as in Table 4.5. The lowest h value, and thus the worst

TABLE 4.5

Heat Transfer Coefficient h versus Flow Velocity, for Two Air Temperatures

$$h(V) := \frac{\text{cond}}{d_{\text{wire}}} \cdot \left[\left(\frac{\text{den} \cdot d_{\text{wire}} \cdot V}{\text{vis}} \right)^{0.51} \cdot 0.48 \cdot 1.0 \right]$$

V	$h(V), T_f = 293\,K$	$h(V), T_f = 293\,K$
1	3169.9	2953.3
3	5551.1	5171.7
5	7203.2	6710.9
7	8551.6	7967.2
9	9721	9056.7
11	10768.6	10032.6
13	11726.3	10924.9
15	12614.1	11752
17	13445.5	12526.6
19	14230.3	13257.7
21	14975.5	13952

self-heating error, will be found at the lowest velocity (1.0 m/s) and highest temperature (493 K); it is h = 2953 W/(m²-s). The smallest self-heating error corresponds to h = 14975. We can solve Equation 4.3 for T_w and then do the following Mathcad calculation:

$$h1: = 14975. \quad Aw1: = 1.5708 \cdot 10^{-9} \quad Tf1: = 293$$

$$Tw1: = \frac{-h1 \cdot Aw1 \cdot Tf1 - 1.5017 \cdot 10^{-5}}{5.0875 \cdot 10^{-8} - h1 \cdot Aw1}$$

$$Tw1 = 294.27$$

$$h1: = 3169. \quad Aw1: = 1.5708 \cdot 10^{-9} \quad Tf1: = 293$$

$$Tw1: = \frac{-h1 \cdot Aw1 \cdot Tf1 - 1.5017 \cdot 10^{-5}}{5.0875 \cdot 10^{-8} - h1 \cdot Aw1}$$

$$Tw1 = 299.07$$

$$h1: = 13952. \quad Aw1: = 1.5708 \cdot 10^{-9} \quad Tf1: = 493$$

$$Tw1: = \frac{-h1 \cdot Aw1 \cdot Tf1 - 1.5017 \cdot 10^{-5}}{5.0875 \cdot 10^{-8} - h1 \cdot Aw1}$$

$$Tw1 = 494.83$$

$$h1: = 2953. \quad Aw1: = 1.5708 \cdot 10^{-9} \quad Tf1: = 493$$

$$Tw1: = \frac{-h1 \cdot Aw1 \cdot Tf1 - 1.5017 \cdot 10^{-5}}{5.0875 \cdot 10^{-8} - h1 \cdot Aw1}$$

$$Tw1 = 501.74$$

At low velocities, we see that the self-heating error is about 1% or 2%, while at high velocities it is only about 0.3%. Because heat transfer coefficients such as h are notoriously difficult to predict (even though they are based on correlations of *experimental* data), our self-heating error predictions are only estimates. If our thermometer system is calibrated against a suitable standard, and then used under conditions close to calibration conditions, this error can, in principle, be eliminated. In any case, our theoretical analysis has made us aware of the effects of system parameters on this error and given us at least an order of magnitude estimate of its significance.

If one has access to a *numerical equation solver,* Equation 4.3 can be solved without the approximation (replacement of the term $\left(1 + \dfrac{T_w - T_f}{2T_f}\right)^{0.17}$ by 1.0), that we used above. Mathcad has such a solver and we show below how it would be used to solve for R_w when flow velocity V and air temperature T_f are both given. Being numerical rather than analytical, such solvers give approximate answers, but the approximations can be made very good (better than our above results), so they can be used to check whether such results are sufficiently accurate. The example shown uses a fluid temperature of 293 K and a flow velocity of 20 m/s. For these values, we find that the self-heating error is 1.47 K, which shows that our approximate method (which gave 1.27 K) is probably acceptable.

I will now try to use the "solver" to compute the wire resistance and temperature more correctly than earlier. We begin by choosing an air temperature and a flow velocity.

$$\text{Tf} := 293. \quad V := 20.$$

$$\text{den} := 3.23155 - 0.0106688 \cdot \text{Tf} + 1.50741 \cdot 10^{-5} \cdot \text{Tf}^2 - 7.70731 \cdot 10^{-9} \cdot \text{Tf}^3$$

$$\text{vis} := 3.42892 \cdot 10^{-6} + 5.81618 \cdot 10^{-8} \cdot \text{Tf} - 3.03056 \cdot 10^{-11} \cdot \text{Tf}^2$$
$$+ 1.08778 \cdot 10^{-14} \cdot \text{Tf}^3$$

$$\text{cond} := 3.24662 \cdot 10^{-3} + 9.41753 \cdot 10^{-5} \cdot \text{Tf} - 6.28511 \cdot 10^{-8} \cdot \text{Tf}^2$$
$$+ 3.48159 \cdot 10^{-11} \cdot \text{Tf}^3$$

$$\text{den} = 1.20582 \quad \text{vis} = 0.000018 \quad \text{cond} = 0.02632 \quad \text{dwire} = 1 \cdot 10^{-6}$$

$$\text{Aht} = 1.570796 \cdot 10^{-9}$$

$$C := \frac{0.48 \cdot \text{cond}}{\text{dwire}} \left(\frac{\text{den} \cdot \text{dwire} \cdot V}{\text{vis}}\right)^{0.51} \qquad C = 14607.458285 \qquad \text{Aht} \cdot C = 0.000022945$$

I can now use Mathcad's "solver." You first have to give your "best guess" for the unknown (Rw). This guess is usually not critical, and you can refine it once the solver produces an answer: just use the "answer" as your new "guess." There is also a "tolerance" value which has a default value of 0.01. Setting it to a smaller value gives better accuracy, but the calculation takes longer.

You have to define a function which would be zero when the unknown is at the solution value. For example, for the equation X^3–4X^2 = sin(X), the function would be X^3–4X^2–sin(X). In the example below, the function is called Z(Rw).

Rw : = 121.0786   I started with 120.0 and changed the "guess" to agree with the value that "answer" came up with.

TOL : = 0.00001   I overrode the default value of 0.001

$$Z(Rw) := \left[\frac{7.070}{(14000+Rw)}\right]^2 \cdot Rw - Aht \cdot C \cdot \left[\left[\frac{Tf + \left(\frac{60.0699}{159.9196}\right)}{2 \cdot Tf}\right]^{0.17} \cdot \left(\frac{Rw-60.0699}{0.20346} - Tf\right)\right]$$

answer : = root (Z(Rw), Rw)   answer = 119.9837

$$Tw := \frac{answer - 60.0699}{0.20346} \qquad Tw = 294.4746$$

TOL = 1.10^{-5} I typed "TOL" to check whether my requested tolerance value was actually used.

sh_error : = Tw − Tf   sh_error = 1.474639

4.4 Estimating the Sensitivity to Desired and Spurious Inputs

We have already shown that the range of output voltage produced by the full-scale range of fluid temperatures is about 20 mV, giving an *average* sensitivity of about 0.1 mV/K. We call this an average sensitivity because the instrument is not exactly linear. However, since the current is nearly constant, the self-heating errors are not excessive, and the wire resistance is modeled as nearly linear with wire temperature, the variation of sensitivity over the temperature range is not large. That is, the instrument is nearly linear. When we are measuring a certain air temperature, the output voltage we get is unfortunately also sensitive to the flow velocity, a *spurious input*. If we would measure the velocity, then we might compensate our reading to correct, at least partially, for this effect. It would of course be more convenient to avoid this additional measurement and calculation by showing that the velocity effect is negligible. Let us make a few calculations to investigate this by assuming a certain fluid temperature and calculating the change in output voltage over a range of velocities.

We could do this for several different temperatures but it may suffice to check this situation at only the mid-range temperature of 393 K. I used the Mathcad solver program shown just above to find how much the wire temperature changes when the velocity changes from 1.0 to 2.0 m/s and repeated this for a velocity change from 19.0 to 20.0 m/s. For the change from 1.0 to 2.0 m/s the wire temperature changes from 401.31 to 398.82 K. This makes the sensitivity to velocity 2.49 K/(m/s). For the change from 19 to 20 m/s, the wire temperature changed from 394.84 to 394.79 K making the sensitivity 0.05 K/(m/s). We see that both the self-heating error and the error due to velocity improve as we go to higher flow velocities.

4.5 Dynamic Response to Fluid Temperature Fluctuations

Our study so far has been limited to steady-state conditions but the main purpose of cold-wire sensors is to measure rapid changes in fluid temperature. We can modify Equation 4.3 to include dynamic effects by considering the conservation of energy over a short time interval dt. That is, if both T_f and T_w are varying with time, during a time interval dt, the energy generated by electrical heating of the wire, minus the heat lost (or gained) by convection heat transfer between wire and fluid, must equal the additional energy stored in the *thermal capacitance* (product of wire mass and specific heat) of the wire, due to heating or cooling of the wire.

$$i^2 R_w - h A_w (T_w - T_f) = M_w c_w \frac{dT_w}{dt} \tag{4.7}$$

The terms A_w, M_w, and c_w are treated as constants, but i is a function of R_w as is T_w, and h is given by Equation 4.6 which depends on V, T_f, and T_w, making Equation 4.7 a *nonlinear* differential equation which is analytically *unsolvable*. It can, however, be easily "solved" numerically using Simulink simulation, as shown in Figure 4.4. In this diagram I assume that the flow velocity is constant at 10.0 m/s and that T_f has been constant at 393 K long enough that R_w has taken on the value (140.548 Ω) dictated by the steady-state self-heating error. Then T_f makes a step change to 403 K, initiating a transient response of all the variables. To clearly show these events I initiate this step change at $t = 0.0002$ s, so that we can see the initial steady state and then the dynamic response.

While simulations such as Figure 4.4 can get us accurate numerical results, they *never* provide any *formulas* for predicting the effect of system parameters on speed of response. Therefore we always try to get such useful design formulas, usually by making approximations which result in *linearized* versions of the differential equation. In our present example we have earlier noted that the steady-state response is very nearly linear over the entire range of

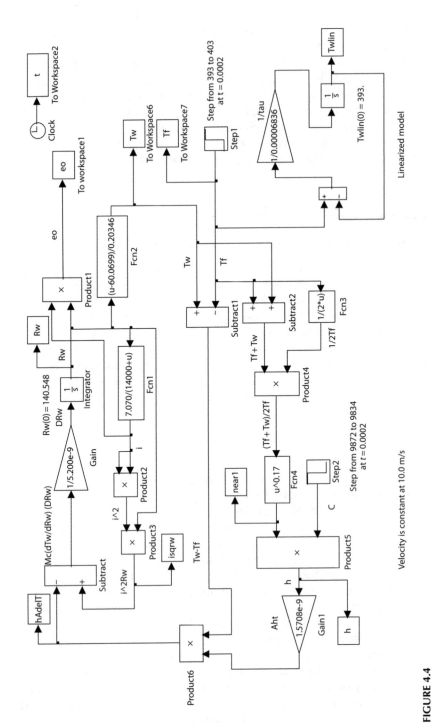

FIGURE 4.4
Nonlinear and linearized simulation of cold-wire step input response.

fluid temperatures, and even when this is not true, most *smooth* nonlinear relations can be closely approximated with linear ones *for sufficiently small changes in the variables*. It is well known that most temperature sensors, when they are used "bare" (not inserted into some kind of protective well) have a dynamic response adequately modeled as a simple linear first-order dynamic system, characterized by a single parameter called the *time constant*. This can be shown by deriving a differential equation that expresses the relation between convective heat flow from fluid to wire (a *thermal resistance* effect) and energy storage in the wire's thermal capacitance.

During time interval dt: energy into the wire = additional energy stored in the wire

$$hA_w(T_f - T_w)dt = M_wc_wdT_w \quad \tau \cdot \frac{dT_w}{dt} + T_w = T_f \quad \tau \triangleq \frac{M_wc_w}{hA_w} \quad (4.8)$$

In this equation, we assume that initially the fluid temperature is steady at some value and that the wire temperature is also steady at that same value, that is, we *neglect* the self-heating error. If we then let T_f undergo a step change, T_w will go through a typical first-order transient and asymptotically approach T_f. Since h depends on T_f, to approximate it as a constant, we use its average value for the initial and final T_f values (393 and 403 K for the simulation example). Using the fixed and known values of M_w, c_w, and A_w and the average h value, the time constant turns out to be 68.36 μs.

The simulation of this linearized model is shown at the lower right of Figure 4.4; the nonlinear model takes up most of the diagram. Figure 4.5 shows the responses. We see that before the step change the linearized model makes

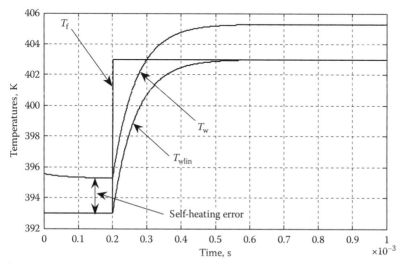

FIGURE 4.5
Nonlinear and linearized step response of cold wire.

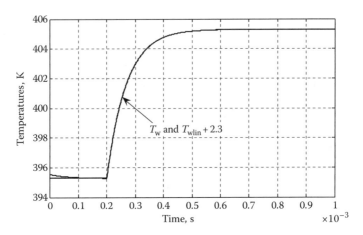

FIGURE 4.6
Dynamics of linearized and nonlinear models agree.

T_{wlin} exactly equal to T_f while the nonlinear model correctly shows the self-heating error. The *dynamic* portions of the graphs are however very nearly identical, as the shifted T_{wlin} curve of Figure 4.6 shows. One advantage of the linearized model is that it is easy to find the effect of the four parameters that affect the time constant on the dynamic response. In our example, the wire size and material are fixed, so the only remaining effect is that of h, which depends mainly on flow velocity, which has a large effect. For first-order systems, knowing the time constant also fixes the bandwidth of the frequency response, which is usually taken as the frequency where the amplitude ratio has fallen by 3 dB. Using the h values computed earlier, we can construct Table 4.6. If temperature changes are not too large and we ignore the effect of

TABLE 4.6

Effect of Velocity on Time Constant
and Bandwidth

V, m/s	tau(V), μs	f3 dB(V), Hz
1	221.186566	719.551
3	126.306851	1260.066
5	97.338364	1635.069
7	81.989536	1941.16
9	72.126535	2206.607
11	65.110117	2444.397
13	59.792627	2661.782
15	55.584308	2863.307
17	52.147056	3052.041
19	49.271344	3230.173
21	46.819514	3399.329

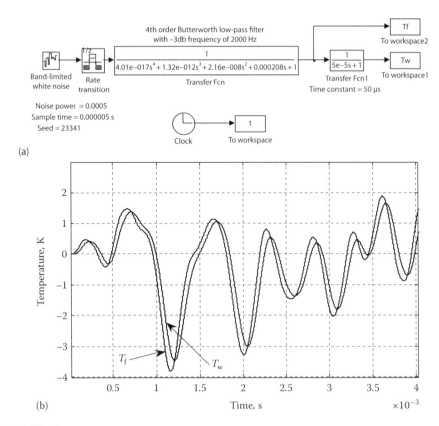

(a)

(b)   Time, s

FIGURE 4.7
Random-signal response of linearized cold wire. (a) Simulation diagram. (b) Results.

the self-heating error, we can use the linearized model to study the response to fluctuating temperature of any waveform of interest. Figure 4.7 does this for a random input with a bandwidth of about 2000 Hz when the time constant is 80 µs. The temperature traces there (which are sometimes negative) should be thought of as fluctuations about some average temperature, such as 393 K for example.

4.6 Use of Current Inputs for Dynamic Calibration

Since our predictions of dynamic response are all based on theories and parameter values which are not exact, it would be helpful to have a method of determining the dynamic response for the actual system based on *measurements*. Such measurements would, it seems, require that we produce a contollable input of fluid temperature with sufficient dynamic content (frequency spectrum) to exercise the thermometer over its useful frequency range. This turns out to be rather difficult, so alternative methods were sought. The method usually used (which has been applied also to hot-wire anemometers for measuring velocity) subjects the wire to a *current input*, causing the wire temperature (and thus the output voltage) to respond. We will show that the dynamic response to a current input has the same time constant as the response to a fluid temperature input, so testing with current inputs will reveal also the response to fluid temperature inputs. Current inputs of almost any waveform and frequency content are relatively easy to produce since the apparatus is totally electronic, rather than involving mechanical or thermal elements.

Since the practical interest is in relatively small temperature changes and the system is somewhat nonlinear, we base the analysis on small changes away from an initial steady-state condition, starting with Equation 4.7. We rewrite this equation using a notation which assigns a subscript *o* to steady values of variables and *p* to small perturbations away from those steady values.

$$(i_o + i_p)^2 (R_{wo} + R_{wp}) - hA_w(T_{wo} + T_{wp} - T_{fo} - T_{fp}) = M_w c_w \frac{dT_{wp}}{dt} \qquad (4.9)$$

Since h varies mainly with flow velocity V, which we assume fixed, h is taken as a constant. The steady-state version of this equation is

$$i_o^2 R_{wo} - hA_w(T_{wo} - T_{fo}) = 0 \qquad (4.10)$$

Since both these equations are true, we can subtract Equation 4.10 from Equation 4.9 to get another true equation:

$$i_o^2 R_{wp} + 2i_o i_p R_{wo} + 2i_o i_p R_{wp} + i_p^2 R_{wo} + i_p^2 R_{wp} - hA_w T_{fp} = M_w c_w \frac{dT_{wp}}{dt} \qquad (4.11)$$

In this kind of *perturbation analysis* it is conventional and mathematically justifiable to neglect *terms of the second order* relative to larger terms, as a simplifying approximation. If a term is a product of two perturbations, it will be small relative to a term which is a product of some constant and a perturbation, and will be neglected. In Equation 4.11 we will neglect $2i_o i_p R_{wp}$, $i_p^2 R_{wo}$ and $i_p^2 R_{wp}$, giving

$$\frac{M_w c_w}{hA_w} \frac{dT_{wp}}{dt} + T_{wp} - i_o^2 R_{wp} = T_{fp} + (2i_o R_{wo})i_p \qquad (4.12)$$

For the small changes we are using, $R_{wp} = 0.2035 T_{wp}$, and $i_o^2 = (0.0005)^2$ so we can replace the term $i_o^2 R_{wp}$ by the term $5.1e{-}8\, T_{wp}$, which is clearly negligible relative to $1.0\, T_{wp}$. This shows that the simple analysis of Equation 4.8 is valid for the given numerical values and also that the responses to inputs

T_{fp} and i_p have exactly the same dynamic behavior (time constant) though of course they have different gains (sensitivities). We can thus dynamically calibrate (find the frequency response of) our thermometer for fluid temperature inputs without actually applying such inputs. Instead we apply dynamic current inputs, which are much easier to achieve.

4.7 Electronic Considerations

While the simple circuit of Figure 4.1 has actually been used, more convenient and versatile circuitry is common. Our focus has been on the basic fluid and thermal aspects of the sensor itself but we want to at least briefly comment on some of the electronic details. A main consideration is the need with our simple circuit for external amplification. That is, the voltage e_o has a full-scale range of only 20 mV; if we want to use a digital data acquisition system, at some point this must be boosted into a range more like 0–10 V or ±10 V since common analog-to-digital converters require volt-level (rather than millivolt-level) input signals. If we use the circuit of Figure 4.1 with an amplifier connected directly to e_o, that amplifier would need to be the *differential* type, since neither end of this voltage is at ground potential. Many practical circuits are based on some form of Wheatstone bridge, and include internal amplification (usually based on operational amplifiers), some kind of constant-current source, and even means to apply the dynamic current inputs we just described (for calibration).

Figure 4.8 shows perhaps the simplest bridge circuit, which still uses external amplification but grounds one end of the amplifier so that a differential

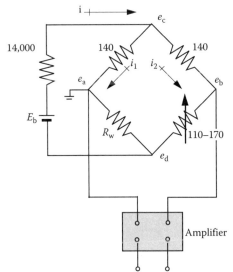

FIGURE 4.8
Basic Wheatstone bridge circuit.

amplifier is not needed. It also gives an output voltage signal that can be set at zero for, say, the mid-range value of temperature (393 K) and then goes positive or negative as the temperature increases or decreases from this value. Suppose the temperature is at the mid-range value ($R_w = 140\ \Omega$) and we balance the bridge by setting the adjustable resistor at 140 Ω. Since we have grounded the left corner of the bridge, voltage e_a is by definition always zero, and for a balanced bridge, e_b is also zero. If the cold wire then changes resistance, the bridge will be unbalanced and e_b will change to some positive or negative value, depending on whether the wire resistance increased or decreased from 140 Ω. To find e_b we can write the following equations:

$$R_{eq} \triangleq \text{bridge equivalent resistance} \triangleq \frac{(140 + R_w)(280)}{140 + R_w\ 280} \qquad i = \frac{E_b}{14,000 + R_{eq}} \qquad (4.13)$$

$$i_2 = \frac{140 + R_w}{280} i_1 \qquad i = \frac{(420 + R_w)E_b}{14,280R_w + 5,919,200} = i_1 + i_2 = \frac{420 + R_w}{280} i_1 \qquad (4.14)$$

$$i_1 = \frac{280}{14,280R_w + 5,919,200} E_b \qquad e_c = 140i_1 \qquad e_b = e_c - 140i_2 = \frac{19,600 - 140R_w}{14,280R_w + 5,919,200}$$

$$(4.15)$$

Since voltage e_a is defined as 0 V (grounded), e_b is the bridge output voltage and the amplifier input voltage. Using Equation 4.15 we see that for $R_w = 140$ the output voltage is zero (bridge balanced) and for the low and high values of R_w (120 and 160 Ω) we get, respectively, output voltages of $+0.0003668E_b$ and $-0.0003413E_b$. If we want to have the same wire current at mid-range that we used in our earlier circuit (Figure 4.1), we would use a battery voltage of $(14280)(0.0005) = 7.14$ V. The full range of output voltages would then be -0.002437 to $+0.002619$ V which is about 5 mV. An external amplifier gain might typically be 1000 V/V, giving the system output voltage a range of ± 2.5 V.

The literature documents a variety of cold-wire circuits that have been used. LaRue et al. (1975)[*] describe such a system and give a circuit diagram with complete details on electronic components and numerical values used. Dynamic current inputs of either sinusoidal or step nature are provided. Both theory and experimental test data are included. A somewhat simpler instrument[†] uses mainly well-known op-amp techniques to build a low-drift voltage source (uses a mercury cell and an op-amp voltage follower), a voltage-to-current converter, and an output-voltage reference circuit.

---

[*] J.C. LaRue, T. Deaton, and C.H. Gibson, Measurement of high-frequency turbulent temperature, *Rev. Sci. Instrum.*, 48(6), June 1975, 757–764.

[†] S. Tavoularis, A circuit for the measurement of instantaneous temperature in heated turbulent flows, *J. Phys. E: Sci. Instrum.*, 11, 1978, 21–23.

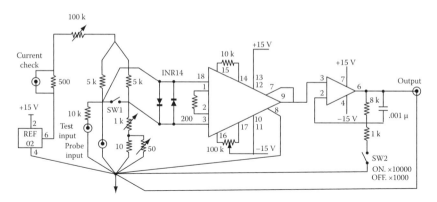

FIGURE 4.9
Cold-wire circuit of Cho and Kim.

An external differential amplifier (PAR 113 used at a gain of 2000 V/V)* is used and the cold-wire probe is a DISA 55P31[†] (still available in 2008!). The PAR 113 is no longer available but amplifiers of this type can be obtained from several sources.[‡] Another relatively simple circuit[§] was experimentally tested using a radiant heat source to heat the wire. A focused beam from a heat lamp was interrupted periodically by a rotating slotted disk (chopper) to provide a dynamic excitation. An analysis of the effect of heat conduction to the prongs supporting the wire when the radiant heat source was applied was carried out. A more recent design[¶] uses a Wheatstone bridge and an internal instrumentation amplifier, as shown in Figure 4.9.

4.8 Effect of Conduction Heat Transfer at the Wire Ends

Usually the wire length/diameter ratio is large enough (500–1000) so that our assumption of negligible heat loss by conduction from the wire to its supports is justified. Experience shows, however, that occasionally this is not the case, and the measured frequency response deviates from the simple first-order model we have used. Smits et al. (1978) discuss this question and include further references. The careful measurement of sensor frequency

---

* www.princetonappliedresearch.com, this firm is still in business but no longer makes this amplifier.
[†] www.dantecdynamics.com
[‡] www.thinksrs.com model SR 560
[§] A.J. Smits, A.E. Perry, and P.H. Hoffmann, The response to temperature fluctuations of a constant-current hot-wire anemometer, *J. Phys. E: Sci. Instrum.*, 11, 1978, 909–914.
[¶] J.R. Cho and K.C. Kim, A simple high-performance cold-wire thermometer, *Meas. Sci. Technol.*, 4, 1993, 1346–1349.

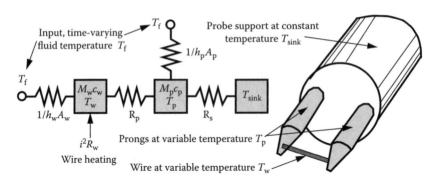

FIGURE 4.10
Lumped-parameter thermal model of prong conduction problem.

response sometimes shows a (relatively small) deviation from the predicted first-order behavior at very low frequencies when the l/d ratio is too small. The observed behavior is that the amplitude ratio drops off from near 1.0 at zero frequency to about 0.95–0.90 over a low-frequency range (typically 0 to about 0.5–5 Hz) and then remains flat at this value until we reach the high-frequency rolloff predicted for any first-order system. This behavior is due to the conduction heat loss which is neglected in the usual analysis.

We want to analyze this behavior using a simple model for our cold-wire system. Figure 4.10 shows the actual apparatus and the simplified thermal model representing it, which uses a *lumped-parameter* method that models heat transfer rates with *thermal resistances* and energy storage with *thermal capacitances*. This is the usual approach not only of system dynamics texts but also many studies by thermal-system specialists. The alternative is to use the more correct *distributed-parameter* approach which leads to partial differential equations, whereas the lumped approach gives ordinary differential equations. For system geometry as complex as the wire/probe system under study, the partial differential equations cannot be solved analytically and one must appeal to finite-element software available for thermal-system analysis. (Some simplified geometries have been solved analytically.) This specialty software is quite complex to use and requires considerable user time and effort to learn and apply. Also, as with any finite-element software, the "answers" are numerical or graphical results, rather than *equations* relating system parameters. Thus the accuracy disadvantage of lumped analysis is balanced by its provision of useful design formulas relating basic parameters to system response. In any specific engineering study of some practical problem, we need to be aware of these trade-offs between lumped- and distributed-analysis models and try to make the best choice at a given stage of the analysis. Often, preliminary lumped studies give useful insights which later lead to more detailed and expensive distributed models.

Conceptually, our model assumes that heat flows from the fluid temperature T_f to the wire through a convective thermal resistance $1/h_w A_w$ and from

T_f to the prongs through a convective thermal resistance $1/h_p A_p$. We will treat the two prongs as a single thermal capacitance $M_p c_p$. Heat leaks from the wire to the prongs through a conductive thermal resistance R_p, and from the prongs to the probe support rod through a conductive thermal resistance R_s. In the usual analysis, these two conduction paths are neglected; in our analysis we will include them but make their thermal resistance values much higher than the convective resistances. The wire carries a small constant current i. Our model is for small changes around a fixed operating point, so we take all the parameters as constants. To avoid the need for conversion factors, we again assume SI units so that heat transfer rates and electrical heating rates are both in watts. We generate two simultaneous differential equations by considering heat transfer into and out of the two thermal capacitances, and energy storage within each, using conservation of energy over a short time interval dt:

$$i^2 R_w + h_w A_w (T_f - T_w) - \frac{(T_w - T_p)}{R_p} = M_w c_w \frac{dT_w}{dt} \qquad (4.16)$$

$$\frac{(T_w - T_p)}{R_p} + (T_f - T_p)h_p A_p - \frac{(T_p - T_s)}{R_s} = M_p c_p \frac{dT_p}{dt} \qquad (4.17)$$

Now, as in Equation 4.9, we rewrite these equations with each variable expressed as a constant steady-state value plus a small dynamic perturbation, using subscripts o for the steady state and p for the perturbation. Subtracting the steady-state equation, we get

$$M_w c_w \frac{dT_{wp}}{dt} + \left(h_w A_w + \frac{1}{R_p} - i^2 K_{RT} \right) T_{wp} + \left(\frac{-1}{R_p} \right) T_{pp} = (h_w A_w) T_{fp} \qquad (4.18)$$

$$M_p c_p \frac{dT_{pp}}{dt} + \left(h_p A_p + \frac{1}{R_s} + \frac{1}{R_p} \right) T_{pp} + \left(\frac{-1}{R_p} \right) T_{wp} = (h_p A_p) T_{fp} \qquad (4.19)$$

The constant K_{RT} is the conversion factor between wire resistance and wire temperature ($0.2035\,K/\Omega$ in our earlier example).

For convenience, we now define the coefficient of T_{wp} in Equation 4.18 to be a constant C_w and the coefficient of T_{pp} in Equation 4.19 to be a constant C_p. It is often useful in first-order equations such as these to define some kind of *time constant* in each equation. We could do this by dividing Equation 4.18 by C_w and Equation 4.19 by C_p, which would then make the coefficients of the derivative terms the two time constants. These would be called the *coupled-system* time constants. We prefer instead to define two *uncoupled* time constants by dividing Equation 4.18 by $h_w A_w$ and Equation 4.19 by $h_p A_p$, giving

$$\frac{M_w c_w}{h_w A_w}\frac{dT_{wp}}{dt} + \frac{C_w}{h_w A_w}T_{wp} + \left(\frac{-1}{h_w A_w R_p}\right)T_{pp} = T_{fp} \qquad (4.20)$$

$$\frac{M_p c_p}{h_p A_p}\frac{dT_{pp}}{dt} + \frac{C_p}{h_p A_p}T_{pp} + \left(\frac{-1}{h_p A_p R_p}\right)T_{wp} = T_{fp} \qquad (4.21)$$

We are interested in the differential equation and transfer function relating response T_{wp} to input T_{fp} so we use our usual determinant method on these two equations to get

$$\frac{T_{wp}}{T_{fp}}(D) = \frac{K(\tau_n D+1)}{(\tau_w \tau_p)\dfrac{h_w A_w h_p A_p R_p^2}{C_w C_p R_p^2 -1}D^2 + \left(\dfrac{\tau_w C_p}{h_p A_p}+\dfrac{\tau_p C_w}{h_w A_w}\right)\left(\dfrac{h_w A_w h_p A_p R_p^2}{C_w C_p R_p^2 -1}\right)D+1}$$

$$\tau_n \underset{=}{\Delta} \tau_p\left(\frac{h_w A_w h_p A_p R_p}{h_w A_w C_p R_p + h_p A_p}\right) \qquad (4.22)$$

Here τ_w and τ_p are the coefficients of the derivative terms in Equations 4.20 and 4.21. Time constant τ_w is the same time constant that we have used earlier, when we neglected the conduction heat loss, so we can take its numerical value from our earlier calculations. It depends mainly on h_w (which in turn depends mainly on flow velocity) and for a mid-range value of flow velocity we take it to be 10,000, which makes $\tau_w = 0.00006735\,\text{s}$.

While the numerical values associated with the wire are easily estimated, the rest of the parameters are much more nebulous, since the prongs and their support do not have obvious and clearly defined geometry and heat transfer properties. One way to choose values is to be guided by the resulting frequency response of the entire system, for which at least some measured data is available, as mentioned above. Clearly, the probe time constant must be much longer than that of the wire, so we choose the mass and heat transfer area to achieve this. Initially I tried making the probe area 1000 times the wire area and the probe mass to be 1 million times the wire mass. It then remains to choose values for the conduction resistances R_s and R_p. Here our main guidance is that these conduction resistances should be somewhat larger than the convection resistances related to h_w and h_p, since in the ideal case, the convection heat transfer is taken as dominant, with the conduction negligible. As initial trial values I took the conduction resistances as five times the respective convection resistances. With all these choices made, we can get numerical values for all the parameters in Equation 4.22, and then check the frequency response against available measured data to see whether our initial choices were reasonable. Equation 4.22 is now

$$\frac{T_{wp}}{T_{fp}}(D) = 0.975 \frac{0.0481D + 1}{3.158e - 6D^2 + 0.05617D + 1} \tag{4.23}$$

Since thermal systems cannot exhibit natural oscillations (unlike mechanical, electrical, and fluid systems, they have *only one* form of energy storage), the denominator second-order term *must* be factorable into two first-order terms, giving

$$\frac{T_{wp}}{T_{fp}}(D) = 0.975 \frac{0.0481D + 1}{(0.05612D + 1) \cdot (0.00005628D + 1)} \tag{4.24}$$

If the conduction loss had been totally neglected, we would have

$$\frac{T_{wp}}{T_{fp}}(D) = 1.00 \frac{1}{0.00006735D + 1} \tag{4.25}$$

The gain of 0.975 in Equation 4.24 simply means that, for a change in fluid temperature, the resulting change in wire temperature is only 97.5% of the fluid temperature change because the conductive heat flow path lets some of the heat flow go to elements other than the wire. The numerator time constant is smaller than 0.05612, so in the amplitude ratio of the frequency response, there will be an initial *downward* trend, but this will be cancelled by an equal upward trend, giving then a *flat* region until we encounter the breakpoint at $1(2\pi)(0.00006735) = 2563\,\text{Hz}$. This *is* the general shape that is found in experimental tests, though the numerical values do not of course agree with our arbitrarily chosen numbers. Figure 4.11 shows that the desired shape is achieved but the effect of conduction is more severe than observed in experiments, so our initial choice of numbers can be improved.

Rather than going to a trial-and-error adjustment procedure, I want to show how, if *measured* frequency-response data is available, we can use a MATLAB® procedure to *fit* to that data a model of our chosen form. Since real experimental data was not available to me, I "manufactured" some. This actually has some advantages since we will now know the correct answer! That is, with *real* experimental data, we never know what the *true* transfer function is, so we have no function to compare with that which the curve-fitting procedure produces. Of course, we will always compare the measured frequency-response curves with those predicted by the fitted model, and if they agree pretty well we will accept the fitted model. The curve fitting procedure requires as input both the set of measured amplitude ratio and phase angle data over a range of frequencies, and your *best guess* as to the *form* of transfer function to be used. That is, you have to choose the *order* of the numerator and denominator, such as, say, a second-order numerator (three coefficients) and a fourth-order denominator (five coefficients). Given this

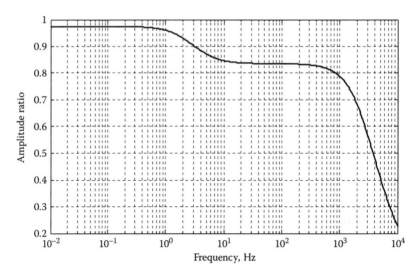

FIGURE 4.11
Effect of conduction heat loss on frequency response.

choice and the set of measured data, the routine finds the best set of numerical coefficients in your chosen model form. When we do not have a good
guess as to the model form, we may have to try several different forms before
getting a good fit. A MATLAB program goes as follows:

```
% computes and plots frequency response of cold-wire
% thermometer with conduction loss
% define gain and numerator time constant
K = 0.975
Taun = 0.0481
% define denominator coefficients
d2 = 3.158e-6
d1 = 0.05617
% compute numerator and denominator of transfer function
num = [K*Taun K];
den = [d2 d1 1];
% compute data and plot curve
f = logspace(-2,4,500);
w = 2*pi*f;
[mag,phase] = bode(num,den,w);
semilogx(f,mag);
% hold on
% now prepare to use the curve fitting method
% "manufacture" some data which will be treated
% as if it were experimental data
Taun1 = 0.0481
Taud1 = 0.0500
Taud2 = 0.000067
```

```
num1 = [K*Taun1 K];
den1 = [Taud1*Taud2 Taud1 + Taud2 1];
ha = freqs(num1,den1,w);
mag1 = abs(ha);
semilogx(f,mag1);hold on; % this is the "experimental" data
% now apply the curve fitting method to the mag1, phase1 data
[g,h] = invfreqs(ha,w,1,2);
% now put results in standard form
gs = g/h(3)
hs = h/h(3)
[mag2,phase2] = bode(gs,hs,w);
semilogx(f,mag2-0.03); % displace the fitted model curve to
% compare with the experimental data curve
```

The "experimental" data is generated from a transfer function with the same form as Equation 4.23, with a gain of 0.975, numerator time constant of 0.0481, and denominator time constants of 0.0500 and 0.000067. The frequency-response data (amplitude ratio and phase angle) for this transfer function is called *ha*. The statement which invokes the curve-fitting procedure is [g,h] = invfreqs(ha,w,1,2), where *ha* is the *measured* data, *w* is the list of frequency points, 1 is the order of the numerator and 2 is the order of the denominator. Note that the curve-fitting procedure does *not* know the coefficients of the transfer function; it will *find* the best values for these, which will be printed out as the polynomials *g* (numerator) and *h* (denominator). In our case, because we *know the right answer*, we can compare these with the polynomials that were used to manufacture the data. In Figure 4.12, the curves for the measured data and the fitted model lay right

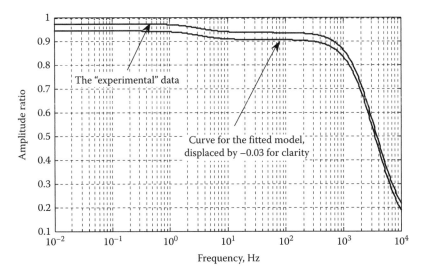

FIGURE 4.12
Use of experimental data to fit a system model.

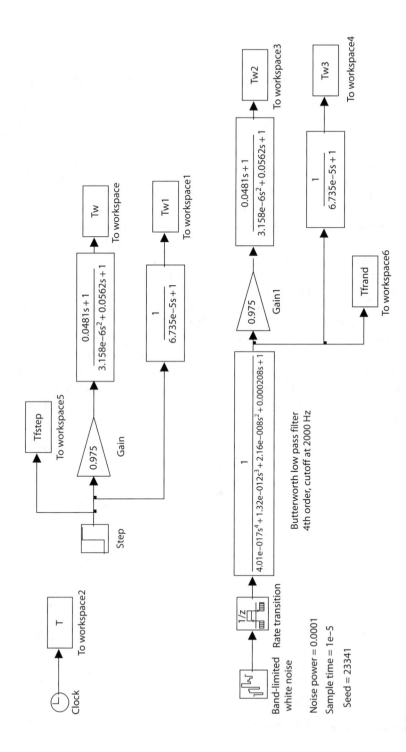

FIGURE 4.13
Simulation for step and random input responses.

on top of each other; to see them separately I displaced the model curve by −0.03. The polynomials returned by the fitting procedure were for all practical purposes identical with those used to generate the measured data. In a real-world example, with actual experimental data, we should not expect such perfect results, however the method often gives good results and is a powerful aid whenever we want to find an analytical transfer function from measured data.

Once we have obtained a valid model for a system (whether from theory or fitting experimental data) we can use simulation to study the time response of our system to any kind of dynamic input. Figure 4.13 shows a Simulink diagram for our present system. Figure 4.14 shows the response of wire temperature to a step input of fluid temperature. The upper graph, which shows only early times, is somewhat misleading since the conduction effect introduces a long time constant. By replotting (in the lower graph) on

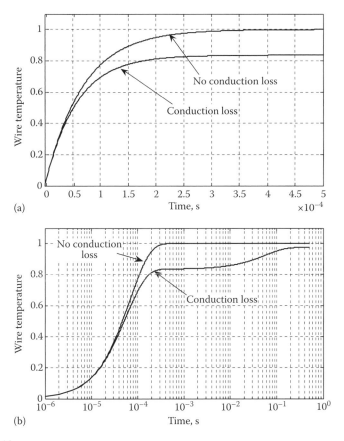

FIGURE 4.14
Step input response for wire temperature.

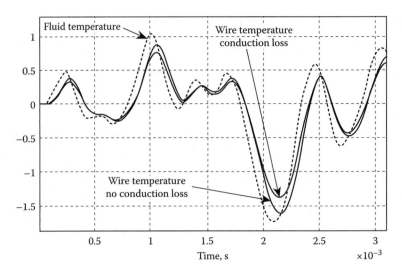

FIGURE 4.15
Response of wire temperature to random fluid temperature.

a logarithmic time scale, we can see both early times and the final equilibrium state. Since cold-wire thermometers are often used to measure small
temperature fluctuations in turbulent flows, the random input of Figure 4.13
is appropriate. I chose a white noise signal with frequency content of about
2000 Hz since this is a little beyond the system's accurate range; Figure 4.15
shows the results. For a signal with lower frequency content, the response
would be more accurate.

5

Piezoelectric Actuation for Nanometer Motion Control

5.1 Introduction

Many instruments and machines today deal with processes at the micrometer and nanometer scale of motion. Examples include manipulation of biological materials (single cells, etc.), focusing and alignment of optical systems, specimen manipulation in microscopes, and manufacture and testing of microcircuits and microelectromechanical devices (MEMS). While larger motions are routinely controlled using electromagnetic actuators (voice coils, stepping motors, galvanometers, rotary dc and ac motors) together with mechanical elements such as ball-screws, gear trains, and lever systems, for the tiniest motions we find piezoelectric actuation very widely used. These actuators are based on the *piezoelectric effect*, which is usefully applied in two basic modes. In force, pressure, and acceleration sensors,[*] we apply a mechanical input to the piezoelectric material and it responds with a change in electrical charge and voltage, which become a measure of the mechanical input. In actuators[†] for micrometer and nanometer motion control, we apply a voltage and the material responds with a motion and force. In addition to catalog type of information, www.physikinstrumente.com/en/products/piezo_tutorial.php provides very detailed and practical tutorials on actuator application to motion control systems. Of the various forms of piezoelectric actuators, we concentrate on the basic *stacked element* version, which provides translational motions up to about 1 mm full scale, with nanometer resolution.

In Figure 5.1, I show a somewhat schematic sketch of a piezoelectrically driven rotation stage from Mad City Labs.[‡] This is a rotational positioner but it uses a translational piezo actuator acting at a lever arm to produce the desired rotary motion. Figure 5.2[§] shows how six translational actuators

---

[*] E.O. Doebelin, *Measurement Systems*, 5th edn., McGraw-Hill, New York, 2004, pp. 284–292, 359–363, 452–455, 503–505; G. Gautschi, *Piezoelectric Sensorics*, Springer, New York, 2002.

[†] www.physikinstrumente.com/en/products/piezo_tutorial.php

[‡] www.madcitylabs.com; B. OBrien, Rotational flexures deliver high precision, *Laser Focus World*, April 2005, pp. 89–92. (Dr. O'Brien is in the R & D group at Mad City Labs) 608-298-0855.

[§] Physik Instrumente.

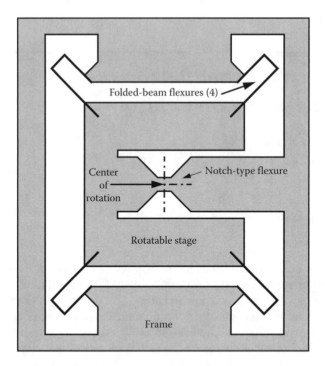

FIGURE 5.1
Rotation stage for nanometer motion control.

FIGURE 5.2
Hexapod piezoelectric actuators of Physik Instrumente.

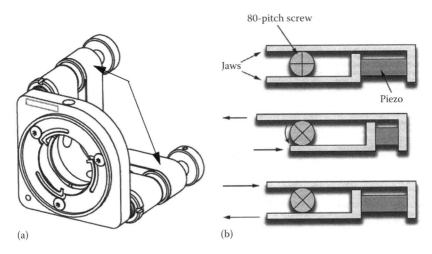

FIGURE 5.3
Rotary piezoelectric motors (PicoMotor™) of New Focus, Inc, San Jose, CA. (a) Two rotary piezoelectric motors tilt the lens mount in two different directions. (b) Two jaws grasp an 80-pitch screw and a piezo actuator slides the jaws in opposite directions. Slow action of the piezo causes a screw rotation while fast action, due to inertia and friction, causes no rotation. Fast expansion and slow relaxation gives counter-clockwise rotation. Slow expansion and fast relaxation gives clockwise rotation.

can be combined in a so-called *hexapod* configuration to provide 6-degree-of freedom (3 translations and 3 rotations) motion control of a mechanical load such as a telescope mirror. Rotary piezoelectric *motors* are available in several forms. Figure 5.3* shows two Picomotors™ from New Focus, Inc., being used to control the angular attitude of a lens mount. These motors use translational piezo actuators to drive the rotation of a fine-pitch screw in a stepwise manner. That is, the screw is driven through a small and fixed step rotation, *slowly* in the desired direction by frictional torque in a piezoelectrically actuated clamp. The actuator is then driven *rapidly* in the reverse direction, so fast that the screw inertia torque exceeds the friction torque and the friction coupling slips, allowing the piezo actuator to return to its *neutral* position, but the screw stays in its advanced location. Each pulse advances the screw about 20 nm and pulses can be applied as fast as 1000/s. In this way, full-scale motions of several inches can be controlled, far beyond the range of conventional translation actuator. Another form[†] of piezoelectric motor uses high-frequency wave motion to produce rotary or translational motion.

---

* www.newfocus.com
† G. Cook, An introduction to piezoelectric motors, *Sensors*, December 2001, pp. 33–36.

5.2 Mechanical Considerations

This chapter will treat a generic rotary actuator as in Figure 5.1 and use data from several commercial products to guide the design as a form of *reverse engineering*. Initial data for my generic actuator comes from the Nano-Theta rotary stage of Mad City Labs, as shown in their catalog page (Figure 5.4; you can download their entire catalog). The rotation stage and frame are all one piece of aluminum alloy (or, at extra cost, *Invar*, to provide smaller temperature effects). The stage has been cut free from the frame using *wire electrical discharge machining*, about the only practical method of producing such parts.

Features

- ▶ Precision rotation: 2 mrad range
- ▶ Accessible and well defined axis of rotation
- ▶ Mount in any orientation
- ▶ High resolution: 4 nrad
- ▶ *pico*™ sensor technology
- ▶ Closed loop control

Typical applications

- ▶ Laser beam scanning
- ▶ Lithography
- ▶ FBG writing

LabVIEW compatible USB interfaces
Examples supplied with Nano-Drive USB interfaces

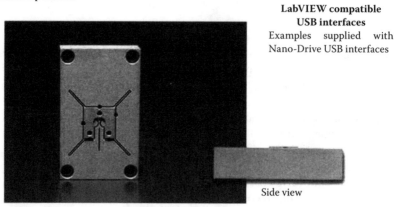

Side view

FIGURE 5.4
Nano-Theta rotational stage constructed from aluminum. (Courtesy of Mad City Labs, Madison, WI.)

Product description

The Nano-Theta is a unique piezoactuated rotational stage having 2 mrad of total motion. With nanoradian resolution, the Nano-Theta is designed for applications in lithography, optical disk manufacturing, and laser beam tracking or scanning. The innovative design of the Nano-Theta incorporates a readily accessible and well-defined axis of rotation which allows a mirror to be mounted so that it is coplanar with the axis of rotation. Internal position sensors utilizing proprietary *pico*™ technology provide absolute, repeatable position measurement with nanoradian accuracy under closed loop control.

Their photo and drawing do not clearly show the details of the five flexure elements that allow the stage to rotate, but that suppress all other degrees of freedom. I made the drawing in Figure 5.1 to clarify these details; however, my drawing is not faithful to the actual geometry, since the machining method allows very fine cuts, rather than the large open spaces needed in my sketch to show flexure details. Because the stage full-scale rotation angle is very small by design (0–2 mrad), these fine cuts still provide enough clearance for the motion. The center of rotation is mainly defined by the *notch-type* flexure, which has some (but not sufficient) stiffness against other undesired motions. To augment stiffness in these other degrees of freedom, the four *folded-beam* flexures at the corners are added. My sketch again is not faithful to the actual geometry; the beams are defined by fine cuts made by the electrical discharge machining method. They are not welded-on separate pieces as my drawing might suggest.

Mad City uses piezoelectric actuators of the *stacked* type to get sufficiently large displacement motions without excessively high drive voltages. They were, for proprietary reasons, not willing to supply me with the details of how the actuators are connected to the stage in order to get rotary motion. Any competitor can, of course, buy a unit and attempt to *reverse engineer* it and discover their practices. *Discovering* manufacturing secrets of a competitor is often easier than actually *duplicating* the product, so companies sometimes avoid the patent process (which does, to some extent, reveal secrets). For protection against pirating their trade secrets they instead rely on the difficulty would-be copiers often encounter in duplicating the *engineering lore* accumulated by the original inventors over years of experience. Stages of this sort nearly always use *feedback* methods to overcome the hysteresis and nonlinearity present in all piezoelectric actuators. This requires high-quality displacement sensors, with capacitance types being quite common. Mad City, however, uses only "piezoresistive position sensors" (probably semiconductor strain gages, but they would not elaborate), and their Web site claims several advantages. When I asked them, they said that the gages were *not* applied directly to the actuators but rather to the flexures, in some way. They would not give details, but again, one could purchase a unit and find out. Their drive electronics appears to use standard proportional-plus-integral (PI) control laws to achieve the desired closed-loop performance.

As an exercise to get a feel for the kind of calculations involved in designing such a nanomotion control system, we next perform a sort of *reverse engineering* on this particular system. While Mad City Labs would not supply me with detailed dimensions, their catalog page does give the natural frequency (2000 Hz), the full-scale rotation (0.002 rad), and the resolution (4 nrad). A catalog drawing gives the thickness as 0.50 in. and the moving part appears from the drawing to be about 0.5 inches square. From the given natural frequency, if we estimate the moment of inertia of the rotating part, we can estimate the combined rotary spring constant of the four-folded beam flexures and the single notch-type flexure. If we then make a guess as to the relative stiffness of the folded-beam flexures and the notch flexure, we can get the stiffness of each, and then design their dimensions to give the desired stiffnesses. As mentioned before, the stage is made from a single piece of metal, with unwanted portions cut away by the wire-electrical discharge machining process. The stage is available in either aluminum or, for smaller temperature effects, Invar; let us assume aluminum for our calculations. For a 0.50 in. cube of aluminum, the moment of inertia about the center of rotation (taken at the center of the cube) is given by

$$I_{xx} = \frac{1}{12} M(0.50^2 + 0.50^2) \quad M = \frac{0.1}{386}(0.50^3) = 3.238\text{e}{-5}$$

$$I_{xx} = 1.35\text{e}{-6} \text{ in.-lbf-s}^2/\text{rad}$$

(5.1)

Given this inertia, we can compute the total rotary (torsional) spring constant that will give the spring/mass system the stated natural frequency of 2000 Hz:

$$\omega_n^2 = (2000 \cdot 6.28)^2 = 1.58\text{e}8 = \frac{K_s}{I_{xx}} = \frac{K_s}{1.35\text{e}{-6}} \quad K_s = 213 \text{ in.-lbf/rad} \quad (5.2)$$

This total spring constant is made up of contributions by the four beam flexures at the corners and the single notch flexure at the center of rotation.

We next want to design the flexural elements of this motion *stage*. (The word stage is commonly used for such mechanical arrangements; we speak of x–y stages, x–y–z stages, x–y–θ stages, etc.) When we want to control very small translational or rotary motions, conventional bearings (rolling contact, sliding, etc.) have too much friction and we usually accommodate the small motions with an elastic suspension called a *flexure*. Aerostatic bearings* also provide near-frictionless motion but are used mainly for larger motions, where their larger size and need for an air supply can be accommodated. When a stage is built up by bolting or welding together discrete pieces, we can choose different materials to suit the needs of the different parts. When we use *monolithic* construction, we start with a single piece of material and then cut away the unwanted portions; so, in the present example we are "stuck with" aluminum as our material. The four-beam-type flexures at the corners would clearly all be identical, so we need to design only one and then multiply by four to get their total rotary stiffness or spring constant. It is not obvious, however, what should be the relative contribution to the total stiffness computed above, of the notch flexure and the four beam flexures. The Mad City Lab designers have years of combined experience in making such decisions, which can involve complicated trade-offs. Without being immersed in the practical environment of the actual manufacturer, it is difficult to duplicate the details of

---

* www.newwayairbearings.com. They offer a downloadable 68-page design guide.

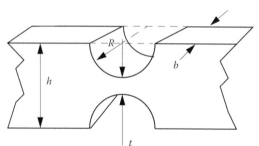

FIGURE 5.5
Notch-type flexure.

their design procedure, and it would be unrealistic to expect them to divulge such details to potential competitors. The best we can do is to show the steps that would need to be carried out, and the type of calculations involved. To this end, we will make the simple assumption that the four beam flexures (together) and the notch flexure each contribute an equal amount to the total rotary stiffness. This might not conform to the actual design, but we will at least be using the same calculation methods that would be employed.

Formulas for the stiffness and stress of various types of flexures are available from a number of papers and books. As far as I know, there is only one book* devoted entirely to flexure design. Professor Smith was educated in England and now teaches at the University of North Carolina, Charlotte, North Carolina.[†] For the notch flexure, we can get the needed formulas from another of Professor Smith's publications.[‡] In Figure 5.1 I drew, for convenience, the notch flexure with straight-line segments, whereas the actual shape is formed from two circular arcs, such as would be obtained by drilling two holes close to each other. The critical dimensions of such a notch are the hole radius R and the minimum thickness t, as in Figure 5.5.

For the notch type of flexure, Smith (2000) offers two alternative analyses relating the angular twist of the elastic hinges to the applied bending moment. The first analysis is quoted from another reference and assumes that the thickness h of the rigid link is related to the circular radius R and the minimum thickness t by

$$\frac{h}{2R + t} \approx 1.0 \quad \text{that is, the notches are nearly semi-circular} \qquad (5.3)$$

If we design to this requirement, then the twist angle θ is related to the applied moment M by

$$\theta = \frac{9\pi R^{0.5}}{2 \cdot E b t^{2.5}} M \qquad (5.4)$$

* S.T. Smith, *Flexures: Elements of Elastic Mechanisms*, Gordon and Breach, Amsterdam, the Netherlands, 2000.
† stusmith@uncc.edu   phone: 704–687–3969
‡ S.T. Smith and D.G. Chetwynd, *Foundations of Ultraprecision Mechanism Design*, Gordon and Breach, Switzerland, 1994, pp. 104–108.

where

    E is the modulus of elasticity

    b the thickness of our stage (see Figure 5.5)

The second analysis is due to Smith and Chetwynd (1994) and gives

$$\theta = \frac{2KRM}{EI} = \frac{24KRM}{Ebt^3} \quad \text{with the requirement,} \quad t < R < 5t \qquad (5.5)$$

where

$$K = 0.565\frac{t}{R} + 0.166$$

These results are based on finite-element studies. The maximum allowable moment M_{max} at the notch, based on a design stress σ_{max} is given by

$$M_{max} = \frac{bt^2}{6K_t}\sigma_{max} \quad \text{where stress-concentration factor } K_t = \frac{2.7t + 5.4R}{8R + t} + 0.325$$

$$(5.6)$$

Smith (2000) does not state a stress equation for the assumptions of Equation 5.4, so the presumption is that one just uses Mc/I, with I computed using the minimum thickness t. I decided to use Equations 5.5 and 5.6 because they include a stress equation and seem to be based on a more comprehensive analysis. Using Equation 5.5 and taking the rotary stiffness of the notch flexure to be half of the total gives

$$\frac{M}{\theta} = \frac{213}{2} = 106.5 = \frac{Ebt^3}{24KR} \qquad (5.7)$$

Since b is known to be 0.50 in. and E for aluminum is about 10^7 psi, only t and R remain to be found. Here the restriction on Equation 5.5 is actually helpful since it can give us a relation between t and R. That is, we could reasonably take $R = 3t$, which also allows us to find K and K_t.

$$K = \frac{0.565t}{3t} + 0.166 = 0.354 \quad K_t = \frac{2.7t + 16.2t}{25t} + 0.325 = 1.081 \qquad (5.8)$$

Equation 5.7 can now be solved for t and then R is also fixed:

$$106.5 = \frac{(10^7)(0.50)t^2}{(24)(3)(0.354)} \quad t = 0.0232 \text{ in.} \quad R = 0.0696 \text{ in.} \qquad (5.9)$$

    Our approach is to *design* R and t to get the desired stiffness, and then *check* to see if the stress is acceptable. Note that the dimension h (which need *not* be equal to $t + 2R$) does not enter into these calculations, but we *would* have to choose it at some point. If h is not made large enough relative to t, what would be the nature of the error in the computed stiffness? What would be the effect on the fixity of the center of rotation? Without any calculation it is clear that the stiffness would be less than we just calculated and the center of rotation would shift more as the stage is rotated. The center of rotation of a flexure is not as obvious and fixed as in, say, a precision, preloaded, ball bearing, but for small deflections its behavior is often acceptable if *carefully designed*. A company that used a lot of flexures would probably invest in detailed finite-element studies, which could answer many design questions, as would the experimental development which is always part of practical design. Since t is so small in our example, there should be no trouble in making h enough larger than t to get good results. Checking the stress, we find

$$\sigma_{max} = \frac{6MK_t}{bt^2} = \frac{(6)(106.5 \cdot 0.002)(1.081)}{0.5(0.0232^2)} = 5260 \text{ psi} \tag{5.10}$$

This stress is well below the fatigue limit for any aluminum alloy, so our choice of t and R meets both the stiffness and the stress requirements. Stress is often not a problem in flexures for such actuators since the motions are so small.

The formulas needed to design the four beam flexures are available from Smith (2000, p. 174, 208, and 209). The model used there is shown in Figure 5.6a. The stiffness which we want to compute is the relation between the applied force F_x and the resulting deflection δ_x. The force F_x will act tangentially relative to the center of rotation, as shown in Figure 5.1, resulting in a torque about that point. We need to compute the torsional spring constant resulting from this arrangement, taking r to be the radial distance from the center of rotation to the attachment point of the beam flexures, as in Figure 5.6b. Equation 4.77 from the reference gives

$$\delta_x = \frac{L^3}{6 \cdot EI} F_x \tag{5.11}$$

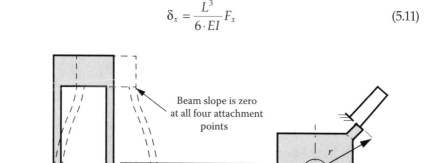

(a)   (b)

FIGURE 5.6
Model for each of the four beam flexures.

where

L is the beam length

I is the area moment of inertia of the beam cross section

Using this result, we can derive the formula giving the total torsional spring constant of all four beam flexures together. The beam length L is estimated from the Mad City drawing to be about 0.38 in. and r appears to be about the same value. The beam width is equal to the thickness of the plate from which the stage is fabricated, which we have already given as 0.5 in. With these assumptions, the only dimension left to *design* is the beam thickness.

$$\delta_x = \frac{L^3 F_x}{6EI} = \frac{M}{r}\frac{L^3}{6EI} = r\,\theta \quad K_\theta \triangleq \frac{M}{\theta} = \frac{r^2 Eb}{2}\left(\frac{t}{L}\right)^3 \tag{5.12}$$

The rotary stiffness K_θ is for a *single* beam flexure so we need to multiply it by 4 and then set this equal to one-half of the total rotary stiffness.

$$106.5 = 2r^2 Eb\left(\frac{t}{L}\right)^3 = (2e7)(0.38^2)(0.50)\left(\frac{t}{0.38}\right)^3 \quad t = 0.01591 \text{ in.} \tag{5.13}$$

As with the notch flexure, we design for stiffness and then check for stress. We thus now need a formula for the maximum bending stress in the beams which form the corner flexures. Smith (2000, p. 175) gives a formula for the maximum bending moment M, from which the stress is easily found.

$$\delta_x = r\theta = (0.38)(0.002) = 0.00076 \quad M = \frac{3EI}{L^2}\delta_x$$

$$\sigma = \frac{Mc}{I} = \frac{3e7}{0.38^2}(0.007955)(0.00076) = 1263 \text{ psi} \tag{5.14}$$

Again we find the stress to be very low and thus acceptable.

5.3 Actuators, Sensors, and Mounting Considerations

Once the mechanical properties of the stage have been defined, one might next consider how piezoelectric translational actuators might be attached to control the stage rotation from electrical command signals. I was again unable to get such mounting details from Mad City Labs, but one of their competitors, Physik Instrumente* (PI) has a catalog which also includes

---

* www.pi-usa.us

Single-axis tilt platforms

Single-axis tilt platforms (θ_x) can be designed in two ways:

(a) Single flexure, single actuator tilt platform

The platform is supported by one flexure and pushed by one linear piezo actuator. The flexure determines the pivot point and doubles as a preload for the PZT actuator.

The advantages of the single flexure, single actuator design are the straight-forward construction and low costs. If angular stability over a large temperature range is a critical issue, the differential piezo drive is recommended.

Several single and multi-axis designs are available.

(b) Differential piezo drive tilt platform
Examples:

Custom designs.

This construction features two PZT linear actuators (operated in push/pull mode) supporting the platform. The case is machined from one solid metal block with FEA (finite element analysis) designed wire EDM (electric discharge machining) cut flexures. The flexures provide for zero friction/stiction and excellent guiding accuracy. The differential design exhibits an excellent angular stability over a wide temperature range. With this design temperature changes only affect the vertical position of the platform (piston motion) and have no influence on the angular position. After the operating voltage is removed the platform returns to the center position.

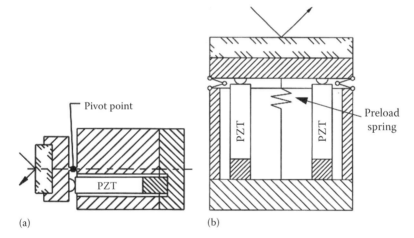

FIGURE 5.7
(a) Construction of a single flexure, single piezo actuator tilt platform and (b) construction of a differential piezo drive tilt platform. (Courtesy of Physik Instrumente, Karlsruhe, Germany.)

extensive technical notes on many such matters. Figures 5.7 and 5.8 are taken directly from their publication. (I took these pages from one of their *older* catalogs. Their current Web site has much useful design data, but Figure 5.7 does not seem to be included any more.) Piezoelectric actuators are usually made from ceramic materials, which are brittle and should not be exposed to side loads (transverse to the direction of actuation) and also are limited in the capability of producing *tension* loads. When tension loads are needed for a particular application, the actuators are generally preloaded with some type of spring effect, to put the actuator in a state of compression without

Mounting guidelines

Piezo actuators must be handled with care because the internal ceramic materials are fragile.

1. PZT stack actuators must only be stressed axially. Tilting and shearing forces must be avoided (by use of ball tips, flexible tips, etc.) because they will damage the actuators.

2. PZTs without internal preload are sensitive to pulling forces. An external preload is recommended for applications requiring strong pulling forces (dynamic operation, heavy loads, etc.).

3. Maximum allowable torque for the top piece can be found in the technical data tables for all PZT stacks and must **not** be exceeded. Proper use of wrench flats to apply a counter torque to the top piece when a mating piece is installed will protect the ceramics from damage.

4. When PZTs are installed between plates a ball tip is recommended to avoid bending and shear forces.

Failure to observe these installation guidelines can result in fracture of the PZT ceramics.

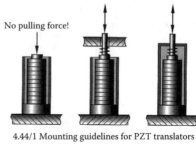

4.44/1 Mounting guidelines for PZT translators

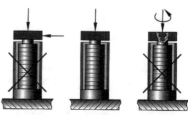

4.44/3 Mounting guidelines for PZT translators

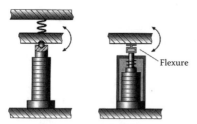

4.44/2 Mounting guidelines for PZT translators

4.44/4 Mounting guidelines for PZT translators

4.44/5 Mounting guidelines for PZT translators

FIGURE 5.8
Mounting guidelines for PZT translators. (Courtesy of Physik Instrumente, Karlsruhe, Germany.)

any electrical signal being applied. (The preload may be internal to the actuator (provided by the actuator vendor) or supplied by the customer (system designer)). When we then apply a voltage in a direction to cause tension, the actuator relaxes some of its preload and the tension force is provided by the spring, not the piezoceramic element. When a motion stage is constrained by flexures, the spring effect of the flexures may sometimes also serve as the preload spring. The lower sketch in Figure 5.7 shows (schematically) the preload spring used in a differential-drive rotary stage.

If you look at a catalog of translational piezo actuators you will find that they are listed by their force capability (usually in newtons) and full-scale stroke (usually in μm). Typical units might range from miniature (about 1 cm square) un-preloaded actuators with rated forces (*T* for tension, *C* for compression) of 150C/30T and stroke of about 10 μm, to preloaded larger (5 cm diameter, 15 cm long) sizes with 30,000T/3,500C and 120 μm stroke. When an actuator drives a stage which has a spring restraint (such as the flexures in our example), Physik Instrumente recommends that the actuator stiffness be about 10 times that of the external spring effect. Typical actuator stiffnesses vary from about 10–2000 N/μm. If the *external* stiffness is too large, available actuator stroke is reduced from the rated value. For example, if the external stiffness is the same as the actuator stiffness, the stroke is cut in half.

Returning to our example rotary stage, the sketches of Figure 5.7 show two possible ways to use translational actuators to drive a rotary stage. Mad City Labs *was* willing to tell me that they use a *single* piezo actuator in this stage, so we will also choose this configuration. Because we wish to choose an actuator whose stiffness is about 10 times that of the stage, we need at this point to compute the stage stiffness at the location where the actuator would be attached. We mentioned earlier the common use of preloading in piezo-driven stages, both to prevent the imposition of tension loads on the piezo element and also to place the desired motion range in a more favorable (linear) region of the piezo force/deflection curve. (We will of course use feedback from a motion sensor to enforce even better linearity.) PI lists three common displacement sensors used with piezo actuators; resistance strain gages, LVDTs, and capacitance pickups, and describes them, as shown in our Table 5.1. Mad City Labs says that they use *only* piezoresistance sensors on all their stages and that they are *not* installed on the piezo actuator itself, but rather "elsewhere," to measure the actual load motion. They were also not forthcoming as to whether *piezoresistance sensor* was just their name for conventional semiconductor strain gages or whether this was something different. An article* written by one of their engineers did not answer these

TABLE 5.1

Feedback Sensors for Piezo Stages

Feature	Strain Gage	LVDT	Capacitance
Resolution	<1 nm	Up to 10 nm	<0.1 nm
Repeatability	0.1% of FS	0.1% of FS	Up to 0.1 nm
Bandwidth	Up to 5 kHz	Up to 1 kHz	Up to 10 kHz

---

* W. O'Brien, Piezoresistive sensors facilitate high-speed nanopositioning, *Photonics Spectra*, July 2003, pp. 90–92.

questions but does provide much useful information. Further details are provided in another article.*

Piezo-driven stages are used in a wide variety of applications, and the details of such applications influence some of the choices made in stage design. Some applications do not require fast response while others need bandwidths of 100 Hz or more. Let's assume that our application *does* need fast response; then PI recommends that some type of preloading be used. This could be either external preloading (using the stage flexures as the preload spring) or internal preloading, where we purchase an actuator which has a built-in preload. To keep our design more simple and compact, let us choose to use a preloaded actuator. The PI catalog lists a number of such actuators, some of which are shown in Table 5.2. The actuators with numbers P841-xx are intended for closed-loop applications and have built-in strain gage displacement sensing. PI makes both high-voltage (1000 V full scale) and low-voltage (100 V full scale) actuators; those in the above list are 100 V full scale. To see whether one of these actuators can meet our needs, we need to do a few more calculations.

The actuator we choose must provide both the desired full-scale motion and also sufficient force to overcome the flexure spring effect as the stage deflects from 0 to 0.002 rad. Let us assume that a single actuator arrangement based on the configuration shown in Figure 5.7a will be used. We first need to estimate the location of the actuator's ball tip on the side of the moving part of the stage (see Figure 5.9). We want to apply the actuator force as far from the center of rotation as we can, since we need to provide sufficient *torque* to rotate the stage against the flexures while minimizing the *force*, which tends to displace the center of rotation from its desired location. That is, if we (foolishly) applied the force *on* the center of rotation, we would get no rotation at all! We *could* provide an extension to the square stage itself to increase the lever arm, but this complicates the machining and enlarges the whole stage, which might have to be kept compact. If we don't use an extension then it appears that the largest available lever arm is about 0.2 in. The maximum translational displacement at this actuator location is $0.20\theta = (0.20)(0.002) = 0.0004$ in. $= 400\,\mu$ in. $= 10.2\,\mu$m. The actuator force needed to move the stage to its maximum rotation can be computed as follows:

$$0.2F_{max} = (213)\theta_{max} = (213)\left(\frac{x_{max}}{0.2}\right) \qquad \frac{F_{max}}{x_{max}} = \text{translational stiffness} = \frac{213}{0.04}$$

$$= 5325 \text{ lbf/in.} = 0.929 \text{ N}/\mu\text{m} \qquad\qquad\qquad\qquad (5.15)$$

The actuator should be at least 10 times as stiff as the driven load, so our actuator must be chosen to provide at least 10.2 μm of displacement and have

---

* J.F. Mackay, Understanding noise at the nanometer scale, *Laser Focus World*, March 2007, pp. 87–89.

TABLE 5.2

Closed-Loop Piezo Actuators from Physik Instrumente

Technical Data	Closed-Loop Models					Units
	P-841.10	P-841.20	P-841.30	P-841.40	P-841.60	
Open-loop travel at 0–100 V	15	30	46	60	90	μm ±20%
Closed-loop travel	15	30	46	60	90	μm
Integrated feedback sensor	SGS	SGS	SGS	SGS	SGS	
Closed-loop/open-loop resolution	0.3/0.15	0.6/0.3	0.9/0.45	1.2/0.6	1.8/0.9	Nm
Static large-signal stiffness[a]	57	27	19	15	10	μm ±20%
Push/pull force capacity	1000/50	1000/50	1000/50	1000/50	1000/50	N
Torque limit (at tip)	0.35	0.35	0.35	0.35	0.35	Nm
Electrical capacitance	1.5	3.0	4.5	6.0	9.0	μF ±20%
Dynamic operating current coefficient (DOCC)	12.5	12.5	12.5	12.5	12.5	μA/(Hz × μm)
Unloaded resonant frequency (f_0)	18	14	10	8.5	6	kHz ±20%
Standard operating temperature range	−20 to +80	−20 to +80	−20 to +80	−20 to +80	−20 to +80	°C
Voltage connection	VL	VL	VL	VL	VL	
Sensor connection	L	L	L	L	L	
Weight without cables	20	23	46	54	62	g ±5%
Material case/end pieces	N-S	N-S	N-S	N-S	N-S	
Length L	32	50	68	86	122	mm ±0.3
Recommended amplifier/controller (codes-explained pp. 1–3)	C, D, G, H	C, D, G, H	C, D, G, H	C, D, G, H	C, D, G, H	

[a] Dynamic small-signal stiffness is about 30% higher.

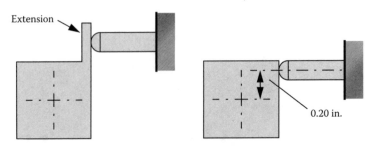

FIGURE 5.9
Configuration to use translational actuator in a rotary stage.

a stiffness of at least $9.29\,\text{N/μm}$. In Table 5.2, actuator P-841.10 easily meets these specifications. Its capacitance is $1.5\,\text{μF}$, its unloaded resonant frequency is $18\,\text{kHz}$, and its total mass is $0.02\,\text{kg}$.

We are now ready to consider the electronic portions of the closed-loop electromechanical system which comprises the complete motion controller. One usually purchases the motion stage and associated electronics from the same source, say Mad City Labs or PI, so that everything fits together properly. Most customers would simply let the supplier choose the hardware to meet performance specifications for the application. In this chapter we want to delve into the details a little more so that you can better understand the operation. A piezo actuator presents electrically a *capacitive* load to the amplifier which drives it; "ordinary" instrumentation amplifiers are not designed for such loads and also usually provide only about $10\,\text{V}$ full-scale output, whereas even "low" voltage piezo actuators need $100\,\text{V}$. If we apply, say, a step input voltage E to a resistive load R, the amplifier output current will jump to $I = E/R$, *assuming* that the amplifier is capable of providing that much current. It should be clear that for a given E, as we reduce R, the current demanded gets larger and larger, at some point exceeding the amplifier's rating, whereupon the current *saturates* at that limiting valued I_s, and the voltage is then not E, but rather RI_s. For a capacitive load, an applied step voltage of *any* size requires an *infinite* current initially (an uncharged capacitor is essentially a *short circuit!*), so the amplifier goes into current saturation, and the capacitor voltage initially rises *from 0* at a rate I_s/C. As the capacitor charges, the voltage rises until it reaches the commanded value, assuming that this voltage does not exceed the amplifier's maximum voltage output. We see that the capacitance of the actuator and the peak output current of the amplifier are critical when evaluating the dynamic response of such stages. For the PI amplifier of Figure 5.10, the peak current (for times less than $0.005\,\text{s}$) is $0.140\,\text{A}$; for times longer than this the peak current is $0.060\,\text{A}$. For the actuator which I selected, the capacitance was $1.5\,\text{μF}$. Suppose we commanded a step input of $10\,\text{V}$ at the amplifier output (actuator input). With a charging current of $0.140\,\text{A}$, the actuator voltage rises at $93333\,\text{V/s}$, so we reach $10\,\text{V}$ in

Notes	Technical data: E-610	
Important calibration information	Function	Power amplifier and sensor/position servo control of PZTs
	Channels	1
	Amplifier	
	Maximum output power	14 W
	Average output power	6 W
	Peak output current <5 ms	140 mA
	Average output current >5 ms	60 mA
	Current limitation	Short-circuit proof
	Voltage gain	10 ± 0.1
	Polarity	Positive
	Control input voltage	−2 to + 12 V
	Output voltage	−20 to + 120 V
	DC offset setting	0 to 100 V with external potentiometer, not included
	Input impedance	100 kΩ
	Input/output connector	32 pin (male) on rear panel (DIN 41612/D)
	Dimensions	One 7T slot wide, 3H high
	Weight	0.35 kg (E-610.00: 0.3kg)
	Operating voltage	12–30 VDC, stabilized
	Operating current	2A
	Position Servo Control (except E-610.00)	
	Sensor types	Strain gage (E-610.S0); LVDT (E-610.L0), capacitive (E-610.C0)
	Servo characteristics	P-1 (analog) + notch filter
	Sensor socket	LEMO ERA.0S.304. CLL (included)

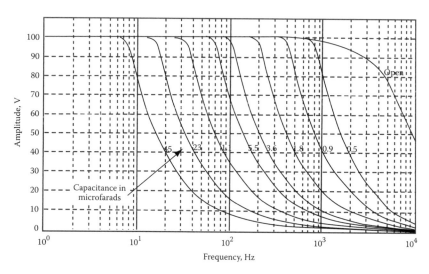

FIGURE 5.10
Typical amplifier for low-voltage (100 V full scale) piezo actuator.

0.000107 s, much less than 0.005 s, so our assumed current is acceptable. If we had commanded 100 V, it would take 0.00107 s, which is still acceptable.

While step inputs lead to the nonlinear amplifier behavior described above, the sine wave inputs used in *frequency-response* studies are more predictable with linear equations. The graph at the bottom of Figure 5.10 shows that our amplifier has a flat amplitude ratio out to about 200 Hz, for full-scale voltage of 100 V. At 1000 Hz it is capable of only about 40 V output even though we are commanding 100. Note that this is amplifier response only; we have not yet dealt with the overall response of the closed-loop electromechanical system. Figure 5.11 shows a block diagram of the complete system, but without the transfer functions needed for analysis using conventional control-system tools. As stated earlier, the basic control mode is the familiar PI control. Also shown is a *notch filter*, a common device for suppressing control system oscillations that are known to occur near a given frequency. Because our mechanical stage does not include any intentional damping, such oscillations can certainly be present if we request rapid changes in position.

Figure 5.12 shows schematically the arrangement of the system components. The preload spring in PI actuators is a conical "washer" type called a Belleville spring. Its stiffness is set at about 10% of the piezo actuator's stiffness, as listed in Table 5.2. The preload force of the preload spring is usually about 20% of the actuator's force rating, again as given in the table. When installing this actuator, we would probably locate it so that the stage flexure system was also slightly preloaded, to guarantee a firm contact between the ball tip of the actuator and the moving part of the stage. The arrangement of Figure 5.12 of course does *not* provide for the actuator exerting any tension forces on the moving stage since the ball tip merely presses against it, it is not "welded" to it. If we want to provide for tension forces (within the

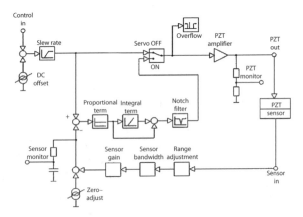

FIGURE 5.11
Block diagram of PI servo controller. PZT amplifiers without position servo control work according to the top branch of the diagram. Also, the slew rate adjustment, servo on/off switch, and overflow indicator are only found on position servo controllers. Note that the DC offset is added to the input voltage before it is amplified.

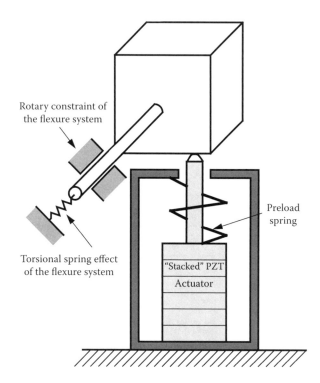

FIGURE 5.12
Mechanical schematic of actuator system.

range allowed by the actuator internal preload spring) then we would have to replace the ball tip with a *flexure tip* firmly attached to both the actuator and the stage moving part. PI offers readymade flexure tips of this sort. While the designer of a piezo actuator motion control system would purchase an actuator to meet the needs of the system design and would not be involved in the design of the actuator itself (this is a very specialized task), it is useful to be aware of some of the details of construction.

The actuators are made from *stacks* of piezoceramic material disks (0.3–1.0 mm thick), whose faces have been metallized to form electrodes to which the actuating voltage will be applied (see Figure 5.13). The disks are cemented together to form the actuator. When a voltage is applied across the two faces of a disk, it expands or contracts, depending on the polarity of the applied voltage. The amount of expansion/contraction (per unit thickness of the disk) is proportional to the *electric field strength* (volts/meter) arising from the applied voltage, and typically is about 0.13% for the maximum allowable field strength. This field strength must be less than about 2000 V/mm to prevent breakdown of the material, which is an electrical insulator (dielectric) in addition to providing the piezoelectric expansion/contraction effect. If a single thick disk (rather than a stack) were used, to get the desired overall motion would require very high voltages. For example, to get, say 15 µm of motion, we would need a disk thickness of 15/0.0013 = 11538 µm = 11.5 mm and the applied voltage would have to be

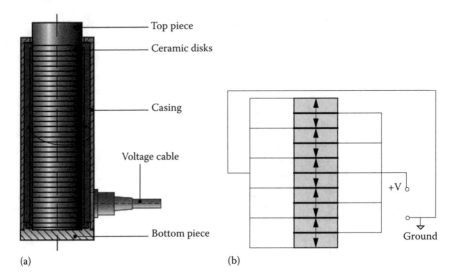

FIGURE 5.13
Details of PI stacked piezo actuator.

about $(11.5)(2000) = 23000$ V. If instead we used, say, n disks of 1 mm thickness, arranged as in Figure 5.13b, each disk would have to provide $15/n$ μm of motion and could contribute, if we applied the maximum allowed voltage of 2000 V, a motion of $(1000)(0.0013) = 1.3$ μm, so we would need $15/1.3 = 11.5$ disks. Lower voltages (such as the 100 V full scale of our actuator chosen from Table 5.2) are achievable using more and thinner disks. While the piezoelectric materials (which are brittle) will produce either tension or compression forces, they are usually used only in compression to prevent mechanical failure of the material and/or the cement bonds. The wiring arrangement of Figure 5.13b shows how all the disks can provide expansion effects which sum to the desired total expansion. When the actuator is moving an external load, this expansion of course puts the disks into compression, as desired.

   The simplest model that represents our system in terms of lumped masses and springs would define a total spring constant made up of the actuator stiffness, the stage flexure total stiffness, and the preload spring stiffness. Since each of the preload spring and stage flexure stiffnesses is intentionally chosen to be about 10% of the actuator stiffness, the system stiffness is dominated by that of the actuator. PI suggests we choose our numbers so that the displacement range of the total system will be quite close to that listed for the actuator in catalog tables. With regard to inertia, PI suggests the effective mass M_{eff} of the actuator is about 1/3 of its physical mass, and since a natural frequency is quoted for each bare actuator, we can find this mass directly from this frequency and the actuator stiffness.

$$\omega_n = (18000)(6.28) = 113097 = \sqrt{\frac{K_s}{M_{eff}}} = \sqrt{\frac{(57e6)\cdot 1.3}{M_{eff}}}$$

(5.16)

$$M_{eff} = 0.005793 \text{ kg}$$

One-third of the physical mass is $20/3 = 6.67$ g, which agrees roughly with our calculation. It will be convenient to replace our actual system, which is partly rotational and partly translational with an equivalent system which is either all rotational or all translational. I chose to use an all-translational equivalent system, so I need to replace the rotary moment of inertia of the stage by an equivalent translational mass M_{eq} attached directly to the end of the piezo actuator.

$$M_{eq} = \frac{I_{xx}}{R^2} = \frac{1.35e-6}{0.20^2} = 3.375 \text{ lbf-s}^2/\text{in.} = 0.00586 \text{ kg}$$

(5.17)

The total mass of our equivalent system is obtained by adding the effective mass of the actuator to the equivalent mass of the rotating stage:

$$M_{total} = 0.005793 + 0.00586 = 0.01165 \text{ kg}$$

The total translational spring constant of our equivalent system is the sum of the actuator stiffness (dynamic small-signal), the preload spring stiffness (usually about 10% of the actuator stiffness), and the translational equivalent of the flexure rotational stiffness.

$$K_{s,total} = 0.929 + (57)(1.3) + (57)(1.3)(0.10) = 82.4 \text{ N}/\mu\text{m}$$

Based on these mass and stiffness values, we can estimate the natural frequency of the complete system:

$$\omega_n = \frac{1}{2\pi} \sqrt{\frac{82.4e6}{0.01165}} = 13,385 \text{ Hz}$$

(5.18)

which we can compare to the "bare" actuator's value of 18,000 Hz.

We started our "design" with some data from the Mad City Labs *Nano-Theta* stage, in particular, the natural frequency of 2000 Hz. We treated this as the natural frequency of the stage *before* attaching the piezo actuator. It turns out that the 2000 Hz number is actually for the *complete* system, including the actuator mass and stiffness. Our selected PI actuator is thus much stiffer than really needed, since the *overall* frequency need only be about 2000 Hz. Mad City Labs seems to concentrate on marketing complete systems and thus offers only a limited selection of actuators, called the Nano-P series, all three of which have natural frequencies of about 2500 Hz. If one of these actuators were used with the stage we designed, it could very easily result in the overall system frequency of 2000 Hz. I used the PI actuator catalog in our design studies since they offer a very wide range of actuators and they give many useful numerical values and design details. However, from here on, we will use the 2000 Hz number as the natural frequency of the *overall* system.

5.4 Control System Design

The basic control mode for many motion control systems (including our piezo actuator system) is the PI, with certain other features added as needed. Some of these added features are derivative control, notch filters, input shaping, and dynamic compensation. Since we assume the reader has only a modest capability in control system design theory, we study our system mainly with simulation, rather than starting with conventional analytical design tools (frequency response, root locus, etc.) and going to simulation only for *fine-tuning*. (Actually, using simulation as the *main* design tool is surprisingly efficient.) We begin with nothing but proportional control and use simulation to show that this is clearly inadequate. This is easily predicted analytically since it is well known that proportional control alone for a spring/mass/damper load will always have a steady-state error for a constant commanded angle, and that a very high loop gain is needed to reduce this error to small values. Another problem with our system is the poor damping, since there is *no* intentional damping and the *parasitic* damping of the flexures and piezo actuator is known to give a damping ratio typically of about 0.01. I do not know of any practical applications where piezo actuator systems have used *added viscous dampers* to improve the response. This is probably because of space restrictions, temperature sensitivity of oil viscosity, and the very small motions involved. Also, most problems can be overcome *electronically*, using one or more of the control system *tricks* mentioned above. Our first simulation uses a loop gain of 100 (to get small steady-state error), and increases the actual damping by 50 times ($\zeta = 0.50$), to show what could be achieved *if* we had good damping. Figures 5.14 and 5.15 show the Simulink block diagram and the step response, respectively. While the dynamic response shows good speed and damping, the steady-state error, even with a gain of 100, is unacceptably large.

We next reduce the damping to the *actual* value ($\zeta \approx 0.01$), and leave the gain at 100, as shown in Figure 5.16. This is clearly an unacceptable dynamic response and if we reduce the gain to reduce the oscillation, the steady-state error will be even larger than in Figure 5.15. While pure proportional control is clearly not adequate, before leaving it, we try, for illustrative reasons, one of the mentioned control system *fixes*; using an input less severe than a perfect step input. That is, our purpose is to move the stage from

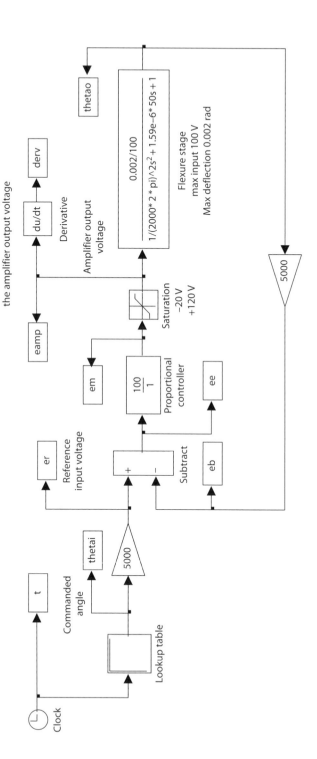

FIGURE 5.14
Simulation of piezo actuator control system.

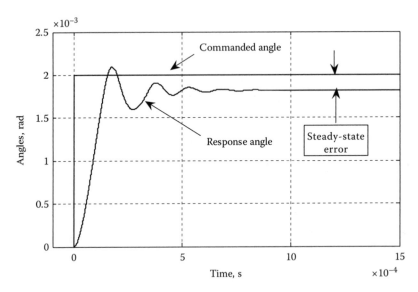

FIGURE 5.15
Step response with 50 times the actual damping.

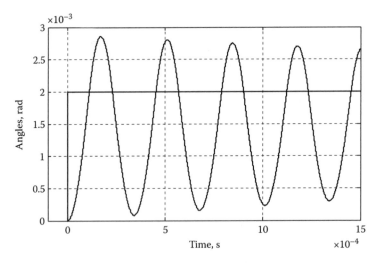

FIGURE 5.16
Step response with realistic (0.01) damping.

one position to another as quickly as possible, which seems to call for a *step* input. We can however achieve our goal by applying a *terminated-ramp* input which ends up at the desired new position, but gets there with a less radical command, hopefully reducing oscillations while not really sacrificing speed. Because of amplifier current saturation for any size step input (as discussed above), the rate of change of amplifier output voltage is

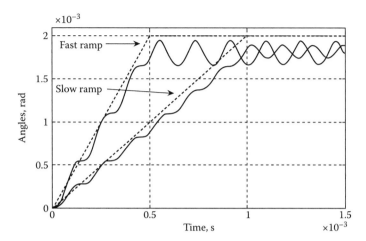

FIGURE 5.17
Terminated-ramp input reduces oscillation.

limited to I_s/C, so step inputs of voltage are really not possible in the actual system, though they are easily implemented in our simulation. It is thus more realistic (and also reduces the apparent oscillation) to apply terminated ramp inputs. The lookup table in our simulation allows use of either steps, or terminated ramps of any desired slope. Figure 5.17 shows that the oscillations, while still unacceptable, are much reduced, so we might try this technique later when we use other control modes. Pure integral control can give a step response without visible oscillation if the gain is not too high. Figure 5.18 shows such a response, which we see is quite slow

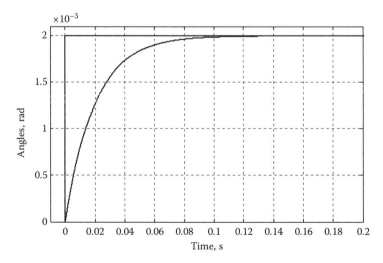

FIGURE 5.18
Pure integral control with gain of 500.

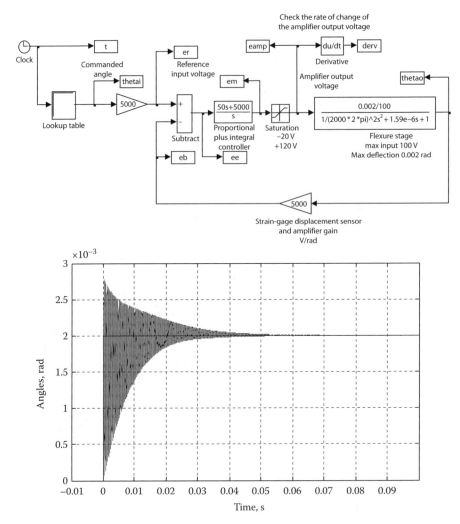

FIGURE 5.19
System with PI control.

relative to the stage natural frequency of 2000 Hz. Higher gain gives faster response but oscillation amplitude is too large, and with even higher gain, we get absolute instability.

Next we try the PI control, which we earlier stated to be widely used for piezo actuator systems from various companies. If we apply an (unrealistic) step input, we get the results of Figure 5.19, which show unacceptable oscillations for the gain values shown there (trial of other combinations of gains showed no acceptable responses). Due to the current saturation that is present in the real systems, we really should not be applying perfect step inputs in our simulations; they cannot actually happen. Note that the simulation *does* include an output voltage saturation effect for the amplifier, but this is *not* the same as the current saturation. The amplifier details needed to properly model the current saturation effect were not available from PI, so I did not make an attempt. We do however get nearly the same effect by applying command voltages that are terminated ramps. By adjusting the slope of the ramps, we can adjust the speed of the response and the degree of oscillation. We should also note that the *size* of any input that we apply is significant; sufficiently *small and slow* inputs to the amplifier will cause neither voltage saturation nor current limiting. *Large and fast* inputs may cause either or both of these nonlinearities to become active. When the system is behaving linearly, changing the size of a given input provides *no* new information; all the responses simply *scale* up or down in proportion to the change in the size of the input. For nonlinear behavior, changing the size of an input may produce a different *form* of response; it doesn't just scale up or down, so we glean new information.

While, as mentioned above, we do not have enough information about the amplifier to exactly model its current saturation effect, we can approximate it by using Simulink's *rate limiter* module. That is, we calculated earlier that, for the actuator's capacitance of 1.5 µF and the current limit of 0.140 A, the maximum possible rate of change of the amplifier output voltage would be 93333 V/s. By placing the rate limiter module after the voltage saturation module (and setting the limits at ±93333) we limit the amplifier output voltage from −20 to 120 V, and its rate of change to ±93333 V/s. At this point we modify our simulation to include these effects and also add the capability of using some derivative control, in the form of a filtered derivative of the strain-gage sensor's output voltage, which is proportional to the output angle. This implements the control mode called *controlled-variable derivative control*. A further modification adds a simulation of the open-loop response of the amplifier/actuator combination, for easy comparison with the closed-loop behavior. Figure 5.20 shows this new simulation. Note that this part of the simulation treats the open-loop actuator as if it had *no nonlinearity or hysteresis*, whereas a main reason for using closed-loop control is to negate these deleterious effects. The reason for neglecting these two features is that they have mainly *steady-state* effects, and we are here concerned mostly with the dynamics.

When I tried several sets of proportional and integral gains, it soon became clear that good response was not going to be possible. This should not have been surprising if we consider the root locus of any system which tries to use only these two modes to control a very lightly damped mass/spring/damper system (see Figure 5.21). We see there that the closed-loop roots associated with the two complex poles are nearly unstable even for very low gains and that their closed-loop damping ratio is *worse* than that of the open loop. There is no set of gains that can achieve a good closed-loop damping ratio, and even small loop gain drives the system into instability. Some types

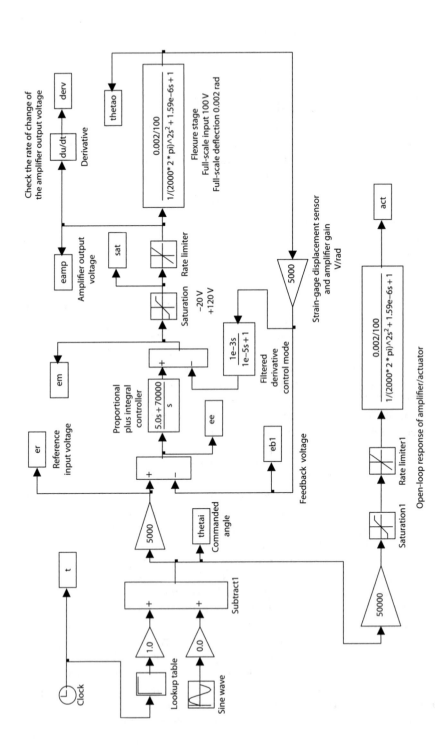

FIGURE 5.20

Simulation including saturation, rate limiting, and derivative control.

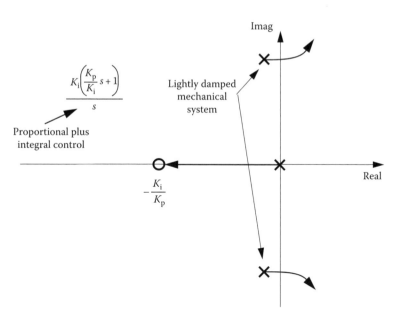

FIGURE 5.21
Root locus of any lightly damped mechanical system with PI control.

of derivative control can usually help in such a situation, and it is routinely recommended by some manufacturers* but I did not see it mentioned in any of the PI or Mad City literature. *The NanoPositioning Book* was written by some engineers at Queensgate who are actually involved in designing and building piezoelectric nanopositioning systems and covers many mechanical, piezoelectric, and control aspects. As far as I know, this is the only book that addresses this subject in a fairly comprehensive way.

If we apply a perfect step command of size 0.0002 rad (10% of full scale) in the system of Figure 5.20, trial-and-error setting of the control mode gains results in the values shown (these *specific* gain values are not *magic numbers*, there are other combinations that also work well). An initial value for the time constant of the low-pass filter used on the derivative control can be *guessed* rather easily since we want the *slowest* filter that will pass frequencies out to about 1 or 2 kHz. That is, the approximate differentiation effect is good at *low* frequencies but gets worse as we go to higher frequencies (the denominator deviates more and more from 1.0). The reason for the filter is its noise rejection (differentiators *accentuate* any noise in our signals), so we want the strongest allowable filter. I tried a time constant of 0.00008 s (−3 dB frequency of 2000 Hz) but the response was too

* Queensgate Instruments Limited; T.R. Hicks and P.D. Atherton, *The NanoPositioning Book,* Queensgate Instruments Ltd., Bracknell, U.K., 1997, pp. 34–41. www.nanopositioning.com

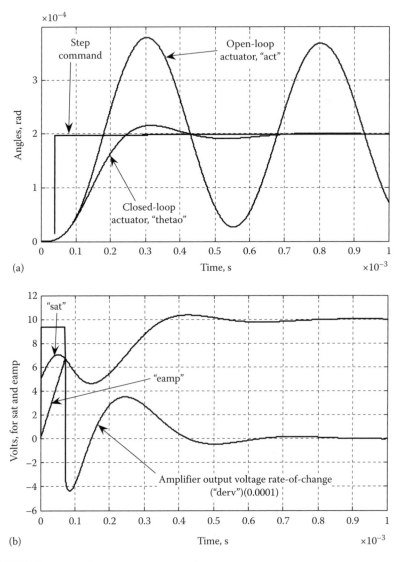

FIGURE 5.22
Small-signal step response.

oscillatory, so I decreased the time constant until I got the good response of Figure 5.22a. Figure 5.22b shows that the amplifier output voltage never saturates, but the current limiting does occur, even with this small step command, since *any* size step must "become a ramp" at the amplifier output until the voltage rate-of-change no longer exceeds the 93333 V/s dictated by the capacitance and the current limit. Our rough simulation of these amplifier features thus seems to be "not too bad."

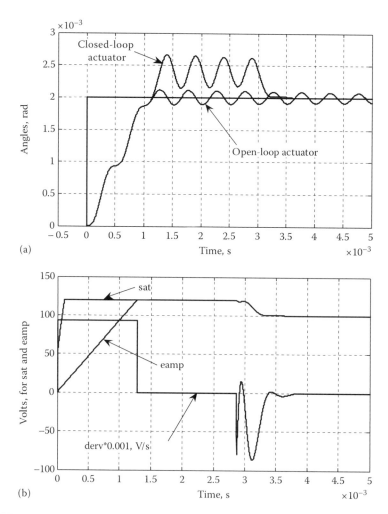

FIGURE 5.23
Large-signal step response.

The large-signal step response is shown in Figure 5.23, where we see that the closed-loop and open-loop responses are identical at first (Figure 5.23a), because both saturation and current limiting are active (Figure 5.23b). With the amplifier "maxed out," the feedback signal in the closed-loop system *can have no effect*. When current limiting ceases at about 0.0013 s, we see that voltage saturation continues until about 0.0028 s; thereafter the system behaves normally. The bad behavior between 0.0013 and 0.0032 s can be charged to a phenomenon called *integral windup*, which regularly occurs in many feedback control systems that are subject to saturation effects and include integral control. It is discussed in some general control literature* and for piezo actuator systems specifically in *The NanoPositioning Book*, pp. 35–40. It is caused by the integrating effect, which, for a large input, quickly builds up a large output signal during saturation, when the system error stays large for some time since stage motion is trying to "catch up" with the large command. Thus the integrator output signal "winds up" to large values (that don't really help the response since the stage is already being driven as hard as possible), and this excessive signal has to be "unwound" later, when things return to normal. We can show this by plotting the signal *em* (see Figure 5.24a), which includes all three modes (P, I, and D) but, under these circumstances, is dominated by the integral portion. We see that the commanded amplifier output greatly exceeds the voltage saturation limit, so it is not really

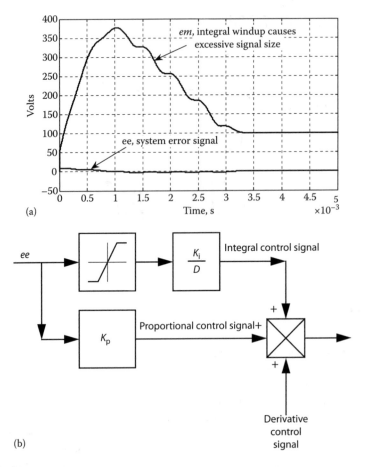

(a)

(b)

FIGURE 5.24
Integral windup and its correction.

* E.O. Doebelin, *Control System Principles and Design*, Wiley, New York, 1985, pp. 352–354.

effective in driving the stage harder, but these excessive values will have to be unwound before we regain control of the stage motion. Controllers can be modified in various ways (called *anti-integral-windup* or *anti-reset-windup* in the literature) to prevent this undesirable behavior. A method suggested in *The NanoPositioning Book* is shown in Figure 5.24b, where the integral and proportional modes have been implemented *separately* and a limiter placed before the integrator. This limiter is set so that the integrator output rate cannot exceed the maximum response rate of the amplifier, as fixed by the actuator capacitance and amplifier current limit. We could easily simulate this "fix" for our system but choose not to at this time; we *will* try it later.

The sinusoidal response of these motion control systems is also of interest and if we ignored the nonlinear effects, we could manipulate the block diagram to get the transfer function relating the response angle to the command angle, and apply our routine MATLAB® frequency-response calculations to plot the usual curves of amplitude ratio and phase angle. We will instead use our simulation (which includes the nonlinear effects) to compute and plot the time histories for a few selected frequencies and for small and large commands. Figure 5.25a shows the small-signal response for three frequencies, 100, 1000, and 2000 Hz. The dotted curve is the command input for 100 Hz and serves as the amplitude reference for all the curves. The solid curves are for the response at 100 Hz and 1000 Hz, with the dashed curve being the response at 2000 Hz. We see that at 100 Hz the amplitude ratio is very close to the ideal 1.0 (with very little phase lag) and even at 1000 and 2000 Hz the amplitude ratio has not fallen appreciably. Figure 5.25b shows the large signal response for 2000 Hz only, which appears to be not much different from the small-signal results. However, Figure 5.26 shows that for the small signal test, neither voltage saturation nor current limiting occurs, while the large signal test exhibits both.

Another possible solution to excessive oscillations in lightly damped systems is the use of a *notch filter* in the forward path of the feedback system. This type of filter has an amplitude ratio of 1.0 for low and high frequencies but at an intermediate frequency of our choice, the amplitude ratio is exactly 0.0 (frequencies near this "notch" are attenuated). For our system, the idea is to set the notch frequency equal to the frequency of the closed-loop system oscillations. For analog signals, this filter is realized in *hardware* (see Figure 5.27); digital software can produce the same effect. The notch filter* has the transfer function

$$\frac{e_o}{e_i}(D) = \frac{R^2C^2D^2 + 1}{R^2C^2D^2 + 4RCD + 1} \tag{5.19}$$

---

* E.O. Doebelin, *System Dynamics: Modeling, Analysis, Simulation, Design,* Marcel Dekker, New York, 1998, p. 573.

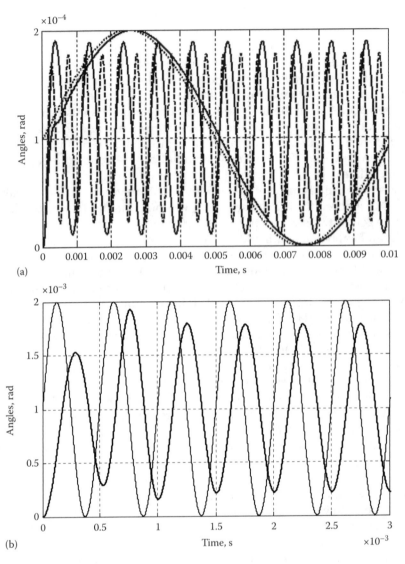

FIGURE 5.25
Sinusoidal responses: (a) small signal and (b) large signal.

where $1/RC$ is the frequency of the notch. We set the notch frequency equal
to the mechanical system natural frequency, in our case 2000 Hz (this is close
to the complete system frequency). Figure 5.28a shows the system with the
notch filter in place, with PI control mode gains set as before, and no derivative control. The step responses of this system with and without the filter
are shown in Figure 5.28b, where we see that without the filter, the system
is absolutely unstable, while, with the filter, we have stability, but excessive

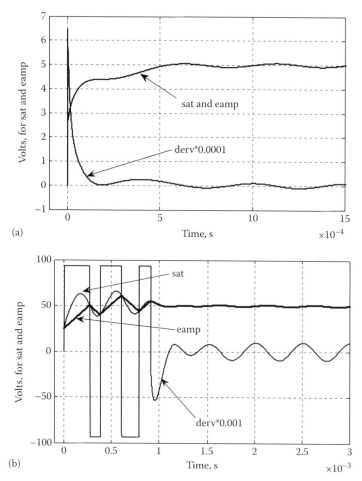

FIGURE 5.26
Sinusoidal response: (a) small signal and (b) large signal.

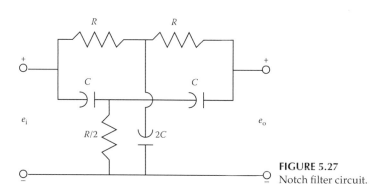

FIGURE 5.27
Notch filter circuit.

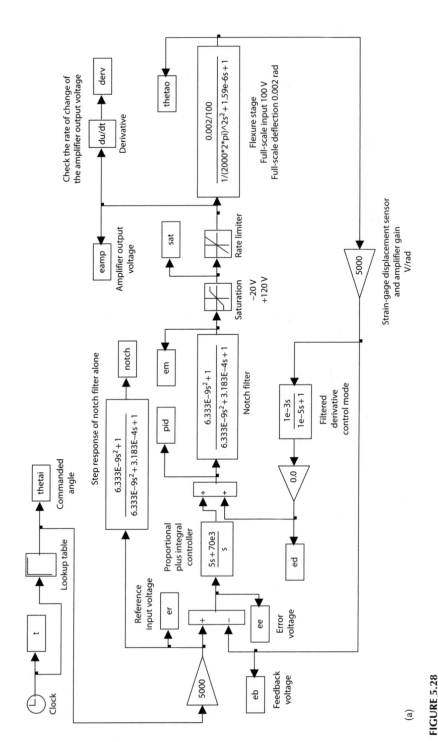

FIGURE 5.28
Effect of notch filter, no derivative control.

(a)

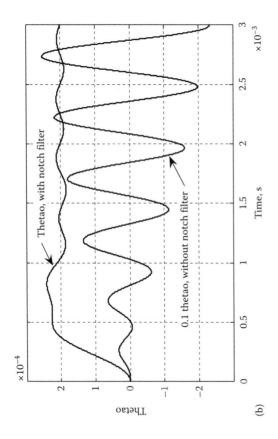

(b)

FIGURE 5.28 (continued)

oscillation. For those unfamiliar with notch filters, I show the step response and frequency response of the filter alone in Figure 5.29. The step response gives some intuitive idea as to how the filter works. At first it applies a positive-going pulse and a little later, at *just the right time*, it applies a negative-going pulse. The first pulse stimulates an oscillation, but the second is timed to be out of phase with the first, so as to nearly cancel its effect, thus reducing the oscillation. We can improve the response of Figure 5.28a by using a suitable terminated ramp command instead of a perfect step, as seen in Figure 5.30. The response for large (full-scale) commands changes radically when the ramp slope is too fast, since the nonlinear effects come into play. For the small (10% of full-scale) commands more typical of actual applications, the terminated ramp response is much improved and certainly acceptable for most practical situations.

Since numerical values of stage frequency and filter notch frequency can never be perfectly known or absolutely fixed, it is important to investigate the effects of a *mismatch* between these two numbers, since that is certain to occur in real applications. Let's try 10% mismatches above and below the desired perfect match, which can be accomplished by changing either the filter dynamics, the stage dynamics, or both. The effect will probably be nearly the same no matter what we change, so let's change just the filter notch frequency, to either 1800 or 2200 Hz. Figure 5.31 shows that for small signals the mismatch has little effect; the response is still quite acceptable. For large signals, where as usual, the nonlinear features come into play, the response becomes unacceptable, and in fact absolutely unstable for the filter with the 2200 Hz frequency. With nonlinearities in action, and theory thus largely lacking, speculation on causes is risky, but one possible explanation can be seen in the frequency-response curves of Figure 5.29. There we see that the filter has a lagging phase angle below the notch frequency and a leading phase angle above. For the filter tuned to 2200 Hz, the lagging phase now extends to higher frequencies, and for most feedback systems, an increase in phase lag is a destabilizing effect, so a system that was stable with the correct filter might become unstable with the mistuned filter. Finally, we might note that any of our results that depend on the nonlinear effects (voltage saturation, current rate-limiting) would *not* be predicted by the available control system theory, since it is largely restricted to linear systems. Using simulation as a major design/analysis tool thus greatly improves our effectiveness as dynamic system designers.

Our fourth, and final, method to deal with the control of poorly damped mechanical resonances is not mentioned by any of the three nanomotion control companies listed before in this chapter, but is discussed in the general control system literature.* To understand the principles behind this technique, it is helpful to have at least a cursory familiarity with the widely used

---

* E.O. Doebelin, *Control System Principles and Design*, Wiley, New York, 1985, pp. 436–440.

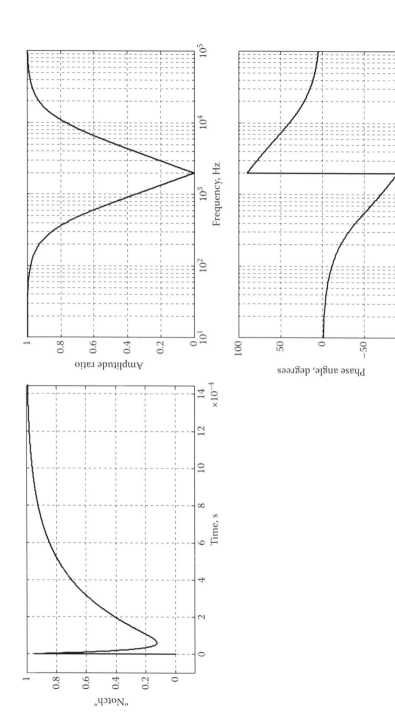

FIGURE 5.29
Step and frequency responses of notch filter itself.

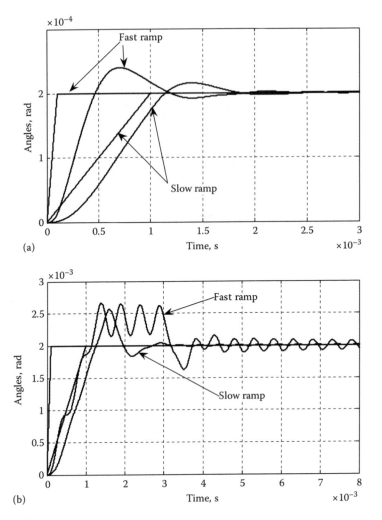

FIGURE 5.30
Terminated-ramp response with notch filter: (a) small signal and (b) large signal.

control system design method called *root locus,*[*] which we used earlier in
Chapter 3 and now expand upon. For those readers with no control systems
background, I now provide the bare essentials. To plot a root-locus graph,
one needs to have all the transfer functions found in the complete control
loop: the "plant" to be controlled, actuators, sensors, and the controller laws.
This combination of transfer functions is called the *open-loop transfer function*.
For the basic stage system of Figure 5.19 (*basic* means *no* notch filter, and *no*

---

[*] E.O. Doebelin, *Control System Principles and Design*, Wiley, New York, 1985, pp. 305–323.

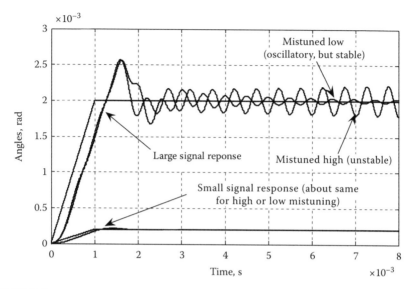

FIGURE 5.31
Effects of filter tuning mismatch.

derivative control), the complete control loop has only the proportional/integral controller dynamics (50s + 5000)/s and the stage dynamics given by

$$2e-5/((1/2000*2*pi)^2\ s^2+1.59e-6s+1)$$

On an *x–y* graph with the horizontal axis being the real part and the vertical being the imaginary, we plot all the poles (using *x*'s as symbols) and all the zeros (using *o*'s as symbols) of these two transfer functions. A control system designer has memorized a few *sketching rules*, which allow quick approximate sketching of the root locus curves. These curves show how the *closed-loop* poles (roots of system characteristic equation) move around in this complex plane as the designer adjusts the *loop gain*, which is a vital design choice in every feedback system. The root loci *always* start at the open-loop poles (the poles of the transfer functions we listed above). If there are as many open-loop poles as there are open-loop zeros, every root locus curve starts on some pole (for loop gain equal to 0) and ends on some open-loop zero, as gain goes to infinity. When there are more poles than zeros, then some of the loci which start at a pole do *not* end at a zero but rather go toward infinity along some straight-line asymptote, which can be sketched from one of the rules. For any particular numerical value of loop gain, there are as many closed-loop roots (poles) as there are open-loop poles, and these roots have specific numerical values. These root values largely control the dynamic response of the entire control system, thus knowing their values is very helpful. In fact, for *every linear dynamic system, whether a feedback system or not*, the location of the roots of its characteristic equation largely determines the system's dynamic behavior. Figure 5.32 reviews the significance of

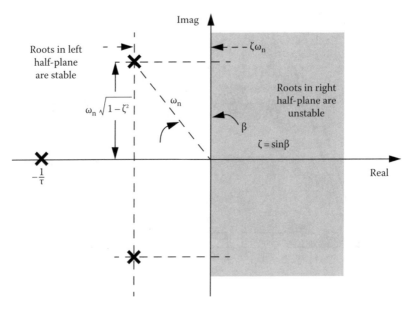

FIGURE 5.32
Significance of root locations.

these root locations. Roots are either real (with *first-order dynamics* and time constant τ), or complex pairs (*second-order dynamics* with natural frequency ω_n and damping ratio ζ). Roots far from the origin are *fast* (large ω_n or small τ). Complex root pairs with small angle β are poorly damped, those with large β are well damped. While the sketching rules are useful in preliminary design, where tentative forms and values of component transfer functions are being chosen, the actual plotting of the root locus curves is done with computer tools such as the control system toolbox of MATLAB. To use the MATLAB tool the *only* things you need to provide are the transfer functions, such as we gave above; MATLAB does the rest.

Returning to our example, Figure 5.21 shows the root locus for that system. Note that when the loop gain goes toward zero, the closed-loop poles (roots of characteristic equation) approach the open-loop poles. Also, as we increase the loop gain, the complex pair of closed-loop poles goes toward *worse* damping than the open loop has, and that if we increase gain enough, these two roots cross over into the right half-plane, denoting instability. Thus applying simple PI control to a poorly damped mechanical load cannot improve the damping. The design "trick" which we now develop suggests that we should somehow add some *zeros* to the open-loop transfer function, and locate these zeros *to the left* of the poorly damped mechanical dynamics, thus "sucking" the root locus paths, that before went toward the right half-plane, leftward toward better damping. Since, for physical realizability, any transfer function must have at least as many poles as zeros, we need the ratio of two second-order terms for our compensator. We choose the zeros to be a complex pair located somewhat to the left of the mechanical dynamics and the poles to be a repeated real pair, located *more* to the left. As in any design problem, there are an infinite number of combinations of such poles and zeros that might work. One simply makes a choice and evaluates the new system response. If it is not satisfactory, we make a second choice, guided by the results of the first. Figure 5.33 shows the general configuration. We cannot show a *general shape* for the root loci since, depending on numerical values, *radically different* root loci are possible. The only thing that is sure is that, because there are five open-loop poles and three open-loop zeros, there *will* be two root loci that go off toward infinity along asymptotic straight lines at ±90°. (This statement is based on one of the root locus rules that we have mentioned but not explained.)

For the first trial, let's put the two compensator real poles at −50,000, well to the left of the PI control zero at −14,000. The polynomial corresponding to this would be $4e{-}10D^2 + 4e{-}5D + 1$. The compensator zeros are chosen to be those of a second-order system with 2000 Hz natural frequency and damping ratio of 0.50. From Figure 5.32 this makes the zeros $-6283 \pm i10883$. The polynomial corresponding to this would be $6.333e{-}9D^2 + 7.958e{-}5D + 1$. A MATLAB program to compute and plot the root locus follows:

```
% crc_book_chap5_comp_rlocus
% computes and plots root locus for nanomotion control system
% first, specify the separate "block" transfer functions
num1 = [5 7e4];den1 = [1 0];
num2 = [6.333e-9 7.958e-5 1];den2 = [4e-10 4e-5 1];
num3 = [2e-5*5000];den3 = [1/(2000*2*pi)^2 1.59e-6 1];
% now combine into one overall function nums,dens
[numa,dena] = series(num1,den1,num2,den2);
[nums,dens] = series(numa,dena,num3,den3);
```

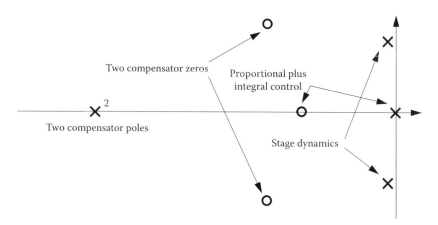

FIGURE 5.33
Compensator pole/zero configuration.

```
% now plot the root locus
rlocus(nums,dens);grid
pause; %you may want to zoom the graph at this point
% now choose selected pole location
% by positioning the cross-hairs
[k,poles] = rlocfind(nums,dens) % this command allows gain
            setting
k % prints the loop gain value to put selected pole where you  wish
poles % prints all the closed-loop poles (roots)for the selected
        gain
```

The statement *rlocus(nums,dens);grid* plots the root locus, with a superimposed grid showing lines of constant closed-loop damping ratio, as an aid
in setting the gain. The statement [k,poles] = rlocfind(nums,dens) also
plots the root locus but provides a mouse-movable crosshair cursor which
you can position at any point on one of the root loci to request that that point
be a closed-loop root. It then computes the loop gain k that will put the root
where you asked, but of course you have no control over where all the *other*
closed-loop roots will be located; you have to accept whatever is dictated by
your chosen root and the system dynamics. The program prints the k value
and all of the closed-loop roots, including the one (usually a *pair* of complex
conjugates) that you requested. For our example, the plots are shown in
Figure 5.34 and the printout is shown below.

```
Select a point in the graphics window
selected_point = -4.0666e+003 + 15.2174e+003i
k =      1.6969e+000
poles =
    -43.2189e+003 + 38.0332e+003i
    -43.2189e+003 - 38.0332e+003i
     -4.0736e+003 + 15.2678e+003i
     -4.0736e+003 - 15.2678e+003i
     -5.6661e+003
```

As the root locus first appears when this program runs (Figure 5.34a), the
scale is automatically set to display portions of all the root loci. This display
usually needs to be zoomed (Figure 5.34b) to expand the region where you
want to position a closed-loop pole, so I put a *pause* in the program to allow this
zooming. To release the *pause* and let the program continue into the statements
that allow you to set gain, type any letter. The zoomed display then appears
with the movable crosshairs and you can position it accurately to select the
desired root. In this example, I choose a point where the loci that leave the stage
dynamics poles intersect a closed-loop damping line of 0.22. This damping is
still fairly low but it is *22 times better* than the 0.01 damping of the stage.
    Having set the gain to get one pair of roots with fair damping, it is of course
necessary to now run simulations of the trial compensator with the suggested

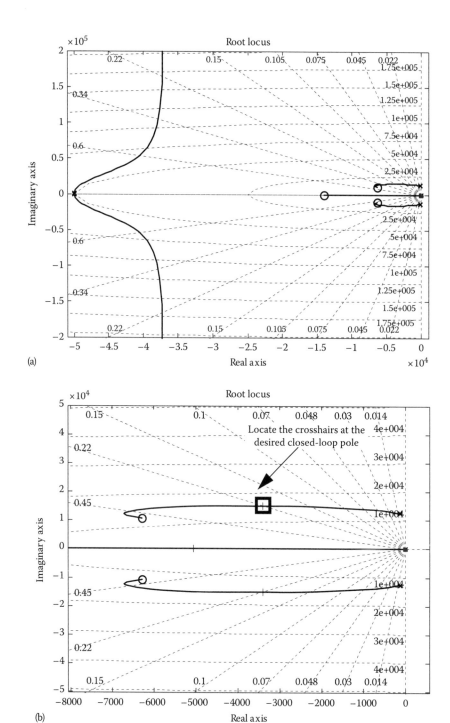

FIGURE 5.34
Root locus for system with compensator.

gain. Because the total response involves *all* the closed-loop roots, not just the one we *designed*, there is no guarantee that we have achieved an acceptable system design. Using the 1.69 gain suggested by the root locus program, and *removing the voltage-saturation and current-limiting nonlinearities*, I ran some simulations. I removed the nonlinearities because the compensator design *assumes* a totally linear system, so it is "only fair" to evaluate the design under those conditions. Figures 5.35 and 5.36 show that the compensated response is really quite good for this "first try" (without the compensator and with the same gain, the system is *unstable*), so this kind of compensator has proved useful. Figure 5.35 shows that if we don't want the brief initial oscillation we can use a terminated-ramp input without sacrificing speed (settling time). In Figure 5.36 the frequency response at 2000 Hz is almost perfect, with amplitude ratio near 1.0 and no noticeable phase shift. At 5000 Hz, for graph clarity, I didn't show the command signal. If I had, a phase lag would be apparent, but this is normal, as is the attenuation which is visible.

While this compensator and gain setting works well in a system without nonlinearities, it must of course work in our real system, which *has* unavoidable nonlinearities. When I tried it under those conditions, if the commands were sufficiently small and slow, so as to not bring the nonlinearities into action, its behavior was the same as for a completely linear system. However, when a command brought the nonlinearities into effect, the system went unstable. That is, the compensator's actions are finely tuned to stabilize an unstable system, and the voltage-saturation and/or current-limiting prevent the corrective signals from being properly implemented. In a search for a "fix" for this problem, I decided to try the anti-integral-windup scheme of

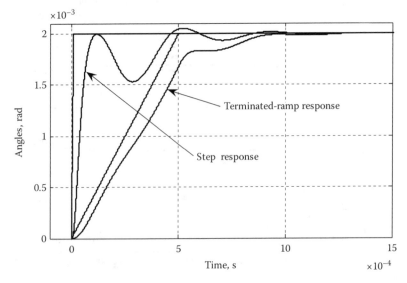

FIGURE 5.35
Full-scale step and terminated-ramp responses with compensator.

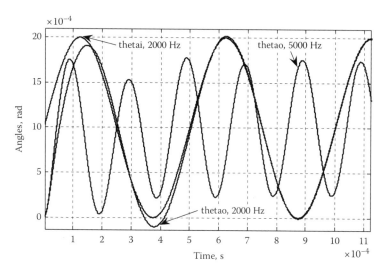

FIGURE 5.36
Full-scale sinusoidal response with compensator.

Figure 5.24b, since it can limit the contribution of the integral control mode to amplifier overload. In the simulation of Figure 5.37 the proportional and integral control modes have been *split up* so that a limiter can be applied to the integrator input. The setting of the limiter involves some trial and error, but an estimate for a first trial is possible by noting that an integrator input of, say, 1.0 will result in an integrator output with a slope of 70,000 V/s, which is a little less than the current-limited amplifier output of 93,333 V/s, so setting the limiter near 1.0 seems reasonable. Trial simulation runs allow this value to be fine-tuned. Note at the upper right that I subject the compensator alone to the same input as the entire system. This allows display of the compensator response for those unfamiliar with such dynamic systems.

Trial runs showed that this system should not be subjected to perfect step inputs since it causes instability due to the nonlinearities. Using appropriate terminated ramps solves this problem and still gives quite fast response, as seen in Figure 5.38. The fact that the large-signal response has a different waveform (not just a larger size) than the small signal means that the nonlinearities are active for the early parts of the response, but the system is able to handle this without instability, and returns to the linear region as the response settles into steady state. In sinusoidal testing we need to note that a sine wave starts out with its largest slope, which is the product of the frequency (in rad/s) times the amplitude. For example, a sine wave given by [0.002 sin (2000*2*pi)], after passing through the gain of 5000 V/rad, produces a slope of 125,663 V/s (at the signal called *ee*), which is greater than the current-limited slope at the amplifier. The signal *ee* is not the amplifier input voltage but we can see the potential problem here. A further detail with regard to sine testing is that the stage is designed for the range 0–0.002

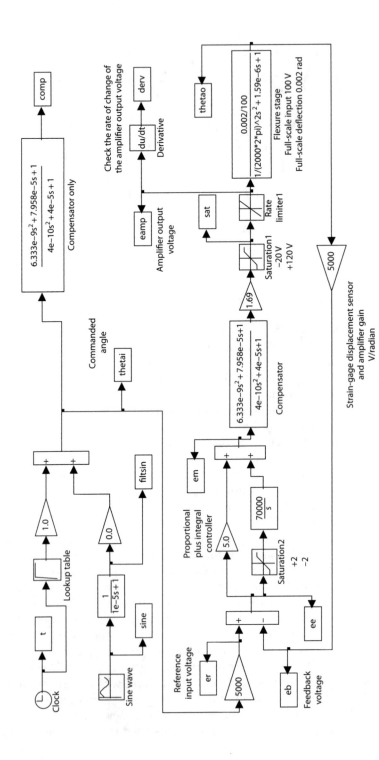

FIGURE 5.37
System with compensator and integral anti-windup.

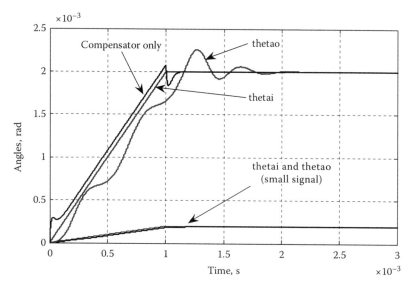

FIGURE 5.38
Compensated system with integral anti-windup.

rad, not a symmetrical range such as ±0.001 rad, thus running a full-scale sinusoidal test involves biasing the ±0.001 sine-wave with a constant 0.001, giving the input signal as 0.001 + 0.001sin (ωt). Such a signal experiences a perfect step change of 0.001 at time = 0, and we said earlier that step changes in this system cause instability. To *smooth out* this sudden change, I added the simple low-pass filter seen in Figure 5.37 at the upper left; its time constant may need to be adjusted depending on the frequency being applied. Figure 5.39 shows early times for a 100 Hz, full-scale sine test with the time constant set at 0.00001 s. The input signal seems to have a step change but the filter has slowed this down sufficiently to prevent instability, but there is some initial nonlinear effect which quickly disappears. A graph (not shown) for longer times documents an almost perfect response with amplitude ratio near 1.0 and very little phase shift.

At 1000 Hz in Figure 5.40 we see significant attenuation and phase shift, but even more interesting is the distortion of the output angle waveform; it is clearly not sinusoidal. Examination of the *sat* and *eamp* waveforms makes clear that for the steady state that occurs at later times, there is no saturation but there is obvious current limiting; the amplifier output signal is essentially a *triangle wave*, not the sine wave typical of a linear system. When we next use small commands we get the results of Figure 5.41. This response shows no visible waveform distortion, and examination of *sat* and *eamp* shows that, in the sinusoidal steady state, neither voltage saturation nor current limiting occur.

While studies of system frequency response are useful diagnostic tools (either in simulation or experimental testing), practical systems are rarely subjected to sinusoidal commands. Thus difficulty in following high-frequency

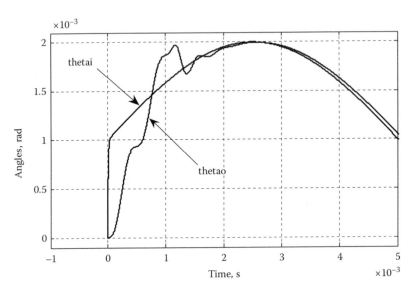

FIGURE 5.39
Sinusoidal response at 100 Hz, input filter time constant is 0.00001 s.

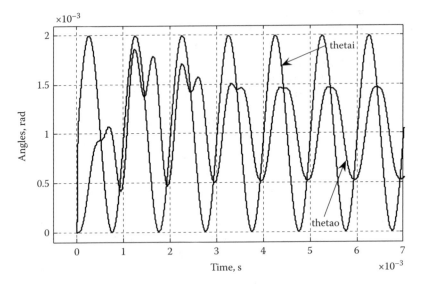

FIGURE 5.40
Sinusoidal response at 1000 Hz, input filter time constant is 0.00001 s.

sine waves of large amplitude may not indicate any real problems with the system's capability of fulfilling its practical specifications. Also, our "first guess" at the compensator dynamics seemed to give usable results, but further trials might result in improvements. Since we didn't describe an actual practical application with detailed information on the needed motions, we

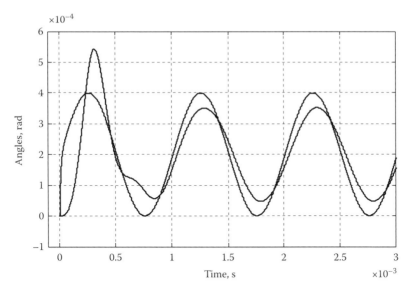

FIGURE 5.41
Small-signal sinusoidal response at 1000 Hz.

are unable to judge whether the various versions of our system would be successful as a working machine. Our real goal was to illuminate the available system configurations and the analytical and simulation tools that can be used to attack this general class of design problems.

A final design technique, which is mentioned in *The NanoPositioning Book* and actually provided as a commercial product by Physik Instrumente,* is the signal-shaping scheme offered for license by Convolve.† We discussed this briefly in Section 2.6, but will not pursue it further here.

---

* http://www.physikinstrumente.com/en/products/prdetail.php?sortnr = 400705
† www.convolve.com

6

Preliminary Design of a Viscosimeter

6.1 Introduction

To begin, note that a viscosity-measuring instrument may be called a *viscometer, viscosimeter, or rheometer;* all these names are used in the technical literature and manufacturers' publications. Most of the material covered in measurement texts focuses on individual *sensors* for measuring some specific quantity such as pressure, temperature, velocity, etc. There are however many "instruments" that might use a combination of sensors and other apparatus to measure some physical quantity of interest. One category of this type would be instruments intended to determine some *physical property* of a solid or fluid material, such as tensile strength, density, thermal conductivity, etc. Such information is vital when we do theoretical calculations that involve that material. In this chapter, you will be familiarized with certain instruments used to measure the *viscosity* of liquids.

6.2 Definition of Viscosity

For our purposes, we use the definition found in most introductory fluid mechanics texts. This definition is given in terms of a *mental experiment* involving the shearing of a liquid film in an apparatus, as shown in Figure 6.1. Viscosity, μ, is defined as

$$\mu \underline{\Delta} \frac{\text{shearing stress}}{\text{rate of shear}} = \frac{\dfrac{F}{A}}{\dfrac{dV}{dy}} = \frac{\dfrac{F}{A}}{\dfrac{V}{r}} \underline{\Delta} \frac{\tau}{\dot{\gamma}} = \frac{\text{psi}}{\dfrac{(\text{in.}/\text{s})}{\text{in.}}} = \frac{\text{Pa}}{\dfrac{(\text{m}/\text{s})}{\text{m}}} \tag{6.1}$$

The *rate of shear* is also called the *rate of change of shearing strain*, with symbol $\dot{\gamma}$. The units of viscosity can of course be simplified by *cancellation* to get psi-s or Pa-s, but this obscures the physical meaning, so I prefer to "leave in the

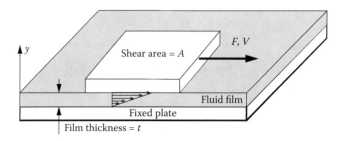

FIGURE 6.1
Definition of viscosity.

details." The quantity defined by Equation 6.1 is also called *dynamic viscosity*, and Greek letter eta (η), rather than mu (μ) is sometimes used for its symbol. Because it appears *naturally* in many fluid mechanics studies, the *kinematic viscosity* υ (nu) is also defined as the ratio (dynamic viscosity)/(fluid density). In the figure, we have *assumed* that the velocity gradient (rate of shear) is linear with distance y, so it can be set equal to V/t. Fluids that behave in this way are called *Newtonian fluids* and this assumption is reasonable for many fluids and applications. Other fluids (*non-Newtonian*) have a nonlinear relation between shearing stress and rate of shear. For a Newtonian fluid, if we made the velocity V in Figure 6.1, say, 1.54 times larger, the shearing stress would also increase by the same factor, as would the force F which causes the plate to move at constant velocity V. For non-Newtonian fluids, an increase in V would cause a change in stress and force, which was *not* proportional to the change in velocity. Thus, such fluids have a viscosity that *changes* when the velocity gradient changes, while Newtonian fluids have a *constant* viscosity, irrespective of the velocity gradient. This linear or nonlinear behavior of the two types of fluids will influence the complexity of differential equations which might be written for systems involving fluid phenomena, the nonlinear (non-Newtonian) model giving rise to less tractable equations. The nonlinearity just described can in general appear in two forms, as shown in Figure 6.2.

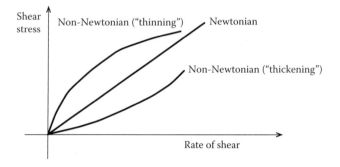

FIGURE 6.2
Newtonian and non-Newtonian fluids.

6.3 Rotational Viscosimeters

Viscosity can be measured with several different types of instruments. Because they can closely approach the ideal flow situation used in the definition of Figure 6.1, *rotational viscosimeters* are widely used. Three versions of this type are shown in Figure 6.3. The concentric-cylinder version is perhaps the most basic since, by making the radial clearance between the rotating and stationary cylinders quite small (typically 1 mm), there is very little variation of shear rate in this gap, approaching a *point measurement* of the viscosity at a specific and known shear rate. This is very important for non-Newtonian fluids because we measure only the rotational speed, ω, and the total torque being exerted to cause this rotation, as a means to get the viscosity value. If there were a significant change in shear rates across the fluid gap, we would be measuring some sort of *average* viscosity rather than the point value for a specific shear rate. This form of rotational viscometer is thus useful for both Newtonian and non-Newtonian fluids. The plate-and-plate version, on the other hand, is most used for Newtonian fluids since there is a very large change in shear rate as we go from the center to the edge of the disk. The cone-and-plate version is designed to give a *constant* shear rate over the entire cone and plate by making the film thickness a linear variation with radius. Since the local velocity also varies with radius in this linear fashion, the shear rate is constant everywhere, and we can thus associate the measured viscosity with a specific shear rate.

To show how viscosity values can be obtained from torque and speed measurements, one must analyze the desired configuration to get the required formulas. All three of the versions shown can be critically analyzed starting with the general Navier–Stokes partial differential equations and making suitable simplifying assumptions. Such analyses have been made and are useful in documenting ranges of applicability and predicting errors that

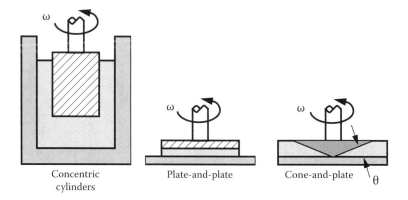

FIGURE 6.3
Three versions of rotational viscosimeters.

might arise in simpler analyses. They are, however, somewhat complicated, so we instead perform only some simple studies, which lead, however, to the formulas actually used in practice. Applying the viscosity definition and neglecting the effects of curvature, the concentric-cylinder instrument is easily analyzed to predict the torque needed to rotate the inner cylinder at a fixed angular velocity, ω. The inner cylinder peripheral surface area is $2\pi RL$ and the shearing velocity is $R\omega$, so the total tangential force is $(2\pi RL)(\mu)(R\omega)/t_r$, where t_r is the radial gap between the two cylinders. The viscosity can then be obtained from the formula

$$\mu = \left(\frac{4t_r}{\pi LD^3} \right) \frac{T}{\omega} \tag{6.2}$$

where T is the torque due to the tangential force. Thus by measuring the inner cylinder diameter D, length L, and radial gap, and by measuring T and ω when the inner cylinder is rotated at constant speed, we can get the value of the viscosity. For the plate-and-plate version, an integration is needed, and using Figure 6.4, we can write

$$\text{incremental tangential force } \underline{\Delta} \, dF = \frac{(2\pi r dr)(r\omega)\mu}{t_a}$$

$$T = \int_0^R \frac{2\pi\omega\mu r^3}{t_a} \, dr \qquad \mu = \left(\frac{32t_a}{\pi D^4} \right) \frac{T}{\omega} \tag{6.3}$$

where t_a is the axial gap between the plates and we assume a Newtonian fluid. For the cone-and-plate version we have

$$dF = \frac{2\pi r \, dr}{\cos\theta} \cdot \frac{r\omega}{r\tan\theta} \cdot \mu \qquad T = \int_0^R \frac{2\pi\omega\mu r^2}{\sin\theta} \, dr \qquad \mu = \left(\frac{12\sin\theta}{\pi D^3} \right) \frac{T}{\omega} \tag{6.4}$$

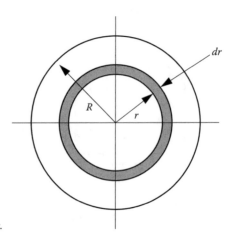

FIGURE 6.4
Analysis sketch for plate-and-plate viscosimeter.

The angle θ is typically small, less than 5°. We see that in each case, the measurement of certain dimensions, rotational speed, and torque, allows us to compute a value for the viscosity.

For the concentric cylinder case, the *bottom* of the cylinder actually corresponds to a *plate-and-plate* configuration and would be an error source when we use Equation 6.2 to find viscosity. We can use Equations 6.2 and 6.3 to compute the torque due to both effects, and thus get a formula for viscosity that includes both. Let us take $L = nD$:

$$T = \frac{\mu\pi\omega D_4}{32\,t_a} + \frac{\mu\pi\omega LD^3}{4\,t_r} = \mu\pi\omega D^4\left(\frac{1}{32\,t_a} + \frac{n}{4\,t_r}\right) \qquad \mu = \frac{1}{\pi D^4}\left(\cfrac{1}{\cfrac{1}{32\,t_a} + \cfrac{n}{4\,t_r}}\right)\frac{T}{\omega} \quad (6.5)$$

Typically, n is about 2 or 3, and t_a can be designed to be much greater than t_r (which is typically about 1 mm). The use of Equation 6.3 for *thick* gaps such as t_a in this case, is open to some question since the velocity gradient may not be uniform, but we will use it as an estimate. Choosing $L = 2D$ and designing to make the error in using Equation 6.2 *alone* in viscosity calculations as 1.0%, we can show that the axial gap should be more than 12.5 times the radial gap (just set the axial gap torque equal to 0.01 times the radial gap torque in Equation 6.5).

6.4 Measurement of Torque

While Figure 6.3 shows the upper member rotating, this is not the only possibility. We could rotate either member, measuring its speed, and hold the other member motionless, reacting its torque into some kind of torque sensor. To get a feel for the torque magnitudes involved, we can assume some typical values for dimensions, viscosity, and speed. Table 6.1 shows viscosity values

TABLE 6.1

Typical Viscosity Values

| Temp, °C | Viscosity, 1e–6 (lb$_f$/ft²)/(ft/s/ft) | | |
	Heavy Oil	Light Oil	Water
0.0	66100.	7380.	36.6
20	9470.	1810.	20.9
40	2320.	647.	13.6
60	812.	299.	9.67
80	371.	164.	7.33
100	200.	102.	5.83

(taken from *Marks' Standard Handbook for Mechanical Engineers*[*]) for a few different liquids. For the concentric cylinder type of instrument the radial gap is typically about 1 mm, but rotary speed and viscosity vary widely, so we must choose some numbers to get a feel for how large the torque might be, and thus what kind of torque sensor might be suitable. Among liquids, water is actually at the low end of the viscosity range, so we choose instead a *machine oil* for our example calculation. The table shows that for *light and heavy* machine oils the viscosity varies widely, and for a given oil, is very temperature sensitive. Let us arbitrarily choose a viscosity value between the *light* and *heavy* types (and at "room" temperature (20°C)) of 5000e−6 psf/((ft/s)/ft). Cylinder dimensions vary among manufacturers, and for a given model, one can select cylinders from a set provided with the instrument. Typical values might be a diameter of 1.0 in. for the inner cylinder, a radial gap of 0.04 in., and a length of 2.5 in. Available rotary speeds cover a wide range; a *midrange* value might be about 100 RPM. Using Equation 6.2 we can now estimate a typical torque value.

$$T = \frac{\mu\omega\pi LD^3}{4t_r} = \frac{(5e-3)\left(\dfrac{100\cdot 2\pi}{60}\right)(\pi)\left(\dfrac{2.5}{12}\right)\left(\dfrac{1.0}{12}\right)^3}{4\left(\dfrac{0.04}{12}\right)}$$

$$= 0.00149 \, \text{ft-lb}_f = 0.00200 \, \text{N-m} \tag{6.6}$$

We see that a very sensitive torque transducer must be used. For typical "top-of-the-line" viscosimeters, torques are measurable in the range 0.2 μ N-m to 150 m N-m.

Some rotational viscosimeters rotate the outer cylinder at the desired speed and constrain rotation of the inner cylinder with a torsion spring of some kind. A given torque will thus cause a certain angular deflection of the spring, which can be measured with an appropriate displacement sensor. One method would mount a translational LVDT[†] (linear variable differential transformer) at the end of a radial arm. For an LVDT with a small full-scale travel (say, 0.005 in.), and an arm length of, say, 3 in., the angular motion would be small enough to avoid binding of the LVDT core in its bore. Figure 6.5 is a schematic sketch of the essential features of a viscosimeter which uses a metal wire in torsion as the elastic torque-sensing element. To get a translational motion of 0.005 in. at a 3 in. radius, the angular displacement would be 0.001667 rad. Let us take the torque computed in Equation 6.6 as the *full-scale* torque of our viscosimeter. We can then find the torsional spring constant needed to give the

---

[*] *Marks' Standard Handbook for Mechanical Engineers*, 9th edn., McGraw-Hill, New York, 1987, pp. 3–37.
[†] E.O. Doebelin, *Measurement Systems*, 5th edn., McGraw-Hill, New York, 2003, pp. 252–262.

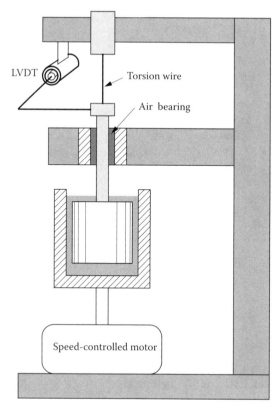

LVDT

Torsion wire

Air bearing

Speed-controlled motor

FIGURE 6.5
Schematic diagram of viscosimeter.

0.005 in. deflection at this torque. For a simple rod of circular cross section as our spring, the torsional spring constant K_s is given by

$$K_s = \frac{\pi r^4 G}{2L} = \frac{0.00149}{0.001667} = 0.894 \, \text{ft-lb}_f/\text{rad} = 10.73 \, \text{in.-lb}_f/\text{rad} \qquad (6.7)$$

where
  r is the wire radius
  L is its length
  G is the torsional elastic modulus of the wire material

Materials used for instrument elastic elements include the Ni-Span-C which we chose in Chapter 4 for the cylinder of the pressure transducer, and it would be suitable for our torsion wire. However, to expand our familiarity with materials we here choose another typical spring material called Elgiloy. This cobalt/chromium/nickel alloy was originally (shortly after World War II) developed for watch springs by the Elgin Watch Company, which had noted corrosion problems with earlier materials. Its G value is 11.2e6, so using

Equation 6.7 we can get a relation between wire radius r and length L for our example torsion spring:

$$\frac{r^4}{L} = \frac{10.73 \cdot 2}{\pi \cdot 11.2e6} = 6.099e - 7 \quad \text{if } L = 2, \text{ then } r = 0.03323 \, \text{in.} \tag{6.8}$$

Available wire diameters can be found* on several Web sites; www.suhm. net/springdesign/materials lists 0.065, 0.067, and 0.072 in. The minimum tensile strength is about 240,000 psi, with a typical design stress given in this Web site as 45% of this value. The small deflection of our spring will probably result in very small stresses:

$$\text{maximum shear stress} = \frac{2T}{\pi r^3} = \frac{2 \cdot 0.00149 \cdot 12}{\pi \cdot 0.0335^3} = 303 \, \text{psi} \tag{6.9}$$

if we decide to use the available 0.067 diameter wire.

6.5 Dynamic Measurements

In addition to the usual steady-state measurements, some studies require measurement of oscillatory angular velocities and torques. Sometimes the oscillations are around zero velocity and other times the oscillations take place around some nonzero average velocity. We can make up a *mass/spring/damper* (J, B, K_s) rotational mechanical model for our viscometer, as shown in Figure 6.6. We assume that the input is an angular velocity, ω_i, applied to the outer cylinder and that the output is the angular rotation θ_o, of the inner cylinder. We can then derive the differential equation relating this input and output, and the associated transfer function $\frac{\theta_o}{\omega_i}(D)$. Note that J is the moment of inertia of the inner cylinder, and that the inertia of the outer cylinder has no effect because we postulate a *motion* ω_i (rather than a torque) as the input. If we were studying the motor control system which produces ω_i as its controlled variable, then the outer cylinder inertia *would* matter. We also neglect any inertial effects associated with the fluid in the gap between the inner and outer cylinders.

FIGURE 6.6
Analysis model of viscosimeter.

---

* www.suhm.net/springdesign/materials

$$(\omega_i - D\theta_o)B - K_s\theta_o = JD^2\theta_o \quad (JD^2 + BD + K_s)\theta_o = B\omega_i$$

$$\frac{\theta_o}{\omega_i}(D) = \frac{K}{\dfrac{D^2}{\omega_n^2} + \dfrac{2\zeta D}{\omega_n} + 1} \tag{6.10}$$

where

$K \triangleq B/K_s$

$\omega_n \triangleq K_s/J$

$\zeta \triangleq B/2(K_sJ)^{0.5}$

While θ_o is the measure of *fluid torque* for a *steady* ω_i, when ω_i is changing there will be some error in taking this angle as proportional to fluid torque and thus the viscosity value will also be incorrect. We will shortly investigate this problem.

The configuration of Figure 6.5 is sometimes called the *Couette* (M. M. Couette, 1888) type of viscosimeter and is considered to be less versatile and accurate than the so-called Searle (Searle, 1912) type, where the outer cylinder is stationary and the inner cylinder is rotated. Figure 6.7a* shows such an instrument with a *plate-and-plate* measurement fixture in place; one can substitute a cone/plate unit or a concentric-cylinder unit, as desired. The Searle instrument has a number of advantages and, due to refinements in several aspects of operation, is the preferred type for the most sophisticated, accurate, and versatile measurements of viscosity and viscoelastic material properties. These advantages of course come at the price of higher cost and complexity. In these instruments, the torque measurement is not made with an elastic element but is instead derived from a *current* measurement of the driving motor. This leads to the need for specially designed motors with integral air bearings to reduce torque measurement errors due to friction. Rotary speed and angular position are measured with shaft-angle encoders.[†] Figure 6.7b[‡]is from a patent assigned to Anton Paar, a large German manufacturer of viscosimeters and rheometers, and shows some air bearing details. This company uses mostly air bearings but has some models that use instrument ball bearings, which are more rugged and cheap, but have higher friction. A table from their catalog gives a useful comparison of these two bearing types; a ball bearing instrument has a torque resolution of 100 nN-m while the air bearing has 0.2 nN-m. Instrument ball bearings have starting torques ranging from 14 to 110 μN-m, depending on size and looseness of fitting.[§] Since the fluid properties to be measured are often quite sensitive to temperature,

[*] TA Instruments, www.tainstruments.com
[†] E.O. Doebelin, *Measurement Systems*, 5th edn., McGraw-Hill, New York, 2003, pp. 327–334.
[‡] Anton Paar GmbH, U.S. patent 6,167,752 (get this patent most easily from Google Patents).
[§] www.myonic.com (A German bearing manufacturer, formerly RMB Bearings).

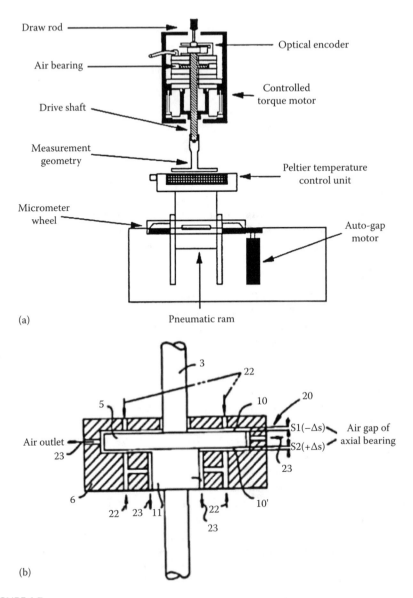

Draw rod

Optical encoder

Air bearing

Controlled
torque motor

Drive shaft

Measurement
geometry

Peltier temperature
control unit

Micrometer
wheel

Auto-gap
motor

(a)   Pneumatic ram

3

22

5

10   20

S1(−Δs)   Air gap of
axial bearing
Air outlet
23

S2(+Δs)

23

6

22   23   11   22   10'

23

(b)

FIGURE 6.7
(a) Schematic diagram of *Searle*-type viscosimeter/rheometer and (b) air bearing of Anton Paar.

feedback-type temperature control is necessary. This is eased by the *nonro-tating* lower member (plate or outer cylinder) since water-jacketing or ther-moelectric heating/cooling requires connections to stationary water and/or electrical facilities. When testing viscoelastic materials using the plate/plate or cone/plate arrangements, it may be necessary to apply a *normal force* to the specimen; the pneumatic ram shown provides this feature. Instruments

that are capable of testing both fluid viscosity and also viscoelastic properties are usually called *rheometers*, rather than viscosimeters. From the above
brief description, it is easy to see that these are generally quite complex and
expensive devices.

Returning to our simple Couette instrument, since Equation 6.10 shows it to
be a standard second-order type, we can apply any of the standard design criteria to aid us in designing it for good dynamic response. The spring constant
and damper coefficient are already fixed to meet the earlier criteria for static
measurement, so the only design variable left to choose is the inertia of the
inner cylinder. For any second-order instrument, fast response requires a large
enough natural frequency, and damping, for most applications, strives for a
damping ratio of about 0.55–0.75, say 0.65. From the definition of the damping
ratio, and given the fixed values of the spring constant and damper coefficient,
there is only one J value that will produce the desired 0.65 damping ratio:

$$B = T/\omega \text{ in Equation 6.6} \qquad \zeta = 0.65 = \frac{B}{2\sqrt{K_s J}} = \frac{0.001707}{2\sqrt{10.73J}} \qquad (6.11)$$

$$J = 1.607\,e{-}7\,\mathrm{lb_f\text{-}in.\text{-}s^2}$$

The inner cylinder has a radius of 0.50 in. and is 2.5 in. long; let us make it of
aluminum and of hollow construction, to minimize inertia. For a thin-walled
cylinder, the inertia may be approximated as the product of the total mass
and the square of the mean radius of the cylinder. Solving for the needed
wall thickness, we find that it must be about 0.000316 in., much too thin for
convenient manufacture, durability, and handling, so designing for optimum damping does not seem to be an option. It is also obvious that since the
instrument will be used with a variety of fluids, achieving optimum damping for the fluid we have used in our calculation would *not* give that result in
general. Thus, we need to explore some other way to get adequate dynamic
response that does not require *perfect* damping.

When we earlier mentioned the desire to work with oscillatory motions, we
were not specific as to the frequency range that might be of interest. Suppose
that our highest frequency will be about 5.0 Hz. Again, for second-order
instruments it is well known that, even with very small damping, if we make
the system natural frequency about 5 times higher than the highest operating frequency, the error in amplitude ratio will about 5%, which is generally
tolerable for most dynamic measurements. Also, for transient inputs, even
though the response to a perfect step input is very oscillatory, response to
more realistic inputs will be quite acceptable if the natural frequency (in
rad/s), times the transient rise time T, is large (say 10 or more) relative to 1.0,
and this result is not very sensitive to the amount of damping. This fact is
easily proven* for *terminated-ramp* inputs and can be easily checked for *any*
input using simulation. Since we want a natural frequency of about 25 Hz

---

* E.O. Doebelin, *Measurement Systems*, 5th edn., McGraw-Hill, New York, 2003, pp. 135–137.

(5 times the highest operating frequency), and the spring stiffness is fixed at 10.73 in.-lb$_f$/rad, we can compute the needed J value for our cylinder, now using a more correct formula for the moment of inertia of a hollow cylinder:

$$\omega_n^2 = \frac{K_s}{J} = (50\pi)^2 = 24674 = \frac{10.73}{J} \qquad J = 0.0004348 = \left(\frac{\pi\rho L}{2}\right)(R_o^4 - R_i^4) \qquad (6.12)$$

Using aluminum, with $R_o = 0.5$ in., $L = 2.5$ in., we find that $R_i = 0$ (*solid* cylinder) gives a J value *smaller* than required for the 5 Hz natural frequency, so we can easily get natural frequencies *higher* than 5 Hz, which is an improvement. We thus choose a reasonable wall thickness, compute the J value for this hollow cylinder, and then find the corresponding natural frequency. A wall thickness of about 0.1 in. is easily manufactured and gives a J value of 3.752e−5 and a natural frequency of 85.1 Hz, so this system would have a "flat" amplitude ratio out to about 17 Hz.

We now have all the numerical values needed to perform a simulation. Earlier we said that using the spring deflection as a measure of the fluid viscous torque is correct for steady input velocities, but not when the velocity is changing. The true fluid torque is of course $B\left(\omega_i - \dfrac{d\theta_o}{dt}\right)$, a quantity which is *not* available in the actual instrument but is easily accessible in our simulation, so we can compare the *measured torque* $K_s\,\theta_o$ with the correct value to see how much error is caused under given measurement situations.

In the simulation of Figure 6.8, at the lower left we provide for a variety of rotational velocities wi; a terminated ramp with adjustable rise time, a sinusoidal variation with adjustable amplitude and frequency, and a ramp of adjustable slope. Other forms are of course easily simulated if the need arises. At the center right, we provide for a time-varying viscosity (if that should be of interest) through the function module $f(u)$. We set $f(u) = 1.0$ when we want the fixed viscosity 3.472e−5. Though not possible in the actual instrument, we can easily get the true fluid torque (trutork) and compare it with the value produced by the actual instrument (mestork) from the spring deflection. The conversion from angular rotation to output voltage in the LVDT is not modeled since we assume it to be "perfect." We also show a signal for the measured viscosity; in the real instrument this would be provided by the data acquisition system, using the LVDT voltage as an input related linearly to spring torque and thus viscosity. Figure 6.9 shows that if we accelerate the input velocity too rapidly (0.01 s) this lightly damped system shows large initial oscillations. A gentler startup (0.10 s) prevents these large vibrations, though of course if we *wait* a few seconds before taking viscosity readings (the usual situation), the results are identical for both cases. While the poor damping is seen to be of little consequence for steady input velocities, it might become important for varying velocities and/or varying viscosity. In Figure 6.10, we see that the performance is essentially perfect for our maximum sinusoidal velocity of 10 rad/s, and frequency of 5.0 Hz.

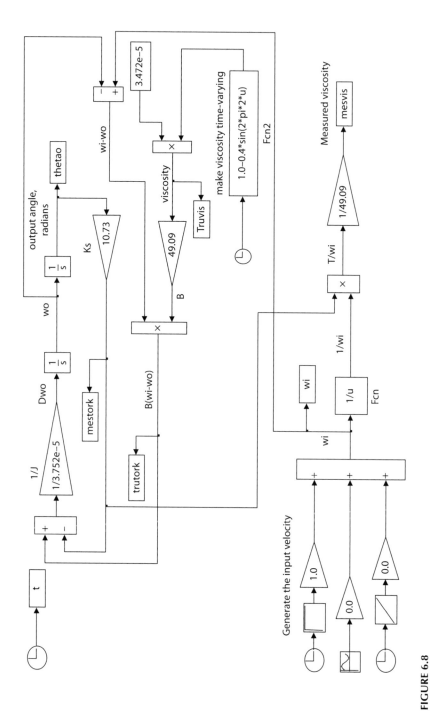

FIGURE 6.8
Simulation of Couette viscosimeter.

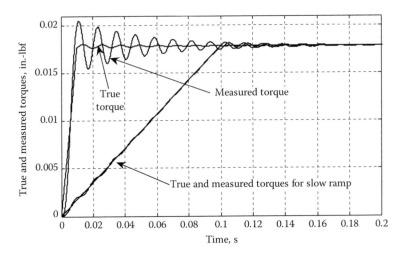

FIGURE 6.9
True and measured torques for slow and fast ramps.

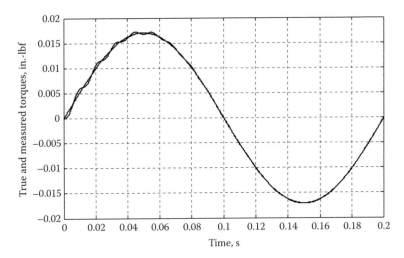

FIGURE 6.10
True and measured torques for 5 Hz sinusoidal velocity.

6.6 Velocity Servos to Drive the Outer Cylinder

The simulation of Figure 6.8 *assumes* that we can perfectly produce input angular velocities of any form we wish. This motion is of course provided by a speed-controlled electric motor system, so we need next to show how such a motor system would *actually* respond to the kinds of commands we have assumed. A typical electric drive of this sort might use a small, brushless DC motor under proportional plus integral control. Brushless DC motor dynamics are essentially the same as those of a brush-type armature-controlled motor, whose analysis is well known, so we assume this simpler system in our analysis and simulation.

While some high-performance rheometers use a brushless DC motor, most employ a *drag-cup** servomotor since its very low inertia allows the fast response (up to about 100 Hz) needed in some measurements on viscoelastic materials. This is an AC two-phase induction motor,[†] a class which includes three types based on the construction of the rotor: *squirrel-cage, solid iron,* and *drag cup*. These all use the same stator arrangement; a two-phase winding excited by alternating voltage of a fixed frequency, most commonly 60 or 400 Hz, although other frequencies are possible. One phase has a fixed-amplitude voltage, say, 110 V, 60 Hz. The other phase is called the *controlled phase* and is supplied with a variable-amplitude voltage of the same frequency, and with a 90° phase shift with respect to the fixed phase. When the controlled phase amplitude is zero, the motor produces no torque; non-zero amplitude produces torque in proportion. To reverse the motor (always necessary in servosystems), the controlled phase voltage must undergo a 180° phase change, say from 90° leading to 90° lagging. In a servosystem the controlled phase is supplied from a simple AC amplifier. Induction motor dynamics are much more complex than those of DC motors, but a linear approximate transfer function[‡]relating shaft rotation angle θ to controlled phase voltage amplitude E has the form $(\theta/E)(s) = K/(s(\tau s + 1))$. While in DC motors the torque is proportional to the current, in AC induction motors such as the drag-cup type, torque is proportional to the square of the current, but we can still use current as the torque measuring scheme. We will not pursue the details of drag-cup motors any further since their application to rheometers apparently requires some computer-aided techniques that are not discussed in the open literature, as seen from the following quote from Barnes and Bell (2003): "The drag-cup motor came into its own when operated by microprocessor electronics capable of dealing with all its quirks and technical difficulties."

The feedback sensor of a velocity servosystem is traditionally a DC tachometer generator, but most recent systems will use an incremental tachometer encoder. Also, the control laws might be implemented digitally, but the sampling rates are often fast enough that the equivalent analog system is a good model for design. We thus analyze a totally analog system model. First we need to model the mechanical system which the motor drives. This consists of the motor inertia rigidly connected to the outer

---

[*] H.A. Barnes and D. Bell, Controlled-stress rotational rheometry: An historical review, *Korea-Aust. Rheol. J.,* 15(4), December 2003, 187–196.

[†] J.E. Gibson and F.B. Tuteur, *Control System Components,* McGraw-Hill, New York, 1958, Chapter 7.

[‡] W.R. Ahrendt and C.J. Savant, *Servomechanism Practice,* McGraw-Hill, New York, 1960, p. 289.

cylinder with this inertia J_{tot} coupled directly to the bottom of the damper in Figure 6.6, giving the model of Figure 6.11. To include these dynamics as part of the motor control system, we need to get the transfer function relating motor (and thus outer cylinder) velocity to motor torque T_m.

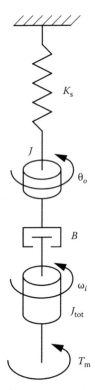

$$T_m - B(\omega_i - \dot{\theta}_o) = J_{tot}\frac{d\omega_i}{dt} \qquad B(\omega_i - \dot{\theta}_o) - K_s\theta_o = J\ddot{\theta}_o \qquad (6.13)$$

Manipulation of these two equations leads to

$$\frac{\omega_i}{T_m}(D) = \frac{JD^2 + BD + K_s}{JJ_{tot}D^3 + (BJ_{tot} + BJ)D^2 + (J_{tot}K_s + B^2)D + BK_s}$$

$$(6.14)$$

The motor inertia can be estimated from catalog data; Doebelin (1998)* lists parameters for a family of 19 such motors with output power ranging from 0.15 to 20 kW. The smallest of these motors should be adequate for our application; it has an inertia of 7.5e−5 in.-lb$_f$-s^2 and a torque constant of 1.95 in.-lb$_f$/A. (The actual procedure for selecting a motor for a motion-control system can be somewhat involved, and usually comes down to making sure that the motor does not *overheat* when subjected to a worst-case duty cycle. We do not here want to get involved in that level of detail, but the text by Doebelin (1998) will supply useful information for those interested.) The inertia of the outer cylinder is calculated for a wall thickness of 0.1 in., a length of 2.5 in., and aluminum as the material; it turns out to be 8.418e−5 in.-lb$_f$-s^2. The output voltage from the PI controller is sent to a *transconductance amplifier*, which puts out a *current* proportional to the input voltage. Such amplifiers are often used to suppress unwanted electrical lags due to inductance in a circuit. The gain of such amplifiers will have dimensions of amps/volt. Motor torque is instantaneously proportional to motor current, with a gain of 1.95 in.-lb$_f$/A.

FIGURE 6.11
Mechanical load for motor control system.

A root locus graph of this control system requires that we state the complete open-loop transfer function, which is a *series combination* of Equation 6.14 and the transfer function of the PI controller, which has the form $K_I[(K_P/K_I)s + 1]/s$. We can use a MATLAB® program to compute some needed numerical values and then implement the root locus procedures which we have used in earlier chapters.

---

* E.O. Doebelin, *System Dynamics*, Marcel Dekker, New York, 1998, pp. 423–426.

```
% crc_book_chap6_couette_servo
% calculates parameters for a velocity control servo
% calculate B for the concentric cylinder configuration
mu=3.472e-5;  L=2.5;  D=1.00;  tr=0.04
mu
L
D
tr
B=mu*pi*L*D^3/(4*tr)
% compute J for inner cylinder
Do=1.0; Di=0.80; Li=2.5; rho=0.1/386
Do
Di
Li
rho
J=(pi*L*rho/2)*((Do/2)^4-(Di/2)^4)
% compute Jo for outer cylinder
di=Do+2*tr;  do=di+0.2;  Lo=2.5;
Jo=(pi*Lo*rho/2)*((do/2)^4-(di/2)^4)
% compute torsion spring constant
rs=0.0335;  Ls=2.065;  G=11.2e6;
Ks=pi*rs^4*G/(2*Ls)
% compute damping ratio of inner cylinder system
z=B/(2*sqrt(Ks*J))
% give the motor inertia
Jm=7.5e-5
% compute coefficients in mechanical system transfer
function
denm=[J*(Jo+Jm) B*(Jo+Jm) Ks*(Jo+Jm)+B^2 B*Ks]
numm=[J B Ks]
% get roots, natural frequencies, and damping ratios
% of system dynamics
denmrts=roots(denm)
[wn,z]=damp(denm)
nummrts=roots(numm)
[wn1,z1]=damp(numm)
sys=tf(numm,denm)
% put transfer functions in standard form
sys1=tf(numm/numm(3),denm/denm(4))
K=numm(3)/denm(4)
% now get servo system root locus
% state PI controller transfer function
numc=[1.5 70]
denc=[1 0]
% get complete open-loop transfer function
[nums,dens]=series(numc,denc,numm,denm)
% plot root locus and set a trial gain
rlocus(nums,dens);grid;pause %you may want to zoom the graph
[k,poles]=rlocfind(nums,dens)
```

Let us look at some of the results of this program run. *For only the mechanical load*, the open-loop zeros are obtained from the statements:

```
nummrts=roots(numm)
[wn1,z1]=damp(numm)
```

with the results:

```
nummrts = -22.6997e+000  + 534.1426e+000i
           -22.6997e+000 -534.1426e+000i

wn1 = 534.6248e+000
      534.6248e+000

z1 = 42.4592e-003
     42.4592e-003
```

The open-loop poles are

```
denmrts = -17.3484e+000  + 534.4506e+000i
          -17.3484e+000 -534.4506e+000i
          -10.7027e+000

wn = 10.7027e+000
     534.7321e+000
     534.7321e+000

z = 1.0000e+000
    32.4431e-003
    32.4431e-003
```

Note that the pairs of complex poles and zeros are very close to each other, which can also be seen graphically on the root locus plot of Figure 6.12a. When poles and zeros are this close to each other, the root locus which starts at one of these poles will almost certainly end on the nearby zero, as we see on the graph. Also, when we set the loop gain, the closed-loop root must fall on this very short path, so its value has little dependence on loop gain. Also, the closed-loop damping *must* be very close to the open-loop values of 0.032 and 0.042, which are very poor (small). What is not obvious, but can be shown,* is that while these poorly damped roots definitely contribute a term to the solution of the differential equation, the *coefficient* of the poorly damped term is so small relative to the other terms in the solution, that its effect on the total solution is usually negligible. If the open-loop pole and zero were not just *close* but actually *identical* (a design trick called *cancellation compensation*), then these dynamics *disappear* from the closed-loop solution. That is, the solution does *not* contain this poorly damped term. If perfect *cancellation* has this effect, then it is plausible that *nearly perfect* cancellation would greatly reduce the influence of these poorly damped terms. This plausibility is converted to a *proof* by numerical examples given in Doebelin.

---

* E.O. Doebelin, *Control System Principles and Design*, Wiley, New York, 1985, pp. 320–321.

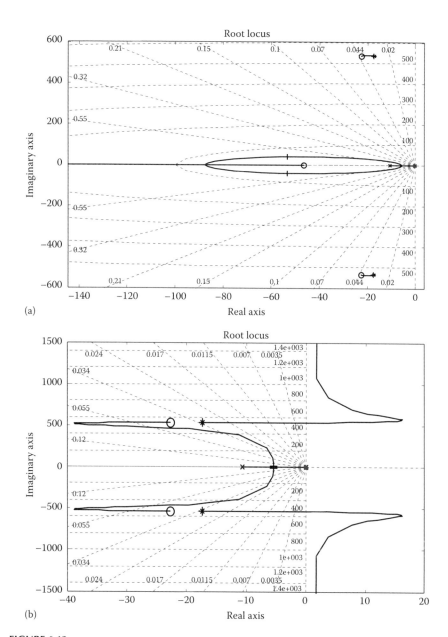

FIGURE 6.12
Root locus of velocity servo system: (a) PI control and (b) integral-only control.

The root locus of Figure 6.12a has some other features that are of general interest. Of the four paths that start at the open-loop poles, there are two that begin at the real poles at 0 and −10.7. As loop gain is increased, the closed-loop roots move along the real axis until they become a repeated root at about −7. They then split up, with one traveling along the upper, and one along the lower part of the "oval" shaped path. They then come together as another repeated root at about –90, whereupon they split up again, each traveling along the real axis, one toward the zero located at –70/1.5 (contributed by the PI controller) and the other heading toward minus infinity along the negative real axis. Note that *none* of these paths ever crosses over into the unstable (right half) of the complex plane, indicating that this feedback system will *not* go unstable, even for very high loop gain! If we build this system and raise the gain too much, it *will* go unstable, as we generally believe about all feedback systems. The apparent discrepancy is that our *model* (as is true of *any* math model) has *neglected* some dynamic features of the real system; *no* model can capture perfectly the behavior of any real system. The neglected dynamics are usually some effect that we have considered *so much faster* than the included dynamics, that we have chosen to not take it into account. There are a number of such effects present in any real system. In the present example, the transconductance amplifier does not really convert an input voltage *instantly* into an output current as we have assumed; there is a small lag. Also, every spring/mass system has an infinite number of natural frequencies, not just one as we have assumed.

When loop gain is increased, the system generally gets faster and faster, and at some point the system speed approaches that of the neglected dynamics, and they are no longer negligible. Thus our simplified model will produce usable predictions so long as we do not raise the loop gain *too much*. When we get to the *experimental development stage* of system design and manufacture, any defects in our modeling will become apparent and adjustments can be made. Of course, designers having long experience with a particular class of systems usually know what can be neglected and what must be included in the early analytic models and they are thus less often *surprised* when the first prototypes are experimentally evaluated. Figure 6.13 shows a zoomed root locus plot (of Figure 6.12a) that is used with the MATLAB gain-setting procedure to place some closed-loop roots at desired locations. Since the root locus paths associated with the lightly damped mechanical dynamics offer little choice of behavior, we instead chose the gain to place a pair of roots at a "good" location on the oval paths. This choice is of course tentative, but we strive for a good damping ratio and a modest open-loop gain, since we saw above that the model predictions for high gain may not be very reliable with regard to the actual (not model) system behavior. The results of this choice are printed out by MATLAB.

```
selected _ point = -49.4076e+000 + 39.1304e+000i
k = 9.4056e-003

poles = -17.4541e+000 + 535.3355e+000i
        -17.4541e+000 - 535.3355e+000i
        -49.5624e+000 + 40.7976e+000i
        -49.5624e+000 - 40.7976e+000i
```

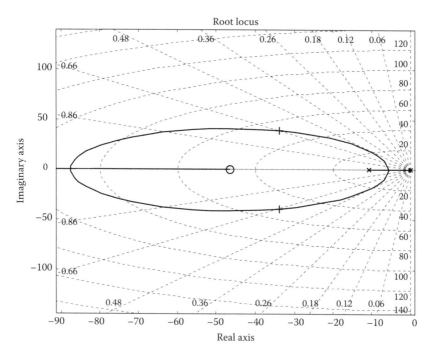

FIGURE 6.13
Zoomed root locus used to set tentative loop gain.

The *selected point* is the position on the oval path where I set the crosshairs; the gain k is the gain which places a root in that location. The *poles* are the four roots of the closed-loop differential equation which result from that gain setting. Two of these are located where I chose them, the other two must be accepted wherever they occur, they cannot be directly chosen. Note that the *selected point* does not agree perfectly with the corresponding poles. This is due to the screen resolution when you position the crosshairs.

While the root locus graphs give us some insight into the influence of the mechanical system dynamics and controller dynamics on the closed-loop behavior, we always use simulation to do the final gain setting once we have chosen the other parameters. In fact, the numbers I used above for the PI controller were not just *guesses*; I did a simulation *first* and used trial and error to choose the P and I values and set the loop gain. This trial and error starts with using a *pure* integral control, which gives the root locus of Figure 6.12b, which is quite different from that of Figure 6.12a (which is PI control) and shows instability for high loop gain. (We *do not* start with pure proportional control because we know it allows steady-state errors.) Using small amounts of pure integral control with step inputs quickly shows that the response is *very* slow, so we increase the amount of integral, which gives the desired speed up, but causes too much oscillation. At this point, we add in some proportional control and increase it until we get *both* good speed and acceptable oscillation (the amount of integral may also need adjustment). This is how the *amounts* of proportional (1.5) and integral (70) control were arrived at.

We see again that using simulation we can design systems with only a modest control theory background. The simulation for this system is shown in Figure 6.14. The command *wv* to this servosystem (generically called the *desired value*) is the desired angular velocity of the outer cylinder which we called *wi* in our earlier studies. This command could take various forms; I provide a selection consisting of a perfect step input, a sinusoidal wave, and a series of terminated ramps (seen in the lookup table display). The transconductance amplifier is shown with a gain of 0.1 amp/V and the motor has a torque constant of 1.95 in.-lb$_f$/A. Motor torque is the input to the mechanical dynamics relating *Tm* to the angular velocity *wc* (the *controlled variable*) of the outer cylinder. This motion causes the motion of the inner cylinder, but note that these dynamics (*wc* to *thetao*) are *outside* the feedback loop and do not affect its behavior directly. (The dynamics from *Tm* to *wc*, of course, do include *all* the mechanical elements in the system of Figure 6.11.) A velocity sensor with a gain of 0.1 V/(rad/s) measures the outer cylinder angular velocity.

The input from the lookup table, which is the series of terminated-ramp commands, can be thought of as a test for Newtonian behavior of the fluid since it subjects it *stepwise* to a range of shear rates, so that we can see if the measured viscosity is the same for all of them. Our fluid model is of course perfectly Newtonian, so we should get the same viscosity no matter what shear rate is applied. This would require that the spring deflection *thetao* be directly proportional to the outer cylinder angular velocity, which we see to be true in Figure 6.15b. This graph is a *cross-plot* of the time history data in Figure 6.15a. Both these graphs show very good servosystem performance with no visible overshooting or oscillating. For perfect step changes in commanded velocity, *wc* responds well but *thetao* has large and long-lasting oscillations which are unacceptable since we use this signal to compute the viscosity, showing again the utility of terminated-ramp inputs. Sinusoidal commands of 1.0 and 5.0 Hz show *perfect* response at 1.0 Hz and a slight *peaking* at 5.0 Hz, the highest frequency the system requires (see Figure 6.16). Since the actual velocity *wc* is *measured* in the practical system, and this is the value used in computing viscosity, the fact that *wc* at the higher frequency is a little larger than the *wv* command is of no consequence. One can use the Simulink *chirp* module to generate a frequency sweep signal for *wv* and thus examine the amplitude response (you cannot get the phase angle this way) of the system. I did this using a sweep from 1.0 to 30.0 Hz in 10 s and it showed only the slight *resonance* that was observed in Figure 6.16 for the 5.0 Hz signal, thus the frequency response is acceptable. Finally, the loop gain of the system shown in Figure 6.14 is (70)(0.10)(1.95)(586.7)(0.10) = 801. The loop gain corresponding to the root locus study where I somewhat arbitrarily chose a root location was (0.009406)(586.7)(70) = 386. I reran the simulation using this lower value of loop gain and got results very similar to those for the 801 gain, so there is quite a bit of flexibility in choosing this system parameter. This is

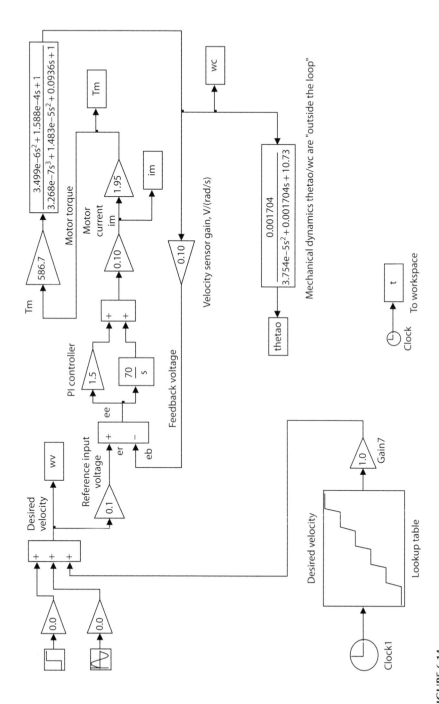

FIGURE 6.14
Simulation of viscometer velocity servodrive.

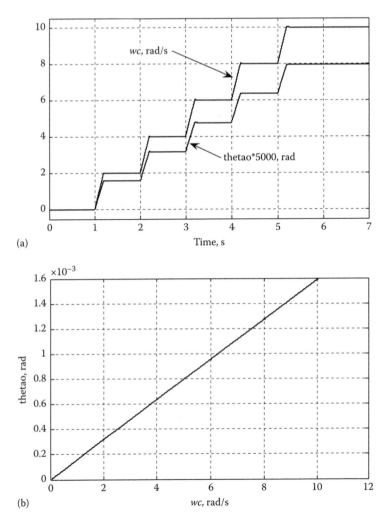

FIGURE 6.15
Performance of velocity servodrive.

also true for the PI controller parameters; the 1.5 and 70 are not *magic numbers*, good response can be achieved with other combinations. Fast simulation software allows us to quickly explore any parameter combinations that are of interest.

The velocity control system we have been studying was originally introduced in this chapter as a drive for a Couette viscosimeter, but the identical scheme will serve also to drive the *inner* cylinder in the Searle configuration. Recall that here no *mechanical* torque sensor is needed; we instead measure the motor current (which is proportional to torque) and use this, together with the measured rotational speed, to compute viscosity. As mentioned earlier, we can still use a brushless DC motor but now (for the highest performance)

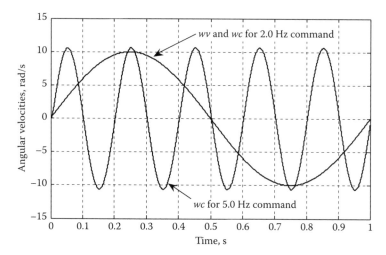

FIGURE 6.16
Sinusoidal response of velocity servodrive.

it must be provided with (essentially frictionless) *air bearings* since at steady speed, the motor torque is the sum of bearing friction (which causes error) and viscous fluid drag which we wish to measure. (Low-friction instrument ball bearings can be used if we accept poorer torque measurement resolution.) Even with air bearings, if the velocity is not steady, there will be some error due to inertia effects. However we have seen, for example in Figure 6.10, that these inertia effects may be quite small relative to the viscous drag, since the inertia elements involved are numerically small.

6.7 Calibration

The theoretical formulas we have used to relate viscosity to apparatus dimensions and measurements of torque and speed are, as usual, *models* of the real physical behavior of our instrument, and are thus not exact. In most *machine* design, theoretical predictions with relatively large errors are regularly accepted because it is routine to provide rather large *safety factors* to allow for the uncertainty of our calculations. In *instrument* design this luxury is generally not available since measuring devices are required to have accuracies of a few percent or less. Thus we *strive* for accurate theoretical results but always, if possible, *calibrate** the instrument before trying to actually make measurements with it. Calibration is essentially a refined form of measurement; we compare our instrument's readings with those of a more accurate

---

* E.O. Doebelin, *Measurement Systems*, 5th edn., McGraw-Hill, New York, 2003, pp. 41–76.

standard. If possible, the standard should be about 10 times as accurate as the instrument to be calibrated, but a 4:1 ratio[*] is acceptable if that is necessary or desirable.

A convenient calibration method for viscosimeters is based on the availability of *standard viscosity fluids.*[†] The *ultimate* standard fluid is freshly distilled water, which is internationally accepted to have a *kinematic viscosity* of 1.0038 mm²/s at temperature 20.00°C. Recall that kinematic viscosity is the *dynamic* viscosity (defined in Figure 6.1) divided by the fluid density. *Capillary* viscosimeters are used to *transfer* the known viscosity of water to other fluids. A capillary viscosimeter is simply a length of glass tubing with a length-to-diameter ratio of at least 30 (preferable 50) oriented vertically, so that gravity will cause the fluid to flow from a reservoir, downward through the tubing into a catch vessel which is used to determine the volume of fluid that is passed through over an accurately timed interval. Such devices are useful only for Newtonian fluids, but are much more accurate than even the most sophisticated rotational viscosimeters.[‡] For capillary viscosimeters, the kinematic viscosity v is given by

$$v = Ct \qquad\qquad (6.15)$$

where
   t is the time for a given volume of liquid to flow through the capillary
   C is the viscosimeter constant

Typical capillary lengths are 400 mm, with the inside diameter chosen so the flow time t is greater than 500 s (Daborn, 1985, p. 226). If we use a capillary viscosimeter with freshly distilled water at 20.00°C, and catch an accurately measured volume over an accurately measured time interval, the instrument's calibration coefficient C can be calculated from Equation 6.15. If we then use a different fluid, but the *same* caught volume, the fluid's kinematic viscosity, at the temperature of the experiment, can be computed from Equation 6.15. Measurement of fluid density then provides a number for the dynamic viscosity. Since water has a rather low viscosity, a capillary size (inside diameter) that is appropriate for water is much too small for, say, oil. According to Daborn (1985), the NPL (National Physical Laboratory, the United Kingdom's National Metrology Institute) has calibrated a series of capillaries (about 10) to cover kinematic viscosities from that of water up to about 100,000 times that of water. A series of 11 oils, with kinematic viscosities ranging from about 1.7–78000 mm²/s is available, with corresponding uncertainties from 0.02% to

---

[*] E.O. Doebelin, *Measurement Systems*, 5th edn., McGraw-Hill, New York, 2003, p. 42.

[†] J.E. Daborn, The NPL reference oils for viscosity, *Measurement and Control*, Vol. 18, July/August 1985, pp. 226–232.

[‡] G. Schramm, *A Practical Approach to Rheology and Rheometry*, Haake GmbH, Karlsruhe, Germany, 1994, p. 153.

0.23%. (When I visited the NPL Web site in 2008, I could not find any reference to these standard oils. Perhaps the lab has discontinued this service, as part of a *privatization* program which substitutes private *for-profit* labs for government facilities. In any case, Daborn (1985) is still a useful resource in that it provides numerical values for "what can be done.") I also visited the NIST Web site and also found no reference to *standard viscosity fluids*. Fortunately, commercial sources of such fluids do exist, in the United States, for example, Brookfield Engineering Laboratories, Inc.* provides a wide range of fluids, with viscosity accuracy of about ±1% of the viscosity value. The use of such fluids for the calibration of a concentric-cylinder rotational viscometer that was designed and built as part of an MSc thesis is discussed in a 1994 paper by Jimenez and Kostic.[†] The thesis by Jimenez, "Design, development, and fabrication of a computerized viscometer/rheometer,[‡] gives even more useful details, including viscosity measurements over a range of shear rates for two Newtonian fluids and one non-Newtonian (see Figure 6.17). The two *standard* Newtonian fluids have nominal viscosities of 51 and 445 mPa-s and the measurements over a range of shear rates seem to agree pretty well with these numbers. The measured viscosities were computed from measured values of torque and speed, together with known parameters of the viscometer, so the agreement with the given standard viscosity indicates that the theoretical formulas for this instrument are quite accurate. The standard non-Newtonian fluid shows, as expected, a large change in measured viscosity as the shear rate is changed.

High-performance viscometers/rheometers are capable of greater accuracy than provided by standard viscosity fluids, so other means of calibration are usually used. Page 156 of *A Practical Approach to Rheology and Rheometry* discusses one such technique, as shown in Figure 6.18a. *Frictionless* (air bearing) pulleys are arranged to transfer the motor torque through thin strings to an accurate standard mass resting on an electronic balance. Such balances[§] can have seven-digit displays with resolution as fine as 2 μg. When the motor torque is set at zero by the motor control system, the balance reads the gravity force on the standard mass, whose weight (gravity force) is larger than the maximum required. When motor torque is set at some nonzero value by increasing motor current (which is accurately measured), the motor stalls as it exerts its torque, putting tension in the strings and reducing the force read by the balance by an amount equal to the string tension. Accurate measurements of pulley radii then allow calculation of motor torque. The reference assumes the motor to be the drag-cup type, where motor torque is nominally proportional to current squared. A log–log plot of motor torque versus motor current is used to fit the best straight line and determine the actual relation

---

* www.brookfieldengineering.com
[†] J.A. Jimenez and M. Kostic, A novel computerized viscometer/rheometer, *Rev. Sci. Instrum.* 65(1), January 1994, 229–241.
[‡] J.A. Jimenez, Design, development, and fabrication of a computerized viscometer/rheometer, Department of Mechanical Engineering, Northern Illinois University, DeKalb, IL, 1992.
[§] E.O. Doebelin, *Measurement Systems*, 5th edn., McGraw-Hill, New York, 2003, p. 439.

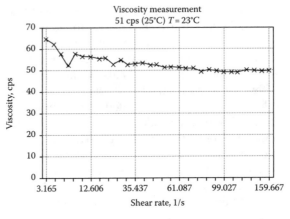

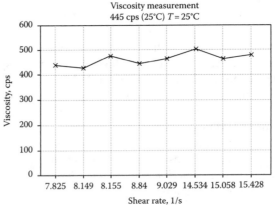

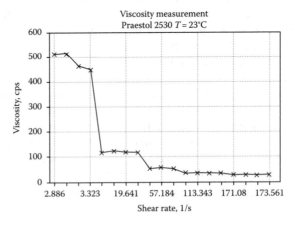

FIGURE 6.17
Measured viscosity values for a range of shear rates.

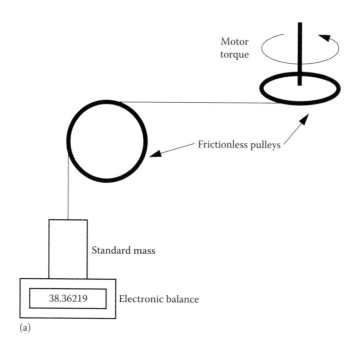

Standard mass

| 38.36219 | Electronic balance

(a)

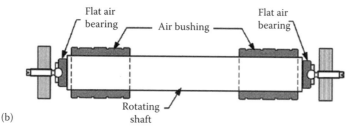

(b)

FIGURE 6.18
(a) Torque calibration scheme and (b) frictionless pulley arrangement.

of current and torque, which in the reference example gave a power of 2.14 rather than the theoretical 2.0. This torque/current relation is then entered into the system computer to be used for converting current measurement to torque, as needed in any viscosity or rheology calculations. When I talked to Haake personnel in 2008 (the reference is dated 1994) they told me that they no longer use the electronic balance, but instead use a series of standard masses attached to the strings to develop the needed torque/current graph. (Haake GmbH had at that time become part of Thermo Fisher Scientific*). Figure 6.18b shows one way of implementing frictionless pulleys using air bearings from NewWay Air Bearings.†

---

* www.thermo.com
† www.newwayairbearings.com

6.8 Corrections to the Simplified Theory

While Equation 6.2 is useful for simple preliminary design, more rigorous analyses and operating experience provide some corrections which get the predictions closer to actual behavior. In Equation 6.2, we need to correct both the diameter D and the length L used to compute viscosity from the measured torque and speed. Equation 6.2 assumes that we can use the definition of viscosity, which is in terms of a purely *translatory* motion, for the concentric-cylinder viscometer, where the fluid has a rotary motion. The rotary case *can* be analyzed, but requires that one assume some fluid model which relates shearing stress and shearing strain rate. The simplest model is of course the Newtonian one which assumes that stress and strain rate are linearly related. In Figure 6.19, the inner cylinder is stationary and the outer one rotates at a constant angular velocity ω_2. We want to compute the shear rate in the fluid gap at any radius r. The radial velocity gradient dv/dr is obtained by differentiating $v = r\omega_r$ with respect to r. Note that the local fluid angular velocity is ω_r, and goes from zero at the inner cylinder surface to ω_2 at the outer cylinder. Carrying out the derivative, we get

$$\frac{dv}{dr} = r\frac{d\omega_r}{dr} + \omega_r \tag{6.16}$$

To get the shear rate, we need to properly interpret this result. If the fluid body were a *rigid solid*, the velocity gradient would just be the angular velocity; that is, the linear velocity is proportional to the fixed angular velocity. In our fluid, *this* change of velocity with radius does *not* cause a shearing of the fluid, so the term ω_r at the right in Equation 6.16 is ignored, and the shear rate is simply the first term, $r\dfrac{d\omega_r}{dr}$. Using the definition of viscosity as shear stress divided by shear rate, and applying this at the location r, we can get an equation for $d\omega/dr$, and integrate this between R_1 and r to get ω_r as a function of r.

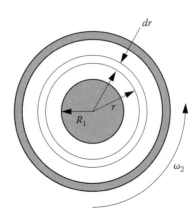

FIGURE 6.19
Analysis sketch for fluid angular velocity.

$$\mu = \frac{\tau}{\dot{\gamma}} = \frac{T/r}{2\pi r L} = \frac{d\omega_r}{dr} = \frac{T}{2\pi r^3 \mu L} \tag{6.17}$$

$$\int_0^{\omega_r} d\omega_r = \frac{T}{2\pi\mu L} \int_{R_1}^r \frac{1}{r^3} dr \tag{6.18}$$

$$\omega_r = \frac{T}{4\pi L\mu} \left(\frac{1}{R_1^2} - \frac{1}{r^2} \right) \tag{6.19}$$

If we let $r = R_2$, the inside radius of the outer cylinder, we get

$$\omega_2 = \frac{T}{4\pi\mu L} \left(\frac{R_2^2 - R_1^2}{R_1^2 R_2^2} \right) \tag{6.20}$$

If we solve this formula for viscosity, it becomes a more correct way to compute viscosity from measurements of torque and angular velocity than our earlier Equation 6.2, which neglected the effects of curvature. For comparison, we can rewrite Equation 6.2 as

$$\mu \approx \frac{4(R_2 - R_1)}{8\pi L R_1^3} \frac{T}{\omega} \quad \text{whereas the more correct version (Equation 6.20) is}$$

$$\mu = \frac{(R_2 + R_1)(R_2 - R_1)}{4\pi L R_1^2 R_2^2} \frac{T}{\omega} \tag{6.21}$$

The error in the approximation is the difference between the correct term $(R_2 + R_1)/R_1^2 R_2^2$ and its approximation $2/R_1^3$. If we let $R_2 = kR_1$, then $(R_2 + R_1)/R_1^2 R_2^2$ becomes $2/kR_1^3$ and the error in μ is a multiplying factor $1/k$. For our earlier typical dimensions ($R_1 = 0.50$, $R_2 = 0.54$), $k = 1.08$, and the multiplying factor is 0.9259. To get an expression for the shear rate at any radius r we first differentiate Equation 6.19 and then multiply this by r:

$$\frac{d\omega_r}{dr} = \frac{T}{2\pi\mu L} \cdot \frac{1}{r^3} \quad \text{shear rate} = \dot{\gamma}_r = r \frac{d\omega_r}{dr} = \frac{T}{2\pi\mu L} \cdot \frac{1}{r^2} = \frac{2(R_1 R_2)^2}{(r^2)(R_2^2 - R_1^2)} \cdot \omega_2 \tag{6.22}$$

The shear rates at the inner and outer cylinders are then

$$\dot{\gamma}_1 = \frac{2R_2^2}{R_2^2 - R_1^2} \cdot \omega_2 \qquad \dot{\gamma}_2 = \frac{2R_1^2}{R_2^2 - R_1^2} \cdot \omega_2 \tag{6.23}$$

For the Searle type of viscosimeter the outer cylinder is stationary and the inner cylinder rotates. Thus the fluid angular velocity now goes from ω_1 at the inner cylinder to zero at the outer cylinder. Whereas the rate of change of fluid angular velocity with respect to radius r was positive for the Couette type, it is now negative. However, for practical instruments (where the gap is very small), the shear *rate* is well estimated from the above formulas for either case, so we will not pursue a detailed analysis of the Searle arrangement. Of course, when computing the shear *stress* from a measured torque, since both types measure torque on the inner cylinder, we use *its* surface area in this calculation.

6.9 Non-Newtonian Fluids

When a fluid does not have its shear rate and shear stress related by a fixed proportionality constant, some other mathematical model must be formulated to try to fit the observed behavior. Of the various models[*] that have been invented and used, the so-called *power law* is perhaps most common and useful. Here the shear rate and shear stress are related as follows:

$$\text{shear rate} = \dot{\gamma} = k_1(\text{shear stress})^n = k_1 \tau^n \qquad (6.24)$$

where k_1 and n are constants chosen to best fit experimental data. (Some references use a different form of the power law: $\tau = k\dot{\gamma}^n$.) Newtonian fluids of course have $n = 1.0$, while *thinning* fluids have $n < 1$ and *thickening* fluids have $n > 1$. The value of n is mainly in the range 0.1–3.0, with most fluids being of the thinning type. Thickening behavior however, is found, for example, in polymer melts, which are of great economic significance. Thinning fluids include human blood ($n = 0.29$) and yogurt ($n = 0.1$). Actually, any model, including the Newtonian and power law, will only fit the data accurately within a certain range of shear rates, perhaps a range of 100 to 1 or 1000 to 1. Outside this range, the behavior deviates from the assumed model. If the fluid *application* is restricted to a modest range of shear rates, one of these simple models may be quite useful in predicting the behavior of some fluid machine or process. For thinning fluids (the most common kind), there is often a range of very low shear rates where the viscosity is nearly constant (Newtonian behavior) and another range of very high shear rates where behavior is again Newtonian (but with a different viscosity than for the low range). Between these two regimes, the behavior follows a power law. Figure 6.20 shows the effect of varying n over the range 0.20–3.0. When the fluid model is the power law form, we need to reanalyze the situation to get a relation analogous to Equation 6.19. We are particularly interested in the shear rate because non-Newtonian fluids do not really have a *viscosity* (as a given

---

[*] C.W. Macosko, *Rheology*, VCH Publishers, New York, 1994, p. 85.

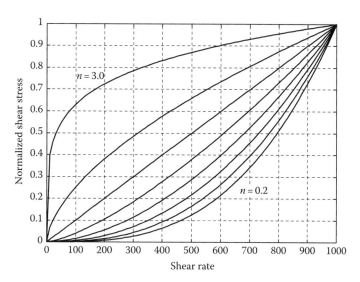

FIGURE 6.20
Power-law model for non-Newtonian fluids.

number) but rather are described by a graph of shear stress (obtained from the torque measurement) versus shear rate (obtained from the velocity measurement and instrument dimensions).

To find the shear rate, we use the power law and the definition of shear rate:

$$\text{shear rate} = r \cdot \frac{d\omega_r}{dr} = k_1 \left(\frac{T}{2\pi r^2 L} \right)^n \qquad \frac{d\omega_r}{dr} = k_1 \left(\frac{T}{2\pi L} \right)^n \frac{1}{r^{2n+1}} \qquad (6.25)$$

Integrating this equation leads to

$$\omega_r = k_1 \left[\frac{T}{2\pi L} \right]^n \left[\frac{1}{2n} \left(\frac{1}{R_1^{2n}} - \frac{1}{r^{2n}} \right) \right] \qquad (6.26)$$

Evaluating this at $r = R_2$ gives

$$\omega_2 = k_1 \left(\frac{T}{2\pi L} \right)^n \left(\frac{1}{2n} \left(\frac{1}{R_1^{2n}} - \frac{1}{R_2^{2n}} \right) \right) \qquad (6.27)$$

Since $k_1 (T/2\pi L)^n = r^{2n} \dot{\gamma}_r$, Equation 6.27 becomes

$$\dot{\gamma}_r = 2n\omega_2 \cdot \left(\frac{1}{r^{2n}} \right) \left(\frac{R_1^{2n} R_2^{2n}}{R_2^{2n} - R_1^{2n}} \right) \qquad (6.28)$$

Returning to Equations 6.23, our earlier simple Newtonian analysis (Equation 6.2) gives *the same* shear rate at both locations 1 and 2. For our assumed dimensions, we can compute how much error is caused by our assumption. Assume, as for a Couette viscosimeter, that the inner cylinder is stationary and the outer rotates at a constant angular velocity ω and that we use our earlier dimensions ($D = 1.0\,\text{in.}$, $L = 2.5\,\text{in.}$, and $t_r = 0.04\,\text{in.}$). Our simple analysis says that the outer cylinder shear rate is $0.54\omega/0.04 = 13.5\,\omega$. The more rigorous analysis gives instead the value $[2(0.54)^2/((0.54)^2 - (0.50)^2)](\omega) = 14.02\omega$, and for the inner cylinder, 12.02ω. We could, of course, decide to use the *average* radius in Equation 6.2, and would then get a shear rate of $(0.52/0.04)\omega = 13.00\omega$.

We also need to correct the shear stress value. For a constant-speed rotation of the outer cylinder, the torque at *any* radius in the fluid film or at the cylinder walls will be the same. This is because, since there is *no* acceleration, there will be no inertial effects on any "cylindrical shell" of fluid within the film. That is, the sum of the torques exerted by fluid viscosity on the inner surface of a *shell* of fluid and on the outer surface must be zero, thus the two torques must be equal and opposite. While the *torques* are the same, the shear *stresses* will be different at different radii because the shear *area* is different. If we measure torque on the stationary inner cylinder (the usual case), then the shear stress there will be the torque divided by the area, divided by R_1. In Equation 6.2, the meaning of D was ambiguous since we there assumed the gap to be very small, making the inner and outer cylinders of "equal" diameter. When we *use* Equation 6.22 we must of course choose an actual number for D. Torque is usually measured on the inner cylinder, so the shearing stress there is (torque/R_1)/(shear area) $= T/(2\pi R_1^2 L)$. If in Equation 6.2 we choose to use the value $D = 2R_1$, then we of course get the same shear stress for both analysis methods. At this point we can replace Equation 6.2 by a more correct one:

$$\mu = \frac{\dfrac{T}{2\pi R_1^2 L}}{\dfrac{2R_2^2}{R_2^2 - R_1^2}\omega} = \frac{R_2^2 - R_1^2}{4\pi R_1^2 R_2^2}\left[\frac{T}{\omega}\right] = \left(\frac{1}{R_1^2} - \frac{1}{R_2^2}\right)\frac{1}{4\pi L\omega} = \frac{0.04541}{L}\left[\frac{T}{\omega}\right] \qquad (6.29)$$

In Equation 6.2, if we were to take D to be the *average* diameter, we get the constant 0.04528.

6.10 The Concept of the Representative Radius

Some concentric-cylinder viscometers are designed on the basis of German industry standards DIN53018 and 53019, as discussed in the literature.[*] These standards use a so-called *representative radius*, defined as the radius where a *representative shear stress* exists. This shear stress is defined as the

---

[*] J.A. Jimeniz and M. Kostic, A novel computerized viscometer/rheometer, *Rev. Sci. Instrum.* 65(1), January 1994, pp. 229–241.

average of the shear stresses at the inner and outer cylinder surfaces. The standards state that the shear rate at the representative radius is virtually independent of the fluid type (Newtonian or non-Newtonian). Using this concept, the standards develop a formula for computing viscosity from the measured angular velocity and torque:

$$\mu = \frac{R_2^2 - R_1^2}{4\pi R_2^2 R_1^2 L_{eff}} \frac{T}{\omega} = \frac{0.04541}{L_{eff}} \frac{T}{\omega} \quad \text{for our dimensions} \qquad (6.30)$$

We see that Equation 6.30 (which uses the *representative radius* concept) is exactly the same as Equation 6.29 (which does not), except for the use of the *effective length* L_{eff} (which we will shortly discuss). In fact, the formula relating μ to T and ω would be the same *no matter what* we use for the radius r. This is because the formulas for shearing stress τ and rate of shear $\dot{\gamma}$ contain r is in such a way that r *cancels*. Once we see this it is natural to ask, "what is the concept of representative parameters good for if it has *no effect* on the measured value of viscosity?" It turns out that the concept *is* useful, but mainly for *non-Newtonian* fluids. When studying such fluids, the usual procedure is *not* to calculate μ, but to instead measure shearing stress τ and rate of shear $\dot{\gamma}$ *separately*, and make a plot of τ versus $\dot{\gamma}$. For non-Newtonian fluids the *purpose* of such a graph is to find an empirical *model* which relates τ and $\dot{\gamma}$ for the fluid under study. While such models could take various forms, the *power-law* model discussed earlier often allows a good curve fit. If we did *not* use the representative radius to compute $\dot{\gamma}$, but chose instead some other radius, then the number for $\dot{\gamma}$ would be *different* for different n values. Since we do not yet *know* n (that is the purpose of the graph) we would be *unable* to calculate a meaningful value for $\dot{\gamma}$. Thus use of the representative radius allows us to generate the desired plot and then discover what value n should have for the particular fluid. This assumes that a *good* curve fit can be obtained using the power-law model and adjusting n.

6.11 The Concept of Effective Length

When we derived our original simple formulas, we did not discuss the concept of effective length, but rather used, without any concern, the *actual* cylinder length, which seemed reasonable at the time. We did recognize that the *bottom* of the inner cylinder contributed some torque, and suggested how this might be minimized. Even if the bottom caused *no* additional torque, there would still be *end effects*. That is, a more rigorous theory really should have been related to cylinders of *infinite* length, with our calculation giving the *torque per unit length*, rather than the total torque. Just below the bottom of the inner cylinder, the shear rate will be quite different from that in the narrow gap above. This abrupt change will actually be felt *above* the cylinder bottom; that is, for the lower portion of the narrow gap, the shear rate will *not* be the same as for an infinite-length cylinder. These deviations from our simple theory are *not* easy to analyze, and the usual approach is to study them experimentally, and/ or use calibrations with known viscosity fluids to determine the corrections. Another approach is to modify the cylinder configuration[*] to minimize any end effects that might occur. One such scheme makes the bottom of the inner cylinder a *cone*, with the cone angle chosen so that the shear rate in the conical gap is the same as what we would compute with the simple theory in the radial gap between the two cylinders. This matching of the shear rates would hopefully reduce the end effect because we would no longer have the sudden change in flow at the cylinder bottom. Figure 6.21 shows some "tricks" used in the Ferranti portable viscometer[†] to reduce end effects. As the cylinders are lowered into a beaker full of the test liquid, an air bubble is trapped as shown. Since gases have much lower viscosities than liquids, there is negligible drag torque at the bottom of the rotating inner cylinder, largely eliminating this error. At the top, a stationary *guard disk* prevents rotating liquid from exerting a torque on the inner cylinder.

In place of (or in addition to) measures to reduce end effects such as those shown in Figure 6.21, one can experimentally estimate, and then correct for, the existing end effects. Figure 6.22a shows one such approach.[‡] By filling the cylinders to different lengths, and then measuring the torque required to rotate at a given speed with a given liquid (constant temperature), one can plot a curve of torque versus length, which can be extrapolated to zero torque, as shown in

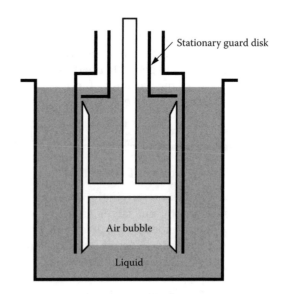

FIGURE 6.21
Cylinder design of Ferranti viscometer.

[*] R.W. Whorlow, *Rheological Techniques*, Wiley, New York, 1980, pp 136–144; C.W. Macosko, *Rheology: Principles, Measurements, and Applications*, VCH Publishers, New York, 1994.
[†] www.ravenfield.com
[‡] R.W. Whorlow, *Rheological Techniques*, Wiley, New York, 1980, p. 138.

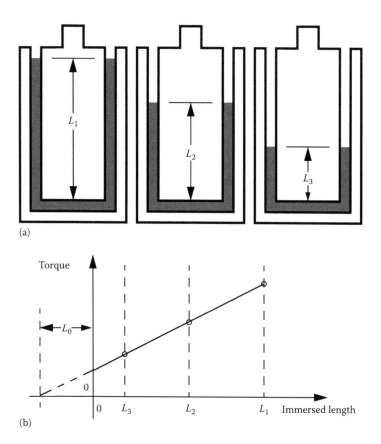

(a)

(b)

FIGURE 6.22
Experiment to find the effective length.

Figure 6.22b. This curve should be nearly a straight line, as shown. The torque goes essentially to zero when the liquid level drops below the bottom of the inner cylinder, since air viscosity is so small. Just *above* this point, the torque measured must be due to the shearing of the fluid below the inner cylinder. Extrapolating the curve to zero torque gives an intercept which defines the length L_0 which needs to be added to the physical cylinder length to get the effective length, which is then used in the viscosity formula. Unfortunately, such end corrections are usually different for different liquids, so the *correction experiment* must be done for every liquid to be studied.

6.12 Cylinder Design according to German Standards

We mentioned earlier the existence of German (DIN) standards which give guidelines for viscometer design, including end effects. If we design according to these rules, correction experiments may not be needed. This situation is analogous to the ASME Fluid Meters guidelines for designing orifice plates for flow measurement. Orifice plates also need corrections (discharge coefficients) which, in principle, must be found from experiment. However, if we design according to the ASME rules, *then no experiments are needed*; we can use the discharge coefficients provided in the handbook. In both cases, the standards organizations have collected and generalized the results of many *past* experiments so that *we* do not have to use experiments to characterize our specific apparatus. The DIN configuration* for concentric-cylinder viscometers is as shown below in Figure 6.23. The design rules are as follows:

$$\delta \triangleq \frac{R_a}{R_i} \leq 1.10, \quad \text{preferably } 1.0847 \qquad \frac{L}{R_i} \geq 3.0, \quad \text{preferably } 3.0 \quad (6.31)$$

$$\frac{L'}{R_i} \geq 1.0, \quad \text{preferably } 1.0 \qquad \frac{L''}{R_i} = 1.0 \qquad \frac{R_s}{R_i} \leq 0.30 \qquad (6.32)$$

$$90° \leq \alpha \leq 150° \quad \text{preferably } 120° \pm 1° \qquad (6.33)$$

We are here using the symbols of the standard, rather than our own. If we dimension our viscometer to conform to Equations 6.31 through 6.33, then the correction for the *end effects* consists of simply multiplying the actual length L by the factor 1.10, which is called in the standard a *drag coefficient* C_L.

We can now make more clear the utility of the representative shear stress concept. This stress is defined as the *average* of the stresses at R_1 and R_2 (using *our* notation).

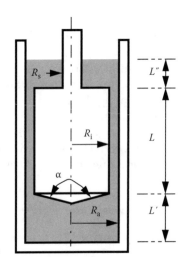

FIGURE 6.23
Cylinder design according to DIN Standards.

---

* DIN 53018 (part 1 and 2), DIN 53019 (part 1).

$$\tau_{\text{rep}} \triangleq \left(\frac{T}{2\pi R_1^2 L} + \frac{T}{2\pi R_2^2 L} \right) \frac{1}{2} = \frac{T}{4\pi L} \left(\frac{R_1^2 + R_2^2}{R_1^2 R_2^2} \right) \tag{6.34}$$

The r value which gives this stress is called R_{rep}:

$$R_{\text{rep}}^2 = \frac{T}{2\pi L \tau_{\text{rep}}} = 2\left(\frac{R_1^2 R_2^2}{R_1^2 + R_2^2} \right) \qquad R_{\text{rep}} = \frac{\sqrt{2} R_1 R_2}{\sqrt{R_1^2 + R_2^2}} \tag{6.35}$$

The shear rate at the representative radius is

$$\dot{\gamma}_{\text{rep}} = 2n\omega_2 \left[\frac{1}{R_{\text{rep}}^{2n}} \left(\frac{R_1^{2n} R_2^{2n}}{R_2^{2n} - R_1^{2n}} \right) \right] \tag{6.36}$$

We can now compute and graph the relation between shear rate and radius for a range of n values in the power-law model for non-Newtonian fluids, using Equation 6.36 with R_{rep} replaced by r. We take $R_i = R_1 = 0.5$ and all the other parameters at the preferred values of the standard, and let n range from 0.2 to 3.0. Since angular velocity is just a multiplying factor in Equation 6.36, we can set it equal to 1.0, or plot $\dfrac{\dot{\gamma}}{2\omega_2}$ rather than γ. Figure 6.24 shows that the representative radius concept is indeed useful in making the measured shear rate independent of the parameter n in the power-law model for non-Newtonian fluids.

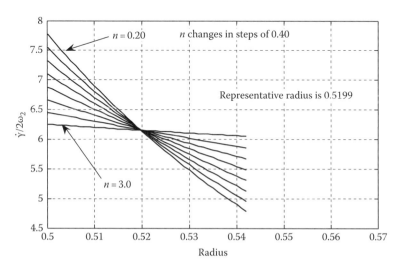

FIGURE 6.24
Utility of the representative radius.

The DIN standard defines a quantity $\delta \triangleq R_a/R_i = R_2/R_1$. Using this nomenclature and the representative radius, we can get a formula for computing the viscosity from measured torque and speed values plus instrument parameters.

$$\mu = \frac{\tau}{\dot{\gamma}} = \frac{\dfrac{T}{2\pi R_{rep}^2 L_{eff}}}{2\omega_2\left[\dfrac{1}{R_{rep}^2}\cdot\left(\dfrac{R_1^2 R_2^2}{R_2^2 - R_1^2}\right)\right]} = \frac{R_2^2 - R_1^2}{4.40\pi L R_1^2 R_2^2}\cdot\frac{T}{\omega} = \frac{\delta^2 - 1}{4.40\pi L R_2}\cdot\frac{T}{\omega} \quad (6.37)$$

We have here used the Newtonian formula for shear rate since n is usually not known for non-Newtonian fluids, and the use of the representative radius concept makes the shear rate largely independent of n. Note that in Equation 6.37 R_{rep} cancels out, as would *any* value of r. This does not, of course, negate the utility of this concept.

6.13 Designing a Set of Cylinders

We next want to explore the design of a set of cylinders for use in a practical viscometer. These would be required to measure viscosity over some chosen range, and also provide for measuring a given viscosity over a range of shear rates, to check for and document non-Newtonian behavior if it is present. Cylinder design depends also on the characteristics of the torque sensor, and the motor drive which rotates the outer cylinder at a desired angular velocity. These two pieces of hardware require their own design considerations, but to simplify our work, let us assume some typical values, so we can then concentrate on the cylinder design. We earlier designed a torsion bar to give the LVDT 0.005 in. of deflection for 0.00149 ft-lb$_f$ of torque. Because accuracies are often given as a percentage of *full scale*, sensors tend to become too inaccurate to use toward the low end of their range (a 1% *of full scale* error becomes 10% of the *reading* when we are at 10% of full scale). Let us thus assume that we do not want to use the torque sensor below 10% of its full-scale value. Let us also assume that our cylinder drive system can be used accurately from 1 to 500 RPM. Again take the inner cylinder diameter to be 0.500 in., but now make the other dimensions conform to the DIN rules ($L = 1.5$ in., $\delta = 1.0847$, $R_a = 0.54235$). To find the *largest* viscosity that we can measure with this setup we consider the *lowest* speed (1.0 RPM) and *highest* torque (0.01788 in.-lb$_f$).

$$\mu = \frac{\delta^2 - 1}{4.40\pi L R_a^2}\cdot\frac{T}{\omega} = \frac{0.1766}{6.097}\cdot\frac{T}{\omega} = 0.02896\cdot\frac{T}{\omega} = \frac{0.02896\cdot 0.01788}{0.1047}$$

$$= 0.004945\,\text{psi-s} \quad (6.38)$$

We have been using viscosities given as psf-s or psi-s. The SI unit of viscosity is the Pa-s. You will also often encounter the unit *centipoise* for viscosity. The conversion is $1\,\text{mPa-s} = 1$ centipoise. One Pa equals 0.02088 psf or 0.000145 psi, so the maximum viscosity of Equation 6.38 would be 34.1 Pa-s or 34,100 centipoise. Since these numbers are for the lowest speed and the highest torque, we could *not* explore non-Newtonian behavior for this fluid since either higher or lower shear rates exceed our instrument's capabilities. To find the *smallest* viscosity we can measure, we use the highest speed (500 RPM) and a torque which is 10% of the full-scale value (0.001788 in.-lb$_f$):

$$\mu = 0.0869 \cdot \frac{T}{\omega} = 0.02896 \cdot \frac{0.001788}{52.36} = 9.889\text{e}{-7}\,\text{psi-s} = 6.820\,\text{centipoise} \qquad (6.39)$$

Recall that water has about 1.0 centipoise and is considered a low-viscosity liquid.

If we design cylinders in accord with the DIN standard, we can show that a formula for the measured viscosity depends *only* on R_i and the measured values of torque and speed:

$$\mu = \frac{0.003620}{R_i^3} \cdot \frac{T}{\omega} \qquad \mu_{min} = \frac{0.003620}{R_i^3} \cdot \frac{T_{min}}{\omega_{max}} \qquad \mu_{max} = \frac{0.003620}{R_i^3} \cdot \frac{T_{max}}{\omega_{min}} \qquad (6.40)$$

Thus if the *span* (lowest value to highest value) of accurate speed and torque measurements is fixed, we can change R_i to get the range of viscosities of interest to us. However, when we are dealing with the fluids which have these highest or lowest viscosities, and R_i is *fixed*, we are *not* able to check for non-Newtonian behavior since this would require running at speeds which would produce torques outside our allowable range. For fluids with viscosities between the highest and lowest allowable, we *can* explore non-Newtonian behavior by running at a range of shear rates, so long as we stay within the capabilities of our instrument. If we wish to test fluids with viscosities, say, 10 times lower than is possible with the present cylinder set, we need to change R_i (and of course $R_a = 1.0847R_i$) in accord with Equation 6.40; $R_{inew} = R_{iold}(10^{1/3}) = (2.154)(0.50) = 1.0772$ in. Most commercial instruments do provide a selection of cylinder sets, but because the best instruments provide much larger accurate torque and speed spans than our simple design, a single cylinder set is usable over larger ranges of viscosity.

6.14 Temperature Effect on Viscosity

Table 6.1 clearly showed the strong effect of temperature on viscosity, especially for highly viscous fluids. This means that all viscometers must be operated under *known* temperature conditions for the fluid being tested. Sometimes just allowing the apparatus to stabilize at "room temperature" and measuring that temperature is sufficient. For critical work, however, the viscometer itself must be provided with some kind of temperature control system for the fluid as it is tested. Three types of such controls are in common use. In one, the outer cylinder is surrounded by a *water jacket* which is supplied with temperature-controlled water from a thermostated supply. Another approach builds electric heaters and temperature sensors into the cylinder system, but this is limited to providing temperature *above* the ambient temperature. Perhaps the most sophisticated controls use thermoelectric (Peltier) heating and cooling to provide regulated temperatures either above or below ambient.

We need to first recognize that the shearing of the fluid under test will *itself* generate heat and raise the fluid's temperature. Complex heat transfer studies have been made of this problem but we again prefer to take a simple approach to get an idea of the order of magnitude involved. Assume that the outer cylinder rotates at a constant speed, and that *all* the heat generated by shearing the fluid goes into the fluid film in the gap. That is, no heat is lost from this shell of fluid. This appears to be a conservative assumption, since certainly *some* heat will actually be lost and the fluid temperature rise will thus be *less* than we predict. We want to derive a formula which predicts the rate of rise of the fluid film temperature in °C/s under these assumptions. The rate of heat generation due to shearing of the fluid is the product of angular velocity and torque, expressed in J/s (W), which is N-m/s. This heating rate is the input to the thermal capacitance of the fluid film, which is its mass times its specific heat c_p(J/(kg-°C)). The rate of change of the fluid film temperature (°C/s) is the heating rate (J/s) divided by the thermal capacitance (J/°C):

$$\frac{dT}{dt} = T\omega \cdot \frac{1}{\pi \rho L_{\text{eff}} c_p (R_2^2 - R_1^2)} \tag{6.41}$$

where
c_p is the specific heat
ρ is the density of the fluid

The *worst-case* situation involves the maximum torque and maximum speed. Using our original dimensions and typical density and specific heat values for "oil" we get

$$\frac{dT}{dt} = 0.002 \cdot 52.36 \left(\frac{1}{3.142 \cdot 895 \cdot 0.0635 \cdot 1810(0.0001898 - 0.00004032)} \right)$$

$$= 0.002168 \, K/s \tag{6.42}$$

While this seems to be a rather small rate of temperature change, we now need to estimate how much change in *viscosity* this causes.

To evaluate the need for temperature control, we must first know how much viscosity changes with temperature. If we want to associate our measured value of viscosity with a certain temperature (± some small tolerance band), then our temperature control system must be capable of holding fluid temperature within that range. Table 6.1 earlier listed the variation of viscosity with temperature for a few typical liquids. The temperature values in the table are rather widespread and few in number, making it difficult to estimate the rate of change of viscosity with temperature. One way to deal with this is to fit a suitable mathematical curve to the data and then use this math model for further calculations. Since curve fitting is a widely useful technique in engineering, this will give us some practice in this skill. The *heavy oil* example in the table exhibits a strong variation of viscosity with temperature (high viscosity fluids tend to act this way), so it will make a good candidate for our curve fitting exercise, since large variations are harder to accurately fit than small variations. That is, if we can get a good fit for the heavy oil data, we can probably fit most other fluids.

In any curve fitting problem, we should always use any available theoretical knowledge which might give us clues as to what kind of functions to try. Lacking such knowledge, we often start with polynomial functions, since by adding higher power terms, we can progressively fit more complex curve shapes. Recall also that curve fitting is a special case of the general statistical tool called *multiple regression,* which we have used in some earlier chapters. This technique is of great utility in finding empirical math models that relate a dependent variable to any number of independent variables.[*] If you need a short tutorial on these methods, consult the Minitab statistics manual in the appendix. The manual shows how to use the Minitab statistics software for this and other tasks. The viscosity data in Table 6.1 covers the range from 0°C to 100°C but is rather sparse, having only six entries. It is well known that a polynomial fit (including a constant term) requires data with at least as many values as the degree of the polynomial, plus one, so the highest degree we can use in our example is the fifth. We also know that when the number of data points, plus one, is exactly equal to the polynomial degree, we get a *perfect* fit at each of the data points (six linear equations in six unknowns in our example). While this at first seems a good result, most readers will already know that it may be *really bad,* since the *curve* of the fitted points is often oscillatory (whereas the physical relation may be monotonic) and thus completely misleading when used to interpolate *between* the data points. Figure 6.25 shows that in this case the fifth-degree polynomial fits the data points exactly (as expected) but does *not* noticeably oscillate and is actually a better fit than the fourth-degree function. Near 360 K, however, it does show an *increase* in viscosity as temperature rises, which is not the true physical behavior. These defects in fitting are at least partly due to the small number of data points available, but may also be related to the limitations

---

[*] E.O. Doebelin, *Engineering Experimentation*, McGraw-Hill, New York, 1995, Chapter 4.

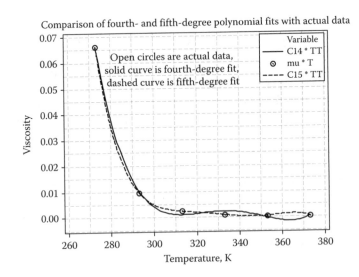

Comparison of fourth- and fifth-degree polynomial fits with actual data

Open circles are actual data,
solid curve is fourth-degree fit,
dashed curve is fifth-degree fit

FIGURE 6.25
Fourth- and fifth-degree polynomial fits to viscosity data.

of polynomial functions as related to the physical behavior of the viscosity/ temperature relationship.

There actually *is* some theory that predicts what the viscosity–temperature relation could be. This approximate theory says that the equation might be of the form

$$\mu = e^{\left[C_1 + \frac{C_2}{T} + \frac{C_3}{T^2}\right]}$$ (6.43)

where
 the Cs are constants
 T is the absolute temperature

The multiple regression (curve fitting) software just used to fit the polynomials is actually *linear* regression, which means that it only works with equations that are linear in the fitting coefficients. Equation 6.43 does not fit this pattern, but by taking natural logs, we can force it into the required form:

$$Ln(\mu) = C_1 + \frac{C_2}{T} + \frac{C_3}{T^2}$$ (6.44)

Figure 6.26 shows that this form of function seems to better suit the viscosity/temperature phenomenon; Table 6.2 shows the numerical values. The coefficients C_1, C_2, and C_3 are respectively −4.37572e−01, −9.51608e+03, and 2.42733e+06. Using these in Equation 6.43 we can differentiate to find the

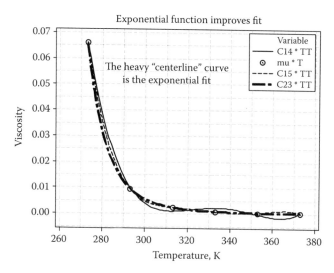

FIGURE 6.26
Exponential function improves curve fit.

TABLE 6.2

Quality of Curve Fit

Exponential Fit	Actual Data
0.0654746	0.066100
0.0096455	0.009470
0.0023253	0.002320
0.0008050	0.000812
0.0003650	0.000371
0.0002028	0.000200

rate of change of viscosity with temperature as $\mu(-C_2/T^2 - 2C_3/T^3)$. It is clear from Figure 6.26 that this derivative (slope) has its maximum negative value at the lowest temperature (273 K), and calculation shows it to be -0.007262 psf-s/K. The viscosity at this point is 0.06610 (curve fit gives 0.065476), so a 1 K change in temperature causes a 10.9% change in the viscosity. At the highest temperature (373 K) we have a viscosity of 0.000203 and the slope is -0.0000051 psf-s/K, giving a 2.5% change per K. While these calculations are based upon a curve fit which is only approximate, it is clear that if we want to associate a measured viscosity with a certain temperature, we need to know that temperature rather accurately.

6.15 Temperature Control Methods

As mentioned earlier, when temperature control is required, the Searle (outer cylinder stationary) configuration is most convenient because of the necessary electrical or fluid connections that are needed between the cylinder and the rest of the rheometer. For concentric-cylinder configurations,

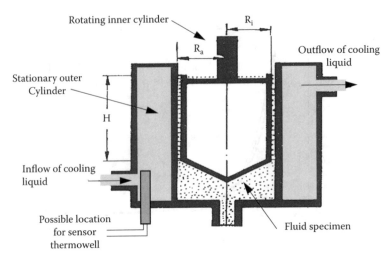

FIGURE 6.27
Water-jacket temperature control (Haake).

the usual approach seems to be to jacket the cylinder region with a surrounding chamber supplied with temperature-controlled liquid or gas,
thus one can use an *external* thermostated source of fluid. Figure 6.27 shows
such a water-jacket arrangement depicted in a Haake brochure. Two platinum resistance thermometer (Pt 100) sensors are used, one for indication
and one for control. The locations of these sensors are not shown in the
brochure but they are probably attached to the outside of the jacket, which
is made of nickel-plated copper (low thermal resistance) to reduce the temperature drop between the specimen fluid in the shearing gap and the metal
of the jacket. Alternatively the sensors might be in protective thermowells
immersed in the cooling/heating fluid. Typical temperature ranges might
be −50°C to 100°C; the lower temperatures would require cooling liquids
other than water.

Figure 6.28 is from a patent (U.S. 7,367,224 B2, May 6, 2008) where a *gas*
is used for the heating medium and the measuring setup is a plate/plate
or cone/plate configuration. (This patent was assigned to Thermo Electron,
which had acquired Haake, and is now Thermo Fisher Scientific.) The gas
medium here actually contacts the plates or cone/plate and the exposed
edges of the specimen fluid, so it must be selected so as to not contaminate or
otherwise influence the specimen. A resistive heater 22 is used to control the
gas temperature with a sensor and feedback scheme (not shown). For very
sensitive measurements, the gas flow does cause some undesired *forces* on
the apparatus, so it should be shut off when measurements are actually being
made, but the temperature control is then lost. To regain this control, other
resistive heaters 23 are imbedded in the chamber walls and are controlled to
give the desired specimen temperature.

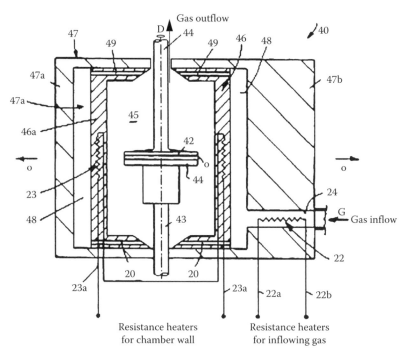

FIGURE 6.28
Gas-jacket temperature control (Thermo Electron).

For plate/plate and plate/cone configurations, *thermoelectric* heating/cooling is widely used. Figure 6.29a (Anton Paar/Physica)* shows a typical arrangement, while Figure 6.29b shows gas cooling with electrical resistance heating. A very useful paper by Lauger et al.[†] discusses several aspects of rheometer temperature control, such as methods to reduce the temperature *gradient* within the fluid sample. That is, a proper control system must not only produce the desired *average* temperature but also assure that there are not large *variations* of temperature within the sample. Figure 6.30 shows three levels of measures available to control such variations. When heating/cooling is applied only at the bottom of the sample as in Figure 6.30a, large vertical gradients will exist if the set temperature is not very close to the ambient temperature. Addition of an insulating hood as in Figure 6.30b will reduce such gradients somewhat, but to minimize them we must actively control both the bottom and the top areas of the sample, as is done with two sets of thermoelectric modules in Figure 6.30c. To evaluate the effectiveness of any proposed control scheme, one must be able to *measure* the temperature gradients. Lauger et al. explain the use of thin plastic disks with embedded

---

* www.anton-paar.com
[†] J. Lauger et al., New developments in accuracy for temperature control of rotational rheometers, Anton-Paar/Physica, www.anton-paar.com

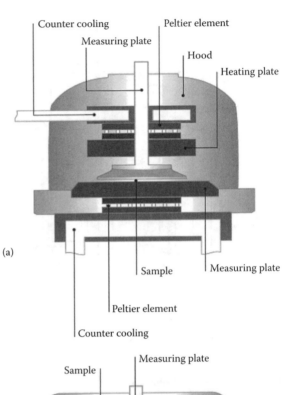

(a)

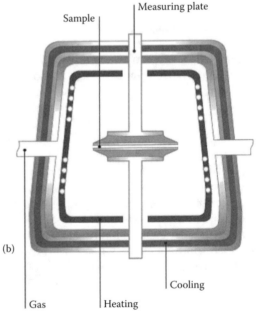

(b)

FIGURE 6.29
Temperature control schemes
(Anton Paar/Physica): (a) thermoelectric and (b) gas cooling,
resistance heating.

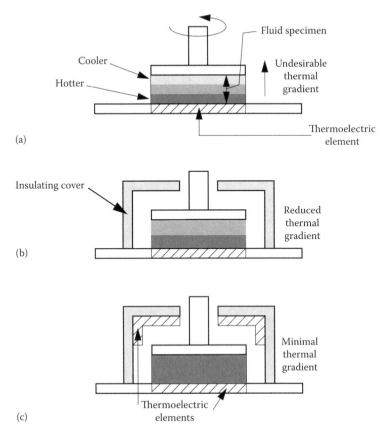

(a)

(b)

(c)

FIGURE 6.30
Improving specimen temperature uniformity using passive shielding and active heating.

thermocouples for this purpose (see Figure 6.31). These disks are made of a solid plastic material (Ertaxel) with thermal properties similar to the thin layer of specimen material that they simulate, though of course they do not allow the convective flows that might occur in a liquid specimen. The thermal parameters are such, however, that the disks *overestimate* the measured gradients and are thus conservative sensing devices. The disks are placed between the plates of a plate/plate setup to simulate the fluid layer and then the temperature control system is exercised with various commands, such as a sequence of step changes from the lowest to the highest temperatures, with a dwell time of about 15 min at each level, to allow equilibrium. Taking temperature differences between the top/bottom and inside/outside thermocouples allows estimation of the vertical and horizontal gradients. Systems with only a passive cover (Figure 6.30b) allow gradients as high as 10 K, while the systems with thermoelectric elements at both top and bottom locations (Figure 6.30c) reduced these to about 0.2 K.

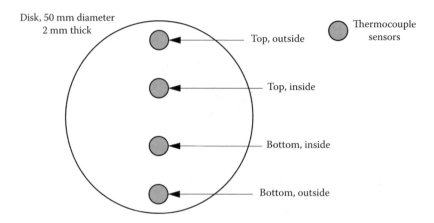

FIGURE 6.31
Sensor disk to check temperature gradients in vertical and horizontal directions.

While we cannot access the detailed rheometer data needed to realistically design thermoelectric temperature controls for that application, we are able to explain the general approach to such problems. Thermoelectric heating/cooling modules are made from ceramic semiconductor materials such as bismuth/telluride alloys, and take the form of thin plates about 40 mm square and 4 mm thick, with two attached wires for electrical connections. If larger areas are needed one must use an array of such units since thermal stresses prevent the successful fabrication of larger modules. The input electrical power of a single module is modest; say 15 V and 4 amp maximum, with a heating/cooling power of about 20 W. While the *cold side* of a module can be directly mounted to an object which we wish to cool, the *hot side* must be attached to some sort of *heat sink*, to dissipate the total heat load. While commercial fin-type heat sinks with a fan to provide forced convection are sometimes adequate, most rheometer applications use a water-cooled heat sink. Using commercial electronic controllers, we can build feedback systems which use a suitably located temperature sensor to provide the feedback signals and PID control modes to manipulate the thermoelectric module's input voltage.

We can explain the operation of thermoelectric modules with the help of Figure 6.32* which is for a specific module but will serve as an example for any module; they differ mainly in the maximum amount of heating/cooling that they can provide. Two major considerations are the maximum temperature difference between the hot and cold sides of the element, and the maximum amount (W) of heating or cooling provided by the element. The referenced vendor offers nine different modules. They all have the same maximum temperature difference (71°C), but the maximum heating/cooling rate ranges from 16 to about 80 W. Sizes range from about

---

* www.electracool.com

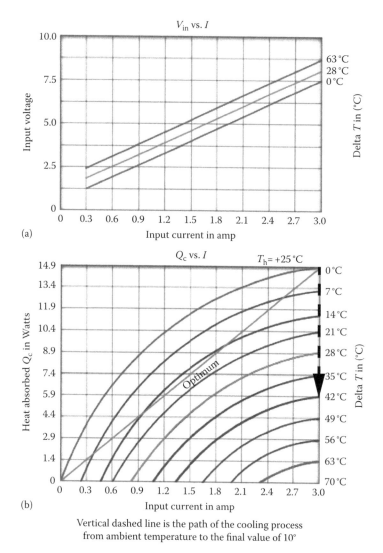

FIGURE 6.32
Thermoelectric module characteristic curves (Advanced Thermoelectic ST-71-1.0-3.0).

20 mm by 20 mm (3 mm thick) to 40 mm by 40 mm (4 mm thick). Reversing the input voltage, which also reverses the current, will reverse the element's heat flow, and the object which we had been cooling is now subject to heating. For a given module, if we require a large temperature difference, then the amount of heating/cooling available will be small. If we require a large amount of heating/cooling, then we cannot get a large temperature difference. The data of Figure 6.32b is for an ambient temperature (T_h) of 25°C, and shows that the maximum possible heating or cooling rate is about 14.9 W, with an input current of 3.0 A, but then the temperature difference would be

zero. If we follow the curve that starts at the upper right of this graph, down and to the left, we see that if we reduce the current, the heating/cooling rate drops off, reaching zero at 0 A. When temperature differences larger than what is possible with a single module are needed, *multistage* systems are used. Here several modules are connected *thermally in series*, so that the total temperature difference is the sum of the individual module values. Design of multistage systems becomes more complicated, so we will consider only the simple single-stage devices.

When one considers some kind of cooling system, we will have in mind a *coldest necessary* temperature for the cooled object, say, 10°C (see Figure 6.33). Assuming that the object's thermal environment is atmospheric air at *room temperature*, say 25°C, we first need to do a heat transfer study to find out how much heat flow (W) will be absorbed by the object when it is at the

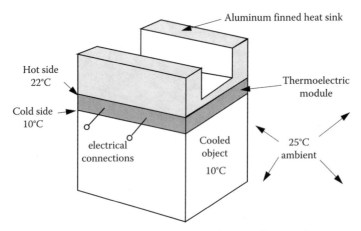

Thermoelectric cooling couples are made from two elements of semiconductor, primarily bismuth telluride, heavily doped to create either an excess (n-type) or deficiency (p-type) of electrons. Heat absorbed at the cold junction is pumped to the hot junction at a rate proportional to current passing through the circuit and the number of couples.

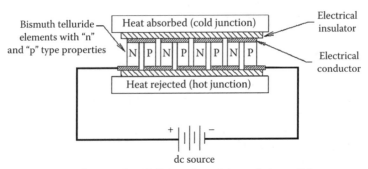

Elements electrically in series and thermally in parallel

FIGURE 6.33
Thermoelectric heating/cooling system.

coldest temperature. If, for example, the major heat flow is convection from the ambient air to the object, and the thermal resistance is, say, 1.25 °C/W, then the heat flow will be (25–10)/1.25 = 12.0 W. We would thus require a thermoelectric module that can provide at least 12 W of cooling. The module of Figure 6.32 can provide about 15 W maximum, so it would be a candidate for our application. However, when we examine Figure 6.32b (visually interpolate between the curve labeled 7°C and 14°C) we see that when the module provides 12 W of cooling, it has only about 12° of temperature difference between its hot and cold sides. This means that the hot side is about 22°, which is *less* than the ambient temperature (25°C). The finned aluminum heat sink is thus *colder* than ambient and thus cannot reject any heat into the ambient air. If it were replaced by a water-cooled cold plate at, say, 10°, then heat rejection from the hot side of the thermoelectric element is possible, but we would have to calculate whether *enough* heat can be rejected. This kind of water cooling appears to be used in the rheometer system of Figure 6.29a, where it is called *counter cooling*.

If our original calculation of the heat flow to the cooled object (12 W) had been, due to a higher thermal resistance, a lower number, say, 6 W (the cooled object thermal resistance would have to be 2.50), then we could get a temperature difference of about 35° with an input current of 2.3 A, or 42° with 3.0 A, which is the maximum allowed current. Using the 42° figure, the thermoelement hot side would now be at 52°, well above ambient, so an air-cooled finned heat sink would have a (52–25) = 27° temperature difference available for heat rejection to the ambient air. To see whether we can get *enough* heat rejection in the heat sink, we have to first calculate the *total* heat flow that it must get rid of. That is, the heat flow *pumped* by the thermoelement is just *part* of the total heat that must be rejected by the heat sink. The other part is the *input* heat, which is just the product of the volts and amps at the thermoelement electrical terminals. *This input heat flow always has the same direction as the Peltier pumped heat flow,* so in this example it goes into the heat sink. We already know that the current must be 3.0 A; to get the corresponding voltage we use Figure 6.32a, which shows the voltage to be about 8 V, making the input power 24 W. The total power is thus 24 + 12 = 36 W. To check whether a particular heat sink is capable of dissipating this heat flow with the available temperature difference of 27° we need to know the thermal resistance of that heat sink. To meet our needs this resistance has to be less than 27/36 = 0.75°C/W. (If we were to *start* the process (with everything initially at 25°C) by turning on the 3 A current at time zero, in Figure 6.32b we would be starting at the top right, with a thermoelement delta *T* of 0°, and the maximum amount of heat pumping, 14.9 W. As the controlled object gradually cools down over time, we travel down a vertical line on the graph at (3 A current) until we reach a delta *T* of 42°, at which point we stop because now the pumped heat flow exactly matches the heat gain, giving equilibrium. Since the heat pumping rate diminishes as the temperature difference increases, this process is an asymptotic approach (roughly an exponential curve) with respect to time.)

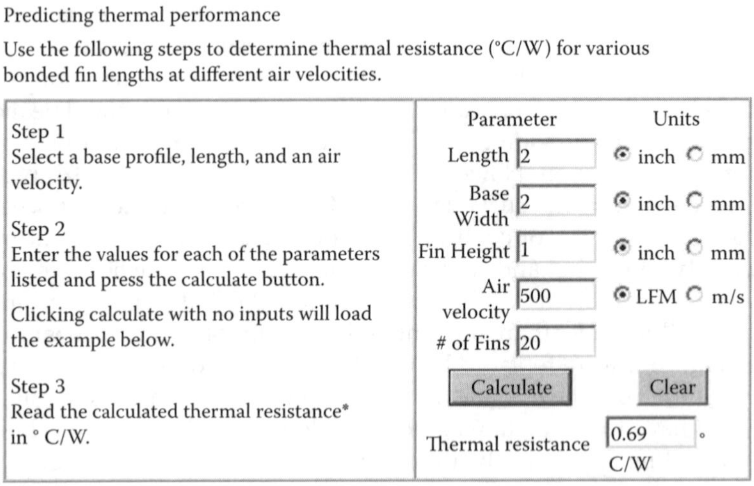

Predicting thermal performance

Use the following steps to determine thermal resistance (°C/W) for various bonded fin lengths at different air velocities.

Step 1 Select a base profile, length, and an air velocity. Step 2 Enter the values for each of the parameters listed and press the calculate button. Clicking calculate with no inputs will load the example below. Step 3 Read the calculated thermal resistance* in ° C/W.	Parameter	Units
	Length 2	⊙ inch ○ mm
	Base Width 2	⊙ inch ○ mm
	Fin Height 1	⊙ inch ○ mm
	Air velocity 500	⊙ LFM ○ m/s
	# of Fins 20	
	Calculate	Clear
	Thermal resistance 0.69	C/W

FIGURE 6.34
Thermal resistance calculator for bonded fin heat sinks (Aavid Thermalloy).

Many heat sink manufacturers provide Web sites with product performance listings and helpful design and selection guides.* Figure 6.34 shows a thermal resistance calculator available on this Web site for a particular type of heat sink called *bonded fin*, with some example numbers that I entered. Note that I entered an air velocity of 500 ft/min, which means that some kind of *fan* must be provided to give forced (rather than free) convection heat transfer. If one enters much lower air velocities, the thermal resistance will be much higher than the 0.75 value which is needed for our example problem. This Web site provides similar information on other types of heat sinks, such as water-cooled, together with technical papers in PDF form.

Once one has found a suitable combination of thermoelectric module and heat sink for a given application, the next step is to select an electronic *controller* that will manipulate the module input voltage so as to provide the range of temperatures needed in the application, say a rheometer. Our simple example considered only a single temperature, which required cooling, to illustrate the procedure for module and heat sink selection. When a *range* of temperatures is needed, some of which require heating and some cooling, one must check *all* these conditions to make sure that the module and heat sink can meet them. In the system of Figure 6.33, what happens when we need to *heat* the controlled object, say to 15° *above* ambient (40°C)? Whatever input voltage polarity was used when cooling was desired, we now need to *reverse* that polarity (this also reverses the current's direction), and then the hot and cold sides of the module will be reversed from the situation in Figure 6.33. The controlled object now *loses* heat at the rate of (40−25)/(2.50) = 6 W,

---

* http://www.aavidthermalloy.com/products/heatsinks.shtml

and the thermoelement must supply this heat. However, since the input heat flow direction reverses when the pumped heat flow reverses, these two heat flows now *help* each other in supplying the needs of the controlled object. Because the heat sink is now *colder* than ambient, an additional heat flow from ambient to the heat sink occurs and this *also* goes to supply the controlled object. With *all* these heat flows being effective in heating the controlled object, it will not take a very large reversed current and voltage to meet its needs. We see that if we have designed a system to provide adequate cooling, there will *generally* be no problem in that same system providing an equal amount of heating.

At this point, I want to analyze a thermoelectric cooling/heating system similar to what might be used in a rheometer. To do a simulation of such a system it is helpful to have *equations* for the thermoelectric module behavior and these are available in the literature and also from some manufacturers' Web sites. The Melcor* site has the equations and also a convenient software program called AZTEC, which you can download at no cost. It gives the equations and also does all the steady-state calculations needed to choose one of their thermoelectric modules. It does *not* do dynamic simulation, so we will use Simulink for this. The available thermoelectric equations are for *steady-state* conditions, but the devices are *electrically* mainly resistive, so the input current follows the input voltage without any significant delay. Any *thermal* delays in the devices will be neglected (we assume that pumped heat flow follows input current instantly). This is usually valid since module manufacturers state that the response of heated or cooled objects should, by design, be kept slower than about 1 °C/s, to prevent mechanical damage due to excessive thermal stresses in these brittle materials. Time lags in the controlled object and the heat sink will *not* be neglected and are easily modeled with combinations of thermal resistances and capacitances. Since we do not have the *thermal details* of rheometers available to us, we will use the simplest thermal models to demonstrate the general approach. These thermal models will use one resistance and one capacitance each for the controlled object and the heat sink.

To calculate the thermoelectric *heat pumping rate* Q_p we use the following formula:

$$Q_p = 2N\left(\alpha i T_c - \frac{\rho i^2}{2G} - \kappa G \Delta T \right) \tag{6.45}$$

where

N is the number of thermocouples in the module
α is the Seebeck coefficient in V/K
i is the current through the module

---

* www.melcor.com

T_c is the cold-side temperature in Kelvin
ρ is the material resistivity in Ω-cm
G is area/length of the thermoelectric element
κ is the thermal conductivity in W/(cm-K)
ΔT is the temperature difference $T_h - T_c$ across the thermoelement

For any specific module, the manufacturer will provide values of N (typically about 10–200) and G (typically 0.02–1.2). For a specific material (usually bismuth telluride) the values of α, κ, and ρ depend on the average temperature $(T_h - T_c)/2$ and numerical values can be found in references, such as the AZTEC software. To activate a module, one must apply a voltage to its terminals, which then causes a current i to flow, according to the module's electrical resistance. The formula for the resistance is

$$R = \frac{2N\rho}{G} \, \Omega \qquad\qquad (6.46)$$

These two equations are sufficient to model the thermoelectric actions in the module and we will combine them with models of the controlled object, heat sink, temperature sensor, and control system to simulate a complete system. To get specific values for the thermoelectric module, one must select a specific module from those available from a vendor (we will use the Melcor modules because the AZTEC software makes the selection relatively easy). To select a module, one must specify the ambient temperature, the coldest desired temperature of the controlled object, and the total heat flow to that object when it is at this coldest temperature. Let us suppose that this coldest temperature is 10°C and that a heat transfer study has computed the maximum heat flow of 10 W to the controlled object at an ambient temperature of 25°C. We also need to choose a heat sink thermal resistance by selecting from a range of maximum values displayed by the software. Remember that *low* heat sink thermal resistance is beneficial to the thermoelectric module but will require more weight and space and will cost more. I choose a maximum value of 1.0 °C/W. The software also gives one a choice of the most efficient module or the lowest cost module; I chose the most efficient.

Figure 6.35 shows a results page from the software. At the top left we see the module which has been selected, a PT8-7-30, which is shown to have $N = 71$, the number of *couples*. We asked for a heat-sink thermal resistance less than 1.0 and the software shows that 0.784 is what is needed for this application. We could then go to heat-sink Web sites and try to find a commercial product close to this value. To get the cooling we asked for, this module will require an input voltage of 4.08 V, with a current of 3.51 A. For the heat sink suggested, the hot-side temperature would be 44.09, the heat pumped out of the controlled object would be 10.02 W (we had entered a requirement of 10.0), and the total hot-side heat dissipated (the sum of the heat pumped and the electrical input wattage) is 24.35 W. With the module selected, we can

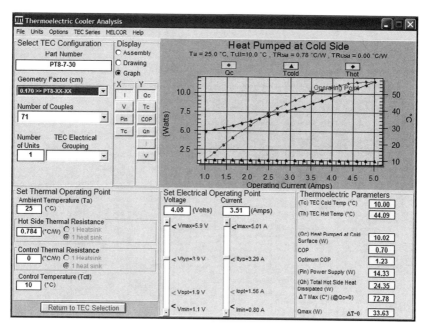

FIGURE 6.35
Results page from AZTEC software.

look up in the software the values of the needed parameters, for an average temperature of $(10.00 + 44.09)/2 = 27.05°C = 27.05 + 273 = 300.05\,\text{K}$. The G value is 0.171, $\kappa = 1.51e{-}2$, $\rho = 1.1e{-}3$, and $\alpha = 2.02e{-}4$. We now have enough data to formulate a Simulink simulation of a complete system, as in Figure 6.36. The controlled object and heat sink are each modeled as single thermal capacitances C_c and C_h, and thermal resistances R_c and R_h. The heat-flows into the thermal capacitances are summed to produce the terms of form $C\,dT/dt$, which are multiplied by the $1/C$ values to get dT/dt, which is then integrated to get the temperatures. The heat sink model is at the top of the figure, the controlled object model at the bottom. The values of C_c and C_h for an actual application would have to be estimated from dimensions and material properties; in our example I arbitrarily took both as 0.01. The resistance R_h is given by the software, as noted above. The controlled object resistance R_c would have to be estimated from a heat transfer analysis of the specific object; I here took it as 2.5.

A temperature sensor at the lower right measures the controlled object temperature Tc with a sensitivity of $1.0\,\text{V}/°$. The desired temperature Tv does not exist as a physical signal (this is usually the case in all feedback systems); the *physical* command is the reference input voltage er, which is related to Tv through the *reference input element gain*, which must be the same numerical value as the sensor gain, 1.0 in our example. For a sensor with a different sensitivity, such as, say, $0.0135\,\text{V}/°$, we would have to use

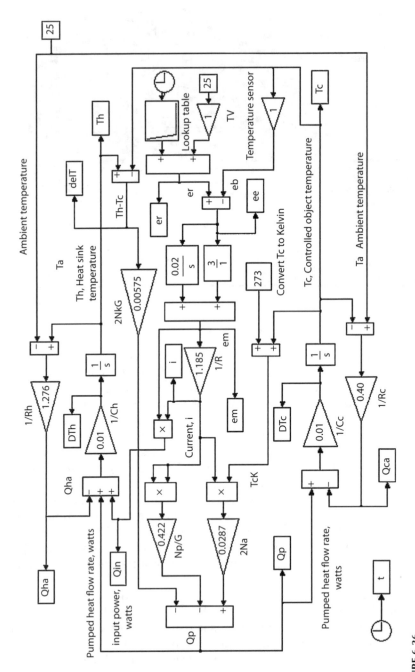

FIGURE 6.36
Simulation of thermoelectric heating/cooling system.

that same value for the reference input gain. The command *er* is formed as the sum of the ambient temperature (25°C) and a terminated ramp formed in the lookup table. Step inputs are not used since they cause temperature changes that are so rapid (more than about 1°/s) that excessive thermal stresses arise in the module. The rise time of the ramp input is adjusted by trial and error to meet this requirement. Note that in the thermal simulation of the heat sink and controlled object, we measure their rates of change *DTc* and *DTh* to allow this adjustment. System error voltage *ee* is obtained as the difference between the command *er* and the feedback (sensor) voltage *eb*, and this error voltage is then processed through a conventional proportional plus integral controller to produce the module input voltage *em*. Module current *i* is just this voltage divided by the resistance *R*. Pumped heat flow Q_p can now be obtained from available signals, according to Equation 6.45.

To see how the temperature *Tc* responds to cooling commands, I made the terminated ramp have a rise time of 100 s with a final value of −15, which commands *Tc* to go to 10°C. To check the heating, the ramp had a final value of +15, which commands *Tc* to go to 40°C. The PI controller was *tuned* in the usual way by starting with pure *I* control and increasing the value until oscillations were excessive, whereupon we gradually add some proportional to regain stability. These adjustments were also guided by watching the temperature rates of change, as discussed above, to prevent excessive thermal stress. As usual, the *final* controller settings of *I* = 0.02 and *P* = 3.0 are not "magic numbers"; good results are obtained with a range of tunings. Figure 6.37a shows the commands and responses; the responses are very good, the curves are nearly identical and thus difficult to distinguish on the graph. Note that for heating, the heat-sink temperature at steady state is *cooler* than ambient, which means that the heat sink is contributing to the total heat flow to the controlled object. Figure 6.37b (which is for heating the controlled object) shows that the temperature rates are quite low and thus meet the thermal stress guideline. The simulation is easily modified to check other performance criteria. For example, in a viscosimeter, we calculated earlier the heating that occurs due to shearing of the fluid at a steady velocity. This heating would tend to change the fluid temperature, but the control system will counteract this. To study this effect we would just add this given heating rate as an additional input to the summer in the controlled object thermal model, with the desired temperature set at a constant value, say, the ambient temperature of 25°. Another study would replace the fixed ambient temperature with some time-varying function, say, a slowly changing random signal, to see how changes in *room* temperature affect the controlled object temperature. Figure 6.38a shows the response to a random ambient temperature change around 25°, and Figure 6.38b the response to a step change in fluid-shear heating rate of 1.0 W, applied at time = 10 min. Both results show good performance of this temperature control system.

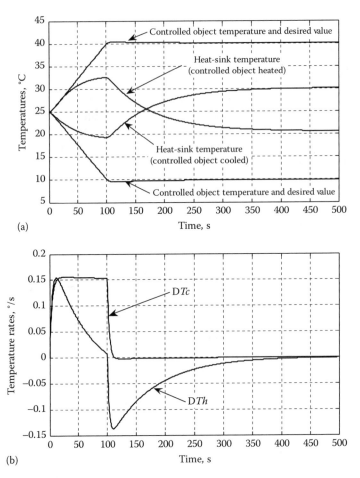

FIGURE 6.37
Response of system to heating and cooling commands.

6.16 Uncertainty Analysis

Equation 6.30 can be expanded to relate torque T to the LVDT output voltage e actually measured, the torsion spring dimensions and shear modulus G, the LVDT arm length L_a, and the LVDT sensitivity K_v (V/in.). Similarly, assuming that we use an incremental encoder with resolution R_{en} counts/rev and a digital *clock* with resolution R_{clk} microseconds to measure angular velocity ω, we can include these in the formula for viscosity. This formula can then be used to estimate the uncertainty in viscosity caused by individual uncertainties in each of the parameters in the formula. The conventional

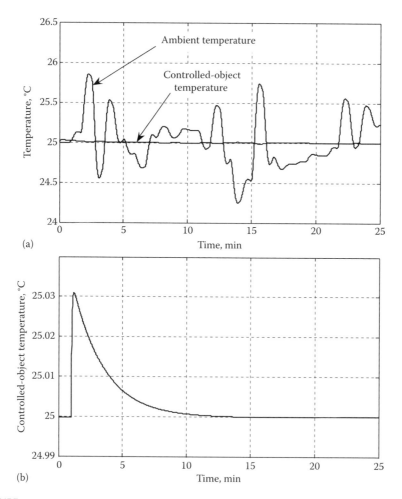

(a)

(b)

FIGURE 6.38
System response to: (a) random ambient temperature and (b) fluid-shear heating rate.

way of doing this uses the *root-sum-square* method* where the uncertainty U_y in a dependent variable y is related to the individual uncertainties u_{x_i} in each independent variable x_i on which it depends by the relation

$$U_y \approx \sqrt{\left(\frac{\partial f}{\partial x_1} \cdot u_{x1}\right)^2 + \left(\frac{\partial f}{\partial x_2} \cdot u_{x2}\right)^2 + \left(\frac{\partial f}{\partial x_3} \cdot u_{x3}\right)^2 + \cdots + \left(\frac{\partial f}{\partial x_n} \cdot u_{xn}\right)^2} \quad (6.47)$$

where $y = f(x_1, x_2, x_3, \ldots, x_n)$ and the variation in each independent variable is assumed to follow the Gaussian (Normal) distribution. Equation 6.47 is

---

* E.O. Doebelin, *Measurement Systems*, 5th edn., McGraw-Hill, New York, 2003, pp. 67–72.

an exact relation when y depends on the x's linearly; for nonlinear relations where the individual uncertainties are small relative to the base values, and/or for non-Gaussian distributions it is a close approximation. It also assumes that the individual x's are statistically *independent*, that is, if one of them changes, it has no effect on the others. The individual partial derivatives in Equation 6.47 are useful in identifying those parameters which are most influential in causing changes in the dependent variable. That is, if $\partial f/\partial x$ is particularly large for a certain x, then changes in that x will cause relatively large changes in y. Thus in manufacture or use of the device, we may want to concentrate design efforts on reducing the variability of that particular parameter.

Returning to Equation 6.30, it can be written as

$$
\mu = \left[\frac{R_2^2 - R_1^2}{4\pi R_1^2 R_2^2 L_{\text{eff}}} \right] \cdot \left[\frac{\pi R_s^4 G}{2 L_s} \cdot \frac{e}{L_a K_v} \right] \cdot \left[\frac{1}{\dfrac{N_{\text{counts}}}{\Delta t} \cdot \dfrac{1}{R_{\text{en}}} \cdot 2\pi} \right]
\tag{6.48}
$$

where the second bracketed quantity from the left is the measured torque, and the third is the measured speed. For the torsion-wire spring, the radius is R_s, the length L_s, and shear modulus G. For the LVDT, the arm length is L_a, the measured voltage is e, and the LVDT sensitivity is K_v V/in. For the incremental encoder, N_{counts} is the number of counts measured over the time interval Δt, and R_{en} is the encoder resolution in counts/rev. We can analytically compute the various partial derivatives of viscosity with respect to each parameter, assign some typical values to each parameter to get numerical values of the partial derivatives, and then make, say, a 1% change in each parameter to see the individual effects on viscosity. In taking the partial derivatives analytically, it is convenient to define three groups of parameters as the three bracketed terms in Equation 6.48. From left to right we will call these Cr, Ct, and Cw. A Mathcad program for doing these calculations is shown below. For parameters which appear in Equation 6.48 in linear fashion (G, e, delT, Ren) a 1% change results, as expected, in a 1% change in computed viscosity. For parameters that appear as *reciprocals* (Le, Ls, La, Kv, Ncts), the change is −1%. For parameter Rs, which appears to the fourth power, we get a 4% change, while R1 gives −13.3% and R2 11.3%.

Some of the parameters display *inherent* variation but also can vary with *temperature*. For example, R1 measured at *room* temperature may deviate from the value specified on the manufacturing drawings, but there is

R1 := 0.500   R2 := 1.0847R1   Rs := 0.0335   Le := 5.0R1   G := 11.2 · 10⁶

Ls := 2.00   La := 3.00   Kv := 1000.0   e := 5.00   Ren := 10⁷   Ncts := $\dfrac{10^9}{60}$

$$\text{delT} := 1.0$$

$$\text{Cr} := \frac{R2^2 - R1^2}{4 \cdot \pi \cdot R1^2 \cdot R2^2 \cdot \text{Le}} \qquad \text{Cr} = 0.019$$

$$\text{Ct} := \frac{\pi \cdot Rs^4 \cdot G \cdot e}{2 \cdot Ls \cdot La \cdot Kv} \qquad \text{Ct} = 0.018$$

$$\text{Cw} := \frac{1.0}{\dfrac{\text{Ncts}}{\text{delT}} \cdot \dfrac{1}{\text{Ren}} \cdot 2 \cdot \pi} \qquad \text{Cw} = 0.095$$

$$\text{mu} := \text{Cr} \cdot \text{Ct} \cdot \text{Cw} \qquad \text{mu} = 3.369 \cdot 10^{-5}$$

Now compute the various derivatives of viscosity with respect to a parameter

$$\text{dudR1} := \frac{\text{Ct} \cdot \text{Cw}}{4 \cdot \pi \cdot \text{Le}} \cdot \frac{-2}{R1^3} \qquad \text{dudR1} = -8.98 \cdot 10^{-4}$$

$$\text{dudR2} := \frac{\text{Ct} \cdot \text{Cw}}{4 \cdot \pi \cdot \text{Le}} \cdot \frac{2}{R2^3} \qquad \text{dudR2} = 7.036 \cdot 10^{-4}$$

$$\text{dudLe} := \frac{\text{Ct} \cdot \text{Cw}(R2^2 - R1^2)}{4 \cdot \pi \cdot R1^2 \cdot R2^2} \cdot \frac{-1}{\text{Le}^2} \qquad \text{dudLe} = -1.348 \cdot 10^{-5}$$

$$\text{dudRs} := \text{Cr} \cdot \text{Cw} \cdot \frac{4 \cdot \pi \cdot Rs^3 \cdot G \cdot e}{2 \cdot Ls \cdot Kv \cdot La} \qquad \text{dudRs} = 4.023 \cdot 10^{-3}$$

$$\text{dudG} := \frac{\pi \cdot \text{Cr} \cdot \text{Cw} \cdot Rs^4 \cdot e}{2 \cdot Ls \cdot Kv \cdot La} \qquad \text{dudG} = 3.008 \cdot 10^{-12}$$

$$\text{dudLa} := \frac{-1}{La^2} \cdot \frac{\pi \cdot \text{Cr} \cdot \text{Cw} \cdot Rs^4 \cdot G \cdot e}{2 \cdot Ls \cdot Kv} \qquad \text{dudLa} = -1.123 \cdot 10^{-5}$$

$$\text{dudKv} := \text{Cr} \cdot \text{Cw} \cdot \frac{\pi Rs^4 G \cdot e}{2 \cdot Ls \cdot La} \cdot \frac{-1}{Kv^2} \qquad \text{dudKv} = -3.369 \cdot 10^{-8}$$

$$\text{dudRen} := \text{Cr} \cdot \text{Ct} \cdot \frac{\text{delT}}{2 \cdot \pi \cdot \text{Ncts}} \qquad \text{dudRen} = 3.369 \cdot 10^{-12}$$

$$\text{dude} := \frac{\pi \cdot \text{Cr} \cdot \text{Cw} \cdot Rs^4 \cdot G}{2 \cdot Ls \cdot Kv \cdot La} \qquad \text{dude} = 6.738 \cdot 10^{-6}$$

$$\text{dudLs} := \frac{\pi \text{Cr} \cdot \text{Cw} \cdot Rs^4 \cdot G \cdot e}{2 \cdot Kv \cdot La} \cdot \frac{-1}{Ls^2} \qquad \text{dudLs} = -1.685 \cdot 10^{-5}$$

$$dudNcts := \frac{Cr \cdot Ct \cdot delT \cdot Ren}{2 \cdot \pi} \cdot \frac{-1}{Ncts^2} \qquad dudNcts = -2.022 \cdot 10^{-12}$$

$$duddelT := \frac{Cr \cdot Ct \cdot Ren}{2 \cdot \pi \cdot Ncts} \qquad duddelT = 3.369 \cdot 10^{-5}$$

Now make 1% changes in each parameter and compute the resulting percent change in viscosity.

$delR1 := 0.01 \cdot R1$ $delR2 := 0.01 \cdot R2$ $delRs := 0.01 \cdot Rs$

$delLe := 0.01 \cdot Le$ $delG := 0.01 \cdot G$ $dells := 0.01 \cdot Ls$

$delLa := 0.01 \cdot La$ $delKv := 0.01 \cdot Kv$ $dele := 0.01 \cdot e$

$delRen := 0.01 \cdot Ren$ $delNcts := 0.01 \cdot Ncts$ $deldelT := 0.01 \cdot delT$

$dmuR1 := 100 \dfrac{dudR1 \cdot delR1}{mu}$ $dmuR1 = -13.327$ $dmuNcts := 100 \dfrac{dudNcts \cdot delNcts}{mu}$

$dmuR2 := 100 \dfrac{dudR2 \cdot delR2}{mu}$ $dmuR2 = 11.327$ $dmuNcts = -1$

$dmuRs := 100 \dfrac{dudRs \cdot delRs}{mu}$ $dmuRs = 4$ $dmudelT := 100 \dfrac{duddelT \cdot deldelT}{mu}$

$dmuLe := 100 \dfrac{dudLe \cdot delLe}{mu}$ $dmuLe = -1$ $dmudelT = 1$

$dmuG := 100 \dfrac{dudG \cdot delG}{mu}$ $dmuG = 1$ $dmue := 100 \dfrac{dude \cdot dele}{mu}$

$dmuLs := \dfrac{100 dudLs \cdot dells}{mu}$ $dmuLs = -1$ $dmue = 1$

$dmuLa := 100 \cdot \dfrac{dudLa \cdot delLa}{mu}$ $dmuLa = -1$ $dmuRen := 100 \cdot \dfrac{dudRen \cdot delRen}{mu}$

$dmuKv := 100 \cdot \dfrac{dudKv \cdot delKv}{mu}$ $dmuKv = -1$ $dmuRen = 1$

additional variation when the temperature differs from the assumed ambient. These two components of variation are statistically *independent*, so the total variation in R_1 is the sum of the two:

$$\Delta R_1 = \Delta R_{1,\text{dim}} + \Delta R_{1,\text{temp}} \tag{6.49}$$

If we use Equation 6.47 to compute the statistical uncertainty in viscosity, when entering $u_{R_1}^2$ we would enter it as

$$(u_{R_1,\text{dim}}^2 + u_{R_1,\text{temp}}^2) \qquad \text{not as} \qquad (u_{R_1,\text{dim}} + u_{R_1,\text{temp}})^2 \tag{6.50}$$

since the two components of uncertainty are independent. All the other parameters which are functions of temperature can similarly have their uncertainties represented as a sum of *inherent* and *temperature* contributions. Having raised this issue we will, however, not pursue the details into actual computations. We will point out an additional effect on viscosity uncertainty, the *eccentricity* of the inner and outer cylinders. That is, when the viscosimeter is assembled, the axes of these two cylinders cannot be perfectly aligned; there will be a radial displacement of these axes. This is expressed as the ratio c of the axis displacement to the nominal radial gap. This effect has been theoretically studied* and the result predicts (for Newtonian fluids) a torque increase on the rotating cylinder by a factor $(1 + 2c^2)$ and a torque decrease on the other cylinder by a factor $(1 - c^2)$. For example, if the gap is 0.04 in. and the axis misalignment 0.001 in., then $c^2 = 0.000625$, and the effect is probably negligible. For narrower gaps, which some instruments use to get more accurate results for non-Newtonian fluids or for very low-viscosity fluids, the correction might be significant.

While the RSS uncertainty analysis is needed to find the effect of each uncertain parameter on the viscosity uncertainty, another approach may also be useful. Here, we use statistical software (such as Minitab) to make each parameter a statistical variable, with its own mean value (average) and standard deviation. We can even choose the statistical *distribution function* from a list of available models, although we usually choose the *Gaussian* (normal) distribution unless we have specific information to the contrary. We request the software to generate an n-item sample of each of the parameters. We then enter these values into the formula (use Equation 6.48) for viscosity and the software calculates n values of viscosity, which will exhibit statistical scatter. We can then ask the software to calculate the mean value and standard deviation of the viscosity values, which will show us how

---

* R.W. Whorlow, *Rheological Techniques*, Wiley, New York, 1980, p. 168.

accurate our viscosity values are. To see how this works out, let us take for each parameter mean value the same numbers we just used above. Let the standard deviation be such that two standard deviations (95% probability) equal 1% of the mean value. (In practice we would of course assign each parameter *its own* standard deviation, rather than using 1% for all of them.) We will use the Gaussian distribution and a sample size of 20 for each parameter.

If the formula for viscosity were *linear* in the various parameters, then theory says the viscosity will also have a Gaussian distribution. Since our formula is *nonlinear* in some of the parameters, viscosity will *not* follow a Gaussian distribution. When the nonlinearities are not too severe and/or the parameter deviations from their mean values are sufficiently small, then the viscosity distribution will be *close* to Gaussian. The Minitab software allows one to visually judge whether a data set clearly deviates from Gaussian. Go to **graph > probability plot** and then ask for analysis and plotting for the viscosity sample. An *infinite-size* sample of perfectly Gaussian data will plot on this graph as a straight line. This graph will show whether your data sample is near Gaussian or far from Gaussian, but one can *never* prove that any data is *exactly* Gaussian. To *educate* our visual judgment on such matters, we can ask for this graph when the data is *known* to be Gaussian. We can try this on any or all of the *parameter* data samples, since we *generated* these using a random number generator which is known to be very close to Gaussian. The problem, of course, is that a *small* sample of perfectly Gaussian data will *not* plot as a straight line. By looking at samples of the *same size*, both Gaussian and *questionable*, we can improve our evaluation of any real data set that we might have to deal with.

Figure 6.39 shows results when the parameters all have Gaussian distributions with mean values as stated in the Mathcad program above, and standard deviations equal to 0.5% of the mean values. This makes two standard deviations equal to 1.0%, and thus 95% of a parameter sample's values will be found within ±1% of the mean value. While the nonlinearity of Equation 6.48 makes the distribution of viscosity non-Gaussian (even though the parameter distributions were all Gaussian), Minitab's *probability plot* (Figure 6.40a) shows that the distribution of viscosity is quite close to Gaussian. We reach this conclusion by comparing Figure 6.40a (the viscosity graph) and 6.40b (a similar graph for the parameter R_1). The straight line shown in each graph is where the points would fall if we had a perfect Gaussian distribution *and an infinite-size sample*. For samples of size 20, the graph of R_1, *even though this data has a nearly perfect Gaussian distribution* (because it was generated by a software random number generator that is nearly perfect) does not, and cannot, fall on the *perfect straight line*. Since these two graphs look quite similar, we conclude that the viscosity distribution is close to Gaussian. Remember that it is always *impossible* to prove any real-world data has *exactly* a Gaussian (or any other analytical) probability distribution; the best we can do is to say it is "close to" or "obviously not" Gaussian.

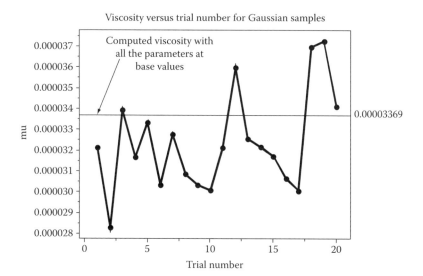

FIGURE 6.39
Effect of parameter variation on computed viscosity (Gaussian parameter distributions).

The viscosity numbers for most of the 20 samples in Figure 6.39 show quite large deviations from the base value of 0.00003369, which might be taken as indication of unacceptable uncertainty in actual viscosity measurements. This large predicted uncertainty, for rather small uncertainty of the individual parameters, is in fact not a proper interpretation. The calculation performed refers to a situation where the base value of viscosity corresponds to the parameters all being exactly at the nominal values of design specifications. We then allow each parameter to vary according to its specified standard deviation, leading to the predicted standard deviation (0.00000236) of the measured viscosity as seen in Figure 6.40a. The situation to which such a calculation applies is one where we manufacture all the system parts and then *accept* these parts with whatever manufacturing variation occurs, and then assemble them directly into a viscosimeter, each of the 20 *trials* of our calculation referring to assembly of one viscosimeter. In actuality, however, we take each component part and carefully *measure* its significant parameters with a measurement uncertainty that is usually much *less* than the manufacturing uncertainty. We then use these measured values in our formula for calculating viscosity, which results in a much smaller (and acceptable) practical uncertainty. Also, *calibrations* of individual components or subassemblies can further reduce the actual uncertainty. A detailed discussion of rheometer uncertainty for practical measurements is available in the literature.*

---

* G. Schramm, *A Practical Approach to Rheology and Rheometry*, Haake, Karlsruhe, Germany, 1994, pp. 154–168.

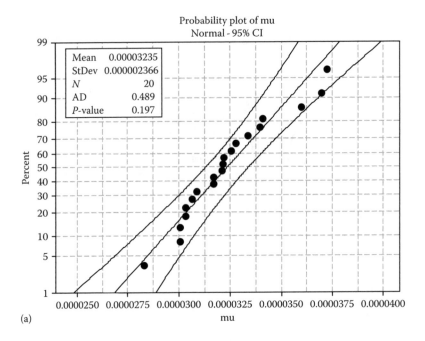

(a)

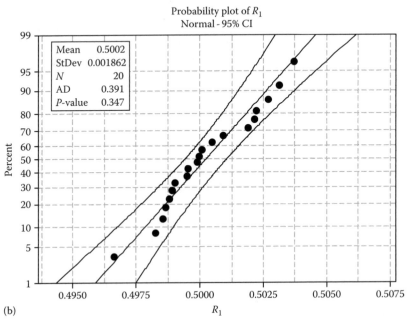

(b)

FIGURE 6.40
Check for Gaussian distribution of viscosity.

Versatile statistical software such as Minitab allows us to explore another version of these uncertainty calculations. The Gaussian distribution may not be the best choice when we measure individual components and then use these numerical values in our calculations. That is, the Gaussian distribution allows for the *occasional* occurrence of values that are rather far from the average (mean) value for a parameter. When we measure, say, R_1 with a conventional micrometer, it is easy to read this to about 0.0001 in., so if the nominal (average) value is taken as 0.5000, then a more realistic description of the possible variation is that it might lie between 0.5001 and 0.4999. A more suitable distribution function for such situations is the *uniform distribution*, not the Gaussian. When specifying a particular uniform distribution, one gives the lowest value (0.4999 in our example) and the highest value (0.5001). The random sample generated by the software will have *all* its values within this range, with any particular value being as likely as any other. Our Mathcad program showed that R1, R2, and Rs contributed disproportionally to the overall uncertainty, so I assumed they could each be measured to the nearest 0.0001 in. and was uniformly distributed over a range of ±0.0001 in. All the other parameters were also taken as uniformly distributed, with a range of ±0.5% of their mean values. This analysis resulted in the graph of Figure 6.41, which shows much smaller uncertainty for the computed viscosity than Figure 6.39. Most of the parameters can probably be measured more accurately than 0.5%, so if we had available these more realistic numbers, the viscosity uncertainty would be even less.

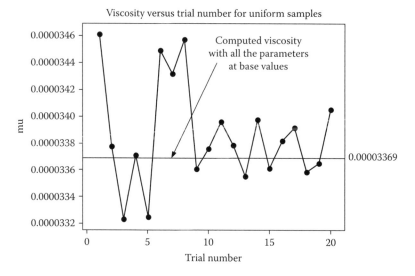

FIGURE 6.41
Effect of parameter variation on computed viscosity (uniform distributions and reduced uncertainty in R_1, R_2, and R_s).

6.17 Encoder Angular Position and Speed Measurement

We had earlier mentioned the use of incremental shaft angle encoders for accurate angular position and speed measurements, but some further details may be useful for those unfamiliar with this technology. Sophisticated rheometers require such measurements over rather large ranges. Incremental encoders for these applications are usually the highest resolution devices available; the order of 1 million counts per revolution. Figure 6.42 gives some details of a Canon optical encoder (Model X-1M). Its output signal is of sinusoidal form, with 225,000 sine wave cycles per revolution. Such signals can be further processed in an electronic *interpolator** to provide even

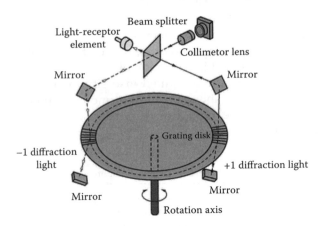

As illustrated in the diagram, laser beams are applied to two points equidistant from the grating disk's center of revolution. One diffraction beam is positive first order (+1) and the other is negative first order (−1).

For each 1 pitch that the grating disc revolves, the ±1 diffraction light will change each phase by ±2π. Reflecting the ±1 diffraction light into respective mirrors and then reapplying it to the grating disk changes the phase by ±4π.

In this way, each time the grating disk revolves 1 pitch, the brightness interference signals for 4 cycles can be obtained, making highly accurate angle sensing possible.

FIGURE 6.42
Canon optical incremental shaft angle encoder.

---

* E.O. Doebelin, *Measurement Systems*, 5th edn., McGraw-Hill, New York, 2003, p. 330.

Encoder Rotary encoder	Signal type	Interpolator	Counter[+1]	Final resolution (")
R-1O	⊓⊔⊓ Balance		Counter	4
R-1L	⊓⊔⊓ Balance		Counter	4
M-1	⊓⊔⊓		Counter	6.48
K-1	∿	CI16-2 / IU16		1
X-1M	Balance ∿	CI-200GA		0.0288
		CI-1000GA		0.00576
KP-1Z	Balance ∿	CI-200GA	Counter	0.08
		CI-1000GA		0.016

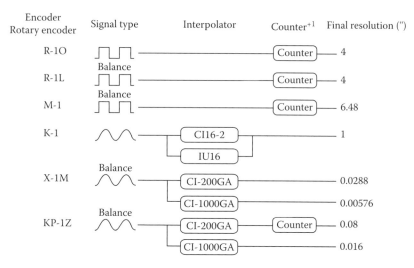

FIGURE 6.43
Resolution of various Canon encoders and interpolators.

more *counts* per revolution. In Figure 6.43, we see that the X-1M can be used with two different interpolators, which accept the sinusoidal encoder output signal and produce many rectangular voltage pulses (counts) for each sine wave cycle. The CI-200GA interpolator divides each sine wave into 200 parts, giving 200 pulses/sine wave, and thus an angular resolution of 0.0288". The CI-1000GA divides each sine wave into 1000 parts, giving 5 times finer resolution. Whether such extreme resolutions can be usefully applied in a rheometer or other mechanism depends on careful design and use of the overall system. The Paar Physica MCR 501 rheometer brochure states the angular resolution as 10 nanoradians, which is 5.73e–7° or 0.00206", a little better than the Canon X-1M/C1-1000GA combination, so it appears that these very fine resolutions can be practically realized.

When a viscometer/rheometer is used to test viscoelastic materials using sinusoidal test signals, the angular amplitude is usually very small, 1° or less. This is because the experimenter is often interested in the *linear* rather than *nonlinear* regime of material behavior, and materials generally go nonlinear for large stresses or strains. When operating with such small motions, the motion transducer (digital encoder) must have sufficient resolution to get accurate data (amplitude ratio and phase angle) for the sine waves of stress and strain. The Paar Physica instrument just mentioned claims to work with amplitudes as small as 100 nrad, which means that a sine wave would be "plotted" with about 20 points per cycle, which seems reasonable.

When the same incremental encoder used to measure angular displacement is used to measure angular *velocity*, two approaches employed in general measurements might be applied. Perhaps the most common is to count the number of displacement pulses over an accurately measured time interval and compute the ratio as the average angular velocity over that time interval. Using the same Paar Physica instrument, the claim for the lowest usable velocity is 1e–7 RPM. One RPM with this instrument produces 1.75e6 pulses/min, which is 10.05 pulses/s for 1e–7 RPM. Since pulse counting is usually uncertain by about 1 count, for 1% velocity accuracy we would need to accumulate 100 pulses, which would take about 10 s to get a velocity measurement, which again seems reasonable for such a slow rotation. The other approach to velocity measurement with an incremental encoder is one that is also used in a commercial *frequency-to-voltage converter*.[*] Here an accurate digital clock is used to time the interval between every two adjacent pulses and the angular resolution (angle between pulses) is divided by the time interval to get a velocity value. This has the advantage of getting velocity updates very quickly but the disadvantage of not providing the *averaging* associated with the other method. For the minimum velocity of 1e–7 RPM, the time between two adjacent pulses is about 0.1 s, so for 1% accuracy we must be able to time this interval to the nearest 0.001 s, which is not difficult with, say a 1 MHz digital clock. It appears that either of these methods might be used in a rheometer; the sales brochures of course do not usually reveal this level of detail.

6.18 Practical Significance of the Shear Rate

While this chapter is not designed to provide even an introduction to the science of rheology, I did want to provide at least a little insight into the practical utility of the kinds of measurements we have discussed. We earlier mentioned that it is often of interest to study the effect of shear rate on fluid viscosity, but did not explain in detail *why* this kind of information is important. Many important industrial processes involve fluids in situations where the success of the process requires that a fluid exhibit properties whose numerical values must lie in a certain range. All fluids are somewhat non-Newtonian and many are strongly so, which means that viscosity varies with shear rate, and a process fluid may be required to have specific viscous (or viscoelastic) properties within a fairly narrow range of shear rates. These properties can only be established through accurate rheological measurements using the kinds of instruments that we have described in this chapter. Papermaking involves fluid processes of various kinds at different stages of manufacture. Figure 6.44 shows how one can estimate the shear rate associated with the *paper coating process* using some simple assumptions. Such coating operations are

---

[*] E.O. Doebelin, *Measurement Systems*, 5th edn., McGraw-Hill, New York, 2003, pp. 343–344.

Quality papers are coated to get a smooth and often glossy surface.

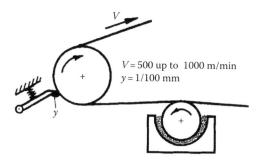

FIGURE 6.44
Schematic of a coating machine. (Adapted from Schramm, G., *A Practical Approach to Rheology and Rheometry*, Haake, Karlsruhe, Germany, 1994, p. 146.)

Typical production conditions in a continuous coating process are:

Paper speed $\qquad v = 500$ [m/min] (up to 1000 m/min)

$$v = 50{,}000 \text{ (cm/min)}$$

$$v = \frac{50{,}000}{60} \text{ (cm/s)}$$

Gap size y = distance between the nip of the scraper blade and the roll surface

$$y = \frac{1}{100} \text{ [mm]} = \frac{1}{1000} = 10^{-3} \text{ (cm)}$$

$$\text{Shear rate } \dot{\gamma} = \frac{50{,}000}{60} \cdot 10^{3} = 8.3 \cdot 10^{5} \text{ (s}^{-1})$$

Maximum shear rates in paper coating can exceed values of 10^{6} (s^{-1}).

successful only if the coating fluid has the proper viscous behavior at the shear rates existing in the process. Figure 6.45 addresses the process of brushing paint onto a surface. When paint is applied to a vertical surface we encounter the problem of *sagging* due to gravity forces, which is also analyzed. Quoting from Schramm (1994): "To have good application characteristics, good paints are obviously non-Newtonian liquids which are highly shear rate dependent. Very important is whether paints show a yield stress value, which requires measurements close to zero shear rate. To classify the quality of paints in rheological terms, one must consider not just one shear rate but a shear rate range covering sometimes more than 6 decades." The term *yield stress value* refers to a viscoelastic property

In normal processing, paints encounter a wide shear rate range. When paint is scooped out of a can, the shear rate can be estimated as $\dot{\gamma} = 10 \, (\text{s}^{-1})$.

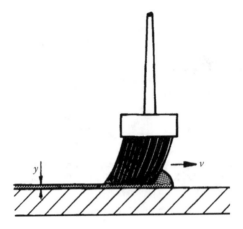

FIGURE 6.45
Brushing paint onto a flat surface. (Adapted from Schramm, G., *A Practical Approach to Rheology and Rheometry*, Haake, Karlsruhe, Germany, 1994, p. 146.)

Estimation of the shear rate of brushing:
During application, the brush can be drawn across a board at a speed of $v = 1 \, \text{m/s} = 1000 \, \text{mm/s}$
The thickness of the paint layer is estimated as
$y = 0.2 \, \text{mm}$
then the shear rate of brushing is approximately:

$$\dot{\gamma} = \frac{1000}{0.2} = 5.000 \, (\text{s}^{-1})$$

Paint sprayed out of the nozzle of an air gun is subjected to shear rates exceeding $\dot{\gamma} = 50.000 \, (\text{s}^{-1})$.

that we have not discussed in this chapter, but is routinely measured with rheometers. Its measurement requires accurate rheometer operation at very low angular velocities, which we have shown possible in the previous section's treatment of incremental encoders. Figure 6.46 (same reference) analyzes the engine processes of piston sliding and lubricant pumping. Table 6.3* shows typical shear rates associated with a range of industrial processes, while Table 6.4 from the same reference (Mezger 1998, p. 12) illustrates the vast range of viscosities for some fluids of practical importance.

---

* T. Mezger, A little course in rheology, *Paar Physica*, 1998, part 1, p. 8.

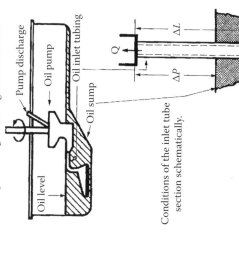

Oil pump provides vaccum to draw oil out of the oil sump of an engine through a tube section.

Conditions of the inlet tube section schematically.

Typical dimensions are:

y = Gap size = 0.03 (mm)
n = Rotational speed of the crankshaft = 4000 (min^{-1})
v_{max} = Maximum speed of the piston = 20 (m/s)
v_{mean} = Mean speed of the piston = 14 (m/s)

This leads to:

$$\dot\gamma_{max} = \frac{V_{max}}{y} = \frac{20.000\,mm/s}{0.03\,mm} \approx 6.7 \cdot 10^5 \;(s^{-1})$$

or

$$\dot\gamma_{mean} \approx 4.7 \cdot 10^5 \;(s^{-1})$$

Shear rates between piston rings and the cylinder obviously reach extremely high levels.

Typical dimensions are:

d = Tube diameter = 0.7 cm = 7 (mm)
R = Tube radius = 0.35 cm = 3.5 (mm)
Q = Flow rate = 30 cm^3/min = 500 (mm^3/s)

$$\dot\gamma = \frac{4}{\pi} \cdot \frac{Q}{R^3} = \frac{4}{3.14} \cdot \frac{500}{3.5^3} \approx 15\,(s^{-1})$$

Shear rates encountered by engine oils in the inlet tube section of the oil pump are very low. Shear rates are even lower in the oil sump itself.

FIGURE 6.46
Shear rate estimates for two engine processes. (Adapted from Schramm, G., *A Practical Approach to Rheology and Rheometry*, Haake, Karlsruhe, Germany, 1994, p. 146.)

TABLE 6.3

Process Shear Rates

Process	Shear Rate Range D (1/s)	Examples for Applications
Sedimentation of small particles in a suspension	10^{-6}–10^{-4}	Coatings, paints, varnishes, pharmaceutical solutions
Sedimentation of bigger particles in a suspension	10^{-4}–10^{-1}	Ceramic suspensions
Leveling due to surface tensions	10^{-2}–10^{-1}	Coatings, paints, inks
Sagging under gravitational forces	10^{-2}–10^{1}	Coatings, paints
Extrusion	10^{0}–10^{2}	Polymers
Press	10^{0}–10^{2}	Polymer molding material
Calender	10^{0}–10^{2}	Dispersions, polymers
Dip coating	10^{0}–10^{2}	Dip coatings, food
Chewing/swallowing	10^{1}–10^{2}	Food
Tube or pipeline flow	10^{0}–10^{3}	Pumped fluids (e.g., blood)
Mixing, stirring	10^{1}–10^{3}	Process fluids in chemical plants
Injection molding	10^{1}–10^{4}	Polymers
Painting, brushing, spraying, blade coating	10^{2}–10^{4} 10^{3}–10^{6}	Paints, spray coatings, fuel injection
Wet grinding of pigments	10^{3}–10^{5}	Printing inks
Rubbing	10^{4}–10^{5}	Skin creams, sun lotion
Rolling	10^{4}–10^{6}	Printing inks
High-speed coating	10^{5}–10^{6}	Paper coatings
Lubrication of machine parts	10^{3}–10^{7}	Mineral oil, lubricating grease

6.19 Fitting a Power-Law Model for a Non-Newtonian Fluid

Equation 6.24 shows the form of the power-law model that is quite commonly used to describe the viscosity behavior of non-Newtonian fluids. To see how well this model works with some actual data, I obtained some measurements on whole human blood, which is well known to be non-Newtonian. (Blood *plasma* is nearly Newtonian but the red blood cells contained in it give whole blood the non-Newtonian behavior.) The data in the first two columns of Table 6.5 comes from Rheologics Inc.,* a company founded by Kenneth R. Kensey, M.D., an alumnus of Ohio Wesleyan University who got his MD degree from the Ohio State University. This company markets a unique (patented) form of *capillary* viscometer which can obtain the data needed for a non-Newtonian model in about 2 min. This speed is significant, since blood will start to coagulate for longer times, making measurements unreliable. I want to thank Nathan Krieger (proj-

---

* www.rheologics.com

TABLE 6.4

Viscosities of Some Fluids (at 20°C unless Otherwise Stated)

Substance (Temperature)	Shear Viscosity h (mPa-s)	Order of Magnitude
Gases	0.01–0.02	0.01
Liquid hydrogen	001	
Air	0.018	
Pentane	0.23	0.1
Acetone	0.32	
Benzine	0.54–0.65	
Water		
20°	1.00	1.0
0°	1.79	
40°	0.65	
Ethanol, alcohol	1.2	
Mercury	1.55	
Blood plasma		
20°	1.7	
37°	1.2	
Whole blood		
20° (in healthy subject)	5–120 ($D = 0.01$–1000 1/s)	
37° (in healthy subject)	4–15 ($D = 0.01$–1000 1/s)	
Wine, fruit juice (undiluted, without additions)	2–5	
Milk, cream	5–10	10
Glycol	20	
Sulfuric acid	25	
Sugar solution (60%)	57	
Motor oil	ca. 50–1000	100
E.g., SAE 5W (50°)	17	
SAE60	100	
SAE 250 (50°)	520	
SAE 10W/30 (50°/100°)	40/11	
Gear lubricant oil	ca. 300 bis 800	
Olive oil	ca. 100	
Castor oil	ca. 1000	1000
Glycerin	1480	
Honey	ca. 10.000	10.000
Syrup	ca. 1000 bis 10.000	
Polymer melt	ca. 10^4 to 10^8	
Tar/silicone caoutchouc/bitumen	ca. 10^5–10^8	100.000
Glass melt	ca. 10^{15}	($\geq 10^6$)
Glacier ice	ca. 10^{16}	
Aluminum melt	ca. 10^{20}	
Earth crust	ca. 10^{24}	

TABLE 6.5

Blood Viscosity Data for Power-Law Curve Fit
(Two Human Subjects)

1	2	3	4	5	
Vis, MPa-s	Shear Rate	Shear Stress	Log Rate	Log Stress	
18.00	1.28	23.13	0.10890	1.36418	
8.31	8.11	67.39	0.90902	1.82862	
6.26	27.70	173.40	1.44248	0.23905	
6.29	37.50	235.88	1.57403	2.37268	Subject 1
5.38	75.00	403.50	1.87506	2.60584	
5.06	94.50	478.17	1.97543	2.67958	
4.80	150.00	720.00	2.17609	2.85733	
4.31	300.00	1293.00	2.47712	3.11160	
3.89	750.00	2917.50	2.87506	3.46501	
3.65	1500.00	5475.00	3.17609	3.73838	
10.22	1.28	13.13	0.10890	1.11835	
6.17	8.11	50.04	0.90902	1.69931	
5.00	27.70	138.50	1.44248	2.14145	
4.97	37.50	186.38	1.57403	2.27039	
4.31	75.00	323.25	1.87506	2.50954	Subject 2
4.28	94.50	404.46	1.97543	2.60688	
3.81	150.00	571.50	2.17609	2.75702	
3.58	300.00	1074.00	2.47712	3.03100	
3.26	750.00	2445.00	2.87506	3.38828	
3.11	1500.00	4665.00	3.17609	3.66885	

ect engineer) and Bill Hogenauer (vice president) for providing this data and technical information on their viscometer. Basically we need a set of measurements of shear rate and corresponding shear stress to fit a power-law model. Rheologics supplied a table of viscosity (really *apparent viscosity*) and shear rate for blood samples from 23 different human subjects. (For non-Newtonian fluids, it is conventional to call the ratio (shear stress)/(shear rate) the *apparent* viscosity.) One can of course easily compute the shear stress from this data. I picked two subjects from the 23 available, choosing the high and low extremes of viscosity for analysis.

To carry out a least-squares curve fit for the power-law model, using Minitab's software, we need to take *logarithms* of the power-law equation in order to get a model that is *linear* in the two coefficients, k_1 and n.

$$\text{shear rate} \underline{\Delta} y = k_1(\text{shear stress})^n = k_1(x)^n \quad \log(y) = \log(k_1) + n\log(x) \quad (6.51)$$

The data in columns 2 and 3 of the table must then be converted to logarithms in columns 4 and 5 and submitted to the linear regression routine to find values of $\log(k_1)$ and n (the two coefficients in the equation) which give the best fit. We then get k_1 itself by taking the antilog. Having k_1 and n we can compute the predicted shear rate values for the given shear stress values, and compare these with the measured shear stresses to judge the quality of fit of the model to the data. Figure 6.47 shows results of the curve-fit studies

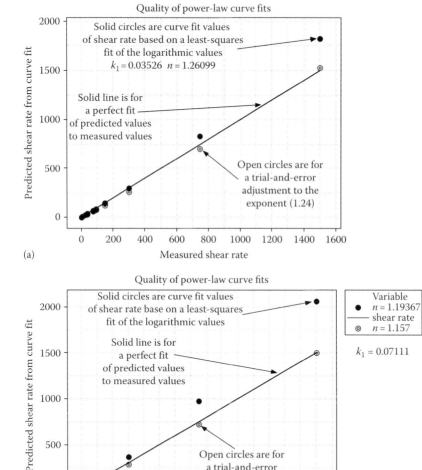

FIGURE 6.47
Power-law curve fits for human blood viscosity: (a) subject 1 and (b) subject 2.

for the blood samples from two different human subjects. For both sets of blood data, we see that the curve fit is not very good for the larger shear rates. Since the least-squares procedure is guaranteed to produce a curve fit which minimizes the sum of the squares of the fit errors, it appears that the power-law model is not well suited to fitting this kind of data, and that we need to try some other kind of equation. Actually, the equation is really adequate; the fault is in the curve-fit procedure we used. That is, to linearize the problem, we took logarithms, so the procedure will minimize the sum of squares of the logarithms, not the shear stress errors themselves, which can lead to poor fits.

To try to improve the fit, I tried adjusting only the value of the exponent n, by trial and error. This turned out to be quite effective for both sets of data. For subject 1's data, I found that changing the exponent from 1.26099 to 1.24 gave a much better fit; for subject 2, the adjustment was from 1.19367 to 1.157. Figure 6.47 shows the improvement. These adjustments were not tedious and gave good results, but are not very "scientific"! A better way is available if one has suitable software; it is called *nonlinear regression*. Here the error minimization is done on the *actual* curve-fit function, not on some distorted version of it, such as our logarithmic transformation. The problem with such procedures is that they are in the nature of iterative algorithms, rather than the "once through" solutions of linear algebra used in Minitab's (and other software's) routines. This means that the procedure may not *converge*, or if it does converge within some stated error limits, the solution may not be the *best* solution. There are a number of nonlinear regression methods available in the literature; we will use MATLAB's Levenberg-Marquardt routine.

To implement this routine one makes up two relatively simple m-files:

```
% an m-file used in the blood curve fitting
function F=bloodfun(x,xdata)
F=x(1).*xdata.^x(2)
% nonlinear least squares fit of blood viscosity
% crc_book_chap6_bloodfit
xdata=[23.3 67.39 173.4 235.88 403.50 478.17 720.0 1293 2917.5 5475]'
ydata=[1.28 8.11 27.7 37.5 75 94.5 150 300 750 1500]'
x0=[1,1]
[x,resnorm]=lsqcurvefit(@bloodfun,x0,xdata,ydata)
plot(xdata,ydata)
hold on
yfit=x(1).*xdata.^x(2)
plot(xdata,yfit)
[ydata yfit]
```

The first m-file simply defines the function that we wish to try for our curve fit and names any parameters that are to be adjusted during the curve-fit process as $x(1)$, $x(2)$, ... $x(n)$. In our case $x(1)$ is k_1 and $x(2)$ is n. The second file gives as "vectors" the measured x and y data that we want fitted, and also provides our best guess as to starting values for the x's. The statement x0 = [1,1] means that I gave the initial values of k_1 and n each as 1.0. This kind of arbitrary choice of the vector x0 usually works well and the algorithm quickly finds more appropriate values. However, if you *do* have a better guess, you should use it, since in some cases, a bad guess may lead to lack of convergence.

The statement *plot(xdata,ydata)* graphs the measured data. The statement yfit=x(1).*xdata.^x(2) takes the *results* (x(1) and x(2)) of the procedure and computes the curve using these results on the same graph as the measured data curve, for comparison. The last statement displays a table of the measured data and the curve-fit values, for more accurate comparison than is provided by the graph. Figure 6.48 shows the good results obtained for one set of data. This shows that the power-law model really is suitable for this blood data when we properly determine k_1 and n.

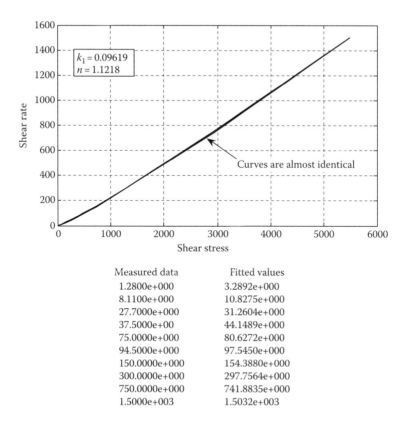

Measured data	Fitted values
1.2800e+000	3.2892e+000
8.1100e+000	10.8275e+000
27.7000e+000	31.2604e+000
37.5000e+00	44.1489e+000
75.0000e+000	80.6272e+000
94.5000e+000	97.5450e+000
150.0000e+000	154.3880e+000
300.0000e+000	297.7564e+000
750.0000e+000	741.8835e+000
1.5000e+003	1.5032e+003

FIGURE 6.48
Nonlinear regression gives excellent curve fit.

Bibliography

H.A. Barnes, J.F. Hutton, and K. Walters, *An Introduction to Rheology*, Elsevier, New York, 1989.

A. Dinsdale, *Viscosity and Its Measurement*, Reinhold, New York, 1962.

J. Ferguson and Z. Kemblowski, *Applied Fluid Rheology*, Elsevier, New York, 1991.

C.W. Macosko, *Rheology: Principles, Measurements, and Applications*, VCH Publishers, New York, 1994.

D.M. Rowe, *CRC Handbook of Thermoelectrics*, CRC Press, Boca Raton, FL, 1995.

G. Schramm, *A Practical Approach to Rheology and Rheometry*, Haake, Karlsruhe, Germany, 1994.

J.R. Van Wazer, J.W. Lyons, K.Y. Kim, and R.E. Colwel, *Viscosity and Flow Measurement: A Laboratory Handbook of Rheology*, Interscience, New York, 1963.

K. Walters, *Rheometry*, Wiley, New York, 1975.

R.W. Whorlow, *Rheological Techniques*, Wiley, New York, 1980.

7

Infrasonic and Ultrasonic Microphones

7.1 Introduction

In the field of acoustic measurements, the basic sensor is some kind of *microphone* (if the measurement is in a liquid, the sensor is usually called a *hydrophone*). Many measurements are related to the auditory response of human beings to the noise or sound, so the frequency range over which the sensor must measure accurately is taken as the *audible range*, nominally about 20–20,000 Hz. Most such microphones are pressure sensors designed to respond to the very small acoustic pressure variations around an average value which is the ambient atmospheric pressure of about 14.7 psia. The basic operation of such microphones is shown in a simplified manner in Figure 7.1. The housing or *cartridge* has a thin diaphragm (typically a stretched nickel film) attached, and a carefully designed *leak* (often a small capillary tube) serves to equalize the inside and outside pressures for pressure changes slower than about 10 Hz. For the higher frequencies of interest (the 20–20,000 Hz range), the leak is ineffective and the diaphragm feels a small pressure difference across it, which causes it to slightly deflect. The equalizing leak protects the very sensitive diaphragm from slow atmospheric pressure changes that are however very much larger than the acoustic signals of interest and would damage the instrument if allowed to act directly on it. To obtain an electrical output signal proportional to the acoustic pressure, the deflection of the diaphragm is measured with some type of displacement transducer, a capacitance method giving the highest accuracy. Further details of such microphones can be found in the literature.*

In this chapter, we discuss two special classes of microphones, *infrasonic* and *ultrasonic*, which are not used nearly as much as the "audio" instruments just described, but have important specialized applications. The names just given relate to the *frequency range* for which the devices are intended, infrasonic being the range below the lowest frequencies of human hearing and ultrasonic, the range above the highest.

---

* E.O. Doebelin, *Measurement Systems*, 5th edn., McGraw-Hill, New York, 2003, pp. 547–568.

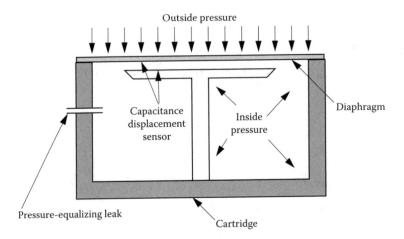

FIGURE 7.1
Microphone basic elements.

7.2 Infrasonic Microphones

Infrasonic microphones are extremely sensitive pressure transducers used to measure very small and low-frequency air pressures associated with the detection and analysis of *geophysical events* such as earthquakes, volcanic eruptions, distant nuclear and conventional explosions, missile launches, thunder, sonic booms, severe weather, and the entrance of meteorites and space debris into the atmosphere. The frequency range of interest is from about 0.001–10 Hz, and the amplitudes are in the range of 0.1 µ bar to 1 mbar (1 µ bar = 1 dyne/cm^2 = 1.45e−5 psi = 0.10 Pa). Because all these events take place within relatively large *background noise* always present in the atmosphere, these microphones will generally be dynamic systems characterized as *band-pass filters*, to screen out so far as possible the background and concentrate the sensitivity in the desired band of frequencies. Mutschlecner and Whitaker (1997)* describe a typical microphone in a simplified manner with the type of diagram seen in Figure 7.2. Another name for this type of instrument is *microbarograph* or *microbarometer* and many useful mechanical and electronic details can be found in the literature.†

The input to the system is the *atmospheric* pressure p_i, which may already have been filtered by a *noise reducer* of some sort, but we take p_i as our input signal. (All the pressures in the analysis are of course *small perturbations* away

---

* J.P. Mutschlecner and R.W. Whitaker, The design and operation of infrasonic microphones, Los Alamos National Laboratory Report LA-13257, UC-706, May 1997.
† R.V. Jones, *Instruments and Experiences*, Wiley, New York, 1988, pp. 145–247. This is a most interesting book, written by a famous Scottish physicist who was involved in many scientific and engineering developments that were critical to the British war effort during World War II.

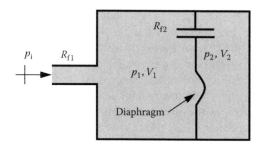

FIGURE 7.2
Generic infrasonic microphone.

from a steady base pressure.) The output signal will be taken as the pressure difference $(p_1 - p_2)$. Actually, this pressure difference deflects the diaphragm, which is transduced to an output voltage with some type of capacitance sensing scheme. Because the diaphragm is very thin and light, it is modeled as a pure spring; its inertia and damping effects are negligible at the low frequencies of interest. The output voltage would be directly and instantaneously proportional to the diaphragm deflection since the capacitance displacement sensor and associated electronics are much faster than the *pneumatic* responses.

The flow resistances, R_{f1} and R_{f2}, can be computed from dimensions and fluid properties, under the assumption of steady laminar flow, since the Reynolds numbers for these flows are *very* small compared to the critical value (2100) for transition from laminar to turbulent flow. Pressure changes and the associated volume changes are related by the adiabatic bulk modulus, B_a, which is assumed constant at the value kP_{base}, where k is the ratio of specific heats for air and P_{base} is taken equal to *standard atmospheric* pressure. At the lowest frequencies (0.001 Hz) isothermal conditions might be more nearly correct, but this only effects certain numerical values (bulk modulus is now P_{base} rather than kP_{base}), not the nature of the analysis. While conservation of *mass* is the fundamental principle, we will use conservation of *volume*, because the density is taken constant as given by the base pressure; the tiny pressure changes have a negligible effect on density. For each of the two volumes we can write over a time interval dt: volume in − volume out = additional volume stored

$$\frac{(p_i - p_1)}{R_{f1}} - \frac{(p_1 - p_2)}{R_{f2}} - C_d(Dp_1 - Dp_2) = \frac{V_1}{B_a}Dp_1 \tag{7.1}$$

$$\frac{(p_1 - p_2)}{R_{f2}} + C_d(Dp_1 - Dp_2) = \frac{V_2}{B_a}Dp_2 \tag{7.2}$$

Here $D \triangleq d/dt$, (derivative operator) and C_d is the diaphragm *compliance* (diaphragm volume displacement $= C_d(p_1 - p_2)$. The flow resistances are defined for *volume* flow rate, as (pressure difference)/(volume flow rate). All the parameters in these equations are assumed to be constant, so we have two simultaneous, linear differential equations with constant coefficients. Our goal is to get the transfer function relating pressure difference

$(p_1 - p_2)$ as an output to pressure p_i as an input. (Those who prefer to use Laplace transform methods can simply replace D's with s's in the above equations.) These equations agree with those given in Mutschlecner and Whitaker (1997), where the output signal is taken proportional to $(p_1 - p_2)$, rather than the (more correct) diaphragm deflection. Since our diaphragm mechanical model is a pure *metal spring* effect (no inertia or damping), diaphragm deflection would seem to follow the pressure difference *instantly*. This would be true except for the fact that when the diaphragm deflects, it *compresses* (or expands) the air on either side, giving an *air-spring* effect, which tends to reduce the amount of deflection and thus the output voltage. We will follow this reference for the time being, but will later modify their analysis to take into account the fact that the diaphragm displacement is influenced not only by the diaphragm *stiffness* but also by the stiffness of the air-spring effect, which occurs when the diaphragm deflection compresses or expands the air in the two volumes.

7.3 Diaphragm Compliance Calculation

We have defined the diaphragm compliance as the ratio (volume change)/ (pressure change) where the pressure is the *difference* in pressure across the diaphragm. To estimate the numerical value of this parameter, we analyze a simple model of the diaphragm. Diaphragms can be treated either as plates in bending or tensioned membranes, depending mainly on the thickness/diameter ratio and the material. For infrasonic microphone diaphragms, which are very thin and often made of plastic film (MYLAR, for example) that has been metallized on one side to serve as a capacitor "plate," the tensioned membrane model is usually used since this material has negligible bending stiffness. Figure 7.3 can be used to derive the differential equation whose solution gives the deflection z as a function of radius r, outer radius R, tension T, and pressure p. Considering a ring of width dr at radius r, a force balance gives

$$(p)(2\pi r dr) = -T(2\pi r)\frac{dz}{dr} + T(2\pi(r+dr))\left[\frac{dz}{dr} + \frac{d^2z}{dr^2}dr\right] \tag{7.3}$$

$$\frac{p}{T} = -\frac{1}{dr}\frac{dz}{dr} + \frac{1}{dr}\frac{dz}{dr} + \frac{1}{r}\frac{dz}{dr} + \frac{d^2z}{dr^2} + \frac{1}{r}\frac{d^2z}{dr^2}dr \tag{7.4}$$

let dr approach zero:
$$\frac{d^2z}{dr^2} + \frac{1}{r}\frac{dz}{dr} = \frac{p}{T} \tag{7.5}$$

This equation is *not* linear with constant coefficients (it has the variable coefficient $1/r$) so a "routine" solution is not available; however, it *is a classical*

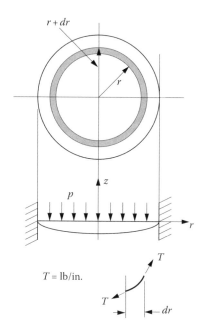

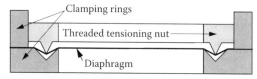

FIGURE 7.3
Analysis sketch of diaphragm and tensioning scheme suggested by Neubert.

equation with a known solution of the form $z = k_1 + k_2 r^2$, which we can substitute into Equation 7.5 to get

$$2k_2 + \frac{1}{r}(2k_2 r) = \frac{p}{T} \qquad k_2 = \frac{p}{4T} \tag{7.6}$$

$$z = k_1 + \frac{p}{4T}r^2 \qquad \text{when } r = R, z = 0 \qquad 0 = k_1 + \frac{p}{4T}R^2 \qquad k_1 = -\frac{p}{4T}R^2 \tag{7.7}$$

$$z = \frac{p}{4T}(r^2 - R^2) \quad \text{note that } z \text{ is negative, since up is positive in Figure 7.3} \tag{7.8}$$

This analysis, which gives a linear relation between pressure and deflection, is valid only for small deflections, but this is usually the case for microphone applications. If a large deflection relation is needed, such is available in the literature.* This same reference (Neubert 1963, p. 275) shows that an

---

* H.K.P. Neubert, *Instrument Transducers*, Clarendon Press, Oxford, 1963, p. 274.

air (permittivity ε) capacitor made up of the diaphragm and a flat plate a distance d from the undeflected diaphragm has a capacitance given by

$$\frac{\varepsilon R^2}{3.6d} + \frac{0.0347\varepsilon R^4 p}{d^2 T} \tag{7.9}$$

Diaphragm compliance (volume change)/(pressure change) is easily computed by integration, using Equation 7.8 as follows:

$$dV = -2\pi r z \, dr = 2\pi r \left(\frac{p}{4T}(R^2 - r^2) \right) \quad \frac{V}{p} = \frac{\pi}{4T} \cdot \int_0^R R^2 r \, dr - \int_0^R r \, dr = \frac{\pi R^4}{16T} \tag{7.10}$$

7.4 Microphone Transfer Function

Returning now to our original set of differential equations, our plan is to get single differential equations for each of the two pressures and then subtract one from the other to get the desired pressure difference. In this way, we will be able to solve for each pressure individually and also for their difference, giving a complete picture of how the system responds. Perhaps the easiest way to proceed now is to use determinants and Cramer's rule to get our single equation:

$$p_1 = \frac{\begin{vmatrix} \dfrac{p_i}{R_{f1}} & -C_d D - \dfrac{1}{R_{f2}} \\[2ex] 0 & \left(\dfrac{V_2}{B_a} + C_d\right)D + \dfrac{1}{R_{f2}} \end{vmatrix}}{\begin{vmatrix} \left(\dfrac{V_1}{B_a} + C_d\right)D + \dfrac{(R_{f1} + R_{f2})}{R_{f1}R_{f2}} & -C_d D - \dfrac{1}{R_{f2}} \\[2ex] -C_d D - \dfrac{1}{R_{f2}} & \left(\dfrac{V_2}{B_a} + C_d\right)D + \dfrac{1}{R_{f2}} \end{vmatrix}} \tag{7.11}$$

This gives the transfer function

$$\frac{p_1}{p_i}(D) = \frac{R_{f2}\left(\dfrac{V_2}{B_a} + C_d\right)D + 1}{(R_{f1}R_{f2})\left[\dfrac{V_1 V_2}{B_a^2} + (C_d)\left(\dfrac{V_1 + V_2}{B_a}\right)\right]D^2 + \left(\dfrac{V_2}{B_a R_{fe}} + \dfrac{V_1}{B_a R_{f2}} + \dfrac{C_d}{R_{f1}}\right)(R_{f1}R_{f2})D + 1} \tag{7.12}$$

To get the transfer function for p_2 we just replace the numerator determinant in Equation 7.11 by

$$\begin{vmatrix} \left(\dfrac{V_1}{B_a}+C_d\right)D+\dfrac{(R_{f1}+R_{f2})}{R_{f1}R_{f2}} & \dfrac{p_i}{R_{f1}} \\[2mm] -C_dD-\dfrac{1}{R_{f2}} & 0 \end{vmatrix}$$

while the denominator determinant stays the same. We then get the transfer function

$$\frac{p_2}{p_i}(D)=\frac{R_{f2}C_dD+1}{(R_{f1}R_{f2})\left[\dfrac{V_1V_2}{B_a^2}+(C_d)\left(\dfrac{V_1+V_2}{B_a}\right)\right]D^2+\left(\dfrac{V_2}{B_aR_{fe}}+\dfrac{V_1}{B_aR_{f2}}+\dfrac{C_d}{R_{f1}}\right)(R_{f1}R_{f2})D+1}$$

(7.13)

In these formulas, the symbol $R_{fe} \triangleq (R_{f1}R_{f2}/(R_{f1}+R_{f2})$. Both transfer functions have a steady-state gain of 1.0, indicating that a steady p_i will result in both p_1 and p_2 becoming equal to that p_i, which is obviously correct. The second-order denominator can be factored into two first-order terms, defining two time constants. As with any other transfer function, if we want the associated differential equation, we just *cross multiply* the numerators and denominators in an expression such as Equation 7.12, and then interpret terms such as D_{p_1}, for example, as dp_1/dt.

To get the response of $(p_1 - p_2)$ to p_i we just subtract the two transfer functions:

$$\frac{(p_1-p_2)}{p_i}(D)=\frac{\dfrac{R_{f2}V_2}{B_a}D}{(R_{f1}R_{f2})\left[\dfrac{V_1V_2}{B_a^2}+(C_a)\left(\dfrac{V_1+V_2}{B_a}\right)\right]D^2+\left(\dfrac{V_2}{B_aR_{fe}}+\dfrac{V_1}{B_aR_{f2}}+\dfrac{C_d}{R_{f1}}\right)(R_{f1}R_{f2})D+1}$$

(7.14)

This is the transfer function of a band-pass filter, defined by its static sensitivity $R_{f2}V_2/B_a$ and the two time constants in the denominator. We could factor the quadratic to get defining formulas for the two time constants, but prefer to do this when numerical values are available.

7.5 System Simulation

We need numerical values to do a simulation study and Mutschlecner and Whitaker (1997, p. 8) provide some typical numbers. The volumes are $V_1 = 40{,}000\,\text{cm}^3$, $V_2 = 200\,\text{cm}^3$, $C_d = 0.0001\,\text{cm}^3/(\text{dyne/cm}^2)$, $R_1 = 0.4\ (\text{dyne/cm}^2)/$

(cm^3/s), and $R_2 = 10,000$ $(dyne/cm^2)/(cm^3/s)$. The fluid resistance values are not given in standard SI units; the SI pressure unit would be the Pascal, however all the parameter units are consistent and the resistance values use a common acoustical unit, the *ohm*, which is the same as 1 $(dyne/cm^2)/(cm^3/s)$. That is, the reference gives the R values in ohms and I have just shown the details for those readers unfamiliar with this terminology. Before running our simulation it will be useful to first compute and graph the *frequency response*, as given by Equation 7.14. A MATLAB® program for this calculation goes as follows:

```
% infsound.m computes frequency response for infrasonic microphone
% first define parameter values
Rf1=0.4; Rf2=1e4; V1=4e4; V2=200; Cd=1e-4; Ba=1.4e6;
Rfe=Rf1Rf2/(Rf1+Rf2);
% next compute transfer function coefficients
num=[Rf2*V2/Ba 0];
d2=(Rf1*Rf2)*(V1*V2/(Ba^2)+Cd*(V1+V2)/Ba);
d1=(V2/(Ba*Rfe)+V1/(Ba*Rf2)+Cd/Rf1)*Rf1*Rf2;
den=[d2 d1 1];
% factor the denominator
[Wn,z]=Damp(den)
tau1=1/Wn(1)
tau2=1/Wn(2)
% estimate the time constants
taua=(Cd+V1/Ba)*Rf1
taub=(Cd+V2/Ba)*Rf2
w=logspace(-2,3,500);
h=freqs(num,den,w);
mag=abs(h);
ph=angle(h);
subplot(2,1,1),semilogx(w,mag,'k');grid;
phi=ph*57.3;
subplot(2,1,2),semilogx(w,phi,'k');grid;
```

The frequency-response graphs produced by this program are shown in Figure 7.4. We see that the passband for this particular set of numbers is from about 0.6–60 rad/s (0.1–10 Hz). The denominator second-order polynomial is factored to give two time constants *tau*1 and *tau*2, which are respectively 2.43 and 0.01145 s. For design purposes it is useful to have analytical estimates *taua* and *taub* of these two time constants, which are obtained from Equations 7.1 and 7.2 by considering them as separate (uncoupled) equations, giving the formulas shown in the program under *estimate the time constants*. The estimates are very good; 2.43 and 0.01147, so we can use these approximate formulas for choosing parameters to achieve a desired passband since the low- and high-frequency ends of the passband are roughly the reciprocals of the time constants, 0.41 and 87.2 rad/s.

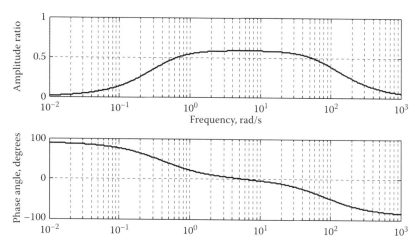

FIGURE 7.4
Frequency response of infrasonic microphone.

Returning to the simulation, we usually prefer to do dynamic system simulation *directly* from the original physical equations, in our case, Equations 7.1 and 7.2, since this makes all the physical variables directly available, and all the basic parameters appear directly in the block diagram. Figure 7.5a shows such a simulation, for an input pressure that contains three components with frequencies within the passband and two "noise" components, one above and one below the passband. The *results* graph of Figure 7.5b shows that the desired signals at 6 and 60 rad/s *get through* the filter, while the noise signal at 600 rad/s is attenuated significantly but still visible. The desired signal at 0.6 rad/s and the noise at 0.06 rad/s cannot be seen because they require longer recording times to become apparent. Using these longer recording times gives an *aliased* graph because the higher frequency components are *squeezed together*. We could make a simulation run that had only these two components in the input signal, if we wanted to see such results. (Since the amplitude ratio in the passband is about 0.6, when plotting the graph of Figure 7.5 I divided the p_o signal by 0.6 to give an easy comparison with the input signal.) Sometimes the microphone output voltage is subjected to additional *electronic* filtering; the 600 rad/s noise signal could be reduced by applying a suitable electronic low-pass filter. At first glance one might think that *all* the filtering could be done electronically, using a pressure sensor that had a *flat* frequency response. This is usually not feasible because, due to the required very high sensitivity, the pressure sensor would be *overranged* by being subjected to the *total* pressure signal. We need to get rid of the undesired frequencies *before* they impinge on the diaphragm.

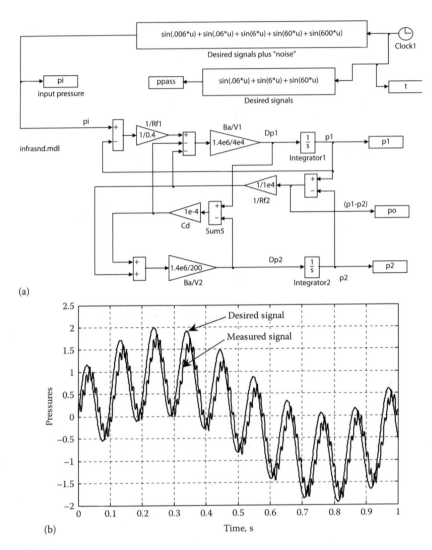

(a)

(b)   Time, s

FIGURE 7.5
Simulation diagram and results.

7.6 Adjusting Diaphragm Compliance to Include Air-Spring Effect

We have so far followed Mutschlecner and Whitaker (1997) in our analysis, but I now want to deviate from that path, with regard to the diaphragm behavior, as mentioned earlier. That reference did have some comments on the diaphragm compliance, but I have been unable to explain their argument. In the reference, page 8, equation 24, it is stated that C_d *must* be less than V_2/B_a, and this was used to set the value of C_d. A somewhat cryptic explanation for this requirement is given in the reference but I could independently find no reason why this relation was required. It is easy to violate it in a numerical example, using either the time histories from the simulation or the frequency-response graphs. I set $C_d = 2.0e{-}4$, which violates the requirement, but the frequency-response graphs, while showing a definite effect, seem perfectly *normal*. This change did not seem to cause anything unusual in the time history simulation either. I have tried to contact the authors for more details but have so far been unsuccessful.

The deviation from the reference, which I propose, is to take into account the *stiffening effect* of the compressed air on the diaphragm deflection. This could be significant since the output voltage depends directly on the diaphragm deflection, not the pressure difference as assumed in the reference and used in our analysis so far. At low frequencies this effect is negligible because the *leak* R_2 prevents any air compression due to diaphragm motion. At high frequencies (but still within the passband), the diaphragm moves rapidly enough that an *air-spring* effect is possible. This effectively stiffens the diaphragm and reduces its deflection, and thus the output voltage. It will be convenient to represent the diaphragm by an equivalent *piston/cylinder* and spring as shown in Figure 7.6. To be "equivalent" the compliance (volume change)/(pressure change) must be the same for both systems. This requires that C_d for the diaphragm be equal to A^2/K_s. Any combination of piston area A and spring stiffness K_s that meets this requirement can be used to define the equivalent piston system. The piston/cylinder is assumed to be frictionless and without inertia, just as we assumed for the diaphragm itself. Our goal is to replace the constant compliance C_d with a *dynamic compliance*, which changes with frequency to account for both the fixed "metal" compliance and the variable air-spring compliance. This will be in the form of a

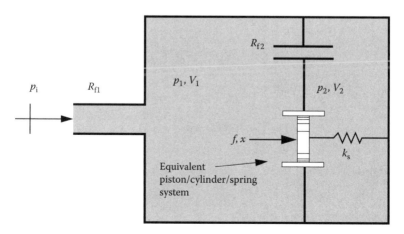

FIGURE 7.6
Analysis sketch for air-spring compliance effect.

transfer function, which we can enter into Equations 7.1 and 7.2 to replace the constant C_d used there.

In Figure 7.6 we can take the input pressure as fixed at the ambient atmospheric pressure, since we are not now concerned with the response to that input. That is, our models are all linear systems, so the superposition theorem holds and we can study any input by itself, neglecting all other inputs. The input of interest will be taken as a fictitious force f acting on the piston. If there were no air-spring effect, this force would cause an instantaneous piston displacement x according to $x = f/K_s$. Actually, the deflection would not be this large because the piston motion will also be opposed by an air-spring effect caused by compression of the air in volume V_2. As an additional simplifying assumption we will take the pressure p_1 to be constant at the ambient atmospheric pressure, unaffected by the motion of the piston. This is not unreasonable since the resistance R_1 is very small, allowing the input pressure p_i to easily and quickly "make up" any attempted momentary small changes in p_1. The large volume V_1 also is helpful in satisfying this assumption. We will find a transfer function relating displacement x to applied force f, thus defining an overall spring effect which includes the metal spring stiffness and the air-spring stiffness. We can then reinterpret this relation in terms of our *real* system as a diaphragm dynamic compliance, which can replace our earlier and simpler model. This result will show that at low frequency, the earlier model (no air spring) is accurate, but as frequency increases, the diaphragm compliance gradually decreases and for very high frequencies becomes constant again, but at a lower value than that for the diaphragm alone.

Under our assumptions, we can write

$$Dp_2 = \frac{B_a}{V_2}\left(\frac{0-p_2}{R_{f2}} + ADx\right) \tag{7.15}$$

and

$$f = K_s x + A(p_2 - p_1) = K_s x + Ap_2 \tag{7.16}$$

Equation 7.15 quickly yields

$$(\tau_2 D + 1)p_2 = R_{f2}ADx \qquad \tau_2 \triangleq \frac{R_{f2}V_2}{B_a} \tag{7.17}$$

Using Equation 7.16

$$(\tau_2 D + 1)\left(\frac{f - K_s x}{A}\right) = R_{f2}ADx \tag{7.18}$$

$$(\tau_2 D + 1)f - (\tau_2 D + 1)K_s x = R_{f2}A^2 Dx \tag{7.19}$$

This equation is easily manipulated to get a transfer function relating input force f to piston displacement x:

$$\frac{x}{f}(D) = \frac{(\tau_2 D + 1)\dfrac{1}{K_s}}{\left(\tau_2 + \dfrac{A^2 R_{f2}}{K_s}\right)D + 1} \tag{7.20}$$

We now need to convert this result for the equivalent piston/cylinder back to the diaphragm compliance form, using the fact that $C_d = V/p = Ax/(f/A) = A^2 (x/f)$. This gives

$$\frac{V}{p}(D) = \frac{\dfrac{A^2}{K_s}(\tau_2 D + 1)}{(\tau_2 + C_d R_{f2})D + 1} = \frac{C_d(\tau_2 D + 1)}{(\tau_2 + C_d R_{f2})D + 1} \tag{7.21}$$

Considering the frequency response of this transfer function, we see that at low frequency it is just the *metal* diaphragm compliance C_d, but as frequency increases, the *total* compliance gets less (system is *stiffer*), leveling off at the value $C_d \tau_2/(\tau_2 + C_d R_{f2})$ at high frequency. Inserting the definition of τ_2, we see that the high-frequency compliance is given by $C_d C_{as}/(C_d + C_{as})$, where $C_{as} \triangleq V_2/B_a$, the air-spring compliance. That is, the high-frequency compliance is the *parallel combination* (product/sum) of the *metal* compliance and the air-spring compliance. If we wanted to work with stiffnesses rather than compliances, the high frequency stiffness would be the sum of the individual stiffnesses. Using the same numerical values as before, we can graph the frequency response of Equation 7.21 as in Figure 7.7, which clearly shows the stiffening effect as we go from low frequency to high. The phase angle approaches zero for both high and low frequencies but shows some lagging effect at intermediate frequencies.

Having Equation 7.21 we can now insert this transfer function everywhere we see C_d in Equation 7.14. Manipulation of this equation leads to a new transfer function $(p_1 - p_2)/p_i$, which will have a *cubic* denominator rather than the second-order form of Equation 7.14.

$$\frac{(p_1 - p_2)}{p_i}(D) = \frac{\dfrac{R_{f2}V_2}{B_a}D}{(R_{f1}R_{f2})\left[\dfrac{V_1 V_2}{B_a^2} + \left(\dfrac{C_d(\tau_2 D + 1)}{(\tau_2 + C_d R_{f2})D + 1}\right)\left(\dfrac{V_1 + V_2}{B_a}\right)\right]D^2}$$

$$+ \left(\dfrac{V_2}{B_a R_{fe}} + \dfrac{V_1}{B_a R_{f2}} + \dfrac{\dfrac{C_d(\tau_2 D + 1)}{(\tau_2 + C_d R_{f2})D + 1}}{R_{f1}}\right)(R_{f1}R_{f2})D + 1 \tag{7.22}$$

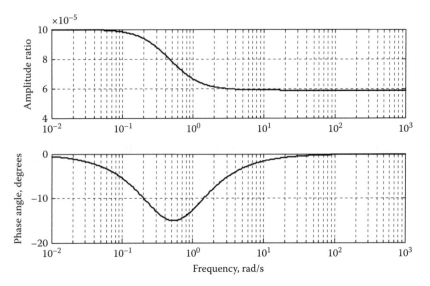

FIGURE 7.7
Frequency response of the dynamic compliance (amplitude ratio units are cm³/(dyne/cm²)).

Manipulation leads to

$$\frac{(p_1 - p_2)}{p_i}(D) = \frac{a1(a9D+1)}{(a2a3a9 + a2a4a8C_d)D^3 + \left(a2a3 + a2a4C_d + a5a7a9 + \dfrac{C_d a7 a8}{a6}\right)D^2}$$

$$+ \left(a5a7 + \frac{a7a9C_d}{a6}\right)D + 1 \tag{7.23}$$

Definitions of the a's will be given in an upcoming MATLAB program. The actual output voltage of a capacitance displacement sensor is proportional to the diaphragm *displacement*, usually taken as the displacement at the center of the diaphragm. We thus need to multiply the amplitude ratio of Equation 7.23 by the amplitude ratio of Figure 7.7, since volume and center displacement are proportional ($z_{center} = 4V/\pi R^2 T$). This operation is included in the program below.

```
% crc_book_chap7_infraphone_dyn_compliance.m computes and plots
% infrasonic mike with dynamic compliance
% first define parameter values
Rf1=0.4; Rf2=1e4; V1=4e4; V2=200; Cd=1e-4; Ba=1.4e6;
Rfe=Rf1*Rf2/(Rf1+Rf2);
% now compute parameter combinations
```

```
a1=Rf2*V2/Ba
a2=Rf1*Rf2
a3=V1*V2/Ba^2
a4=(V1+V2)/Ba
a5=V2/(Ba*Rfe)+V1/(Ba*Rf2)
a6=Rf1
a7=Rf1*Rf2
a8=Rf2*V2/Ba
a9=a8+Cd*Rf2
% next compute transfer function coefficients
num2=a1*a9
num1=a1
num0=0.0
den3=a2*a3*a9+a2*a4*a8*Cd
den2=a2*a3+a2*a4*Cd+a5*a9*a7+a8*Cd*a7/a6
den1=a5*a7+Cd*a7/a6+a9
den0=1.0
% now define numerator and denominator
num=[num2 num1 num0]
den=[den3 den2 den1 den0]
% get the 3 breakpoint frequencies of cubic denominator
[wnd,z1]=damp(den)
% get the breakpoint frequency of the numerator
[wnn,z2]=damp(num)
% plot frequency-response curves
w=logspace(-2,3,500);
h=freqs(num,den,w);
mag=abs(h);
ph=angle(h);
subplot(2,1,1),semilogx(w,mag,'k');grid;
phi=ph*57.3;
subplot(2,1,2),semilogx(w,phi,'k');grid;pause
% % now compute and plot the dynamic compliance
numc=[Cd*Rf2*V2/Ba Cd]
denc=[(Rf2*V2/Ba)+Cd*Rf2 1]
hdync=freqs(numc,denc,w);
magdync=abs(hdync);
phdync=angle(hdync);
subplot(2,1,1),semilogx(w,magdync,'k');grid
phidync=phdync*57.3;
subplot(2,1,2),semilogx(w,phidync,'k');grid;pause
% now multiply the two transfer functions
magt=mag.*magdync;
phit=phi+phidync;
subplot(2,1,1),semilogx(w,magt,'k');grid;
subplot(2,1,2),semilogx(w,phit,'k');grid
```

Figure 7.8 shows that the air-spring effect actually slightly *increases* the amplitude ratio of the pressure difference in the passband. To see the effect on

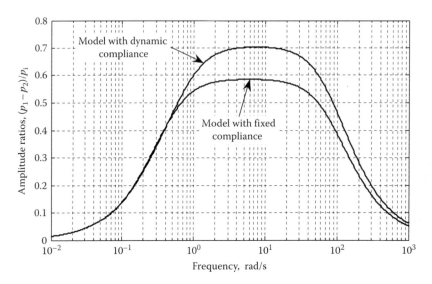

FIGURE 7.8
Effect of air-spring stiffness on response of pressure difference.

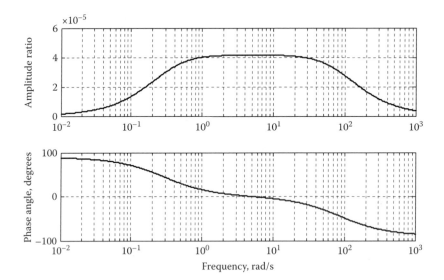

FIGURE 7.9
Response of diaphragm volume with air-spring effect.

output voltage however, we must multiply this amplitude ratio by that of the dynamic compliance, which has dimensions of $V/(p_1-p_2)$, so the amplitude ratio of Figure 7.9 has dimensions of V/p_i. If we had numerical values for diaphragm tension and radius, we could get actual values of diaphragm center displacement, which would be useful in the design of the capacitance sensor:

$$z_{center} = \frac{pR^2}{4T} = \frac{R^2}{4T}\left[\frac{16T}{\pi R^4} \cdot V\right] = \frac{4}{\pi R^2} \cdot V \qquad (7.24)$$

If we wanted to get the response to time-varying signals of arbitrary form, we can set up a simulation based on Equation 7.23, using the numerical values *num* and *den* provided in the program. We can do the same for the dynamic compliance itself, using *numc* and *denc*. Figure 7.10 shows such a simulation with the inputs taken as step functions of size 1.0. Equation 7.21 for the dynamic compliance is a form generically called a *lead–lag element* since its numerator contributes a leading phase angle and its denominator a lagging phase angle. In our case the time constant in the denominator is larger than that in the numerator, so the net effect is a lagging phase angle, as seen in

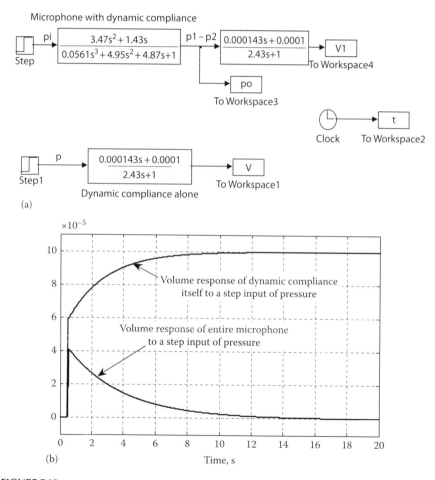

(a)

(b)

FIGURE 7.10
Simulation of entire microphone and dynamic compliance alone. (a) Simulation diagram. (b) Results.

Figure 7.7. The step response of our lead-lag element is shown in Figure 7.10b. Note that the volume suddenly jumps up when the step input is applied at $t = 0.5\,\text{s}$ to a value corresponding to the sum of the *metal* stiffness and the air-spring stiffness. Thereafter the volume gradually increases as the leak causes the air spring to become ineffective, leaving finally only the *metal* compliance of the diaphragm itself. The volume response of the entire microphone shows the initial jump, but then decays to zero as the pressure equalizes on each side of the diaphragm.

7.7 Calibration

An apparatus as in Figure 7.11 is sometimes used to calibrate infrasonic microphones. This is essentially a low-frequency version of the *pistonphone* used to calibrate audio microphones. The piston is oscillated with a fixed amplitude of displacement, generating a dynamic change in volume, which in turn produces the pressure oscillation desired. The piston motion is easily measured accurately, and from this we can compute the pressure change from the known bulk modulus of the air in the large chamber:

$$\Delta p = \frac{B_a}{V} \cdot \Delta V \qquad (7.25)$$

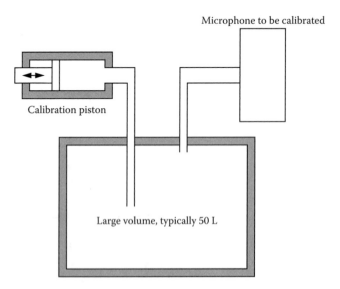

Microphone to be calibrated

Calibration piston

Large volume, typically 50 L

FIGURE 7.11
Calibration apparatus.

Often it is assumed that the compression process is adiabatic, so we use the adiabatic bulk modulus B_a, which is the chamber pressure (taken as atmospheric) times the ratio of specific heats, 1.4 for air, giving $B_a = 1.4e6$. For a chamber of volume 50 L (50,000 cm³) and a typical piston volume change of 10 cm³ the pressure change would be 28 dyne/cm² = 28 μ bar = 2.8 Pa.

A recent paper by Starovoit et al. (2006)* points out that the simple interpretation just given may lead to calibration errors due to neglected heat transfer effects. A more correct treatment is presented in the paper and we now go through their analysis. The analysis is based on the perfect gas law, conservation of energy, and heat transfer through the walls of the volume chamber, driven by the temperature difference between the inside gas and the ambient air outside.

$pV = mRT$ where m is the total mass of air, V is the total volume

$$pv = RT \quad v = V/M \tag{7.26}$$

Using the *specific volume v* allows us to work on a *per kilogram* basis in this equation and those coming up. Conservation of energy, again on a per unit mass basis gives, for a time interval dt:

$$\text{change in internal energy} = -\text{work} + \text{heat transfer}$$

$$c_v dT = -P_0 dv + \frac{(T_a - T)}{R_t} dt \tag{7.27}$$

where
  c_v is the specific heat at constant volume
  P_0 is the (assumed constant) initial pressure in the vessel
  T is the time-varying gas temperature
  T_a is the constant ambient temperature
  R_t is the thermal resistance of the vessel wall (includes conductive resistance of solid wall plus convective resistances of inside and outside air films)

From Equation 7.26, taking p as the variable part of the vessel pressure (deviation from P_0) we get

$$P_0 \cdot \frac{dv}{dt} + V_0 \cdot \frac{dp}{dt} = R \frac{dT}{dt} \tag{7.28}$$

---

* Yu. O. Starovoit et al., About dynamical calibration of microbarometers, Inframatics, #14, June 2006. (available at www.inframatics.org)

From Equation 7.27

$$c_v \cdot \frac{dT}{dt} = -P_0 \cdot \frac{dv}{dt} - \frac{T}{R_t} \tag{7.29}$$

Here, since initially $T_a = T$, and our analysis considers *changes* from the initial state, T_a does not appear in Equation 7.29. This equation directly gives a relation between T and v:

$$\left(c_v D + \frac{1}{R_t}\right) T = -P_0 D v \qquad \frac{T}{v}(D) = \frac{-R_t P_0 D}{\tau_1 D + 1} \qquad \tau_1 \triangleq R_t c_v \tag{7.30}$$

Combining this with Equation 7.28 gives

$$P_0 D v + V_0 D p = RDT = \frac{-RR_t P_0 D^2}{\tau_1 D + 1} \qquad \frac{p}{v}(D) = \frac{P_0}{V_0}\left[\frac{\tau_2 D + 1}{\tau_1 D + 1}\right] \qquad \tau_2 \triangleq (\tau_1 + RR_t) \tag{7.31}$$

Now, using the fact that $R = c_p - c_v$:

$$\tau_2 = \tau_1 + RR_t = R_t(c_v + R) = R_t(c_v + c_p - c_v) = R_t c_p \tag{7.32}$$

For air, $c_p = 1.4 c_v$ so the lead-lag element of Equation 7.31 has a net lagging phase angle and an amplitude ratio that starts out at P_0/V_0 at low frequency and gradually decreases, leveling off at $P_0/V_0 \,(c_p/c_v)$.

In Equation 7.25 we could use either the adiabatic bulk modulus $(c_p/c_v)(P_0)$ or the isothermal P_0, giving

$$\left|\frac{\Delta p}{\Delta v}\right| = \frac{c_p P_0}{c_v V_0} \qquad \text{adiabatic} \qquad \text{or} \qquad \left|\frac{\Delta p}{\Delta v}\right| = \frac{P_0}{V_0} \qquad \text{isothermal} \tag{7.33}$$

We now see that the *conventional* practice of choosing one of these to use at *all* frequencies is incorrect since the detailed analysis shows that this ratio changes with frequency, approaching the isothermal value at low frequencies and the adiabatic at high. Thus, during calibration, when converting measured volume changes to predicted pressure changes, we must compute the two time constants and use Equation 7.31 to adjust the pressure change amplitude and phase angle according to frequency. The main difficulty here is getting a good value for the thermal resistance R_t, which would usually be calculated by measuring chamber and insulation dimensions and then using theoretical formulas and standard material property values. The reference mentions (page 4) that for typical numerical values, the isothermal assumption is good from low frequencies up to about 10 Hz, which is often near the upper end of a microphone's passband, suggesting that we might design the apparatus for such operation. They argue against this since an isothermal design makes the chamber walls very *conductive* (adiabatic requires good insulation), allowing any fluctuations in the *ambient* temperature to affect the pressure in the chamber, leading to calibration errors. I e-mailed the authors with a suggestion, which so far, they have not responded to. This suggestion arises from a recent paper* on pneumatic valve dynamics that discussed a design for an isothermal chamber needed in that study. The idea is to *fill* the chamber with metallic material such as steel wool or fine copper wire. Actually, the metallic material takes up only a small fraction of the chamber volume, but its total thermal capacitance is so much greater than the surrounding air that it tends to maintain the air temperature steady by absorbing or giving up heat whenever the air *tries* to change its temperature. In the quoted study, this technique was very successful, recommending it to others who might need such isothermal behavior in some experimental investigation. In the microphone calibration apparatus, the chamber would still be heavily insulated so as to prevent any influence from ambient temperature changes, since the isothermal behavior is provided by the internal high thermal capacitance, *not* by a poorly insulated chamber wall.

7.8 Wind Noise Filtering with Pipes and Spatial Arrays

In many applications of infrasonic microphones the instrument's *pneumatic* band-pass filtering needs to be augmented with a pre-filter of some sort. The need for this additional filtering is often due to *wind noise*, which unfortunately may have frequency content that lies within the passband of the microphone and thus will not be removed by that filtering effect. That is, the passband would be designed to suit the particular type of signal to be studied and often the existing ambient wind noise has strong frequency content in that same range. Fortunately, there *is* a usable difference between the desired signals and the wind noise which can be the basis of a practical filtering effect. This difference is based on the wind signals being essentially of a random nature, whereas the desired signal is more deterministic. Also, the wind-associated pressures appear as turbulent cells or *packets* that move over the surface at typical wind speeds (say 5–20 mph, (7–29 ft/s)) whereas the acoustic pressure signals of interest are *waves* that propagate at the speed of sound (about 1100 ft/s). These waves are produced by geophysical

---

* K. Kawashima et al., Determination of flow rate characteristics of pneumatic solenoid valves using and isothermal chamber, *ASME J. Fluid Eng.*, 126, March 2004, 273–279.

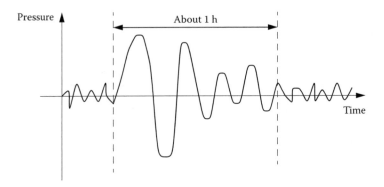

FIGURE 7.12
"Generic" infrasonic signal.

phenomena such as volcanic eruptions, meteors entering the earth's atmosphere, earthquakes, etc., and man-made events such as explosions of atomic and nuclear bombs. In fact, much of the activity in infrasonics has been associated with international treaties concerning nuclear proliferation and the remote detection of bomb explosions that might be in violation of a treaty. All these types of acoustic events produce some sort of pressure pulse or oscillation that can be distinguished from the background noise (such as the local wind) with suitable filtering. Figure 7.12 (*not* actual data) shows what such a signal might look like. Actual records can be found in the literature. A wave (similar to that of Figure 7.12) associated with a Russian bomb test (estimated at 50 megatons of TNT) that occurred on October 30, 1961 had a peak-to-peak amplitude of about 600 μ bar when it was detected at Aberdeen, Scotland* after traveling about 21,000 mi around the earth from Novaya Zemlya. Natural events such as the Krakatoa volcanic eruption of August 27, 1883 and the Siberian meteorite explosion of June 30, 1908 create similar pressure signals which were detected by microbarographs around the earth at those historic times.

While the various infrasonic signals of interest are somewhat complicated transient pressure signals, we can, as usual, using Fourier transform concepts, analyze them in terms of their frequency spectrum. This is also true of noise effects, such as the wind-associated pressures. The spatial filtering provided by a suitably-designed *pipe* or array of pipes was perhaps first suggested and analyzed by Daniels.[†] His scheme used a pipe made up of sections of different diameters, with a properly-designed *hole* at each location where the pipe changes diameter. The example described in the reference used a pipe of total length 1980 ft, with 100 openings and

---

* R.V. Jones, *Instruments and Experiences*, Wiley, New York, 1988, p. 177.
† F.B. Daniels, Noise-reducing line microphone for frequencies below 1 cps, *J. Acoust. Soc. Am.*, 31(4), April 1959, 529–531.

pipe sizes from 0.5 in. diameter to 6 in. diameter. Daniels' theory predicts that the improvement in signal-to-noise ratio when the noise is a random white noise will be proportional to the square root of the number of *holes* in the pipe, in the example case, an improvement of 10-to-1. A figure in the paper shows actual recordings when the wind was 25–30 mph that validate this significant improvement. When I found *recent* references* to these types of *pipe* filters, I noticed that none of them used the multi-diameter configuration of Daniels but rather used pipes of uniform diameter for the entire length. Also, instead of providing discrete *drilled holes* at locations along the length, they often used simple porous *garden hose* as the pipe. I tried to find out when and why the filter design philosophy changed from the multi-diameter pipe to the uniform diameter pipe. My literature search finally turned up a paper[†] which I believe documents this important change. Burridge provides an analysis and numerical examples to show that performance equaling that of the Daniels configuration can be achieved with the simpler uniform pipe system. Most current infrasound work uses, instead of a single *linear* pipe, multiple pipes, which can be arranged in various configurations. Figure 7.13a[‡] shows a microphone with multiple hoses arranged in a *star* pattern. An infrasonic "observatory" might consist of four such systems, each with its own microphone, using 50 ft hoses bent back on themselves, and situated at the corners of a square of size about 200 ft on a side. Such *arrays* (Figure 7.13b) are capable of measuring, with suitable signal processing, not only the pressure signal but also its direction and speed of propagation. Figure 7.13c[§] shows a setup for a comparison of two commercial microphones, using three samples of each. I tried to find a Web site for the manufacturer of the MB2000 microphone, but was unsuccessful. The Chaparral microphones would be considered "high end" (most versatile model costs about $7000), but have excellent performance and are widely used. Their Web site[¶] gives complete specifications for their various models but does not reveal any details of construction.

The details of array signal processing are beyond the scope of our introductory treatment, but it will be useful to explain in some detail how the wind noise reduction works in a single linear pipe, since this concept has application beyond our current interest. The acoustic signals that are of interest

---

* M.A.H. Hedlin and J. Berger, Evaluation of infrasonic noise reduction filters, *23rd Seismic Research Review: Worldwide Monitoring of Nuclear Explosions*, October 2–5, 2001. Defense Threat Reduction Agency Contract DTRA01-00-C-0085.

† R. Burridge, The acoustics of pipe arrays, *Geophys. J. R. Astr. Soc.*, 26, 1971, 53–69.

‡ A.J. Bedard, Jr. et al., The Infrasound Network (ISNET): Background, design details, and display capability as an 88D adjunct tornado detection tool, NOAA Environmental Technology Lab, Boulder, CO.

§ R.P. Kromer and T.S. McDonald, Infrasound sensor models and evaluation, Sandia National Lab, U.S. Dept. of Energy, Contract DE-AC04-94AL85000.

¶ www.chaparral.gi.alaska.edu/

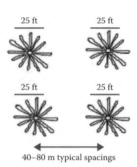

40–80 m typical spacings

(a) Photograph of a spatial filter for reducing wind-induced pressure fluctuations in the infrasonic frequency range.

(b) View of a typical infrasonic observatory array configuration. The exact positioning of the twelve porous irrigation hoses radiating outward from each of the four sensors is not critical.

(c) MB2000 and Chaparral 5 Comparison Test at Sandia FACT Site.

FIGURE 7.13

(a) Microphone with spatial filter using porous hose. (b) Infrasonic observatory array configuration. (c) Test setup for comparison of two different microphones.

have rather long wavelengths; the lowest frequencies (about 0.001 Hz) having a wavelength of (1100 ft/s)/(0.001 cycles/s) = 1.1e6 ft = 208 mi. The lower the frequency, the lesser the attenuation as such waves propagate over the earth's surface at a typical sound speed of 1100 ft/s, some traveling completely around the earth, as mentioned earlier. The guidelines for designing a spatial (pipe) wind noise filter suggest that the pipe total length be as long as possible (relative to the size of the wind's atmospheric turbulence cells), but short compared with the wavelength of the propagating pressure signals of interest. The holes drilled into the pipe at regular intervals should be spaced at intervals whose length is the same order as the length of the wind cells. These "holes" need to have proper acoustic impedance, typically a brass plug with a 0.8 mm diameter hole and 6 mm long might be screwed into the pipe at each location. The impedance of this flow passage should be large relative to the impedance of the length of pipe that it feeds, but small enough so that the pressure signals of interest are not attenuated too much at the higher frequencies of interest. While the pressure signals of interest travel *through* essentially stationary air by elastic wave propagation at the speed of sound (nominally 1100 ft/s), the undesired wind noise travels *with* the air as a gross motion of a turbulence packet or cell at the local wind speed, which is a few feet per second. Also, the desired signals are *coherent* over distances the order of the signal wavelength (hundreds or thousands of feet), while the random wind noise becomes incoherent over distances the order of the size of the turbulent cell (usually less than 50 ft).

Coherence can be measured by a signal's autocorrelation function.* Here we take a signal, delay it by a certain amount (called the *lag*), multiply the original and delayed signals together, and then average this product over time. We do this computation for a range of lags (starting at zero lag) and then plot the averaged product against lag. For a lag of zero, we see that the signal is just multiplied by itself, giving the square of the signal; the subsequent averaging is then nothing but the computation of the *mean-squared value* of the signal, the square root of which is the so-called *root-mean-square* (RMS) value. When we now *delay* the signal by a small amount, we begin to "misalign" the positive and negative portions, so the averaged product is *smaller* than the undelayed (zero lag) value. For a random signal, as we increase the lag, the smaller becomes the autocorrelation function, which means the signal is losing coherence with itself. As the lag is further increased, the autocorrelation function may go through zero and then oscillate around zero. The lag (delay time) value where the function starts oscillating is an indication of the signal's coherence. Slowly changing signals have a large time value for the first zero crossing and start of oscillation; fast signals have a small value.

When our perforated pipe (Figure 7.14) is exposed to the total (desired plus noise) signal, the "samples" picked off at each of the "holes" will tend to add up within the pipe, as they travel toward the microphone. If the holes are spaced properly, the wind noise components will be time-shifted enough that they tend to cancel each other, reducing the undesired random signal at the microphone. The desired signal is coherent for much longer distances, so its components at each hole will tend to reinforce each other, leading to an improved signal-to-noise ratio. As mentioned earlier, analysis has shown that the improvement is proportional to the square root of the number of holes, thus to get a, say, 10-to-1 improvement, the pipe needs to have about 100 holes, leading in most cases to rather long pipes. We have left the proof of all these statements to the references, but I want to do a simulation that will hopefully demonstrate the validity of the concept. This simulation does not duplicate the actual situation in all its details, but does capture the essential features of the method. Figure 7.15 shows

---

* E.O. Doebelin, *Measurement Systems*, 5th edn., McGraw-Hill, New York, 2003, pp. 183–185.

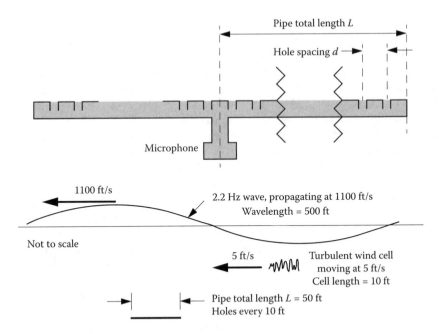

FIGURE 7.14
Explanation of spatial wind noise filter.

the simulation diagram (a) and some graphical results (b). Since the noise reduction goes as the square root of the number of holes, and we have only 5, we should expect only a modest improvement, but the graphical results do show it to be noticeable.

Some details of the simulation are in order. The random noise signal was designed to have frequency content in the same range as the desired signal, which is a 2.2 Hz sine wave of amplitude 1.0. This was accomplished by first producing a random square wave with *flat* frequency content to about 10 Hz (see Appendix A). This signal was then sent through a band-pass filter with a passband of 1–5 Hz. The *size* of the random signal (noise power) was adjusted by trial-and-error to give a strong random component relative to the desired sine wave. To keep the simulation simple, I chose the desired signal frequency rather high, to give a short wavelength and thus a short pipe length, and a small number of "holes." Simulink *transport delay* modules were used to delay the desired signal according to its propagation velocity and the hole spacing. The noise signal was similarly delayed, but the delay times are much longer since the wind moves over the holes at only 5 ft/s. The final gain block was set by trial-and-error to make the *out* signal equal to the *in* signal when the random portion was turned off.

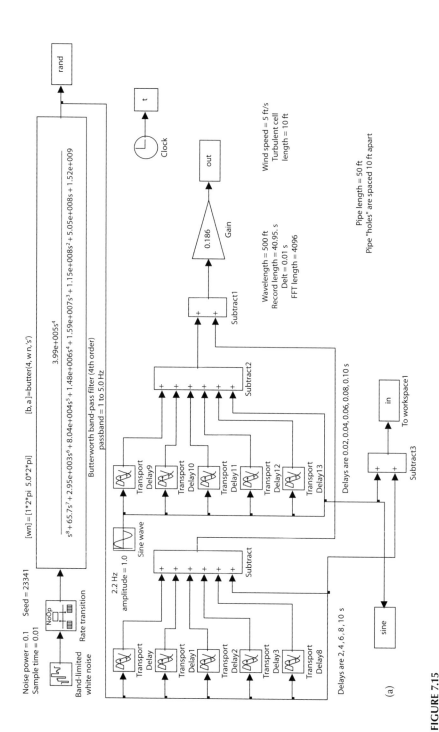

FIGURE 7.15
Simulation to demonstrate the principle of spatial filter.

(continued)

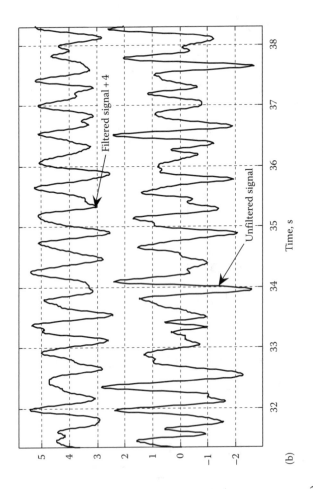

FIGURE 7.15 (continued)

In addition to the simulation, I also wrote a MATLAB program to demonstrate some other features of interest, such as FFTs of some of the signals, and an autocorrelation function calculation for the random signal, to illustrate the coherence concept.

```
% crc_book-chap7_spatial_filter  computes and plots
% characteristics of a simplified spatial filter
% plot frequency response of bandpass filter, 4th order
% Butterworth  [wn]=[1*2*pi 5*2*pi] [b,a]=butter(4,wn,'s')
bode(b,a);grid;pause
% use Hanning window on wind noise signal rand
randh=rand.*hann(4096);
% get FFT of hanned noise
ftran=fft(randh,4096);
ftran=ftran(1:2048);
delt=0.01;
tfinal=40.95;
mag=delt.*abs(ftran);
fr=(0:1/tfinal:2047/tfinal);
plot(fr,mag);pause
axis([0 10 0 3]);pause
% compute and plot autocorrelation function of wind noise
autocor=xcorr(rand,rand,'unbiased');
autocor1=autocor(4096:8191);
autocor2=autocor1(1:500);
lag=[0:0.01:499*0.01]';
plot(lag,autocor2);pause
axis([0 1.0 -.4 1]);pause
% compute and plot FFT of unfiltered total signal
inh= in.*hann(4096);
ftranin=fft(inh,4096);
ftranin=ftranin(1:2048);
magin=delt.*abs(ftranin);
plot(fr,magin);pause
% compute and plot FFT of filtered total signal
outh=out.*hann(4096);
ftranfilt=fft(outh,4096);
ftranfilt=ftranfilt(1:2048);
magfilt=delt.*abs(ftranfilt);
plot(fr, magfilt);pause
axis([0 20 0 11]);pause
% compare FFT of input and output signals
hold on
plot(fr+10,magin);pause
% get FFT of 2-Hz sine wave for a check
hold off
ftsin=fft(sine,4096);
ftsin=ftsin(1:2048);
```

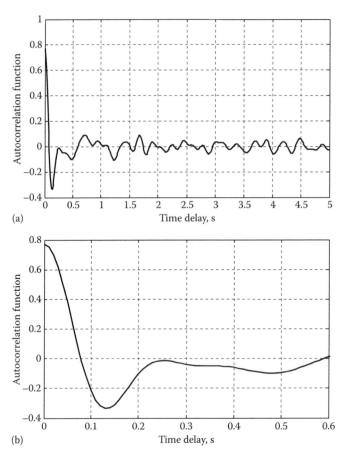

FIGURE 7.16
Autocorrelation function of random signal.

```
magsin=delt.*abs(ftsin);
plot(fr,magsin)
```

Figure 7.16 shows the autocorrelation function of the signal called *rand*, at two different time scales. We see that for delays longer than about 0.2 s, this signal becomes *incoherent*. Thus in the spatial filter, the holes need to be spaced at least (0.2 s)(5 ft/s) = 1.0 ft apart, so our choice of 10 ft is clearly adequate to provide a good averaging of the random fluctuations. The MATLAB autocorrelation routine always computes about *twice* as many lag values as the number of points in the time record. In this example, I ran the simulation for 40.95 s with a computing increment of 0.01 s, giving a total of $n = 4096$ points, convenient for the upcoming FFT calculations. The *xcorr* statement always calculates autocorrelation values for $2N - 1 = 8191$ points.

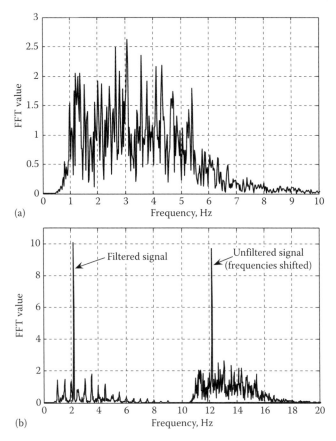

FIGURE 7.17
(a) FFT of random signal and (b) comparison of filtered and unfiltered FFT's of total signal.

The sequence of $2N-1$ points will always be an odd number, so it has a *center point*, in our case the 4096th point. We actually want only this point and those after it, in our case, points 4096–8192. Of these 4096 points, only the first 5%–20% will be statistically reliable, so in this example I decided to take the first 500. These comments apply to *all* autocorrelation calculations, not just this example. Figure 7.17a shows the FFT of the random part (*rand*) of the signal. This spectrum is quite *ragged*, as we would expect for a *single* spectrum of any random signal. To get a more representative spectrum one must always *average* a number (often a *large* number) of spectra, as demonstrated in the Appendix A. The *total* signal (random plus sine wave) before filtering has the spectrum shown (with frequencies shifted 10 Hz for easy comparison) at the right in Figure 7.17b, while the filtered signal's spectrum is shown at the left. Both show a prominent spike at 2.2 Hz but the filtered signal exhibits less randomness and some small periodic spikes.

7.9 Ultrasonic Microphones

Our intention is not to survey the whole field of ultrasonics or even that of ultrasonic microphones, but to focus on one type of microphone that is used in a number of important industrial applications. Our incentive here is partly to illustrate a technique called *frequency translation*, which is employed in many measurement systems, using as a vehicle some instruments designed for the detection of *fluid leaks* and incipient *mechanical component failure*. A small leak in a truck's air brake system may be inaudible to the human ear but produces ultrasonic noise which can be detected with suitable instruments. Production of automotive parts such as fuel tanks, radiators, etc. requires some kind of leak inspection to insure a leak-free component before it is assembled into the vehicle. Here an ultrasonic "loudspeaker" may be placed inside the vessel and one or several microphones placed outside can detect even tiny leaks. Such testing can often be automated to screen out bad components without materially slowing the production line. Fluid leaks can occur with either laminar or turbulent flow; ultrasonic noise generation requires turbulent flow. Preventive maintenance of vital machinery may involve a monitoring system based on ultrasonic microphones and signal processing equipment. While a failing ball bearing will produce noise audible to human hearing when the bearing is about to fail catastrophically, ultrasonic noise may occur much earlier, when action can be taken to prevent expensive unscheduled down time.

While fluid leaks and bearing deterioration produce ultrasound over a wide range of frequencies, experience has shown that an optimum frequency range for many applications is in the neighborhood of 40 kHz, so many commercial systems are designed for this range. While electronic equipment has no trouble dealing with such frequencies, many systems are intended for human operation. That is, when searching for air leakage in a trailer-tractor rig's air brake system (which has many components distributed over considerable distances), the best approach may be to provide the mechanic with a sensing probe of some sort, together with earphones to listen for unusual sounds. For such applications it is necessary to *translate* the frequency range of the signals picked up by the microphone to the lower frequencies that can be heard by humans. A similar situation occurs for biologists studying the echo-location behavior of bats, whose emitted sounds are also in the ultrasonic range. There are various methods for translating the frequency content of a time-varying signal to a lower or higher range while preserving the information content. If the signal were recorded on a tape recorder we could *play it back* at a different tape speed. A digitally recorded signal can be played back at almost any speed of our choice. A relatively simple method used in many ultrasonic leak detectors is that of *amplitude modulation*,* also called *heterodyning*. Here the microphone signal

---

* E.O. Doebelin, *Measurement Systems*, 5th edn., McGraw-Hill, New York, 2003, pp. 167–178.

with a bandwidth of about 36–44 kHz is multiplied electronically with a fixed-amplitude sine wave (called the *carrier*) of, say, 45 kHz, producing so-called sideband frequencies which are the sum and difference of the carrier and signal frequencies. That is, a 36 kHz signal produces components at two frequencies, 45 + 36 = 81 kHz and 45–36 = 9 kHz. For a 44 kHz signal, the two side frequencies are 89 and 1 kHz. We see that the information that was originally in the ultrasonic range 36–44 kHz has now been translated down to the audible range 1–8 kHz. Most readers will recognize the amplitude modulation process as that used in transmitting AM radio signals. Music or voice signals are too low in frequency to be efficiently transmitted by radio, so they are *heterodyned* up to the radio frequency range for transmission by the broadcaster. The listener's radio receiver than *demodulates* (see Doebelin's *Measurement Systems*, pp. 167–178) the signals to restore them to the audio frequency range for human hearing.

The microphones needed for the typical leak detection system can be quite simple and inexpensive piezoelectric devices, using a diaphragm to transduce the sound pressure to a small force and motion, with a flexible piezoelectric element converting the diaphragm center motion to a voltage. A *general purpose* microphone requires a flat frequency response over a wide frequency range, which is difficult to attain and leads to complicated and expensive designs. Our requirements are much easier to meet since we *want* a very peaked (not flat) response, which is easily accomplished with a lightly damped mechanical structure. Figure 7.18 shows two possible configurations for piezoelectric transducer elements, and some typical characteristics for a transmitter or receiver (microphone) by one of the large ultrasonic manufacturers.* I wanted to show the *actual* physical details of the TR-89B40 transducers, but Massa does not provide such information, so my sketches are *guesses* as to what arrangement *might* be used. Massa is basically a supplier to OEMs (original equipment manufacturers) and does not offer for sale single copies of off-the-shelf products. Irrespective of the details of construction, such microphones have fairly simple dynamic models; they are essentially lightly damped resonant mechanical structures with the usual piezoelectric dynamics for the transduction from mechanical displacement to electrical output voltage. The Massa circuit diagrams of Figure 7.18 show that the frequency response of transmitters and receivers can be influenced by adding external resistors and inductors to the circuit. If no such circuit elements are added, there is a single resonant peak. Adding only an inductor produces two peaks since this inductor forms an electrical resonant circuit which interacts with the mechanical resonance. Adding also a resistor augments the slight mechanical damping with electrical damping, producing a lower and much broader peak. Because ultrasonic leak detection systems work best in a narrow range of frequencies near 40 kHz, and a high microphone

---

* www.massa.com

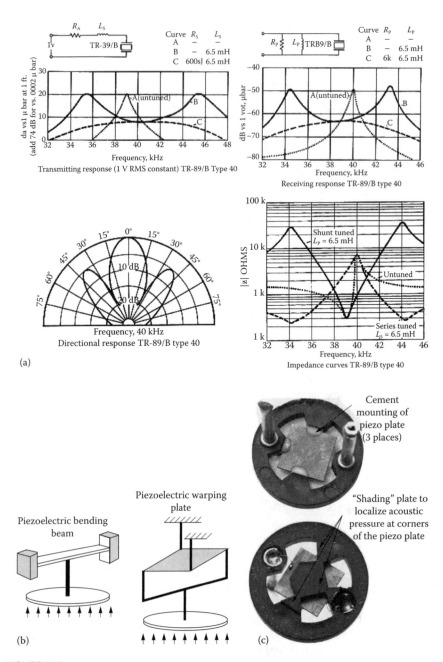

(a) Characteristics of Massa TR-89B40 transmitter/receiver, (b) possible transducer configurations, and (c) Gulton piezoelectric microphone (circa 1975).

FIGURE 7.18

sensitivity is desirable, the simplest circuit (no added resistor or inductor) can be employed.

When I was unable to get from Massa or any other manufacturers the constructional details of this class of microphones, I recalled that back around 1975 I had purchased for a student lab demonstration an ultrasonic transmitter and also a receiver (microphone) from the Electronic Components Division of Gulton Industries Inc. A Web search for Gulton in 2008 found that a company of that name was still in existence but apparently no longer in the microphone business. My files still contained the data sheets for these components and since I no longer needed them for my classes I disassembled the microphone and took digital photos of the internal construction (see Figure 7.18c). These microphones have *no* diaphragm but rather use a square piezo plate as the receiver of the acoustic pressure. (There was a simple protective metal screen over the piezo element, but it serves no other function and is not shown.) The piezo plate is cemented to a plastic case at 3 of the midpoints of the square; these are nodal (stationary) points for the vibrational mode shape of the plate, the fourth midpoint is left free, apparently to increase the element deflection (and voltage output) due to pressure. In Figure 7.18c (lower photo) I show the reverse side of the upper view; the acoustic pressure would be applied from this side. A *shading plate* blocks the pressure from acting on the central portion of the piezo plate, so the pressure effect is concentrated at the four corners, where it is most effective. What you see in Figure 7.18 is *all there is* to this microphone, except for the protective screen and a pair of electrical terminals to pick off the piezo voltage. These simple microphones sold for $8.00 in 1975 and were used in various applications by many customers over the years. I strongly suspect that the Massa microphones have a similar construction. I believe Panasonic made similar units at one time but I could not find any for sale in 2008. A Japanese company* offers very similar microphones (manufactured in China), but would not supply any constructional details.

A simple model which produces the type of response seen in the experimental curves of Figure 7.18 is given by

$$\frac{e}{p}(D) = \frac{K\tau D}{(\tau D + 1) \cdot \left(\dfrac{D^2}{\omega_n^2} + \dfrac{2\zeta D}{\omega_n} + 1 \right)} \tag{7.34}$$

To get a sensitivity comparable to a practical microphone we can use, the data in the Massa curve of Figure 7.18 for the *receiver* (microphone) which shows a peak of about $-48\,\text{dB}$ relative to $1\,\text{V}/\mu$ bar. We need to convert $-48\,\text{dB}$ to the actual amplitude ratio AR, using the conversion $1\,\text{dB} = 20\log_{10} AR$, giving $AR = 0.00398$. The piezo transfer function $\tau D/(\tau D + 1)$ will have

---

* www.fuji-piezo.com

a time constant which might be around 0.01 s, so its amplitude ratio will be very near 1.0 for any frequencies beyond 1000 rad/s and thus near the 40 kHz peak, so we need to choose the mechanical damping ratio and K so as to get the peak value of 0.00398 at 40 kHz. For simple mechanical second-order systems the peak amplitude ratio is given by $K/(2\zeta\sqrt{1-\zeta^2})$, so there is an infinite number of combinations of K and ζ that will give us the desired peak value. Since there is no intentional damping in the piezo element, we can chose a damping ratio typical of mechanical structures, which might be about 0.005, giving $K/0.01 = 0.00398$, and thus $K = 0.0000398$ V/μ bar. A MATLAB calculation gives us the frequency-response curve as shown in Figure 7.19 which is visually a very close match to the experimental curve of Figure 7.18a. Since amplitude ratios in dB give a distorted view of the relative values at different frequencies. Figure 7.19b shows the *true* values in proper proportion, making clear that the microphone is very frequency-selective, probably *too* selective for our practical application to leak testing. That is, experience has shown that we need to include ultrasonic signals in the range 36–44 kHz in roughly *equal* strength. To get this increased bandwidth in the actual microphone, the circuit labeled "C" in Figure 7.18a could be used; the external resistor and inductor provide the desired broader (and lower) peak, so we gain bandwidth at the expense of sensitivity (which can be regained, if necessary, with amplification). In our simulation, much the same effect can be attained by simply increasing the mechanical system damping to about 0.10, as we see in Figure 7.19b. If we need to recover the original sensitivity, it appears that an amplifier with a gain of about 20 is needed.

We now set up a Simulink simulation as in Figure 7.20 to generate time-response data which is then analyzed for frequency spectra in a MATLAB program:

```
% crc_book_chap7_ultrmike_ampmod computes spectra for amplitude
% modulated ultrasonic mike need to run simulink first
% compute FFT and PSD of raw mike output signal
mikinft=fft(ultrasound,4096);
mikinft2=mikinft(1:2048);
delt=1e-6;
tfinal=0.004095;
delf=1/tfinal;
mag=delt.*abs(mikinft2);
fr=[0:delf:2047/tfinal];
psd=mag.*mag./delf;
plot(fr,psd);axis([2e4 5e4 0 0.8e-15]);pause
axis([3.5e4 4.5e4 0.01e-13]);pause
loglog(fr,psd);pause
psddb=20.*log10(psd);
semilogx(fr,psddb);axis([3e4 5e4 -400 -300]);pause
```

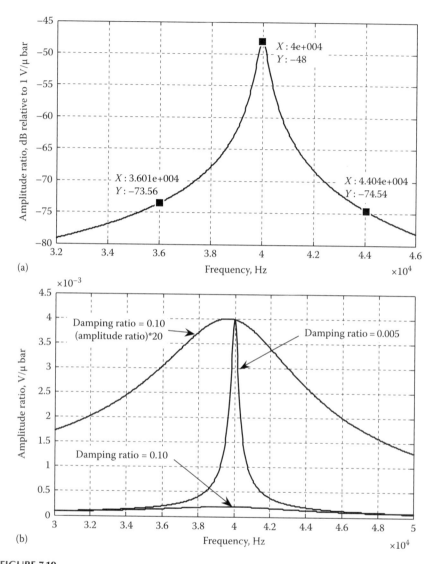

FIGURE 7.19

Frequency response of ultrasonic microphone: (a) amplitude ratio in dB and (b) true amplitude ratio.

```
% compute FFT and PSD of lowpass-filtered
% amplitude modulated signal
mikoutft=fft(lopass,4096);
mikoutft2=mikoutft(1:2048);
magout=delt.*abs(mikoutft2);
psdout=magout.*magout./delf;
plot(fr,psdout);axis([0 10000 0 8e-17])
```

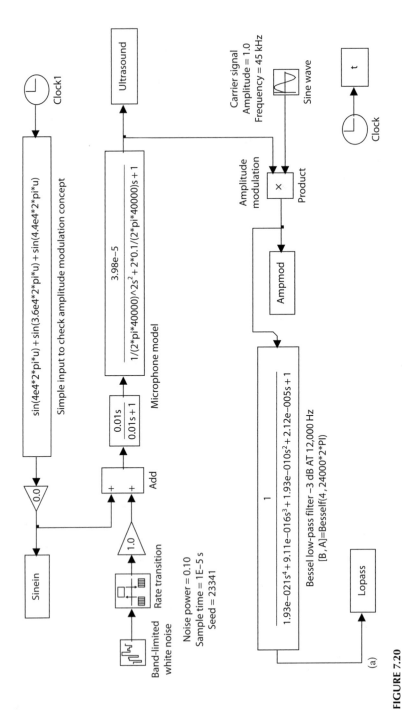

FIGURE 7.20
Simulation of ultrasonic microphone with amplitude modulation.

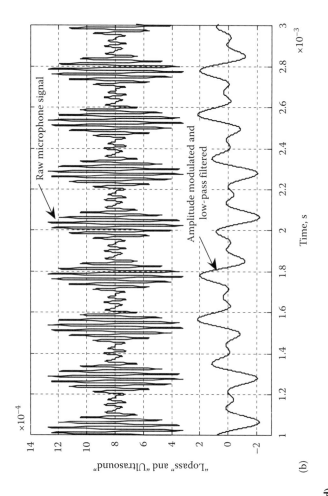

(b)

FIGURE 7.20 (continued)

The simulation is set up with a computing increment of 1e−6 s and final time of 0.004095 s, so as to generate 4096 samples of each signal for subsequent FFT processing. I have provided at the top of the simulation diagram a simple input made up of 3 sine waves of frequency 36, 40, and 44 kHz, as a means of demonstrating that the amplitude modulation process works as claimed in translating the high ultrasonic frequencies down to the lower range audible to human ears. (Once we have seen the verification of the principle, this input will be *turned off* and replaced by a wideband random *ultrasonic* input generated with the Simulink *band-limited white noise generator*.) When the *carrier* frequency of the amplitude modulation process is set at 45 kHz, the process generates *desired* frequencies of 1, 5, and 9 kHz and *undesired* frequencies of 81, 85, and 89 kHz. The undesired frequencies are removed with a Bessel lowpass filter designed to have its *cutoff* (−3 dB) frequency at 12 kHz. In an actual leak detection system this filter may not be needed since the conventional earphones used in such apparatus have negligible response at frequencies of 80–89 kHz. Also, the human ear has no response in this frequency range. A Butterworth low-pass filter might also be used here but I prefer the Bessel since its linear phase shift with frequency will better preserve the waveform of the total signal. (The MATLAB/ Simulink appendix provides some discussion of such filter selection questions.) Figure 7.20b shows the time histories of the *raw* microphone signal and its amplitude-modulated and filtered version, clearly verifying the frequency-translation claimed.

Computing and plotting the frequency spectra of these signals with the above MATLAB program further clarifies the operation of the amplitude-modulation process. Note that one must *first* run the Simulink simulation to make the time-history signals available to the MATLAB program. (It is of course possible to *embed* the Simulink program within the MATLAB m-file, but I kept them separate here to minimize possible confusion. The Appendix A material explains the *embedding* process should you ever need to do this.) Since our main interest here is to display the frequency spectra of interest, we need to compute the FFTs. Rather than then plotting these directly, I chose to compute and plot *power spectral densities* (PSDs), since the ultrasonic signals of practical interest will be random signals, and the PSD (more correctly called *mean-square spectral density*) is the conventional spectral measure for such signals. When I use the sine wave signals to verify the operation of the amplitude-modulation process, those signals would conventionally be processed, since they have *discrete* spectra, with the *Fourier series*, but one can certainly compute their PSDs since the peak frequencies will be properly displayed either way. This is what Figure 7.21 shows, and we see that the frequency translation is achieved as desired. Repeating these calculations for the random ultrasonic signals gives the results of Figure 7.22, where the frequency translation effect is again clearly evident. As explained in Appendix A, frequency spectra of random signals require considerable *averaging* to get accurate values. Since even rough spectra are sufficient, in

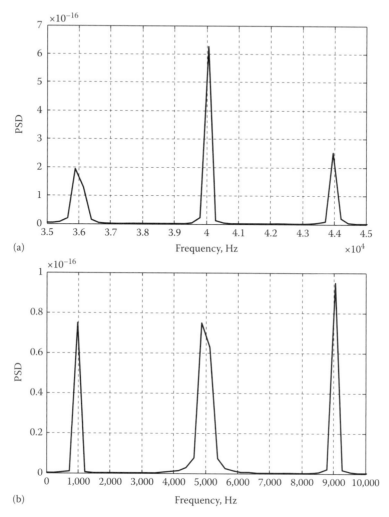

FIGURE 7.21
Power-spectral-density of (a) raw microphone signal and (b) amplitude-modulated and low-pass-filtered signal.

our present example, to convince the reader of the utility of amplitude modulation, I chose to avoid the added complexity of the averaging process.

7.10 Ultrasonic Acoustics Pertinent to Leak Detection

It will be helpful to discuss here briefly some characteristics of ultrasonic signals as they pertain to commercial leak detection and mechanical component failure detection systems, particularly *attenuation with distance* and *beam width*. As an acoustic wave propagates through the air its intensity diminishes, and this reduction in strength is greater for high frequencies than for low. For frequencies below 200 kHz the attenuation in dB/ft is approximately given by *attenuation* = *0.01f*, where frequency *f* is given in kHz. Thus 40 kHz signals will attenuate at the rate of 0.4 dB/ft. This loss in signal is in some ways an advantage for ultrasonic versus audible sound. When using a microphone probe to locate an air leak there may be *several* sound sources that might be picked up, which makes the task of isolating a specific source more difficult. With an ultrasonic signal, as we approach one source and thus move away from others, the near source tends to dominate the microphone response, easing the location task.

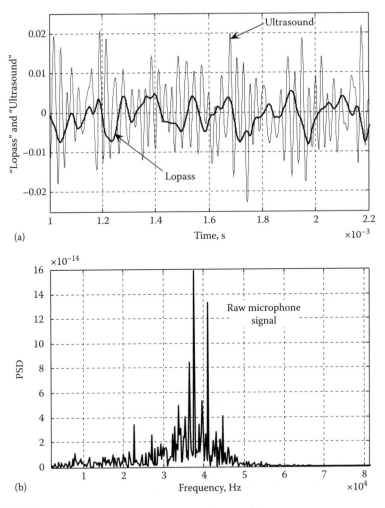

FIGURE 7.22
(a) Time histories. (b) and (c) Frequency spectra of random ultrasound.

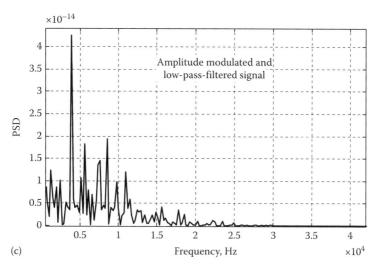

FIGURE 7.22 (continued)

The concept of beam width applies to both transmitters and receivers (microphones), as seen in the polar graph of Figure 7.18a. For both transmitters and receivers, the beam width gets narrower as the frequency gets higher, and as the diameter of the sensor (or actuator) diaphragm gets larger relative to the signal wavelength. Thus a narrow beamwidth requires a *large* diaphragm. Quantitatively[*]:

$$\text{cone angle} = asin\left(\frac{\lambda}{2r}\right) \tag{7.35}$$

At 40 kHz the wavelength λ is 0.325 in., so if the diaphragm radius r were, say, 0.5 in., then the majority of the sound would be found within a cone of included angle 19°. Most ultrasonic microphones of the simple type we are considering have diaphragms somewhat smaller, so their directionality is not this good, but still sufficient to allow accurate location of most leaks. Leaks in unpressurized systems can be found by placing an ultrasonic *transmitter* inside the test object (say an empty fuel tank) and probing the outside with a microphone or array of microphones. The short-wavelength ultrasonic waves can penetrate any small leaks, producing signals that can be picked up outside the vessel and located by scanning the microphone over the surface.

Information on practical leak detection and mechanical failure detection systems based on ultrasonic is available at several Web sites.[†]

---

[*] P.M. Morse, *Vibration and Sound*, McGraw-Hill, New York, 1948, pp. 328–329.
[†] www.ctrlsys.com, www.uesystems.com

8

Some Basic Statistical Tools for Experiment Planning

8.1 Introduction

In Chapter 1, we introduced the topic of statistical design of experiments as a tool for finding mathematical models for industrial processes or engineering experiments, where we wanted to find a relation between a *response* (dependent) variable and a number of independent variables (*factors*) that we assumed affected the dependent variable in some important way. Our main tool there was *multiple regression*. In this chapter, we address questions such as the following:

1. If I want to find a reliable numerical value for some material property, dimension, or parameter of some component or system, *how many samples must I test to get a good average (mean) value?*

2. If I want to get a good estimate of the *variability* of some material property, dimension, or parameter, how large a sample is needed for this?

3. If I want to *compare* the *average values* of two quantities to decide which is best, how large a sample is needed for this?

4. If I want to *compare* the *variabilities* of two quantities to decide which is best, how large a sample is needed for that?

5. If an experiment involves measuring several quantities and then computing a result from these measurements, how accurate must the individual measurements be to guarantee a specified accuracy in the result?

Questions 1–4 can be addressed in two major ways: *hypothesis testing* and *confidence intervals*. In hypothesis testing, we propose a hypothesis such as *aluminum alloy A has a larger tensile strength than aluminum alloy B*. We gather some data and then test this hypothesis statistically to decide whether we accept or reject it at a certain level of *significance*, often 5%. Here, 5% means that in our acceptance/rejection decision, we will be wrong only 5% of the time. Using the confidence interval method on the same problem, we gather some data and compute the difference in the mean values of tensile strength, together with a confidence interval that extends on either side of the difference in average values. For example the difference in average values (A − B) might be 10,673 psi, and the confidence interval might be ±2,473 psi. Confidence intervals always carry an associated *probability*, often 95%. The meaning of the above statements is that our best estimate of the difference in strength is that alloy A has tensile strength 10,673 psi larger than that of alloy B, *and we are 95% sure that the true difference is somewhere between 8,200 and 13,146*. The phrase "95% sure" means that if we use this test routinely in our experimental work, we will be correct 95% of the time (19 times out of 20). About 5% of the time we will be *wrong*. While 95% confidence is often used, we can choose any percentage that we wish, *but with an important trade-off*. If we insist on being wrong only, say, 1% of the time, then for the same sample size, *the confidence interval must be larger*. If we want *both* a higher confidence level *and* a smaller confidence interval, then we *must* use larger sample sizes, which usually costs time and money.

When using hypothesis testing, we say that the result (alloy A is stronger than alloy B) is *significant at the 5% level. Significant* here means *statistically significant*, and that may be where the problem lies. For example, alloy A may be only *slightly* stronger than alloy B, or *much* stronger, but we get no *numerical* indication of this. That is, the *practical* significance is our real concern, and this requires some numerical information (not just the *yes/no* of the hypothesis test). Confidence intervals provide this vital information.

With the above brief comparison of the two possible approaches (hypothesis testing and confidence intervals), most engineers would wonder why anyone would want to use hypothesis testing; the other approach clearly provides much more useful information! Some early literature* made the case for confidence intervals, but these articles seem to not have had much impact on most later statistics texts. I personally developed a preference for confidence intervals some years ago,† and more recent literature‡ seems to also go in this direction. Many earlier medical and drug experiments suffered from an "addiction" to hypothesis testing, and now some journals in that field *insist* that this approach be augmented by (or replaced with) confidence interval methods. While most engineering experiments do not carry the *life or death* aspect of those in the medical field, we still want to use the best tools available. Statistics texts still present considerable material on hypothesis testing (with or without critical evaluation), and it may be that there are some applications where this method might be useful, but I (as a non-statistician) have been unable to justify it for practical engineering work. In the upcoming detailed discussions we will thus present only the confidence interval viewpoint.

* M.G. Natrella, The relation between confidence intervals and tests of significance, *The American Statistician*, 14, February 1960, 20–23; G.J. Hahn, Don't let statistical significance fool you!, *Chemtech*, 4, January 1974, 16–17.

† E.O. Doebelin, *Engineering Experimentation: Planning, Execution, Reporting*, McGraw-Hill, New York, 1995, p 72.

‡ S.T. Ziliak and D.N. McCloskey, *The Cult of Statistical Significance*, University of Michigan Press, Ann Arbor, MI, 2008; D.G. Altman et al., *Statistics with Confidence*, BMJ Books, London, U.K., 2000.

8.2 Checking Data for Conformance to Some Theoretical Distribution

Many of the standard formulas for computing confidence intervals assume that the data comes from some theoretical distribution, the most common being the *normal* or *Gaussian* distribution. This sort of assumption is *necessary* in order to come up with specific formulas. We know, of course, that *no* real data can ever *exactly* follow *any* theoretical distribution, so there is never any hope of proving that real data exactly conforms to some theoretical model. Our check methods then must be satisfied with showing that the data "comes close" to some chosen distribution function. Actually, we will accept methods that only show that our data *does not grossly diverge* from the model; this is the best that one can do in practical work where we cannot economically justify very large sample sizes. These kinds of tests may take a numerical or graphical form; we prefer the graphical approach, implemented by readily available statistical software, such as the Minitab described in Appendix B.

In Figure 8.1a, we see the familiar *bell-shaped curve*, which is the probability density function (distribution function) for the Gaussian (also called *normal*) distribution; I have chosen a mean (average) value of 10, and a standard deviation of 2.0. Minitab will make such graphs for a wide variety of distributions, using the following sequence: *graph>probability distribution plot>view single>distribution*. You then select the distribution you want from the list offered and enter the numerical values of any parameters (such as mean, standard deviation, etc.) associated with that distribution. To decide whether we can treat some real data as if it were Gaussian, we really want to see whether it conforms to the *bell-shaped curve*, but the best we can do is to plot a *histogram* of these data. This is a *bar chart* that shows how the sample values are distributed. If we had an infinite-size sample, this bar chart would become a smooth curve, which could be directly compared with the ideal curve. To see some actual graphs, we can "manufacture" a data sample using the sequence *calc>random data*, which brings up a list of available theoretical distributions, from which we can choose the one we want. These are *random number generators*, which produce numerical samples that almost perfectly follow the theoretical distribution. To get Figure 8.1b, I chose the *normal* distribution, requested a mean value of 10, a standard deviation of 2, a sample size of 100, and asked that these data be stored in column c1 of the worksheet. Once this data is available, we can process it in various ways. To plot the histogram of Figure 8.1b, we use the sequence *graph>histogram>simple*, which brings up a window where you can select column c1, take the defaults, and request the plot. We see that, even with a large (100 item) sample, the histogram is still quite "ragged" compared with the ideal smooth curve, so *it is not obvious* that our data sample is Gaussian, even though it was generated by a "perfect" Gaussian random number generator.

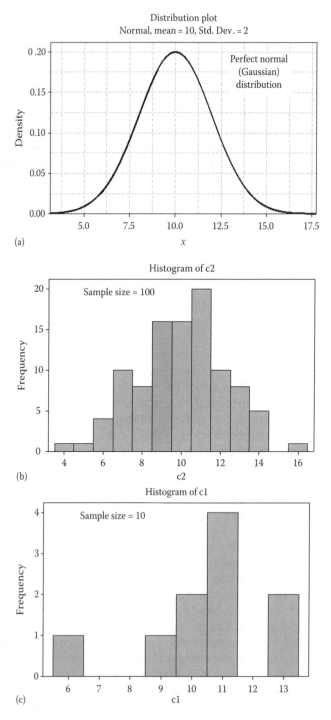

FIGURE 8.1
Comparing finite sample size results to infinite sample size results.

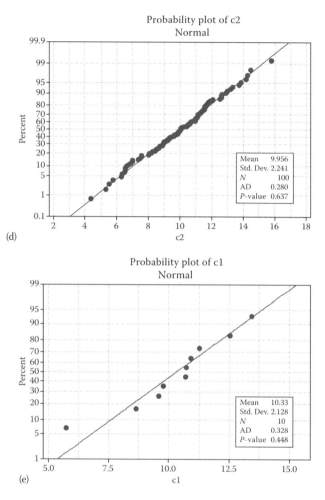

(d)

(e)

FIGURE 8.1 (continued)

To make such decisions easier and more reliable, a *different* graphical display is available, using the technique called *curve rectification*. By properly distorting the plotting scale on a graph, *any* curve can be made to plot as a *straight line*. To make the *perfect* curve of Figure 8.1a plot as a straight line, one uses special graph paper called *normal probability graph paper*. Minitab provides the computerized version of this with its sequence *graph>probability plot>single*, which opens a window where you can choose the data column to be plotted and the distribution to be used for comparison (take the defaults on the other choices). Figure 8.1d shows that this 100-item sample *does* plot very close to a straight line, so we would conclude that these data are *close enough* to the Gaussian distribution that we could use any theoretical formulas that made this assumption. However, in Figure 8.1c and e the histogram looks *very bad* and the "straight-line" graph is not very reassuring, even though this sample came from *a* "perfect" *Gaussian number generator!* These graphs reveal the basic problem in judging whether real data is *close enough* to some theoretical model; *small samples* from even a perfect distribution will not *look like* that distribution. Another aspect of this dilemma is found in the small tables included in Figure 8.1d and e. These show that even though we *requested* a mean value of 10 and a standard deviation of 2, the *samples generated* will not have exactly these values. For *very* large samples, the sample means and standard deviations *will* converge on the requested values, but in practice we rarely can justify such sample sizes.

A practical approach to this problem might go as follows. If you have gathered an *n*-item sample of data and want to check whether these data can be treated as Gaussian, have Minitab generate, say, five *n*-item samples from that distribution, and visually compare the probability (straight-line) plots of your data and the Gaussian data. When you generate the perfect Gaussian samples, be sure to use a *different* "base" value (sometimes called the *seed*) for each sample. If the plot of your data does not appear to deviate radically from the five perfect plots, you can treat your data as Gaussian. This clearly is a judgment call on your part, and may seem somewhat arbitrary, but there really is no foolproof way of making this decision when the sample size is modest. Some people prefer a *numerical* rather than a visual way of making such decisions, the *chi-square goodnesss of fit test*[*] being the most common. In my opinion, this method does not really improve the ease in making, or quality of, the decision, so I prefer our graphical approach.

When our data appears to be clearly non-Gaussian, what do we do? We can either try some other theoretical distribution, or we can try to *transform* our data in some way so that it comes closer to Gaussian. Experience has shown that most engineering data can be modeled with a distribution chosen from this list:

Normal	Lognormal	Weibull (two-parameter)	Weibull (three-parameter)
Exponential	Binomial	Hypergeometric	Poisson

For example, fatigue failure of metals (stress/life data) closely follows the lognormal distribution.[†] Detailed discussion of the applicability of specific distributions to specific types of data is available.[‡] Graphical (straight-line)

---

[*] E.O. Doebelin, *Engineering Experimentation: Planning, Execution, Reporting*, McGraw-Hill, New York, 1995, pp. 55–58.

[†] E.O. Doebelin, *Engineering Experimentation: Planning, Execution, Reporting*, McGraw-Hill, New York, 1995, pp. 58–61.

[‡] C. Lipson and N.J. Sheth, *Statistical Design and Analysis of Engineering Experiments*, McGraw-Hill, New York, 1973, pp. 59–62.

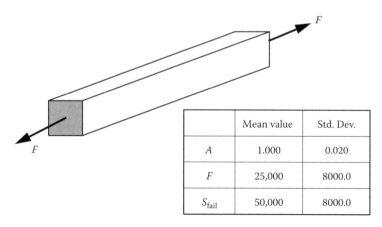

	Mean value	Std. Dev.
A	1.000	0.020
F	25,000	8000.0
S_{fail}	50,000	8000.0

FIGURE 8.2
Simple tension member.

tests are available for most of these distributions. We will here concentrate on the transformation approach, using a method (Box–Cox) that is available in Minitab. An example problem, which also covers some other useful ideas, concerns a simple tension member (see Figure 8.2) with cross-sectional area A, made from a material with failure stress, s_{fail}, loaded with a force F. Each of these parameters is treated as a random Gaussian variable with mean and standard deviation as given in the figure. "Ordinary" machine design uses a safety factor defined by

$$\text{Safety factor } f_{safe} \triangleq \frac{S_{fail}}{F/A} \qquad (8.1)$$

where all the parameters are treated as constants, *not* as statistical variables, giving a single number, 2.0, for the safety factor in our example. Statistical design is not much used because the data needed is difficult and expensive to gather, though it *is* used in those cases where the expense can be justified by the advantages gained. In statistical design we think of each item in the sample having its own (randomly variable) failure stress, area, and applied force, giving each item *its own* safety factor. This viewpoint is of course *more realistic* than the usual assumptions of constant parameters and a fixed safety factor, and is one of the advantages of statistical design. We will now act as if our *manufactured* Gaussian data on area, force, and failure stress is real data and proceed to calculate the individual safety factors of the 100 items in our sample. The mean values and standard deviations shown in Figure 8.2 are "made up" values chosen to suit our demonstration, they are not actual material properties, dimensions, or forces. Using Equation 8.1, Minitab will calculate the individual safety factors and store them in a worksheet column of our choice. Note that a safety factor of 1.0 means that the actual stress and the failure stress are equal, thus any values less than 1.0 correspond to *failures*. These are easily detected by sorting the column of safety factors in increasing (or decreasing) order and placing them in another column, using the sequence *data>sort*. From Table 8.1 we see that two failures are predicted and that some items will have excessive safety factors such as 8.2 and 15.3. These large spreads are of course due to the large standard deviations I assumed for force and failure stress.

When a *dependent variable* depends in a *linear* fashion on several independent variables, and if the independent variables each have a Gaussian distribution, it can be shown that the dependent variable

TABLE 8.1

First and Last 10 Sorted Safety Factors

f_{safe}	f_{safe}
0.89354	3.5859
0.93161	3.6731
1.05226	4.3179
1.15542	4.4150
1.23174	5.0477
1.27191	6.1795
1.27222	6.2279
1.29349	8.2164
1.32133	8.6330
1.32367	15.3083

will also have a Gaussian distribution. Furthermore, if we know the means and standard deviations of the independent variables, we can calculate the mean and standard deviation of the dependent variable.

$$y = a_1x_1 + a_2x_2 + \cdots + a_nx_n \tag{8.2}$$

The mean value of y would be the sum $(a_1x_{1,avg} + a_2x_{2,avg} + \cdots)$, and its standard deviation would be the square root of the sum of squares $[(a_1s_1)^2 + (a_2s_2)^2 + \cdots]$ where the s's are the individual standard deviations. When the relation of y to the Gaussian x's is *nonlinear*, then y's distribution will *not* be Gaussian and we are not allowed the use of any theoretical formulas based on Gaussian behavior for making predictions about y. When the standard deviations are *small relative to the respective mean values* (coefficient of variation $C_v = s/x_{avg}$ is $\ll 1.0$), then a useful approximation goes as follows. First, the distribution of y will be close to Gaussian and we can use formulas based on that assumption. To get the mean value of y, we merely insert the mean values of the x's into the nonlinear formula relating y to the x's. To get the standard deviation of y we compute the partial derivatives of y with respect to each x and use the formula

$$s_y = \sqrt{\left(s_{x1} \cdot \frac{\partial y}{\partial x_1}\right)^2 + \left(s_{x2} \cdot \frac{\partial y}{\partial x_2}\right)^2 + \cdots + \left(s_{xn} \cdot \frac{\partial y}{\partial x_n}\right)^2} \tag{8.3}$$

In our safety factor example, the relation between the dependent and independent variable is linear in s_{fail} and A, but nonlinear in F, so the

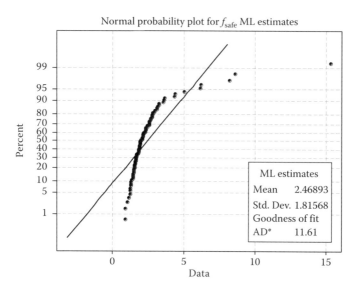

FIGURE 8.3
Nonlinear relation gives non-Gaussian distribution.

overall relation is nonlinear. I intentionally made the standard deviations of F and s_{fail} too large to allow use of the approximate relation based on the partial derivatives, and the distribution of y will probably be quite non-Gaussian. To check this last assertion I subjected the column of f_{safe} data to the probability graph test, giving Figure 8.3, which clearly shows a non-Gaussian distribution. We will now try a Box–Cox transformation on these data to see if it gives a more Gaussian result. This transformation is of the form $y_{new} \triangleq y_{old}^{\lambda}$ where lambda is an adjustable parameter which the software automatically optimizes to get the most nearly Gaussian result. The transformation is invoked with the sequence *stat>control charts>box-cox*, which opens up a window where you can select which column of data is to be transformed and in which column the transformed values are to be stored (enter 1 for *subgroup sizes*). Figure 8.4 shows that the "estimate" lambda value was found to be −0.75. I then asked that the transformed data be plotted on the normal probability plot, giving Figure 8.5, which we see is acceptably close to Gaussian, with a mean value of 0.5773 and a standard deviation of 0.1762 (Figure 8.6). One could thus perform any standard Gaussian calculation with the transformed data, and then convert those results back to the original data values, using the formula. For some reason unknown to me, the *estimate* value of lambda displayed in panels like Figure 8.4 is *not* the one used in transforming the data, and the value that *is* used is not displayed anywhere! The lambda value that *was* used is easily calculated using the definition of the transformation:

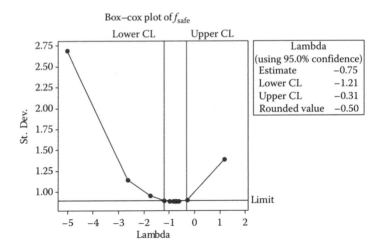

FIGURE 8.4
Box–Cox transformation result panel.

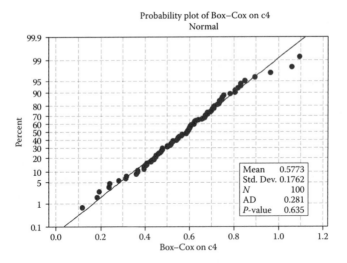

FIGURE 8.5
Box–Cox transformation is successful.

$$\lambda = \frac{\log(y_{\text{new}})}{\log(y_{\text{old}})} \tag{8.4}$$

For our example, the lambda value actually used in the transformation is −0.786017. If we want to convert values of y_{new} to associated values of y_{old}, the formula is

$$y_{\text{old}} = y_{\text{new}}^{\frac{1}{\lambda}} \tag{8.5}$$

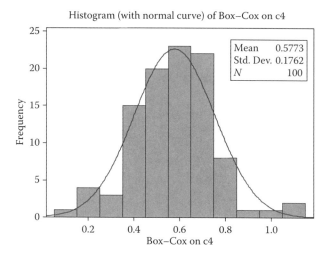

FIGURE 8.6
"Descriptive statistics" display for the transformed data. (Use sequence: *stat > basic statistics > graphical summary*).

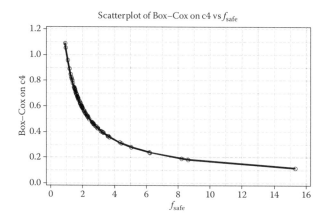

FIGURE 8.7
When untransformed data decreases, transformed data increases.

Suppose we want to calculate the probability that the safety factor will be less than, say, 1.5. This is the same as asking for the probability that the *transformed* value be *more* (see Figure 8.7) than 0.72709. Since the transformed data can be treated as if it were a Gaussian distribution, we can use the standard tables of that distribution to compute the desired probability. Minitab has the computerized equivalent of these tables, using the sequence *calc>probability distributions>normal*. This brings up a screen like Figure 8.8, where I have filled in the necessary numbers. Minitab returns the computed probability as 0.1976 (19.76%) in the session window. Since our original sample of 100 safety factors did include values below 1.5, we could have estimated this

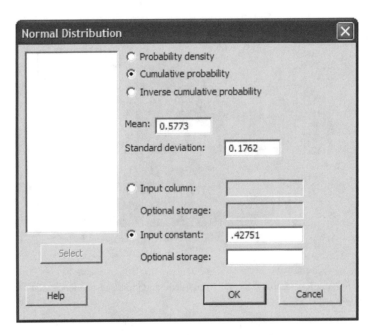

FIGURE 8.8
Calculation of normal (Gaussian) probabilities.

probability *without* going through the transformation process. Examining the sorted list of safety factors we see that the 17th–20th values are 1.4844, 1.4877, 1.5030, and 1.5375, respectively, so this is in close agreement with the Gaussian probability calculation. If our original sample had been smaller, then it might *not* have included values below 1.5, and we could then *not* get the probability of interest from that sample, but we still *could* get it from the transformed data and the Gaussian calculation. Since sample sizes in practical engineering work will often be 30 items or less, the use of transformations has considerable application.

8.3 Confidence Intervals for the Average (Mean) Value

Most readers will be familiar with the calculation of the average value of a set of data. This is a useful number, but has its limitations. If we compute an average from a sample of 5 items and another from a sample of 20, we intuitively know that one of these numbers is more reliable than the other. The concept of *confidence intervals* provides a numerical indication of this difference. When we compute a confidence interval for a mean value, we are then able to make the following sort of useful statement: "My best estimate of the average value is 12.4, and I am 95% sure that the true value is within the interval

12.4 ± 2.6." The meaning of "95% sure" is that if I use this method routinely in my work, I will be right 95% of the time (19 times out of 20), and I will be wrong 5% of the time. The calculation of numerical values for this confidence interval requires use of another well known statistical distribution, the *t-distribution*. This calculation is based on the assumption that the data comes from a Gaussian distribution, so here again we need to check that assumption before proceeding. The formula for computing the confidence interval is

$$\text{Confidence interval } \Delta \ \bar{x} \pm t_{\alpha/2, n-1} \cdot \frac{s}{\sqrt{n}} \tag{8.6}$$

where
   $\bar{x}$ and s are the average value and standard deviation of the data sample
   n is the sample size
   α is (1-decimal value of the *confidence level*)

The confidence level sets *how sure* we are of our statement; it would be 95% in the example stated above, and α would be 0.05. When looking up values of the *t*-distribution in a table, $(n - 1)$ is usually called the *degrees of freedom*. If we had a sample size of, say, 10, we would look up $t_{0.025, 9}$, which turns out to be 2.262. Minitab of course provides the computerized equivalent of the table, using the sequence *calc>probability distributions>t* which brings up the screen of Figure 8.9, where you need to select *inverse*

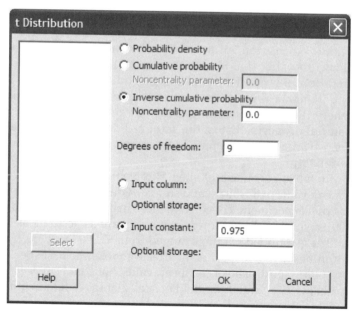

FIGURE 8.9
Using Minitab's "*t*-table."

cumulative probability, and take the default 0.0 for the *noncentricity param-eter*. Our choice of confidence level is entered in the *input constant* box, but note that Minitab uses $(1.0 - \alpha/2)$, not $\alpha/2$. Thus we need to be careful in using statistical tables from books or papers, or statistical software; they do not all use the same conventions and definitions. The use of a 95% confidence level is very common, so we might look at the variation of the *t* value with sample size *n* when we use 95%. For sample sizes from 5 to 120, *t* varies from 2.571 to 1.980, and for an *infinite* sample size, *t* = 1.960. Thus in Equation 8.6, the major effect of sample size will be found in the term $1/\sqrt{n}$. If the mean value and standard deviation were fixed, we see that the confidence interval shrinks with increasing sample size, but only as the square root. Thus to cut the confidence interval in half, we need to quadruple the sample size.

If one wants to design an experiment (choose the sample size) so as to achieve a confidence interval that is some (small) percentage of the mean value (so that we can say, for example, "I am 95% sure the average value is $A \pm 0.1A$"), we have a problem in that we do not know the mean value and standard deviation *before* we run the experiment, and thus cannot use Equation 8.6. To explore this sort of question, it will be helpful to run a *simulation experiment* using Minitab's random number generators. Suppose we were running strength tests on some new metal alloy using a conventional tensile testing machine to find the ultimate (breaking) stress value. Let us also assume that the testing machine and procedure are suf-ficiently accurate that our readings are the *true* values of the material's tensile strength. (This assumption is made to keep the example clearly focused on the question of confidence intervals, without the additional complication of measurement accuracy. There *are* ways to include mea-surement accuracy, but that is not our current interest.) Suppose that we are able to afford a large (100-item) sample, but of course we gather that sample one item at a time. After we test the first specimen, we get some idea of where the mean value might lie, but get *no* idea as to the *scatter* in this material property, that is, the standard deviation. As we test more specimens, we are able to estimate both mean value and standard devia-tion more reliably.

I asked Minitab to generate a 100-item sample from the Gaussian distribu-tion with mean value of 87,400 psi and standard deviation of 1,500 psi, to sim-ulate our actual experiment. For convenience I had the values rounded to the nearest 100 psi since such measurements rarely are as accurate as 1 part in 874 (0.11%). These rounded readings ranged from 83,000 to 90,700. If we were actually running the real experiment, after we broke the first specimen, we would have our first estimate of the mean value, but could say *nothing* about the standard deviation, and thus could not calculate a confidence interval for our mean value. Sometimes there is available *generic* information about the variability of certain physical parameters. Most metals exhibit a *coefficient of variation* (standard deviation divided by mean value) of 2%–7% for tensile

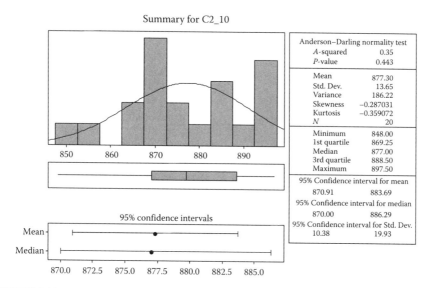

FIGURE 8.10
The basic statistics graphical summary screen.

strength.* Thus we could estimate a standard deviation from data on the first specimen and thus compute a confidence interval. After testing the second specimen, we could calculate a mean value and standard deviation and thus a confidence interval. I used the sequence *stat>basic statistics>graphical summary* to process these data, which gives results, for a sample of size 20, in the form of Figure 8.10. (For some reason, using this sequence, Minitab refuses to calculate standard deviations for samples smaller than 3, so I had to do this calculation *manually* using *calc>column statistics*.) From Table 8.2 we see that if we want to estimate the mean value within 1% (±8.7, 865–883) we need a sample size of about 9.

We can construct a useful graph for estimating the necessary sample size to get a confidence interval that is a selected percentage of the mean value by first defining a *confidence ratio*:

$$\text{Confidence ratio} \underline{\Delta} \text{ confidence interval/mean value} = \pm \frac{t_{\alpha/2, n-1}}{\sqrt{n}} \cdot \frac{s}{\bar{x}} \quad (8.7)$$

$$(\text{Confidence ratio/coefficient of variation}) \underline{\Delta} R_c = \frac{t_{\alpha/2, n-1}}{\sqrt{n}} \quad (8.8)$$

---

* E.B. Haugen, *Probabilistic Mechanical Design*, Wiley New York, 1980, pp. 596–604.

TABLE 8.2

Simulation Experiment for Tensile
Strength

n	Mean Stress, 100 psi	95% Conf. Int. on Mean, 100 psi	Std. Dev., 100 psi
1	887		
2	882	818–956	7.1
3	878	857–899	8.5
4	875	860–890	9.2
5	874	864–884	8.3
6	873	865–881	7.7
7	876	866–886	10.8
8	874	863–884	12.5
9	875	865–884	12.2
10	877	867–887	13.5
20	877	871–884	13.7
40	877	873–881	12.1
100	873	870–876	14.7

We choose the confidence ratio that we want, say, 5%, and must then estimate the coefficient of variation as best we can, from historical data or from the first few specimens tested. The ratio R_c of Equation 8.8 can be plotted versus sample size n for any given confidence level α, giving Figure 8.11. Given the value of R_c, we can pick off the necessary sample size.

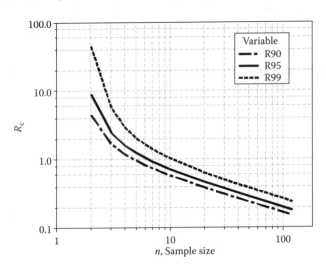

FIGURE 8.11
Choosing sample size to get desired confidence interval for mean value.

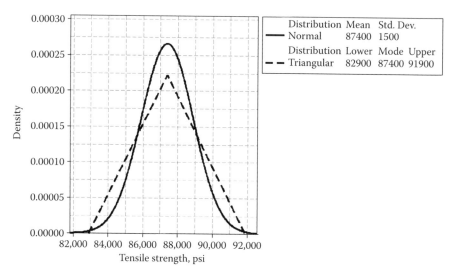

FIGURE 8.12
Comparison of Gaussian and triangular distributions.

Since all the above calculations and graphs are based on the assumption that the data follows a Gaussian distribution, it will be useful to look into how strictly this must be enforced. One way to do this is to have Minitab generate data from a non-Gaussian distribution and then do the calculations as if it were Gaussian. One distribution suitable for such a check is the *triangular*. Here, to define the distribution, we specify the lowest and highest possible value and the mean value. Figure 8.12 compares this with a Gaussian distribution with the same mean value. Since a Gaussian distribution has very few values outside a ± 3*s* range around the mean, I chose the lowest and highest values in my triangular distribution at those locations. When I performed the same calculations (as we just used on Gaussian data), the results from the triangular data were not radically different, so modest deviations from the Gaussian distribution seem to be well tolerated.

8.4 Comparing Two Mean Values: Overlap Plots and Confidence Intervals

Suppose we wanted to choose, based on tensile strength, between the metal alloy used in the previous section and a competing metal. We need to compare the two mean values, but now we know that the confidence interval concept should be included in some way. There are two ways to do this comparison:

1. Graph the individual mean values and their confidence intervals on the same axis and see if they *overlap*.
2. Use an available confidence interval for the *difference of the two means*.

The second method is preferred since it is less ambiguous, but the first method gives a viewpoint useful in explaining results to an audience that is not statistically knowledgeable. We will thus present both methods since they complement each other.

To get the individual confidence intervals we use the sequence *stat>basic statistics>grapical summary* as we did earlier. I generated two 5-item Gaussian samples with different *base* values and different means and standard deviations. I asked for alloy A (874,15) and for alloy B (894,16), but when the generated samples were used to compute the *actual* values they were (873,25) and (886,16). Recall that this deviation is *expected*; small samples of any distribution will *not* exhibit the same parameters as the ideal distribution. Real data, not just simulations, also behave in this way. Figure 8.13 shows the results of this test, arranged to display visually any overlap of the confidence intervals. (Minitab does not provide this display; I had to "piece it together" using Minitab's general graphing tools.) We see that the mean values *alone* show that alloy B is 1300 psi stronger than A but the overlap is extreme; there is a good chance that A might be stronger than B! Note that the larger confidence interval of A is due to its larger standard deviation, even though I had *asked* for them to be almost the same. We see that the small sample size makes it difficult to reach a reliable conclusion. However, even with this small sample

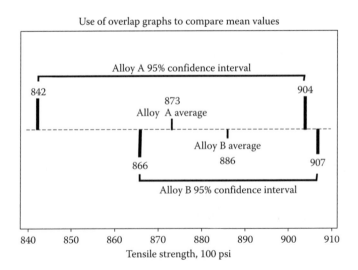

FIGURE 8.13
Comparing two mean values.

size, if the standard deviations had been smaller, we might have been able to make a clear choice.

The overlap display just discussed is often useful when explaining the use of confidence intervals to those unfamiliar with the concept, but it should then be followed up with a presentation of a *single* confidence interval for the *difference* of the two means, which makes the decision a little simpler. Formulas for computing this interval are available,* but not necessary if proper software is available. In Minitab this test is invoked with the sequence *stat>basic statistics>two-sample t*. This brings up the screen of Figure 8.14, which offers three alternatives for entering data. We will usually use the second, where you need only enter the two columns where the data for your two samples is found (mine were in columns 19 and 20). The samples need not have the same number of items. The first method of entering data is for a case where the two samples are located in the same column. The third method allows you to analyze situations where you do not have measured data, but want to stipulate sample size, mean value, and standard deviation for each. This is possible because the formulas for doing the calculating require *only* these

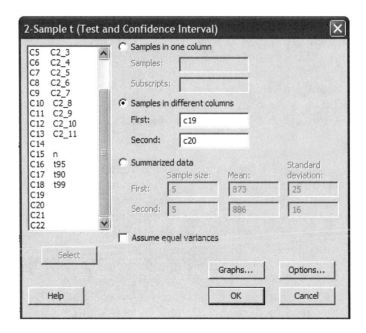

FIGURE 8.14
Minitab screen for comparing mean values.

---

* E.O. Doebelin, *Engineering Experimentation: Planning, Execution, Reporting*, McGraw-Hill, New York, 1995, p. 73.

values. No matter which method you use, the results appear in the *session window* as follows:

Two-sample T-test and CI: C19, C20
Two-sample T for C19 vs C20

    N   Mean St. Dev.
C19 5   873.2   24.9
C20 5   886.1   16.5
Difference = mu (C19) − mu (C20)
Estimate for difference: −12.9
95% CI for difference: (−45.6, 19.8)

We see that our best estimate of the difference in strength is 1290 psi, and we are 95% sure the true value is somewhere between −4560 and +1980 psi. If we insist on being 95% sure, we cannot say that alloy B is stronger; it is possible that A is stronger. If you want to relax the 95% to some lower value, the *options* selection in Figure 8.14 allows you to enter any percentage you like. With our current example, I had to lower the confidence level to about 55% before the confidence interval was entirely in the negative range (−23.7 to −2.1), meaning that we would be 55% sure that alloy B was stronger than A. If these sorts of results seem disappointing to you, it is because of the small sample sizes and *large* standard deviations. Without improvements in either or both of these, one cannot really expect more definitive conclusions. All these discussions should convince you that the direct comparison of raw average values is fraught with danger. We really do need to use confidence intervals.

A particularly significant application of confidence intervals relates to the *experimental validation of theoretical predictions*. All theoretical studies require the postulation of some *simplifying assumptions* that are made to bring the analysis within the scope of the physical/mathematical tools employed. Results obtained in this way are always open to question until they are verified (or refuted) by some experimental tests. This verification ultimately comes down to *comparing* the theoretical prediction with the measurements. As a simple example consider a cantilever beam with an applied end load F. The spring constant at the end of the beam is given *theoretically* by K_s = force/deflection = $Ebt^3/4L^3$. Note that this theoretical result can only be calculated if we have numerical values for modulus E, breadth b, thickness t, and length L. Each of these must be *measured* when we deal with an actual beam, and each measurement has its own uncertainty, leading to an overall uncertainty in the theoretically predicted spring constant, which can be considered a confidence interval around a mean value. Experimental testing might consist of applying a measured force and measuring the resulting deflection, and repeating this for a range of forces, allowing calculation of a set of spring constant values, which would have a mean value and a confidence interval. We could then plot overlap graphs for the theoretically predicted spring constant and that experimentally measured, with the degree of overlap relating to our degree of belief in the theory. The overlap graphs could also be augmented with the single confidence interval for the difference in mean values. Note that we speak of "degree of belief" rather than a flat statement such as "the theory is true" or "the theory is false," since no theory is ever "100% true"; we only require that its predictions are *good enough* for an intended application.

8.5 Confidence Intervals for the Standard Deviation

Knowledge of the standard deviation is useful in a number of situations. If we are studying a manufacturing process which produces a liquid product whose viscosity is important in some application, there will be limits set on the allowable range of the viscosity value. Product outside these limits is *scrap* and represents an economic loss for the manufacturer. The standard deviation of viscosity over some selected time interval is a numerical measure of this loss, and is thus of practical interest. If the process has some adjustable parameters, we might run experiments to try to find the set of operating conditions that minimizes this variability, and would want to *compare* the standard deviations associated with each set of conditions. Sometimes we have *competing* processes for producing the same product, and want to choose the one with the least variability. Such decisions are made more rationally if we know not only the standard deviations, but also the associated confidence intervals.

The calculation of the standard deviation for a set of data is straightforward, but as with the mean value, it is useful to also know how reliable this number is. A confidence interval for a standard deviation may be computed from available formulas,* using tables of the *chi-square* distribution, but Minitab provides this information as part of the *stat>basic statistics>graphical summary* screen, as shown in Figure 8.10. The size of the confidence interval depends on the sample's standard deviation, the sample size, and the confidence level (again often taken as 95%, but selectable at your choice). In Figure 8.10 we see that, in contrast to the mean value's interval (which *is* symmetric), this confidence interval is *not symmetric* about the best estimate; the best estimate is 13.7 and the confidence interval goes from 10.4 to 19.9. The *probabilities* however *are* equal; the true value will be found 47.5% of the time between 10.4 and 13.7, and 47.5% of the time between 13.7 and 19.9. We

---

* E.O. Doebelin, *Engineering Experimentation: Planning, Execution, Reporting*, McGraw-Hill, New York, 1995, pp. 72–73.

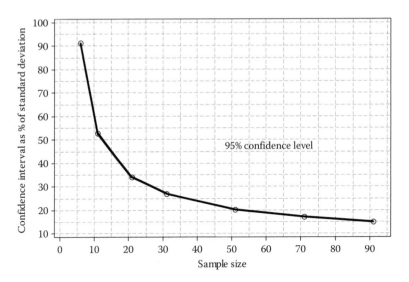

FIGURE 8.15
Choosing sample size for given confidence interval.

can again construct a graph useful for choosing sample size if we want to estimate a standard deviation within an interval which is a chosen percentage of the standard deviation, and at a chosen confidence level (often 95%). Figure 8.15 shows that for a 95% confidence interval, the required sample sizes are very large. To get a confidence interval that is about 20% of the standard deviation, we need a sample size of about 50. Note that the 20% is the *total* width of the interval and thus is comparable to a ±10% interval for a mean value. We cannot *call* it ±10% because it is unsymmetrical. These results show how uncertain small-sample estimates of standard deviation really are, and how dangerous it is to work only with the standard deviation itself, without the confidence interval.

When *comparing* standard deviations for two samples of data, we could again use overlap plots, but it is preferable to work with a *single* confidence interval for the *difference,* just as we did for mean values. This requires use of the *F-distribution,* which again is available in tabular form.* Minitab has a procedure called up by the sequence *stat>basic statistics>two variances,* but as far as I can tell, this does *not* do the calculations that we want, so we have to do this test "manually." We can get the desired *F*-distribution values in Minitab using the following sequence *calc>probability distributions>F,* which produces a screen like Figure 8.16. You need to select *inverse cumulative probability* and take the default (0.0) for *noncentricity parameter.* Numerator and denominator *degrees of freedom* are entered as the respective sample sizes

---

* E.O. Doebelin, *Engineering Experimentation: Planning, Execution, Reporting,* McGraw-Hill, New York, 1995, pp. 75–78.

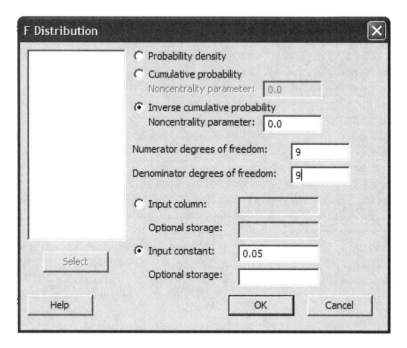

FIGURE 8.16
Getting values from the F-distribution.

minus 1; I used equal sample sizes of 10 for the two samples I was comparing. Select *input constant* and enter (1.0–confidence level)/2; I wanted a 90%
(0.90) confidence level so I entered 0.05. If you wanted to get values for *several*
confidence levels you would select *input column* and put, say, 0.05 and 0.025 in
rows 1 and 2 of some empty column, and enter the column where you want
the results put in *optional storage*. When you hit *OK* the results appear in the
session window or, if you used *optional storage,* the columns you requested.
The value which we will use in our upcoming calculations is *the reciprocal*
of what Minitab produces. For the setup of Figure 8.16 the result in the session window was 0.314575, so the F-distribution number that we will use is
1.0/0.314575 = 3.179.

Whereas the confidence interval for comparing mean values dealt with the
difference of the two numbers, when comparing standard deviations we deal
with the *ratio* of the two numbers. Specifically

$$\frac{1}{F_{\alpha/2,v_a,v_b}} \cdot \frac{s_a^2}{s_b^2} \leq \frac{\sigma_a^2}{\sigma_b^2} \leq \frac{s_a^2}{s_b^2} \cdot F_{\alpha/2,v_b,v_a} \tag{8.9}$$

$$\sqrt{\frac{1}{F_{\alpha/2,v_a,v_b}}} \cdot \frac{s_a}{s_b} \leq \frac{\sigma_a}{\sigma_b} \leq \frac{s_a}{s_b} \cdot \sqrt{F_{\alpha/2,v_b,v_a}} \tag{8.10}$$

Here "a" and "b" refer to the two samples we are comparing. *On the right-hand side* of Inequality 8.9 we treat v_b as the *numerator degrees of freedom*, which is the sample size of sample "b," minus 1; v_a is the *denominator degrees of freedom*, which is the sample size of sample "a," minus 1. *On the left-hand side*, the meanings of *numerator* and *denominator* are *reversed* when we look up the F value. The *true* values of the standard deviation are called σ_a and σ_b, so Inequality 8.9 defines the confidence interval for their ratio. The sample standard deviations, computed in the usual way from the two data samples, are called s_a and s_b. The values of the F-distribution are always less than 1.00, and approach that value as the sample sizes approach infinity. For both sample sizes equal to 500 the value is 1.16. Relation 8.10 shows that for infinite sample size the confidence interval has shrunk to zero, as we would expect. For any finite sample size the left-hand side will always be less than s_a/s_b and the right-hand side will always be greater.

For example, let sample "a" have 10 items and standard deviation of 3.46, while sample "b" has 11 items and standard deviation 4.67. Without any knowledge of confidence intervals we would state that sample "b" has the larger standard deviation. Let us check this assertion using our confidence interval, using a confidence level of 95% ($\alpha/2 = 0.025$). Using the screen of Figure 8.10 to compute the *left-hand side F value*, we enter 9 for the numerator degrees of freedom, 10 for the denominator degrees of freedom, giving the result 0.252279. For the right-hand side F value we reverse the definitions, and Minitab gives 0.264623. Recalling that we use the *reciprocals* of these values, Inequality 8.9 then becomes

$$0.252279\left(\frac{3.46}{4.67}\right)^2 \leq \left(\frac{\sigma_a}{\sigma_b}\right)^2 \leq \left(\frac{3.46}{4.67}\right)^2 3.77896 \qquad 0.372 \leq \frac{\sigma_a}{\sigma_b} \leq 1.440 \quad (8.11)$$

If we think of the two samples as being taken from two competitive production processes we see that at the 95% confidence level, either of the two processes could have the higher variability. This *inconclusive* result may be frustrating, but it does reflect the real uncertainty of the situation. If we invested in larger sample sizes, and/or relaxed our confidence level, a more conclusive result might be obtained, though of course when we use a lower confidence level, we will be wrong a larger percentage of the time. Because confidence intervals for standard deviations have been seen to require unusually large sample sizes, it might be prudent to use a lower confidence level and accept a larger chance of being wrong, to gain the benefit of getting useful conclusions. One of course should not carry this to the extreme. If we go to the 90% level and use sample sizes of 40 for both processes, and assuming that the two sample standard deviations stay the same (this would *not* happen, but it is OK to use it for comparison purposes) the calculations show that the F value is 0.590738 for both the left- and right-hand sides, leading to the result

$$0.569 \le \frac{\sigma_a}{\sigma_b} \le 0.963 \tag{8.12}$$

Now it is clear that, at the 90% confidence level, process "a" has the smaller variability.

8.6 Specifying the Accuracy Needed in Individual Measurements to Achieve a Desired Accuracy in a Result Computed from Those Measurements

Specifying the *uncertainty* associated with measured values, and results computed from those values, is an important consideration in any experimental work. This process is covered in great detail in many texts and other publications, including mine.* I wanted to include in the present text, this brief section on the topic, for those readers whose background in this area is limited, to make them aware of its importance and to provide reference to sources which explain it in more detail. We start with a version of Equation 8.3:

$$u_y = \sqrt{\left(u_{x1} \cdot \frac{\partial y}{\partial x_1}\right)^2 + \left(u_{x2} \cdot \frac{\partial y}{\partial x_2}\right)^2 + \cdots + \left(u_{xn} \cdot \frac{\partial y}{\partial x_n}\right)^2} \tag{8.13}$$

where dependent variable y represents some quantity that is computed, according to a known (linear or nonlinear) formula, from measurements of the n independent variables x_i. The use of 8.13 for *nonlinear y/x* relations is an approximation, but is usually quite accurate since the uncertainties will be small relative to the measured values themselves (instrument errors must be kept small for a valid experiment). The u's are the uncertainties associated with the individual measurements and the computed result. We are interested in specifying the *accuracy* of our results, or alternatively, the *inaccuracy* or *error*. Uncertainties define the error, which is made up of *bias* errors and *random* errors. When *calibrating* an instrument, the word *bias* refers to a *systematic* error whose numerical value is found from the calibration process; this component of error can then be *corrected*. That is, if an instrument's readings are systematically 0.32 units too low, we will *correct* those readings by adding 0.32 to the raw reading. In uncertainty calculations, bias is *redefined* as *bias limit B* and is considered as a *random effect,* so that it is consistent with the *imprecision* portion of the total uncertainty, which is random. Both the bias limit and the imprecision are defined as 95% confidence intervals; we

---

* E.O. Doebelin, *Measurement Systems,* 5th edn., McGraw-Hill, New York, 2004, pp. 40–85.

assume the *true value* will be found within these limits 95% of the time. The basic measure of imprecision is the standard deviation s, and it is called the *precision index*. To attach a certain confidence level to uncertainties, s is multiplied by some constant, called the *coverage factor*. The recommended coverage factor is defined in terms of the *t*-distribution as $t_{95,n-1}$, where n is the number of readings used to compute s. Some specific values for $n = 5, 10, 20$, and infinity are 4.303, 2.365, 2.110, and 1.960. With these definitions we can now give the formula for computing uncertainties:

$$u \triangleq \pm(B + st_{95,n-1}) \tag{8.14}$$

If the numerical value of a standard deviation s is obtained from an actual experiment that is repeated n times, the uncertainty is called a *type A uncertainty*. Bias limits are often based (unavoidably), not on actual experiments, but on things such as expert opinion, past experience, and specification sheets provided by equipment suppliers. In such cases the uncertainty is called a *type B uncertainty*. One reason that bias limits must often be estimated (rather than *measured*) is that most instrument calibrations are performed in a calibration laboratory, *not* with the instrument connected to the experimental apparatus ("*in situ*" *calibration*). (We of course prefer to do *in situ* calibrations, but often it is not possible or economical.) Thus, the uncertainty found in the calibration lab will generally be different from what would be found in an *in situ* calibration. To be conservative we will usually *increase* the uncertainty found in the calibration lab by some amount or percentage, based on our judgment and the factors mentioned above. Since the uncertainty is now estimated rather than measured, we need to give it a name (type B) which distinguishes it from the type A class. When we quote results in written reports or oral presentations, we must always make clear which type of uncertainty is being used. We now go over a simple example* to make these concepts more definite.

When an expensive and complex piece of machinery is purchased, the legal papers will often include the requirement that an *acceptance test* be performed to verify that the agreed upon specifications have actually been met, before final payment is authorized. Such agreements will generally require that a specification value fall within some stipulated range. When we perform the acceptance test, we will compute a confidence interval for the measured value that conforms to the legal document, say a confidence level of 95%. We thus need to *design our experiment* (choose the accuracy needed in each instrument) such that the computed result has the accuracy needed. Figure 8.17 shows a heat exchanger instrumented to measure the variables needed to compute its efficiency, since this parameter was guaranteed to be 0.90 ± 0.05, where the ± 0.05 is to be interpreted as a 95% confidence interval.

---

* E.O. Doebelin, *Engineering Experimentation: Planning, Execution, Reporting*, McGraw-Hill, New York, 1995, pp. 149–152.

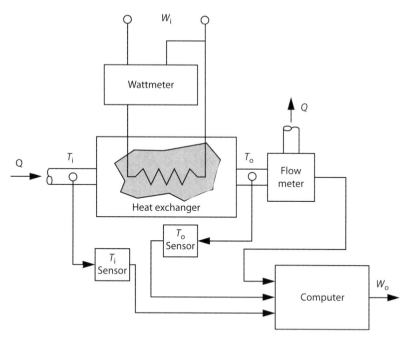

FIGURE 8.17
Heat exchanger acceptance test.

The efficiency is defined as the ratio of the output thermal power W_o watts, divided by the input electrical power W_i watts:

$$\text{Efficiency } E \triangleq \frac{W_o}{W_i} = \frac{\rho c Q(T_o - T_i)}{W_i} \tag{8.15}$$

where
    ρ is the liquid density
    c is its specific heat
    Q is the liquid volume flow rate, cm^3/s

The fluid properties c and ρ are evaluated at the average temperature $(T_o - T_i)/2$, using available tables for the fluid. We *assume* that these fluid properties have 95% confidence limits of $\pm 1\%$, so these are considered *Type B* uncertainties. Tables of fluid properties often do *not* state the uncertainty, and *our* batch of commercial heat exchange fluid will not have exactly the same composition and properties as might have been used when the table was made. An *experiment* to get the needed values and uncertainties for *our* batch of fluid would be prohibitively expensive, so we really must rely on other sources to choose these uncertainties. We consult any *experts* in our company, or lacking these we get in touch with outside competent sources and then make our own judgment as to what numbers to use. If we have been doing tests of this sort for many years, this *experience* may allow a quick choice. Anticipating this sort of problem, we may have stipulated certain numbers in the legal contract that requires the acceptance test. If not, we should notify the heat exchanger vendor of our choice and request approval. *One way or another* a choice must be made.

The temperature sensors and volume flowmeter are conventional instruments which we would rightly be expected to calibrate before proceeding with the test. Calibration is a large subject discussed at length elsewhere in my book *Measurement Systems** so we condense here severely. In general, calibration consists of comparing the readings of our instrument with a more accurate *standard*, over a range which includes all the values that are expected in the test we are to run. These data are plotted as in Figure 8.18 and statistically processed[†] to find the *measurement bias* and the uncertainty (95% confidence interval, using the *t* value). If the instrument is nominally linear (most are) we use least-squares methods to fit the best straight line ($y = mx + b$) to the scattered data. The slope *m* of this line is then the *static sensitivity* and the deviations from this line are used to compute the standard deviation. The *y*-intercept at $x = 0$ is the *measurement bias b*. Typically we take about 10 points going up and 10 going down, giving a sample size of about 20, which provides good statistical reliability. (Figure 8.18 uses 22 points.)

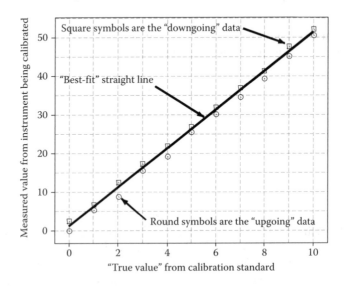

FIGURE 8.18
Generic instrument calibration curve.

* E.O. Doebelin, *Measurement Systems*, 5th edn., McGraw-Hill, New York, 2004, pp. 40–66.
† E.O. Doebelin, *Measurement Systems*, 5th edn., McGraw-Hill, New York, 2004, pp. 54–61.

We now suppose that we have calibrated both temperature sensors and the flowmeter and have numerical values for the measurement uncertainties. Since the instruments were not calibrated *in situ*, we need to provide estimates of the *bias limits*, to add to the measurement uncertainties. Suppose that on a given run of the experiment that we get the following measurements:

$$T_i = (34.1 \pm 1.1)°C \qquad T_o = (87.2 \pm 1.1)°C \qquad Q = 2.92 \pm 0.014 \text{ cm}^3/\text{s}$$

$$W_i = 0.003089 \pm 1\% \text{ W}$$

where all the plus/minus quantities are the total uncertainties as in Equation 8.14. Applying the method of Equation 8.13 to the efficiency calculation of Equation 8.15, we need to first compute the various partial derivatives. These are of course nothing but the "sensitivities" of E to changes in each of the independent variables. If a partial derivative is large, it means that E is particularly sensitive to changes in that variable.

$$\frac{\partial E}{\partial W_i} = \frac{-\rho c Q (T_o - T_i)}{W_i^2} \qquad \frac{\partial E}{\partial \rho} = \frac{c Q (T_o - T_i)}{W_i} \qquad \frac{\partial E}{\partial c} = \frac{\rho Q (T_o - T_i)}{W_i} \quad (8.16)$$

$$\frac{\partial E}{\partial Q} = \frac{\rho c (T_o - T_i)}{W_i} \qquad \frac{\partial E}{\partial T_o} = \frac{\rho c Q}{W_i} \qquad \frac{\partial E}{\partial c} = \frac{-\rho c Q}{W_i} \quad (8.17)$$

We now compute numerical values for all these derivatives, evaluated at the mean values of all the variables, and substitute into Equation 8.13 to get the uncertainty in *E*, using a Mathcad program, which computes partial derivatives for the heat exchanger uncertainty analysis

crc_book_heatexchange
Numerical values

$c := 0.007143 \qquad rho := 0.002543 \qquad Ti := 34.1 \qquad To := 87.1$

$Q := 2.92 \qquad Wi := 0.003089 \qquad Wo := c \cdot rho \cdot Q \cdot (To - Ti) \qquad Wo = 2.811 \cdot 10^{-3}$

Compute the partial derivatives

$DWi := \dfrac{-rho \cdot c \cdot Q \cdot (To - Ti)}{W_i^2} \qquad DWi = -294.612$

$Dc := \dfrac{rho \cdot Q \cdot (To - Ti)}{Wi} \qquad Dc = 127.405$

$$Drho := \frac{c \cdot Q \cdot (To - Ti)}{Wi} \qquad Drho = 357.867$$

$$DTi = \frac{-rho \cdot c \cdot Q}{Wi} \qquad DTi = -0.017$$

$$DTo = \frac{rho \cdot c \cdot Q}{Wi} \qquad DTo = 0.017$$

$$DQ := \frac{rho \cdot c \cdot (To - Ti)}{Wi} \qquad DQ = 0.312$$

State the uncertainties

$$uWi := 0.01 \cdot Wi \qquad uWi = 3.089 \cdot 10^{-5}$$

$$uc := 0.01 \cdot c \qquad uc = 7.143 \cdot 10^{-5}$$

$$urho := 0.01 \cdot rho \qquad urho = 2.543 \cdot 10^{-5}$$

$$uQ := 0.014$$

$$uTi := 1.1$$

$$uTo := 1.1$$

Compute the uncertainty in the efficiency

$$uE := \sqrt{(DTo \cdot uTo)^2 + (DTi \cdot uTi)^2 + (Dc \cdot uc)^2 + (Drho \cdot urho)^2 + (DQ \cdot uQ)^2 + (DWi \cdot uWi)^2}$$

$$uE = 0.0313 \qquad E := \frac{Wo}{Wi} \qquad E = 0.9101$$

We can then quote the efficiency and its uncertainty (95% confidence interval) as 0.91 ± 0.031. Our contract requires 0.90 ± 0.05, so the heat exchanger meets the specification.

Unfavorable results might take several forms. The measured mean value of E might have fallen within the required limits, but the confidence interval might exceed them. This would lead us to try to find more accurate instruments and repeat the experiment. When assigning a required accuracy to instruments, a basic problem is that there is an *infinite number* of combinations of instrument uncertainties that would result in *the same* overall uncertainty in the computed result. One way to deal with this dilemma uses the *method of equal effects*. Here we assume that each instrument contributes *the same* portion to the overall uncertainty in Equation 8.13. Since the partial derivatives are in general *not* the same, this means that the required accuracy of each instrument is *different*. That is, a variable that has a numerically large partial derivative will be required to have a better accuracy than one that has a small partial derivative. To implement the method of equal effects, we compute the required uncertainty of each measurement using

$$u_{xi} = \frac{u_y}{\sqrt{n} \cdot \dfrac{\partial y}{\partial x_i}} \tag{8.18}$$

where u_y is the required uncertainty in the computed result. We could in fact have used this method in *designing* our experiment (assigning required uncertainties to each measurement) and used that information to choose our instrumentation. When one does this, it may turn out that the required accuracy of a specific measurement is *beyond the state of the art* for that type of instrument. This does not mean necessarily that a valid experiment is impossible. Since each of the instruments has been arbitrarily assigned the same accuracy, perhaps some of them are available with *higher* accuracy than that assigned by the method of equal effects. This higher accuracy might compensate for the inadequate accuracy of others. That is, the method of equal effects allows a rational *starting point* for assigning accuracy, but we are not locked into it.

9

Multiaxial Force/Torque Measurement: Thrust Stands for Jet Engines and Rocket Engines

9.1 Introduction

My earlier text,* discusses various methods of measuring vector forces and torques. In general, it is not possible to measure such three-dimensional quantities directly; rather, we design *uniaxial* transducers to selectively measure the *components* of the vectors and then use *computation* to find the vector magnitudes and directions. This chapter will explore those applications where there is sufficient space available to use separate force transducers to measure each component of the vector. One such area of application of major importance to mechanical and aerospace engineers is the measurement of the forces and moments produced by aircraft or spacecraft engines, since knowledge of these quantities is vital to the prediction of vehicle performance. For air-breathing (jet) engines used in aircraft and missiles, the need originally was to measure only the thrust force along the longitudinal axis of the engine. As aircraft (such as the British Harrier) that use thrust vectoring as a control method were later developed, knowledge of *all* the forces and moments became important. For rocket engines, this need was there from the beginning of that technology.

The apparatus used to make all these measurements has come to be known as a *thrust stand*, and many different forms have been developed over the years to meet the needs of flight vehicles of various kinds. In addition to the transducers themselves, the design of thrust stands depends heavily on the intelligent use of different types of elastic elements called *flexures*. These are used to isolate the component forces from the vectors so that the uniaxial transducers can accurately measure the intended component without excessive *crosstalk* from the other force components. Another major consideration is the

---

* E.O. Doebelin, *Measurement Systems*, 5th edn., McGraw-Hill, New York, 2003, pp. 457–464.

treatment of so-called *tare** forces and torques. In the idealized situation used in a textbook presentation focused on *general* force measurements rather than specific applications, we assume that the only forces present are those that we wish to measure. In engine testing, the engines must be provided with fuel, which comes from tanks external to the thrust stand, through suitable piping that runs from the lab foundation to the engine. This, and similar connections

Outdoor test of turbofan engine

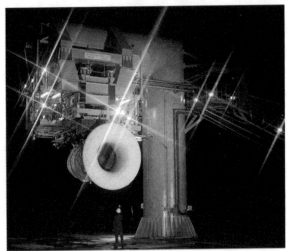

 Rolls-Royce   006563 PF

Measurement and instrumentation

· Amounts and sophistication vary widely

· Detailed performance investigation
pressures and temperatures at virtually every station,
power or thrust, shaft speeds, fuel and air flow, etc.

· Production pass off or endurance testing
ambient conditions,
power or thrust level, and fuel flow

· All engine testing very expensive, must ensure good quality data:
   - The test bed and all instrumentation must be properly calibrated
   - Test planning should include careful specification of instrumentation
   - Key measurements must be checked during the testing. If necessary
     engine removal must be delayed, and testing repeated

· Must understand likely accuracy levels

Rolls-Royce

FIGURE 9.1
Slides from Rolls-Royce presentation. (Courtesy of Rolls-Royce, London, U.K.)

---

* R.B. Runyan, J.P. Rynd, and J.F. Seely, Thrust stand design principles, AIAA-92-3976, 1992; R.B. Runyan et al., Basic Principles of thrust stand design, AFAL-TR-88-060, Arnold Engineering Development Center, Tullahoma, TN, June 1988.

between engine and "ground" will result in forces and torques that we *do not* want to measure being felt by the force transducers, thus corrupting our measurement of the engine-created forces. Such effects are generically called *tare effects* and include things such as instrumentation cables, piping for cooling air flows, etc. These tare effects must be minimized by careful design, and then *calibrated* by experimental testing, to allow their correction at test time. This chapter will lead you through the consideration of these and other aspects of thrust stand design and application. Thrust stands are sometimes designed by instrumentation groups within engine manufacturing companies or government laboratories[*] or by private companies[†] specializing in this field.

Figure 9.1 shows an overall view of a large thrust stand for aircraft jet engines. The design, construction, and operation of such a complex facility is a major engineering undertaking that draws heavily on the accumulated experience of the company involved. Design of the force/moment measurement system is only a small part of the total design task. Many other measurements (pressure, temperature, flow rate, etc.) must be instrumented and facilities provided for fuel and air flow, altitude simulation, etc. Design details are usually closely held by the involved companies and there is little in the open literature on the topic of thrust stand design. A. L. Rowe, corporate operability specialist and engineering fellow of Rolls-Royce kindly provided me with a copy of a Power Point presentation given at an ASME meeting in Stockholm in 1998 by Paul Fletcher (also from Rolls-Royce). This gives a good overview of the topic without going into design details. One of the slides from this presentation is included in Figure 9.1.

9.2 Dynamics of Thrust Stand Force/Torque Measurement

Figure 9.2 shows one of the many different thrust stand configurations that have been developed over the years. Its configuration is actually not typical; most designs use measurement axes that are aligned like a simple orthogonal x, y, z axis set, not the "120°" arrangement of Figure 9.2, which makes the analysis somewhat more complicated. I chose this design since it is geometrically more general and thus its analysis requires some concepts that are of use beyond this particular example. In Figure 9.2, I have included a representation of an *engine* attached to the thrust stand, since its inertial and elastic properties will affect the dynamic performance of the thrust stand. To allow an analytical treatment that matches the intended scope of this chapter, we will consider the engine to contribute only *inertial* effects, and neglect any elastic effects that might be present. That is, we treat the engine as a *rigid body*, with inertial effects comprised of total

---

[*] Sverdrup Technology, Inc., AEDC Group, Arnold Air Force Base, Tullahoma, TN.
[†] Force Measurement Systems, www.forcems.com (formerly Ormond, Inc.)

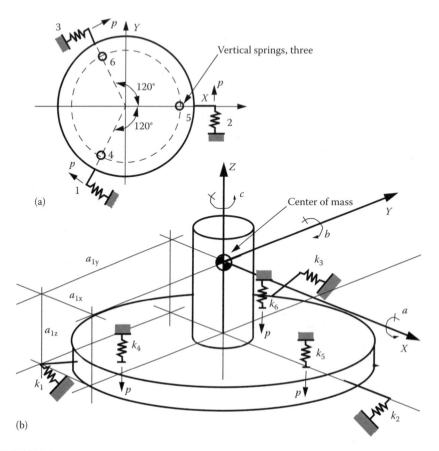

FIGURE 9.2
Thrust stand geometry.

mass, location of center of mass, three mass moments of inertia, and three products of inertia. Such a model would be a reasonable one for some applications, but might be too simplified for others, where engine elastic properties would have to be considered. While damping (frictional) effects are certainly present in the engine, thrust stand, and connections between them, we will neglect these in some of our analytical models. Such effects are minimized in the thrust stand design by using the earlier mentioned flexures (rather than rolling or sliding types of bearings) to provide for the small motions necessary in force measurement with elastic transducers. While *viscous* friction has no effect on the measurement accuracy for *static* force measurement, other types of friction add to uncertainty by causing hysteresis, deadspace, resolution, and related problems. While overall frictional effects are difficult to predict analytically, measurements on many existing thrust stands show an equivalent viscous damping ratio in the range 0.02–0.04 for each mode of vibration.

In Figure 9.2, we consider the engine itself and the thrust stand *plate* to which it is fastened, as a *single* rigid body, whose inertial properties will be needed in modeling the dynamic behavior of the force measurement system. This rigid body is mounted on an immovable foundation (represented by shading in the figure) through springs that represent the uniaxial force transducers. In practice, the force transducers, which themselves are designed to minimize the response to off-axis forces and torques, are further protected by the flexures mentioned earlier. In analyzing a system such as this, one could start with *first principles* or, alternatively, try to find in the published literature a similar analysis. Since many readers of this book will *not* have had a dynamics course beyond the usual undergraduate requirement (which does not address such three-dimensional problems), we appreciate the availability of a published analysis.* The referenced study requires that we locate the center of mass of the rigid body and pass through it the X, Y, and Z axes, as shown in Figure 9.2. Actually, there are *two* such sets of axes, one fixed to the body and one fixed in space. Initially (before any forces are applied to cause motion away from the initial equilibrium position), the two sets of axes are coincident. When motion occurs, the body-fixed axis set moves away from the space-fixed set. By requiring that the translations and rotations of the rigid body are *small*, the rotations can be treated as *commutative*, that is, the *order* in which the three rotations take place does not matter. (For large motions, the order *does* matter and we must deal with the so-called Euler angles,[†] which greatly complicates the analysis.) Furthermore, for small motions, rotations about the body axes can be taken as equal to those about the space-fixed axes, and body inertial properties can be taken as constant and equal to those about the body-fixed axes. All these assumptions are of course not obvious to one who has not studied advanced dynamics, but we often have to rely on the results of studies by *experts* to help us deal with problems that we are not personally able to analyze from "first principles."

Himmelblau and Rubin (1961) provide a complete set of six simultaneous linear differential equations with constant coefficients, which model such systems. There is one equation for each of the six degrees of freedom: the three translations of the system about the center of mass, and the three rotations about the X, Y, and Z axes. The reference discusses several special cases for which the model can be simplified, including some which allow analytical (rather than numerical) solution for system natural frequencies. The general thrust stand case will not allow analytical solution, but one further simplification is usually possible. This has to do with *products of inertia*. These inertial properties are in general not zero and must usually be taken into

---

* H. Himelblau and S. Rubin, Vibration of a resiliently supported rigid body, *Shock and Vibration Handbook*, Vol. 1, C.M. Harris and C.E. Crede (Eds.), McGraw-Hill, New York, 1961, pp. 3-1 to 3-52; C.E. Crede and J.E. Ruzicka, Properties of a biaxial stiffness isolator, *Shock and Vibration Handbook*, Vol. 2, McGraw-Hill, New York, 1961, pp. 30-28 to 30-33.

† E.O. Doebelin, *System Modeling and Response: Theoretical and Experimental Approaches*, Wiley, New York, 1980, pp. 475–496.

account. However, if our rigid body has a mass distribution that is *axisymmetric*, then the products of inertia are all zero and many terms drop out of the equations. While no real body will be *perfectly* axisymmetric, jet engines and rocket motors are often quite close, so we will make this simplification and take all products of inertia to be zero. In Figure 9.2, the Z axis is the axis of symmetry.

Thrust stand natural frequencies are determined by the numerical values of spring stiffnesses, masses, and moments of inertia. The stiffness of a strain-gage load cell depends on its force range; lower ranges will have lower stiffness. We choose the range of a load cell based on the maximum expected value of the engine force component being measured. Engine mass will be a large part of thrust stand total mass. In choosing numerical values for our example calculations, we can be guided by the *thrust/weight* ratio of typical engines. For example, the Pratt & Whitney F117-PW-100 engine* has a thrust of 41,700 lb and weighs 7,100 lb, giving a thrust/weight ratio of about 6. Such data allows us to scale our example stiffness and mass numbers so that the natural frequencies are reasonably typical of actual practice.

9.3 Characteristics of Elastic Elements in Three Dimensions

While you have surely dealt before with springs as parts of dynamic systems, you may have encountered them only in a one-dimensional context, such as a basic spring-mass vibrating system. When a three-dimensional rigid body is connected to a foundation by elastic elements, these elements themselves may often have a three-dimensional behavior. That is, if we apply a force in a certain direction, the deflection caused by this force need not have the same direction as the force. The most general such deflection vector would have three perpendicular components, such as x, y, and z components. Crede and Ruzicka (1961) shows how such elastic elements can be treated analytically. In that reference the *springs* are general elastic elements that are *tied down* at one end but whose other end can move in three dimensions. When we assume that the rigid body has moved away from its original equilibrium position, we need to be able to calculate all the forces and moments produced by the six springs. While the general problem will involve three-dimensional motions and forces, the method of treating the spring effects is most easily first visualized and explained in terms of the simpler two-dimensional case. We follow the reference now in discussing this simpler case. While our "springs" will be uniaxial force transducers, the reference treats the more general case involving rubber and/or metal *shock mounts* used to vibration isolate equipment. Figure 9.3 shows a generic two-dimensional spring

---

* www.pratt-whitney.com

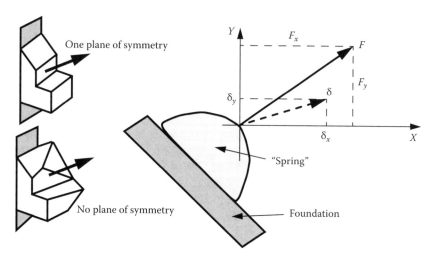

FIGURE 9.3
Two-dimensional elastic element.

attached to a fixed foundation with a force F applied at an arbitrary angle. The spring has a single plane of symmetry, which is the plane of the paper. (If the spring had *no* plane of symmetry, a force lying in the x–y plane could cause a deflection that was *not* in that plane, and we do not want to consider that situation at present.) We see that the deflection δ is *not* in the same direction as the applied force F. To relate the forces and deflections, various *influence coefficients* (k's) are defined:

$$F_{xx} = k_{xx}\delta_x \qquad F_{yx} = k_{yx}\delta_x \qquad F_{xy} = k_{xy}\delta_y \qquad F_{yy} = k_{yy}\delta_y \tag{9.1}$$

Using these definitions, we can express the forces as follows:

$$F_x = k_{xx}\delta_x + k_{xy}\delta_y \qquad F_y = k_{yx}\delta_x + k_{yy}\delta_y \tag{9.2}$$

It can be shown (Maxwell's reciprocity principle) that $k_{xy} = k_{yx}$, so it takes three parameters to define the relation between force and deflection. These k's can be found analytically for simple spring geometries, but for practical shock mounts, experimental measurements are often needed. If we have numerical values for these k's, then, in an application such as our thrust stand, we can find the force components associated with an assumed displacement (translational and rotary) of the rigid body connected to the springs. By rotating the axes in Figure 9.3, it is always possible to find an angular orientation such that an x force produces *only* an x deflection, and similarly for y. These directions are called the *principal elastic axes*. For these axes, only two influence coefficients are needed for us to relate forces and deflections. This situation is in a way analogous to *the principal inertial axes* of a rigid body, where the products of inertia are all zero.

For the three-dimensional case, the geometry and equations are more complex, and there will now in general be six influence coefficients, k_{xx}, k_{yy}, k_{zz}, k_{xy}, k_{xz}, and k_{yz}. Furthermore, any of the springs can have an *arbitrary* orientation with respect to our chosen axes, such as the X, Y, and Z axes in Figure 9.2. Thus when we assume the body to have translated by x, y, and z, and rotated through angles a, b, and c, finding the forces and torques caused by these motions becomes quite confusing. The *Shock and Vibration Handbook* (p. 3-13) assumes that the principal elastic axes (p, q, r) of each spring have been found and that from the location of each spring, we can find the angles between these principal elastic axes and the X, Y, Z axes assumed for the particular problem. Also, the three influence coefficients (k_p, k_q, k_r) associated with the principal elastic axes are also known. The six influence coefficients are then defined in terms of the principal influence coefficients and the cosines(λ's) of the respective angles by the following set of equations, which are taken directly from the reference

$$k_{xx} = k_p \lambda_{xp}^2 + k_q \lambda_{xq}^2 + k_r \lambda_{xr}^2 \qquad (9.3)$$

$$k_{yy} = k_p \lambda_{yp}^2 + k_q \lambda_{yq}^2 + k_r \lambda_{yr}^2 \qquad (9.4)$$

$$k_{zz} = k_p \lambda_{zp}^2 + k_q \lambda_{zq}^2 + k_r \lambda_{zr}^2 \qquad (9.5)$$

$$k_{xy} = k_p \lambda_{xp} \lambda_{yp} + k_q \lambda_{xq} \lambda_{yq} + k_r \lambda_{xr} \lambda_{yr} \qquad (9.6)$$

$$k_{yz} = k_p \lambda_{yp} \lambda_{zp} + k_q \lambda_{yq} \lambda_{zq} + k_r \lambda_{yr} \lambda_{zr} \qquad (9.7)$$

$$k_{xz} = k_p \lambda_{xp} \lambda_{zp} + k_q \lambda_{xq} \lambda_{zq} + k_r \lambda_{xr} \lambda_{zr} \qquad (9.8)$$

For example, λ_{xp} is the cosine of the angle between the X axis and the principle axis p of the spring. The k's defined in Equations 9.3 through 9.8 will appear in the differential equations of system motion. Each spring will have its own set of k values, and will contribute terms to the differential equations, so we begin to see how complicated such analyses become.

In our thrust stand problem, the springs are the uniaxial force transducers (with flexures at each end) and their elastic properties are much simpler than the general case discussed above. When we want to measure one of the forces exerted by the mounted engine on the foundation of the thrust stand, we connect between these two structures a "load column" oriented in the desired direction. Springs 1 through 6 in Figure 9.2 represent such load columns. A load column is made up of a (usually strain gage) load cell and a flexure on each end, as shown in Figure 9.4. The axial stiffness of a flexure is numerically comparable to that of the load cell, so the three springs shown "in series" in Figure 9.4 can be represented as a single spring of about 1/3 the load cell's stiffness. These load cells often use a short column in direct tension/compression (with two "Poisson"

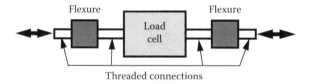

Flexure Flexure

Load
cell

Threaded connections

FIGURE 9.4
Load column configuration.

gages to complete the bridge circuit). A 2000 lb range load cell of this type typically has a stiffness of about 3 million lb/in. In an actual thrust stand design, we would of course get accurate values for the stiffness of the flexures and load cells and properly combine them to get a single stiffness value for the load column. The load columns are then treated as single springs, as in Figure 9.5. Because the purpose of the flexures is to prevent any *sidewise* loads on the load cells, the spring stiffnesses in the q and r directions are taken as zero. The orientation shown for the q and r axes can actually be chosen in many locations, by rotating the pair about the p axis. That is, the p axis is the only one that *must* be chosen along the spring's length, but q and r will give the correct physical behavior for *any* location that lies in the r, q plane shown in Figure 9.5. The availability of this choice for the orientation of the r and q axes may be convenient when we need to find the various angles required in Equations 9.3 through 9.8. These equations are now of course much simplified since any terms involving k_r or k_q will drop out.

$$k_{xx} = k_p \lambda_{xp}^2 \qquad k_{yy} = k_p \lambda_{yp}^2 \qquad k_{zz} = k_p \lambda_{zp}^2 \tag{9.9}$$

$$k_{xy} = k_p \lambda_{xp} \lambda_{yp} \qquad k_{xz} = k_p \lambda_{xp} \lambda_{zp} \qquad k_{yz} = k_p \lambda_{yp} \lambda_{zp} \tag{9.10}$$

While we will rely on the reference for the details of the six equations, we *can* quickly show a simple *physical* analysis of one spring, which makes clear, for example, why Equation 9.9 includes terms that involve the square of the cosine of certain angles. This will remove at least a little of the "black box" nature of our study. In Figure 9.6, we consider only an x displacement of the spring called 3 in Figure 9.2. When we shortly write out the complete set of equations and insert some initial numerical values, the spring constant of springs 1 and 3 will be taken as 1.2e6 lb_f/ft. In Figure 9.6 the angle a is 30°, which makes the force component $K_{s3}\, x \cos^2 30 = 0.9e6\, x$. Since spring 1 will

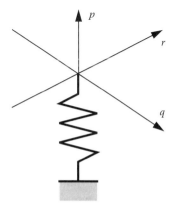

FIGURE 9.5
Spring subject to three-dimensional force.

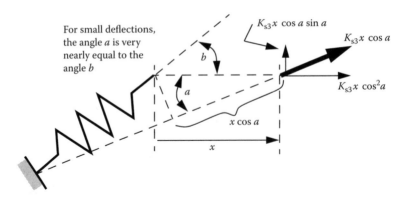

For small deflections, the angle *a* is very nearly equal to the angle *b*

$K_{s3}x \cos a \sin a$

$K_{s3}x \cos a$

$K_{s3}x \cos^2 a$

$x \cos a$

x

FIGURE 9.6
Resolution of vector spring force into components.

contribute an equal effect, the total x force would be $(1.8e6)x$ for these two springs. We will see this same number appear when we use the reference's method. Unfortunately, many of the other spring effects are not as easily analyzed, so we appreciate the availability of the reference.

Let us first establish the three angles needed to compute the three cosines required in Equations 9.9 and 9.10. When we speak of the *angle between the axes*, note that any two lines in space will define a single plane if these lines *intersect*. The axes we speak of will generally *not* intersect, but we can make this happen if we *mentally* translate one of them, parallel to itself, until the two axes *do* intersect. Then they define a plane, and the angle we are looking for is the angle between them in this plane. We then take the cosine of that angle. Note that this operation is similar to the *vector dot product*, which is defined as a scalar which is the product of the magnitudes of the two vectors times the cosine of the angle between them. Multiplying by the cosine amounts to taking the *projection* of one of the vectors on the other. This operation in our case gives the component of the elastic force in the direction of the coordinate axis.

This must be done for all six of the springs. For the three vertical springs (4, 5, and 6), the angle between their p axis and the X axis or Y axis is $-90°$, so these cosines are all zero. The angle with the Z axis is $180°$, so these cosines are all equal to -1.0. Spring 1 has a $150°$ angle with the X axis, $60°$ with the Y axis, and $90°$ with the Z axis, giving, respectively, cosines of -0.866, 0.500, and 0.0. Spring 2 has $90°$ with the X axis, $0°$ with the Y, and $90°$ with the Z, giving cosines of 0.0, 1.0, and 0.0. Spring 3 has $30°$ with the X axis, $-60°$ with the Y, and $0°$ with the Z, giving cosines of 0.866, 0.500, and 0.0. We also need all the a distances for each spring; Figure 9.2 shows only those for spring 1. Note that these "distances" have algebraic signs; some are positive and some negative. This is clearly necessary because the moment of a positive force about an axis could be either positive or negative, depending on which side of the axis it was on.

9.4 Dynamic Response Equations of the Thrust Stand

We will now quote, without any proof, the simultaneous differential equations which describe the three-dimensional motion of a rigid body connected, with any number of elastic elements, to a foundation. The analysis of our reference was *not* specifically intended for thrust stands, but rather for the general situation of vibration isolation for a shock-mounted rigid body. Our thrust stand application will allow some significant simplification, but we will initially give the complete, unsimplified equations, since they have many applications beyond ours.

The equation for x-axis translation is

$$\Sigma F_x + \Sigma k_{xx}(x_f - x) + \Sigma k_{xy}(y_f - y) + \Sigma k_{xz}(z_f - z) + \Sigma(k_{xz}a_y - k_{xy}a_z)(a_f - a)$$
$$+ \Sigma(k_{xx}a_z - k_{xz}a_x)(b_f - b) + \Sigma(k_{xy}a_x - k_{xx}a_y)(c_f - c) = M\ddot{x} \tag{9.11}$$

The equation for x-axis rotation (rotation angle is called a) is

$$\Sigma M_x + \Sigma(k_{xz}a_y - k_{xy}a_z)(x_f - x) + \Sigma(k_{yz}a_y - k_{yy}a_z)(y_f - y) + \Sigma(k_{zz}a_y - k_{yz}a_z)(z_f - z)$$
$$+ \Sigma(k_{yy}a_z^2 + k_{zz}a_y^2 - 2k_{yz}a_ya_z)(a_f - a) + \Sigma((k_{xz}a_ya_z) + k_{yz}a_xa_z - k_{zz}a_xa_y - k_{xy}a_z^2)(b_f - b)$$
$$+ \Sigma(k_{xy}a_ya_z + k_{yz}a_xa_y - k_{yy}a_xa_z - k_{xz}a_y^2)(c_f - c) = I_{xx}\ddot{a} - I_{xy}\ddot{b} - I_{xz}\ddot{c} \tag{9.12}$$

The reference provides similar equations for the y and z axes, but I choose to not reproduce them here. I just wanted to display the general character of the equations and clarify some of the basic concepts needed to understand their application, and that is adequately accomplished by showing the details of just one axis.

These equations allow study of both force/torque inputs and foundation motion (translation and rotation) inputs. The foundation x, y, and z translations are called respectively, x_f, y_f, and z_f, and the foundation rotation angles are a_f, b_f, and c_f. Since our thrust stand foundation is assumed immovable, all these inputs will be taken as zero. In the rotation equation, because of our assumption of axisymmetry, the two terms involving the products of inertia I_{xy} and I_{xz} are also taken as zero. The terms such as a_x, a_y, and a_z refer to the coordinates (lever arms) of the points where the springs are connected to the rigid body. These lever arm distances are needed to convert rotary deflections (radians) into translational deflections (and thus forces) of the various springs. They also convert forces into torques in the rotation equations. Note that the equations are even more complicated than they look at first glance, because of the summation (Σ) terms, which require that we enter the values for all six springs. As an example

$$\Sigma(k_{xz}a_y - k_{xy}a_z) = k_{xz1}a_{y1} - k_{xy1}a_{z1} + k_{xz2}a_{y2} - k_{xy2}a_{z2} + k_{xz3}a_{y3} - k_{xy3}a_{z3} + k_{xz4}a_{y4}$$

$$- k_{xy4}a_{z4} + k_{xz5}a_{y5} - k_{xy5}a_{z5} + k_{xz6}a_{y6} - k_{xy6}a_{z6} \tag{9.13}$$

Also, each k value must be computed from Equations 9.3 through 9.8 using the cosines of the angles between the spring axes and the x, y, z coordinate axes.

Such calculations are best done using *spreadsheet* type software, such as EXCEL or Mathcad. The entry of all six equations (not just Equations 9.11 and 9.12) together with the k calculations is very tedious and prone to error, but once done in letter form, numerical values for different applications are easily entered. Also, for a *given* application, various design alternatives are easily explored by changing dimensions and element values. Since such equations are useful for *any* problem involving vibration of a rigid body supported by some arrangement of elastic elements, the work of entering the general equations can be justified for one who will make extensive use in various application areas. I have done this for my own use but did not want to complicate this chapter with inclusion of this level of detail, especially since our present application leads to many terms in the general equations dropping out. While I do not want to involve you in all the "dog work" just described, it is desirable that we do a few detailed calculations so that you will better understand what is going on. Let us assume some typical numerical values which we will carry along for our first example using ft and lb$_f$ units. While an actual design would choose load-cell ranges to suit the expected maximum force components, our initial example will take all six load cells to have the same range, so the stiffnesses would also be identical. Including the stiffness effects of the two attached flexures, let us take the total stiffness k_p of each load column to be 1.2×10^6 lb$_f$/ft. Using the angles from Figure 9.2, and choosing some typical a values gives the results displayed in Table 9.1. The inertial properties are taken as $M = 250$ slugs, $I_{xx} = I_{yy} = 3333.0$ slug-ft^2, $I_{zz} = 1111.0$ slug-ft^2. Using all the above quoted numerical values, I will now state the six equations used to model the thrust stand dynamics.

TABLE 9.1

Thrust Stand Element Values

Load Column Number	K_{XX}	K_{YY}	K_{ZZ}	K_{XY}	K_{XZ}	K_{YZ}	a_x	a_y	a_z
1	$K_p \cos^2 30$	$K_p \cos^2 60$	0	$-0.5K_p \cos 30$	0	0	-2.5	$-5 \cos 30$	-3.0
2	0	K_p	0	0	0	0	5.0	0.0	-3.0
3	$K_p \cos^2 30$	$K_p \cos^2 60$	0	$0.5K_p \cos 30$	0	0	-2.5	$5 \cos 30$	-3.0
4	0	0	K_p	0	0	0	2.0	$-4 \cos 30$	-2.5
5	0	0	K_p	0	0	0	2.0	0.0	-2.5
6	0	0	K_p	0	0	0	-2.0	$4 \cos 30$	-2.5

$$\Sigma F_x - 1.8 \cdot 10^6 x + (5.4 \cdot 10^6) b = M\ddot{x} \qquad (9.14)$$

$$\Sigma F_y - 1.8 \cdot 10^6 y - 5.4 \cdot 10^6 a = M\ddot{y} \qquad (9.15)$$

$$\Sigma F_z - 3.6 \cdot 10^6 z = M\ddot{z} \qquad (9.16)$$

$$\Sigma M_x - 5.4 \cdot 10^6 y - (4.5 \cdot 10^7) a = I_{xx}\ddot{a} \qquad (9.17)$$

$$\Sigma M_y + 5.4 \cdot 10^6 x - 4.5 \cdot 10^7 b = I_{yy}\ddot{b} \qquad (9.18)$$

$$\Sigma M_z - 9.0 \cdot 10^7 c = I_{zz}\ddot{c} \qquad (9.19)$$

These equations are of course much simpler looking than the general case which we showed earlier. The reasons for this simplicity lie in the *symmetry* of our geometrical configuration together with the equal numerical values of the load column stiffnesses. If springs 1, 2, and 3 were not equally spaced around the periphery and/or did not have equal values, (similarly for springs 4, 5, and 6), we would find more terms present in our equations. This would be due to *elastic coupling* caused by the lack of symmetry. If we had not assumed our rigid body to be axisymmetric, then the terms in the products of inertia would be present, causing *inertial coupling*, and again, equations with more terms present. The general case, of course, would have *every* equation include terms in *every* unknown.

Interpreting our present equations, we see that Equations 9.16 and 9.19 are *totally* uncoupled from the others. They thus can be immediately solved by themselves without any consideration of the other equations. They are clearly the equations of simple *one spring, one mass* systems without damping, and thus would perform free vibrations at a single frequency if given some initial energy (initial displacement and/or velocity). They also define two of the six possible *modes of vibration*. One such mode is a pure translation in the z direction (Equation 9.16) and the other is a pure rotation about the z axis. The other four equations exhibit elastic coupling but the coupling is *pairwise*. That is, Equations 9.15 and 9.17 involve only two unknowns, so this pair could be solved without considering any other equations. A similar case exists for Equations 9.14 and 9.18. Thus what might have been a problem requiring simultaneous solution of six equations in six unknowns has degenerated into much simpler subproblems. For the two equations that are totally uncoupled, we can easily calculate the natural frequencies as 19.1 Hz for the z translation mode and 45.3 Hz for the z rotation mode:

$$f_{nz} = \frac{1}{2\pi} \cdot \sqrt{\frac{3.6\mathrm{e}6}{250}} = 19.1 \qquad f_{nc} = \frac{1}{2\pi} \cdot \sqrt{\frac{9\mathrm{e}7}{1111}} = 45.3 \qquad (9.20)$$

In Equations 9.15 and 9.17, the numbers 1.8e6, 5.4e6, and 4.5e7 are *composite* spring constants that come from complicated equations like Equations 9.11 and 9.12. If we replace these numbers with letter coefficients, respectively, k_{yy}, k_{ay}, and k_{aa}, then these two equations can be combined *in letter form* to get formulas for the natural frequencies associated with this equation pair. Such formulas are useful when designing a thrust stand for a desired speed of response, which is proportional to the natural frequencies. The natural frequencies can be found using the usual determinant approach for reducing simultaneous equations to single equations in single unknowns (see Appendix A).

$$(MD^2 + k_{yy})y + (k_{ay})a = F_y \tag{9.21}$$

$$(k_{ay})y + (I_{xx}D^2 + k_{aa})a = M_x \tag{9.22}$$

$$y = \frac{\begin{vmatrix} F_y & k_{ay} \\ M_x & I_{xx}D^2 + k_{aa} \end{vmatrix}}{\begin{vmatrix} MD^2 + k_{yy} & k_{ay} \\ k_{ay} & I_{xx}D^2 + k_{aa} \end{vmatrix}} \tag{9.23}$$

$$y = \frac{(I_{xx}D^2 + K_{aa})F_y - (k_{ay})M_x}{I_{xx}MD^4 + (Mk_{aa} + I_{xx}k_{yy})D^2 + (k_{aa}k_{yy} - k_{ay}^2)} \tag{9.24}$$

By "cross-mulitplying" Equation 9.24, we get the differential equation relating displacement y to input force F_y and input torque (moment) M_x. The denominator of Equation 9.24 (set equal to 0) is the characteristic equation, whose roots give us the natural frequencies we desire. This is a quartic equation, which usually is not practical to solve in letter form, however, due to the absence of any damping, the first- and third-power terms are missing, which allows a solution for the roots using the familiar quadratic formula. This could be carried out in letter form, but the result is too complicated to be of much use for design studies. Inserting now the numerical values, we get

$$833250D^4 + 1.72494e10D^2 + 5.184e13 = 0 \tag{9.25}$$

Using MATLAB® routine *damp* on this polynomial we get

```
c = [833250 0 1.72494e10 0 5.184e13];

>> damp(c)/(2 * pi)

  9.6131e + 000

  9.6131e + 000

 20.7836e + 000

 20.7836e + 000
```

The quartic polynomial has two pairs of pure imaginary roots, leading to the two natural frequencies of 9.61 and 20.8 Hz. A similar procedure for Equations 9.14 and 9.18 gives *the same* two natural frequencies, giving a total of four *different* natural frequencies for this system, whereas a system of six second-order equations generally has six different natural frequencies. This reduction in the number of distinct natural frequencies is caused by the *symmetry* of our assumed system. There actually *are* six pairs of imaginary roots and six natural frequencies, but two of the pairs are *repeated*. In an actual thrust stand, even if we *designed* it to be symmetrical, unavoidable errors of machining and construction will surely cause at least *some* asymmetry, and then we would get six distinct frequencies, but some of them would be *very close* to others. While we will later compute and plot frequency-response curves for this system, we can even now make use of the above results to estimate the dynamic response of this "instrument." We have found the lowest natural frequency to be about 9 Hz, and since the damping will be quite small (thrust stands traditionally have *no* designed-in damping), accurate measurement will probably be limited to 1 or 2 Hz.

9.5 Matrix Methods for Finding Natural Frequencies and Mode Shapes

In the example discussed above, we were able to find the six natural frequencies rather easily. In the more general case where *all six* of the equations are coupled, we need a more powerful and general approach. In fact, the simplicity of our example depended partly on our assumption that all the load columns were *perfectly aligned* in the angular orientation that we desire. When we actually construct a thrust stand, such perfect alignment is certainly our goal, but it is unrealistic to assume that it can be realized in practice. Thus even our nicely symmetrical design will actually have some small nonzero numbers in equation locations where we had perfect zeros. When these terms are not zero, the six equations are all coupled together and we cannot use the simple means employed above to get our desired results. Fortunately, mathematical tools for dealing with the more general situation are available.

Some readers will already be aware of these methods, but some may not, so I will now present the method* in detail. The method involves first forming a certain matrix of terms and then invoking a software eigenvalue routine

---

* E.O. Doebelin, *System Dynamics: Modeling, Analysis, Simulation, Design*, Marcel Dekker, New York, 1998, pp. 396–399.

(we will use MATLAB's). Associated with the *eigenvalues* (which are the system natural frequencies) are the *eigenvectors*. The eigenvectors supply information about the *vibratory mode shape* associated with each natural frequency. The eigenvalue routines will provide the roots of the system's characteristic equation, which has as many roots as the order of the equation set, in our case, 12 (six second-order equations). If we assume no damping, these roots will be pure imaginary and their magnitudes are the system natural frequencies. If damping is present, the eigenvalues (roots) will be complex, but they can be processed to obtain the undamped natural frequency and the damping ratio associated with each mode of vibration. *Eigenvectors can be calculated whether damping is present or not, but to get the mode shapes the damping must be taken as zero.*

The usual method for doing these calculations is to form the system matrix (sometimes called the *A matrix*). This is easily done once we have the set of equations for the system. For vibration problems, the individual equations will be of second order, since the highest derivative found in our Newton's law equations is the second (translational or rotary acceleration). To form the system matrix, we need to define some new variables so that the equation set will have only *first-order* equations. The new variables will simply be the six *velocities* associated with our six displacements. There will now be 12 unknowns: the 6 displacements and the 6 velocities. The new equation set is given below, where we define the three translational velocities as vx, vy, and vz and the three rotational velocities as wa, wb, and wc. Since in general we might want to consider the presence of damping, we will include a single damping term in each equation, and assume that this damping is proportional to the velocity unknown of that equation. That is, for example, in our x translation equation, the damping term will be taken as $-B_x(vx)$. In the general six degree-of-freedom equations, damping can of course appear in more complicated forms, as we saw for the spring effects, but our special case will adequately explain the treatment of damping without the confusion of the general case.

$$\dot{x} = (0)x + (0)y + (0)z + (0)a + (0)b + (0)c + (1)vx + (0)vy + (0)vz + (0)wa$$

$$+ (0)wb + (0)wc \tag{9.26}$$

$$\dot{y} = (0)x + (0)y + (0)z + (0)a + (0)b + (0)c + (0)vx + (1)vy + (0)vz + (0)wa$$

$$+ (0)wb + (0)wc \tag{9.27}$$

$$\dot{z} = (0)x + (0)y + (0)z + (0)a + (0)b + (0)c + (0)vx + (0)vy + (1)vz + (0)wa$$

$$+ (0)wb + (0)wc \tag{9.28}$$

$$\dot{a} = (0)x + (0)y + (0)z + (0)a + (0)b + (0)c + (0)vx + (0)vy + (0)vz + (1)wa$$
$$+ (0)wb + (0)wc \tag{9.29}$$

$$\dot{b} = (0)x + (0)y + (0)z + (0)a + (0)b + (0)c + (0)vx + (0)vy + (0)vz + (0)wa$$
$$+ (1)wb + (0)wc \tag{9.30}$$

$$\dot{c} = (0)x + (0)y + (0)z + (0)a + (0)b + (0)c + (0)vx + (0)vy + (0)vz + (0)wa$$
$$+ (0)wb + (1)wc \tag{9.31}$$

$$\dot{vx} = (-1.8e6/m)x + (0)y + (0)z + (0)a + (5.4e6/m)b + (0)c$$
$$+ (-B_x/m)vx + (0)vy + (0)vz + (0)wa + (0)wb + (0)wc \tag{9.32}$$

$$\dot{vy} = (0)x + (-1.8e6/m)y + (0)z + (-5.4e6/m)a + (0)b + (0)c$$
$$+ (0)vx + (-B_y/m)vy + (0)vz + (0)wa + (0)wb + (0)wc \tag{9.33}$$

$$\dot{vz} = (0)x + (0)y + (-3.6e6/m)z + (0)a + (0)b + (0)c$$
$$+ (0)vx + (0)vy + (-B_z/m)vz + (0)wa + (0)wb + (0)wc \tag{9.34}$$

$$\dot{wa} = (0)x + (-5.4e6/I_{xx})y + (0)z + (-4.5e7/I_{xx})a + (0)b + (0)c$$
$$+ (0)vx + (0)vy + (0)vz + (-B_a/I_{xx})wa + (0)wb + (0)wc \tag{9.35}$$

$$\dot{wb} = (5.4e3/I_{yy})x + (0)y + (0)z + (0)a + ((-4.5e7)/I_{yy})b + (0)c$$
$$+ (0)vx + (0)vy + (0)vz + (0)wa + (-B_b/I_{yy})wb + (0)wc \tag{9.36}$$

$$\dot{wc} = 0x + (0)y + (0)z + (0)a + (0)b + ((-9.0e7)/I_{zz})c$$
$$+ (0)vx + (0)vy + (0)vz + (0)wa + (0)wb + (-B_c/I_{zz})wc \tag{9.37}$$

(Note that we have not included the input forces and moments in these equations. They have *no* effect on the eigenvalues or eigenvectors, so are *never* included in such calculations.) We see that we now have 12 equations in 12 unknowns, rather than 6 equations in 6 unknowns. We will see shortly that there is no real reason to *write out* this set of equations; I only did this to explain the method. The set of 12 equations can be written compactly in matrix form (Equation 9.38 below), and it turns out we only need the matrix of equation *coefficients* to proceed with the analysis for eigenvalues and eigenvectors.

$$
\begin{bmatrix} \dot{x} \\ \dot{y} \\ \dot{z} \\ \dot{a} \\ \dot{b} \\ \dot{c} \\ \dot{v}x \\ \dot{v}y \\ \dot{v}z \\ \dot{w}a \\ \dot{w}b \\ \dot{w}c \end{bmatrix} =
\begin{bmatrix}
0 & 0 & 0 & 0 & 0 & 0 & 1 & 0 & 0 & 0 & 0 & 0 \\
0 & 0 & 0 & 0 & 0 & 0 & 0 & 1 & 0 & 0 & 0 & 0 \\
0 & 0 & 0 & 0 & 0 & 0 & 0 & 0 & 1 & 0 & 0 & 0 \\
0 & 0 & 0 & 0 & 0 & 0 & 0 & 0 & 0 & 1 & 0 & 0 \\
0 & 0 & 0 & 0 & 0 & 0 & 0 & 0 & 0 & 0 & 1 & 0 \\
0 & 0 & 0 & 0 & 0 & 0 & 0 & 0 & 0 & 0 & 0 & 1 \\
\dfrac{-1.8e6}{m} & 0 & 0 & 0 & \dfrac{5.4e6}{m} & 0 & \dfrac{-B_x}{m} & 0 & 0 & 0 & 0 & 0 \\
0 & \dfrac{-1.8e6}{m} & 0 & \dfrac{-5.4e6}{m} & 0 & 0 & 0 & \dfrac{-B_y}{m} & 0 & 0 & 0 & 0 \\
0 & 0 & \dfrac{-3.6e6}{m} & 0 & 0 & 0 & 0 & 0 & \dfrac{-B_z}{m} & 0 & 0 & 0 \\
0 & \dfrac{-5.4e6}{I_{xx}} & 0 & \dfrac{-4.5e7}{I_{xx}} & 0 & 0 & 0 & 0 & 0 & \dfrac{-B_a}{I_{xx}} & 0 & 0 \\
\dfrac{5.4e6}{I_{yy}} & 0 & 0 & 0 & \dfrac{-4.5e7}{I_{yy}} & 0 & 0 & 0 & 0 & 0 & \dfrac{-B_b}{I_{yy}} & 0 \\
0 & 0 & 0 & 0 & 0 & \dfrac{-9.0e7}{I_{zz}} & 0 & 0 & 0 & 0 & 0 & \dfrac{-B_c}{I_{zz}}
\end{bmatrix}
\begin{bmatrix} x \\ y \\ z \\ a \\ b \\ c \\ vx \\ vy \\ vz \\ wa \\ wb \\ wc \end{bmatrix}
$$

(9.38)

The 12 × 12 square matrix is all that is needed to extract the six natural frequencies and modal damping ratios, but this *must* be done with numerical, not literal coefficient values. If you also want the eigenvectors, which give us the mode shapes of each vibratory mode, then you must enter zeros for all the damping coefficients (the *B*'s). It should then be clear that in any such problem there is no need to write out the equations of form 9.26 through 9.37; we can easily write out the matrix of coefficients directly by simply looking at the *original* set of equations (Equations 9.14 through 9.19).

If you have not done much matrix work in MATLAB, you might consider entering the matrix one column at a time, and then combining all the columns to get the complete matrix. To compute the eigenvalues and eigenvectors for a matrix called, say, A, we enter the command [V, D] = eig (A);. The semicolon is used to *prevent printing* of results at this point. To compute only the eigenvalues we would enter d = eig(A);. In the [V, D] command, *V* is the name given to the eigenvector results. To get numerical values for natural frequencies and damping ratios, once *d* has been calculated, enter [wn, z] = damp(d);. Here, *wn* will be a list of the natural frequencies and *z* will list the damping ratios. In this first example, all the damping terms will be set to zero, so all the damping ratios will also be zero. Since most software computes natural frequencies in rad/s whereas we usually prefer Hz, we enter a command to do this conversion. To show the relation to the system *characteristic equation*, we include in the m-file the commands char = poly(A) and r = roots(char); *char* will be a list of the coefficients of the 12th degree characteristic equation and *r* will be a list of its 12 roots.

The *r* values should be the same as the eigenvalues computed from `eig(A)`. A MATLAB program to do these calculations follows:

```
% thrstmat_crc.m
% matrix operations to get natural frequencies and mode
% shapes for the thrust stand (symmetric version)
% first define the mass and moments of inertia
% the values of the cxx,cxy, etc. will be calculated by Mathcad
% and entered into this Matlab file by hand
m=250;Ix=3333.;Iy=3333.;Iz=1111;format short e;
% define damping values
Bx=0;By=0;Bz=0;Ba=0;Bb=0;Bc=0;
% Bx=1.3e3;By=1.3e3;Bz=1.8e3;Ba=2.3e4;Bb=2.3e4;Bc=1.9e4;
% the spring elements will be given with the signs adjusted
% so that the letters can be entered as positive in the matrix
% I have given the matrix elements convenient names (cxx,cxy,cxz,etc.)
cxx=-1.8e6;cxy=0.0;cxz=0.0;
cxa=0.0;cxb=+5.4e6;cxc=0.0;

cyx=0.0;cyy=-1.8e6;cyz=0.0;
cya=-5.4e6;cyb=0.0;cyc=0.0;

czx=0.0;czy=0.0;czz=-3.6e6;
cza=0.0;czb=0.0;czc=0.0;

crax=0.0;cray=-5.4e6;craz=0.0;
craa=-4.5e7;crab=0.0;crac=0.0;

crbx=+5.4e6;crby=0.0;crbz=0.0;
crba=0.0;crbb=-4.5e7;crbc=0.0;

crcx=0.0;crcy=0.0;crcz=0.0;
crca=0.0;crcb=0.0;crcc=-9.0e7;

% now let's create the matrix which gets us the eigenvalues
% it will be easiest to enter the data as columns
% and then assemble the columns into the A matrix

c1=[0;0;0;0;0;0;cxx/m;cyx/m;czx/m;crax/Ix;crbx/Iy;crcx/Iz];
c2=[0;0;0;0;0;0;cxy/m;cyy/m;czy/m;cray/Ix;crby/Iy;crcy/Iz];
c3=[0;0;0;0;0;0;cxz/m;cyz/m;czz/m;craz/Ix;crbz/Iy;crcz/Iz];
c4=[0;0;0;0;0;0;cxa/m;cya/m;cza/m;craa/Ix;crba/Iy;crca/Iz];
c5=[0;0;0;0;0;0;cxb/m;cyb/m;czb/m;crab/Ix;crbb/Iy;crcb/Iz];
c6=[0;0;0;0;0;0;cxc/m;cyc/m;czc/m;crac/Ix;crbc/Iy;crcc/Iz];
c7=[1;0;0;0;0;0;Bx/m;0;0;0;0;0];
c8=[0;1;0;0;0;0;0;By/m;0;0;0;0];
c9=[0;0;1;0;0;0;0;0;Bz/m;0;0;0];
c10=[0;0;0;1;0;0;0;0;0;Ba/Ix;0;0];
c11=[0;0;0;0;1;0;0;0;0;0;Bb/Iy;0];
c12=[0;0;0;0;0;1;0;0;0;0;0;Bc/Iz];

A=[c1,c2,c3,c4,c5,c6,c7,c8,c9,c10,c11,c12]% assembles the columns
% to form the A matrix
% the Matlab "help" for the needed routines:
```

```
%E = EIG(X) is a vector containing the eigenvalues of a square
%matrix X.
% [V,D] = EIG(X) produces a diagonal matrix D of eigenvalues and a
%full matrix V whose columns are the corresponding eigenvectors so
%that X*V = V*D.
[V,D]=eig(A);
d=eig(A);
[wn,z]=damp(d); % finds natural frequencies (rad/sec) and damping ratios
fn=wn/(2*pi) % natural frequencies in Hz
zeta=z          % damping ratios
ch=poly(A);ch';r=roots(ch);[r d] %gets system characteristic polynomial
% and its roots
V % displays a full matrix whose columns are the eigenvectors
```

Displayed below are some of the results from this program.

fn =	Zeta =	Roots	Eigenvalues	
2.0784e+001	0	2.8422e−014 +2.8462e+002i	0	+1.3059e+002i
2.0784e+001	0	−2.8422e−014 −2.8462e+002i	0	−1.3059e+002i
9.6131e+000	0	−9.0198e−006 +1.3059e+002i	0	+6.0401e+001i
9.6131e+000	0	−9.0198e−006 −1.3059e+002i	0	−6.0401e+001i
2.0784e+001	9.5645e−019	9.0198e−006 +1.3059e+002i	1.2490e−016 +1.3059e+002i	
2.0784e+001	9.5645e−019	9.0198e−006 −1.3059e+002i	1.2490e−016 −1.3059e+002i	
9.6131e+000	−9.2881e−017	3.1974e−014 +1.2000e+002i	−5.6101e−015 +6.0401e+001i	
9.6131e+000	−9.2881e−017	3.1974e−014 −1.2000e+002i	−5.6101e−015 −6.0401e+001i	
1.9099e+001	2.7756e−017	1.4359e−006 +6.0401e+001i	3.3307e−015 +1.2000e+002i	
1.9099e+001	2.7756e−017	1.4359e−006 −6.0401e+001i	3.3307e−015 −1.2000e+002i	
4.5299e+001	−1.2482e−017	−1.4359e−006 +6.0401e+001i	−3.5527e−015 +2.8462e+002i	
4.5299e+001	−1.2482e−017	−1.4359e−006 −6.0401e+001i	−3.5527e−015 −2.8462e+002i	

Note that the natural frequencies are given in Hz, while the roots and eigenvalues use rad/s. The eigenvalues and roots should theoretically all be *pure imaginary* since the system has no damping, but the results give very small nonzero real parts in most cases. This is typical of such calculations. MATLAB prints the eigenvalues to match the sequence of the natural frequencies, but this order is reversed for the roots. The damping ratios (zeta's) should also all be exactly zero, but, as expected, some are quoted as very small positive or negative numbers. When interpreting results, one needs to be aware of these *normal* effects of making computerized calculations. We see that natural frequencies agree with the values we obtained earlier by other methods. These *other methods* of course will *not* work when the thrust stand lacks the symmetry of our simple example, while the matrix approach can handle the set of equations even when all the possible terms are present and nonzero.

I now want to add viscous damping to the system, since all real systems have some kind of frictional affects, whether it is viscous or otherwise.

Actually the damping in a thrust stand is unintentional (I could not find a single example of a thrust stand where damping had been *designed in*), and most of it would not be viscous. However viscous damping is easily modeled and gives linear equations, so we will use it in our study. Nonviscous damping is sometimes modeled as an *equivalent* viscous damping, based on measured transient decay curves of the actual system, once it has been built. Or we might run a frequency-response test on the system and note the *width* of the resonant peaks at each of the modes of vibration, which allows estimation of an equivalent viscous damping ratio zeta. The thrust stand literature indicates typical values might be about 0.03, a very lightly damped system. Our equations can include viscous damping forces, and the easiest way to give the system some damping is to provide a single viscous force such as $B_t(dx/dt)$ or $B_r(da/dt)$ in each of the Equations 9.14 through 9.19 where B_t is a translational damping coefficient and B_r is a rotational damping coefficient. Real damping effects of course do not act in this simple way, but the overall effect will not be unrealistic. To estimate a B value to give the desired 0.03 damping ratio, we can in each equation apply the standard definition of zeta: zeta = $B/2(K_sM)^{0.5}$. For those equations that have more than one *spring* term, we use the one that would be there if the equation were not coupled to any other equations. Again, this is not *physically* correct, but it does produce vibrations that are similar to what would be seen in an experimental test. These B values for axes *x-c* are respectively 1.3e3, 1.3e3, 1.8e3, 2.3e4, 2.3e4, and 1.9e4, as shown in the above program. Rerunning this program gives the following results:

$fn =$	Zeta =	Roots	Eigenvalues
9.6133e+000	4.6777e−002	8.5509e+000 +2.8449e+002i	2.8254e+000 +6.0336e+001i
9.6133e+000	4.6777e−002	8.5509e+000 −2.8449e+002i	2.8254e+000 −6.0336e+001i
9.6133e+000	4.6777e−002	3.2249e+000 +1.3054e+002i	2.8254e+000 +6.0336e+001i
9.6133e+000	4.6777e−002	3.2249e+000 −1.3054e+002i	2.8254e+000 −6.0336e+001i
1.9099e+001	3.0000e−002	3.2249e+000 +1.3054e+002i	3.6000e+000 +1.1995e+002i
1.9099e+001	3.0000e−002	3.2249e+000 −1.3054e+002i	3.6000e+000 −1.1995e+002i
2.0783e+001	2.4696e−002	3.6000e+000 +1.1995e+002i	3.2249e+000 +1.3054e+002i
2.0783e+001	2.4696e−002	3.6000e+000 −1.1995e+002i	3.2249e+000 −1.3054e+002i
2.0783e+001	2.4696e−002	2.8254e+000 +6.0336e+001i	3.2249e+000 +1.3054e+002i
2.0783e+001	2.4696e−002	2.8254e+000 −6.0336e+001i	3.2249e+000 −1.3054e+002i
4.5299e+001	3.0043e−002	2.8254e+000 +6.0336e+001i	8.5509e+000 +2.8449e+002i
4.5299e+001	3.0043e−002	2.8254e+000 −6.0336e+001i	8.5509e+000 −2.8449e+002i

Note that the eigenvalues *match up* with the natural frequencies but MATLAB printed the roots in the *reverse* order, and both are given in rad/s, whereas we displayed the natural frequencies in Hz. As expected, the damping ratios did not all come out at the desired 0.03 value, but they are all acceptably close.

9.6 Simulink® Simulation for Getting the Time Response to Initial Conditions and/or Driving Forces/Moments

The matrix methods get us the natural frequencies, damping ratios, and mode shapes, but do not provide any information on how the system responds to *specific* input forces and/or torques. This kind of information is vital in most applications and can be easily found by a Simulink simulation of Equations 9.14 through 9.19. We can also use this simulation to get some appreciation of the meaning of mode shapes. We have earlier mentioned mode shapes a few times but did not try to explain their meaning or utility. Some readers will already be familiar with the concept of mode shape, but others may not, so a little explanation is in order. For a *system without damping*, any vibrations that might be induced by initial conditions or transient input forces will continue *forever*. If we apply a sinusoidal driving force at exactly one of the natural frequencies, the amplitudes of all the motions will build up toward infinity, but at any given time, the amplitudes of the six motions (x, y, z, a, b, c) will not be equal. Mode shape information tells us these *relative* amplitudes. Instead of using external driving to get our system moving, we could instead apply some set of *initial conditions* (initial displacements and/or velocities). Again, if we use the mode shape information to adjust these initial conditions, when we *release* the system to perform free vibrations, only *one* frequency will be present, and all the motions will occur sinusoidally at that one frequency, with the amplitudes of the relative size given by the mode shape.

To make these statements more meaningful, we now consider the mode shape (eigenvector) information that was generated in our program but not utilized before this. The printout of the MATLAB eigenvector results at first appears quite confusing if you have not encountered it before.

$V =$

Columns 1 through 4

0.0142 − 0.0067i	0.0142 + 0.0067i	**−0.0017** − 0.0034i	−0.0017 + 0.0034i
0.0037 − 0.0022i	0.0037 + 0.0022i	**0.0156** − 0.0032i	0.0156 + 0.0032i
0.0000 − 0.0000i	0.0000 + 0.0000i	**0.0000** − 0.0000i	0.0000 + 0.0000i
−0.0006 + 0.0004i	−0.0006 − 0.0004i	−0.0026 + 0.0005i	−0.0026 − 0.0005i
0.0023 − 0.0011i	0.0023 + 0.0011i	−0.0003 − 0.0006i	−0.0003 + 0.0006i
0.0000 + 0.0000i	0.0000 − 0.0000i	**0.0000** + 0.0000i	0.0000 − 0.0000i
0.4068 + **0.8604i**	0.4068 − 0.8604i	0.2040 − **0.1016i**	0.2040 + 0.1016i
0.1300 + **0.2253i**	0.1300 − 0.2253i	0.1905 + **0.9408i**	0.1905 − 0.9408i
0.0000 − **0.0000i**	0.0000 + 0.0000i	0.0000 − **0.0000i**	0.0000 + 0.0000i
−0.0214 − **0.0370i**	−0.0214 + 0.0370i	−0.0313 − **0.1547i**	−0.0313 + 0.1547i
0.0669 + **0.1415i**	0.0669 − 0.1415i	0.0336 − **0.0167i**	0.0336 + 0.0167i
0.0000 − **0.0000i**	0.0000 + 0.0000i	0.0000 − **0.0000i**	0.0000 + 0.0000i

Columns 5 through 8

−0.0035 + 0.0057i	−0.0035 − 0.0057i	**−0.0008** − 0.0012i	−0.0008 + 0.0012i
−0.0017 − 0.0011i	−0.0017 + 0.0011i	**−0.0052** + 0.0044i	−0.0052 − 0.0044i
0.0000 − 0.0000i	0.0000 + 0.0000i	**0.0000** − 0.0000i	0.0000 + 0.0000i
−0.0008 − 0.0005i	−0.0008 + 0.0005i	**−0.0024** + 0.0020i	−0.0024 − 0.0020i
0.0016 − 0.0026i	0.0016 + 0.0026i	**0.0003** + 0.0006i	0.0003 − 0.0006i
0.0000 + 0.0000i	0.0000 − 0.0000i	**0.0000** − 0.0000i	0.0000 + 0.0000i
−0.7388 − **0.4597i**	−0.7388 + 0.4597i	0.1611 − **0.0988i**	0.1611 + 0.0988i
0.1375 − **0.2273i**	0.1375 + 0.2273i	−0.5713 − **0.6824i**	−0.5713 + 0.6824i
0.0000 + **0.0000i**	0.0000 − 0.0000i	0.0000 − **0.0000i**	0.0000 + 0.0000i
0.0627 − **0.1037i**	0.0627 + 0.1037i	−0.2606 − **0.3113i**	−0.2606 + 0.3113i
0.3370 + **0.2097i**	0.3370 − 0.2097i	−0.0735 + **0.0451i**	−0.0735 − 0.0451i
0.0000 + **0.0000i**	0.0000 − 0.0000i	0.0000 + **0.0000i**	0.0000 − 0.0000i

Columns 9 through 12

0.0000 + 0.0000i	0.0000 − 0.0000i	**0.0000** + 0.0000i	0.0000 − 0.0000i
0.0000 + 0.0000i	0.0000 − 0.0000i	**0.0000** − 0.0000i	0.0000 + 0.0000i
0.0083 − 0.0000i	0.0083 + 0.0000i	**0.0000** + 0.0000i	0.0000 − 0.0000i
0.0000 + 0.0000i	0.0000 − 0.0000i	**0.0000** − 0.0000i	0.0000 + 0.0000i
0.0000 − 0.0000i	0.0000 + 0.0000i	**0.0000** + 0.0000i	0.0000 − 0.0000i
0.0000 − 0.0000i	0.0000 + 0.0000i	**0.0000** + 0.0035i	0.0000 − 0.0035i
0.0000 − **0.0000i**	0.0000 + 0.0000i	0.0000 − **0.0000i**	0.0000 + 0.0000i
0.0000 + **0.0000i**	0.0000 − 0.0000i	0.0000 − **0.0000i**	0.0000 + 0.0000i
0.0000 + **1.0000i**	0.0000 − 1.0000i	0.0000 + **0.0000i**	0.0000 − 0.0000i
0.0000 + **0.0000i**	0.0000 − 0.0000i	0.0000 + **0.0000i**	0.0000 − 0.0000i
0.0000 + **0.0000i**	0.0000 − 0.0000i	0.0000 − **0.0000i**	0.0000 + 0.0000i
0.0000 − **0.0000i**	0.0000 + 0.0000i	−1.0000 + **0.0003i**	−1.0000 − 0.0003i

These results, for our example, are reproduced above. There are six natural frequencies and therefore six mode shapes. The printout has six *pairs* of columns, one pair for each mode shape. The two elements of each pair differ only in the sign of the imaginary part of the numbers. We need only ever consider *one* element of the pair; I chose to use the left element, which has two columns of numbers. The first six real parts give the relative *displacement* amplitudes of the six motions in the downward sequence *x, y, z, a, b, c*. Going now to the bottom half of the second column, those six numbers give the relative *velocity* amplitudes of the six motions in the same sequence as the displacements. In a Simulink simulation of the undamped system, if we want to see only this mode of vibration, we must set the initial displacements at the values 0.0142, 0.0037, 0.0000, −0.0006, 0.0023, 0.0000 (initial velocities all 0.0) and then release the system to perform free vibrations. Alternatively, we could set the initial velocities at 0.8604, 0.2253, 0.0000, −0.0370, 0.1415, 0.0000 (initial displacements all 0.0). (A third way of exciting a mode shape applies *both* the initial displacements and the initial velocities just quoted. All three methods result in pure sinusoidal motion at a single frequency, the motions just start out with different initial conditions.) Note that for the initial velocities, we simply ignore the fact that the printout shows these as *imaginary* parts. By looking at the 3rd, 5th, 7th, 9th, and 11th columns, we can get initial condition information on the remaining five modes of vibration. The initial displacements and velocities are shown in **bold** type (I added this feature *after* the results were displayed.).

If you should use MATLAB to study this or similar problems, I need to warn you of some possible frustrating details. On my own computer I have two versions of MATLAB, an older one and a more current one. I originally did this analysis with the older version, which gave the eigenvector results just explained, which are as they should be. When writing this book I decided to redo the analysis using the newer MATLAB version. The *new* program was actually transferred from the older version, so everything was *exactly* the same. When I ran the newer version everything duplicated exactly, *except* for the eigenvector results, which did not display in the *easy to read* format shown above and also showed numerical differences. I have not yet resolved this difficulty so wanted to warn you of this possibility. I hope the version that you might use will work properly. Even when the eigenvector routine works well, to get the nice format displayed above, you might need to do some simple manipulations on the raw results. If the results are displayed in the default format (1.3178e+002, for example), numbers which are actually zero will show up as a very small number, *not* zero. Also, due to the length of the numbers, the display takes up more space and may be hard to read. The solution to this problem is to change to a fixed-point format (rather than the floating point which is often preferred) as I did in our example above. It may take some trial and error to get the most readable format, but it is well worth the effort.

We next want to use simulation to study how the thrust stand responds to its actual inputs; the forces and torques produced when the engine operates. Let us assume that a rocket engine is being tested and that we want to measure all the forces and moments it produces as it is initially fired and then goes into steady-state operation. We will model the various forces and moments as if they ramped up linearly with time and then leveled off at constant values. This is of course not what really happens but it will serve as a simple introduction. Note also that if the rocket engine is of the solid-fuel type, its mass will reduce as the fuel is burned, changing the weight and also the moments of inertia and the location of the center of mass. If the rocket is mounted with its long axis vertical it will be difficult to distinguish changes in thrust (z-axis force) from changes in weight. These are important considerations that must be dealt with in some way but we will not pursue them here. We will use the damped version of our system to set up the simulation, but the damping is easily *turned off* if we wish. A *lookup table* will be used (with time t as its input) to make up force and torque inputs of any type that we might want to apply to any or all of the axes. Our first study will use the undamped model with *no* externally applied forces, but with the initial conditions given by the eigenvector (mode shape) results, to see whether we do get a single-frequency response with a characteristic mode shape when we use those initial conditions. Figure 9.7 shows the simulation block diagram. To observe the first mode of vibration we set the initial displacements on the integrators according to the first column of the eigenvector results. Physically this means that we deflect each translatory or rotary axis to the specified value and then *let go* at time equals zero. Figure 9.8 shows the response, which *verifies* the eigenvector calculations. That is, even though the system has several natural frequencies, we observe *only one*, and with no damping, the oscillations are never-ending. We can easily measure the frequency from Figure 9.8 and find that it corresponds to the 9.61 Hz entry in our list of natural frequencies. Coordinates x, y, and b oscillate in phase, with the amplitudes shown, while z and c are motionless, and a oscillates 180° out of phase with x, y, and b. With a little visual imagination, one can picture the overall motion as a combination of the individual oscillations. (Some more advanced software* is able to create three-dimensional animations of such motions, making the interpretation much easier.) We could also observe the same mode of vibration by setting the initial displacements all to zero and placing the proper initial velocities on the six integrators whose outputs are those velocities. Or, we could apply *both* the initial displacement and velocities. These three ways of observing a mode shape all show us what we want to see, but the waveforms simply start out (time = 0.0) at different values. The other five modes can be similarly demonstrated, using the proper initial values from the list of eigenvectors.

We next want to study the response of the damped system to input forces and moments. Let us first apply perfect *step* inputs, even though such *instantaneous* changes cannot occur in a real thrust stand. Instead of adding a step input module to our simulation diagram, we can create step inputs using the lookup table that is provided to allow the creation of force/moment inputs of arbitrary shape. This is possible since the lookup table module allows the entry of *double-valued* time points. That is, the time-point *vector* is entered as [0 0.2 0.2 5.0] and the force vector is entered as [0 0 1000 1000]. This creates an input that is zero from $t = 0.0$ to $t = 0.2$ s, but jumps instantly to 1000 lb at $t = 0.2$ and stays there until $t = 5.0$ s. Our transients last less than 5 s, so specifying the force up to 5 s is OK. Rather than specifying an actual set of physical forces and moments, I choose to just enter some arbitrary numbers for force and moment terms in Equations 9.14 through 9.19. The 1000 value (called f_m in the diagram) that comes out of the lookup table is sent through gain blocks which allow us to easily set desired numerical values for each force or moment.

The x, y, and z forces are respectively 200, 150, and 1000 lb$_f$, while the a, b, c moments are 100, 80, and 50 ft-lb$_f$. Figure 9.9 shows the response of each axis

---

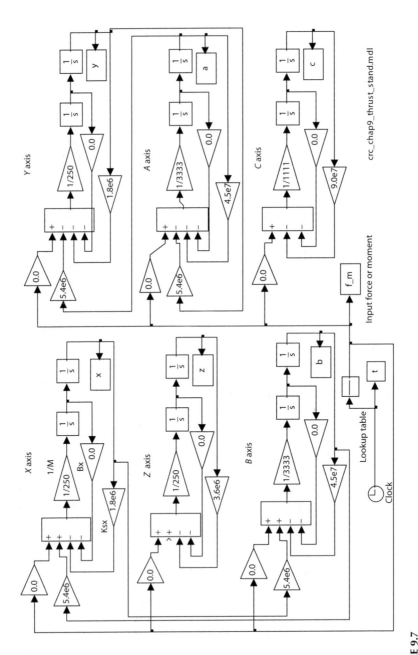

FIGURE 9.7
Simulation diagram with no damping, no applied forces/moments, IC's set for Mode 1.

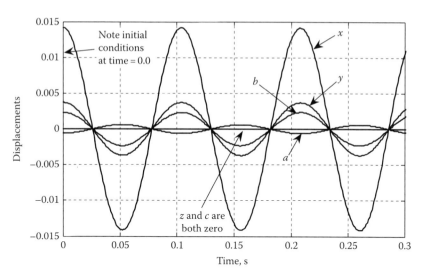

FIGURE 9.8
Responses to initial conditions that give the first mode of vibration.

to these multiple inputs. To the eye, these responses each seem to be at a single frequency, which we easily measure using the cursor facility of MATLAB's graph window. I zoomed each trace until five or six distinct peaks could be seen, and placed cursors at the first and last peaks visible. Counting and timing cycles allow the calculation of the apparent frequency, which I have added to each graph in Figure 9.9. All these measured frequencies are close to the undamped natural frequencies we earlier predicted. The a axis waveform shows a little *distortion*, which is more apparent in a zoomed version. If we recall the *analytical* solution of linear differential equations with constant coefficients (the model that we are using), the response to *any* input includes a transient term for each of the damped natural frequencies, so all these frequencies must be present in our current results. That is, *every* axis's response must include *all* the frequencies, not just certain ones. The explanation of *our* results of course is that each of the six transient solution terms has a *coefficient* in front of it, and some of these coefficients might be much smaller than others, making that frequency's contribution visually *invisible* on our graphs. Only in the a axis trace of Figure 9.9 is this effect subtly visible. Whether all frequencies will be clearly apparent in each response depends on the size and waveform of the six inputs. For the ones I chose, the six axis responses seem to be mainly at single frequencies.

Because of the symmetry of this system, there are only *four* distinct frequencies; two of the characteristic equation's six root pairs are *repeated*. For every complex root pair of the form $a \pm ib$ there is in the transient solution a term of form $C_1 e^{at} \sin(bt + \phi_1)$, and if this root pair is repeated we also get a term of the form $C_2 t e^{at} \sin(bt + \phi_2)$, which of course has *the same* frequency as the other term. Thus only four frequencies can be observed in a transient

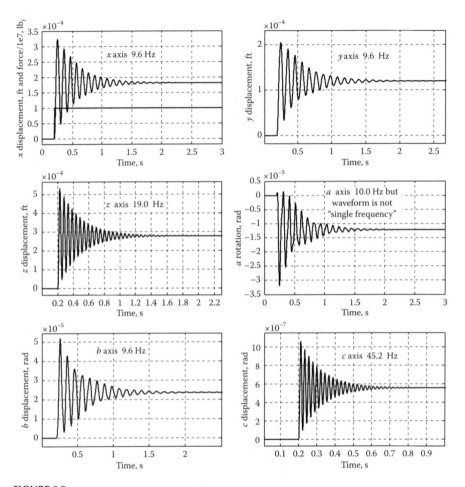

FIGURE 9.9
Response of each axis to combined step inputs on all axes.

response. The t factor in the repeated root term tends toward infinity but the entire term is bounded because the exponential term has a *negative* exponent, which makes that term *overpower* the t term, that is, a is negative in a damped system. In Figure 9.9 the a axis rotation is seen to be in the *negative* direction, even though all our applied forces and moments were entered as *positive*. This can be explained by referring back to Figure 9.2 and noting that a positive y *force* produces a negative a torque. If this torque is greater than the applied a moment (it *is*), then the *net* effect will be a negative a rotation. In Figure 9.10 I applied an arbitrary initial condition to each axis and plotted the response of just the x-axis, showing that in general, the responses are a composite of the various natural frequencies, giving a complicated waveform.

We next want to observe the responses to more realistic force/moment inputs, not the instantaneous *jumps* of a step input. If we knew exactly how

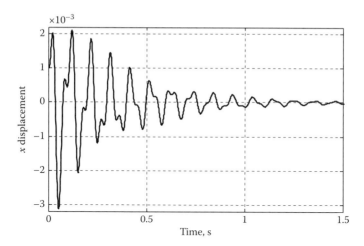

FIGURE 9.10
Waveform showing presence of several frequencies.

the actual forces and moments associated with, say, the ignition of a solid rocket motor built up with time, we could model those with our lookup table by specifying a lot of points on a force versus time graph. Lacking that kind of information we often just use a *terminated ramp* input, which is just a linear ramp from zero to the final value in a specified *rise time*. By adjusting the rise time we can model slow or fast buildups of the force. Fast rise times will cause larger oscillations while slow rise times, even in a lightly damped system, will reduce the oscillations. Let us try a rise time of one second to get a feel for this behavior, using the same force/moment levels we used with the step inputs. Figure 9.11 shows that, even with the slight damping, the various oscillations die out completely after about 2 s, for constant applied forces and moments. In practice, there may however be a problem, since every real engine, even though it may be commanded to produce a steady thrust, will always have some thrust fluctuation, which will continually re-excite oscillations and complicate accurate measurements of the *average* thrust.

Since thrust stands have in general been designed without any intentional damping, and since the *inherent* damping is very small, it is conventional to process the load cell electrical output voltages through some kind of low-pass filter. This filter must be designed to attenuate the offending oscillations but not obscure any short-term thrust changes that might be of interest. Our simulation again provides means for studying this problem. We need to produce a thrust signal, which ramps up as before, but is *contaminated* with some kind of fluctuation. Absent any data on *actual* fluctuations, we might choose to use some kind of random effect, easily produced by the Simulink *band-limited white noise generator* (see Appendix A for details). This appendix also discusses various filters that we might try. Figure 9.12 shows the effect of a rather *strong* filter; it causes a large delay between the actual force and

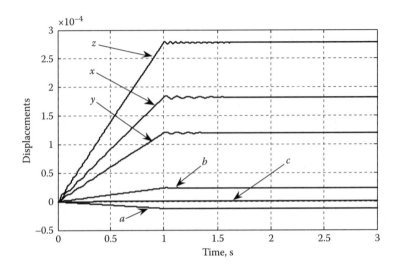

FIGURE 9.11
Terminated ramp response to inputs on all axes.

the filter output voltage, and would mask any force fluctuations of frequency higher than about 1 Hz. It does, however, strongly attenuate the oscillatory force component. In Figure 9.13, a less strong filter has less delay but allows some low-frequency fluctuations to "get through." Note that in both these filters, the time delay is nearly equal to the reciprocal of the cutoff frequency; 1 Hz gives 1 s delay, 5 Hz gives 0.2 s delay.

9.7 Frequency Response of the Thrust Stand

Some jet and rocket engines use various mechanisms (such as movable vanes or gimbaled nozzles) to deflect the exhaust gas flow sideways (or even downward) to control the direction of the thrust vector and thus provide steering or lift forces for the vehicle.* Aircraft using such engines include the British Harrier and the U.S. Joint Strike Fighter (see Figure 9.14). Since frequency-response methods are widely used to design vehicle control systems, the response of thrust forces to sinusoidal directional commands is of interest, and can be measured by the thrust stand if its dynamic response is adequate. Let us do some simple studies of the frequency response of our thrust stand design. One method uses the Simulink model to measure sinusoidal response in the time domain. This is usually an inefficient approach, since a sine wave of each frequency must be applied, transients allowed time

---

* www.harrier.org.uk

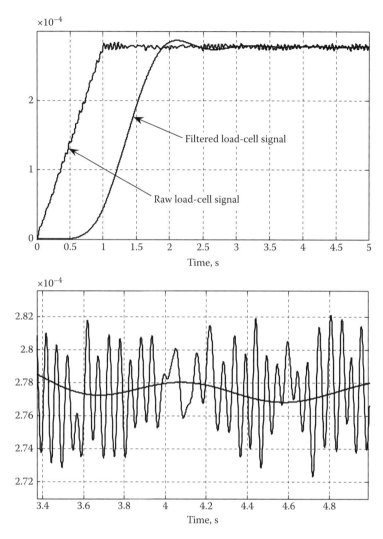

FIGURE 9.12
Effect of filter on load-cell signal (eighth-order Butterworth, cutoff at 1.0 Hz).

to decay, and measurements of amplitude ratio and phase shift finally implemented. These measurements, especially the phase angle, are tedious and prone to inaccuracy.

Another approach derives the *sinusoidal transfer function* relating the desired response variable to the input force. This transfer function allows quick and accurate plots of amplitude ratio and phase angle. If our system did not have the simplifying symmetry observed earlier, the transfer function approach for such a 12th-order system would be quite complicated to perform *in letter form*. If instead we are satisfied with strictly numerical results for assumed

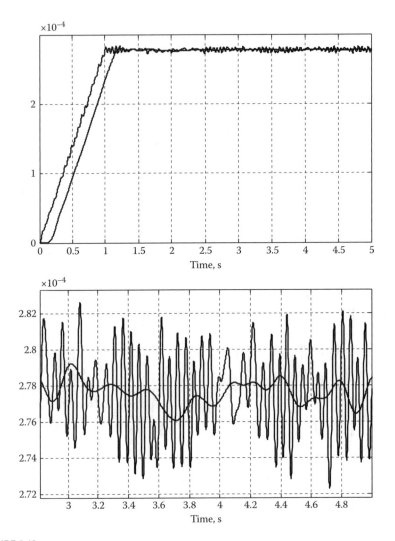

FIGURE 9.13
Effect of filter on load-cell signal (eighth-order Butterworth, cutoff at 5 Hz).

numerical values of system parameters, then an efficient *matrix frequency response* method is available, which gets *all* the output/input transfer functions in "one fell swoop." We will get to this shortly, let us first do the direct Simulink study, but only for a few frequencies. Since sidewise forces are of main interest, let us study the response of y displacement to y forces. We should point out that taking the *response* as the y displacement does not actually correspond to reality. That is, the only output *readings* provided by an actual thrust stand are the voltages coming from each load cell, whose readings are proportional to the *total* force felt by the load cell, not just the y component. We will shortly

FIGURE 9.14
Perhaps the world's largest thrust stand, it accommodates an entire full-scale aircraft. Intended to measure six axes of forces and moments produced by thrust-vectored vehicles such as the British Harrier or the U.S. Joint Strike Fighter. Since such aircraft land and take off vertically, the thrust components are heavily affected by the proximity of the ground, thus testing the isolated engine (the usual method) does not give accurate results. (Courtesy of Carl Schudde, Edwards Air Force Base, California.)

show how we can compute the actual load-cell forces from the computed motions, since that is *necessary* for a study of thrust stand static calibration. For our present study, we consider only the y displacement to be adequate, since we just want to illustrate the general approach rather than getting accurate values. With all initial conditions taken as zero, we apply sinusoidal y forces of amplitude $100\,lb_f$ and frequencies successively of 0.1, 1.0, 5.0 Hz, and at the lowest system natural frequency of 9.613 Hz. For the 0.1 Hz frequency, this is so far below any system natural frequency that the response amplitude will be essentially the same as if the force were applied *statically*. We will define the amplitude ratio (y displacement)/(y force) at this lowest frequency to be in the *flat* (perfect) range and compare the results for other frequencies to this value, so as to *normalize* the data. We can easily do this by ratioing the maximum values of the two signals, which you can get with the MATLAB statement max(y) applied to each signal. Once we have found this ratio (it is about 1.152e6) at 0.1 Hz, we use it as a multiplying factor on displacement y for graphing *all* the runs. This makes the *perfect* amplitude in Figure 9.15 equal to 100; any other amplitude is a deviation from ideal flat amplitude ratio. We see that the 1.0 Hz

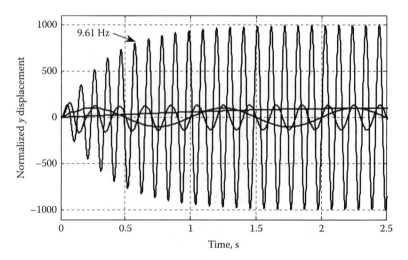

FIGURE 9.15
Simulink frequency-response results: 0.1, 1.0, 5.0, and 9.61 Hz.

trace has the same amplitude as the 0.1 Hz trace (the 0.1 Hz trace peaks at 2.5 s). The 5.0 Hz trace has a slightly larger amplitude, and the 9.61 Hz curve exhibits a large resonant peak with a magnification of about 10. From these sparse results, we conclude that the y-axis measurement might be useable to about 5 Hz.

A better way to get the y-axis frequency response is to use the sinusoidal transfer function relating y displacement to y force. This is available from Equation 9.24, but that model neglected damping, so we need to modify that analysis with the damping values we have been using.

$$(MD^2 + B_y D + k_{yy})y + (k_{ay})a = F_y \tag{9.39}$$

$$(k_{ay})y + (I_{xx}D^2 + B_a D + k_{aa})a = M_x \tag{9.40}$$

$$y = \frac{\begin{vmatrix} F_y & k_{ay} \\ M_x & (I_{xx}D^2 + B_a D + K_{aa}) \end{vmatrix}}{\begin{vmatrix} (MD^2 + B_y D + k_{yy}) & k_{ay} \\ k_{ay} & I_{xx}D^2 + B_a D + k_{aa} \end{vmatrix}} \tag{9.41}$$

$$\frac{y}{F_y}(D) = \frac{I_{xx}D^2 + B_a D + K_{aa}}{MI_{xx}D^4 + (I_{xx}B_y + MB_a)D^3 + (B_a B_y + Mk_{aa})D^2 + (B_y K_{aa} + B_a k_{ay})D + (k_{yy}k_{aa} - k_{ay}^2)} \tag{9.42}$$

We can use a MATLAB program to compute the sinusoidal transfer function and plot the frequency-response graphs.

```
% crc_book_chap9_yaxis_freqres.m computes and plots yaxis frequency
% response for the symmetric, damped thrust stand
% first quote the parameter values
M=250; Ixx=3333.; kaa=4.5e7; kay=5.4e6; kyy=1.8e6; By=1.3e3; Ba=2.3e4
% now compute the denominator coefficients
d4=M*Ixx; d3=Ixx*By+M*Ba;d2=Ba*By+M*kaa+Ixx*kyy;
d1=By*kaa+Ba*kay; d0=kyy*kaa-kay^2
% now define the frequency range
f=0:.001:50;
w=f*2*pi;
% now define the transfer function
num=[Ixx Ba kaa]
den=[d4 d3 d2 d1 d0]
[mag,phase]=bode(num,den,w);
plot(f,mag);pause
plot(f,phase); pause
```

Figure 9.16 shows results that agree with the limited data from the simulation; the amplitude ratio is fairly flat to 5 Hz, with a strong peak at 20.8 Hz. Recall that accurate measurement requires a flat amplitude ratio and a phase shift that is linear with frequency, both within some specified tolerance. Equation 9.42 tells us that the phase angle must be asymptotic to $-180°$ at high frequency (the denominator polynomial is two orders higher than the numerator), whereas Figure 9.16 shows $+180°$. I tried the MATLAB *unwrap* statement on the phase angle, but the results did not change, so Figure 9.16 is probably correct, if properly interpreted. That is, $-180°$ and $+180°$ are really the same angle and the curve between 15 and 21 Hz shows *how we got* to the final $+180°$ value. Questions as to the validity of phase angle plots can usually be answered by graphing the imaginary part of the complex number versus the real part, as in the lower graph of Figure 9.16, which confirms the behavior of the original graph.

9.8 Matrix Frequency Response Methods

When the six thrust stand equations exhibit *complete coupling* (each equation includes nonzero terms in each unknown), the transfer functions needed to compute frequency responses become extremely complicated and tedious to obtain in letter form. If we are satisfied with frequency responses for *numerical* values of all parameters, these difficulties may be avoided by use of *matrix frequency-response** methods. For our thrust stand, complete coupling can be achieved for an infinite number of combinations of basic parameters, if we destroy the symmetry of our original design

---

* E.O. Doebelin, *System Dynamics: Modeling, Analysis, Simulation, Design*, Marcel Dekker, New York, 1998, pp. 640–642.

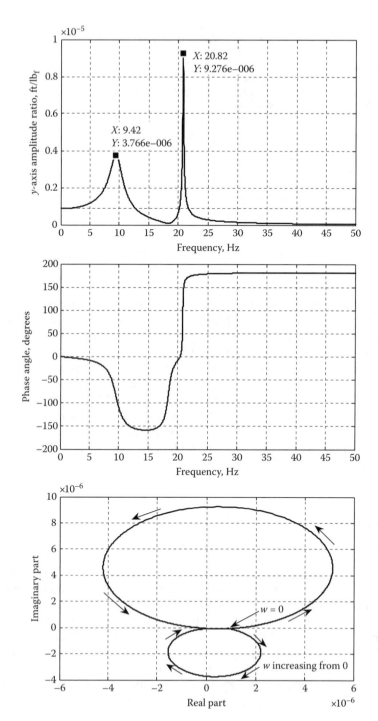

FIGURE 9.16
Frequency response, y/F_y, for the symmetric system.

by allowing angular misalignment of some of the load columns, letting some of the lever arm values that we took to be exactly equal deviate from that perfection, and/or allowing differences in some load column stiffness values which we assumed exactly equal. By relatively minor adjustments of this sort, I obtained the following *completely coupled* set of equations. (I will not give the details of the adjustments, only the equations which resulted.)

$$(250D^2 + 1.3e3D + 1.8e6)x + 8400y + 12000z + 26880a - 5.64e6b - 4.74e4c = F_x \tag{9.43}$$

$$8.4e3x + (250D^2 + 1.3e3D + 1.8e6)y + 12000z + 5.4375e6a - 4.884e4b - 1.71372e5c = F_y \tag{9.44}$$

$$12000x + 12000y + (250D^2 + 1.8e3D + 3.6125e6)z + 1.0524e5a - 6.72a4b + 2.196e4c = F_z \tag{9.45}$$

$$26880x + 5.4395e6y + 1.0457e5z + (3333D^2 + 2.3e4D + 4.5462e7)a - 5.7335e5b - 1.5194e6c = M_x \tag{9.46}$$

$$-5.6394e6x - 4.8838e4y - 6.72e4z - 5.7335e5a + (3333D^2 + 2.3e4D + 4.7762e7)b + 7.261e5c = M_y \tag{9.47}$$

$$-4.73e4x - 1.7137e5y + 2.196e4z - 1.5198e6a + 7.2615e5b + (1111D^2 + 1.9e4D + 8.9291e7)c = M_z \tag{9.48}$$

Note that I have added a damping term to each equation, since we want to compare some results with the frequency-response calculations done above. If we *remove* these damping terms and use our earlier matrix methods to find the system natural frequencies, we find that now there are six *distinct* natural frequencies rather than some repeated ones found for the perfectly symmetrical thrust stand. These new undamped natural frequencies are compared with the *old* ones in Table 9.2.

To use the matrix frequency-response method, in general, the system equation set must first be put in the form shown in our Equations 9.43 through 9.48. That is, we put all the unknowns on the left side and the inputs (driving forces or moments) on the right side. We also write all the derivatives in *D*-operator form and then gather all the terms (derivatives and unknown itself) for each unknown into a single *coefficient*, which we place in parentheses. The method will of course work in the presence or absence of damping.

TABLE 9.2

Frequency Comparison of Symmetrical
and Asymmetrical Geometries

Vibration Mode	Symmetrical	Asymmetrical
1	9.6131	9.5570
2	9.6131	9.6467
3	19.099	19.1324
4	20.784	20.8357
5	20.784	21.3180
6	45.299	45.1225

Doebelin (1998) begins a MATLAB program by stating the frequency values for which we want to compute the sinusoidal transfer function. While the reference sets up *logarithmically spaced* frequencies, we will use linearly spaced ones, since we want to compare this approach with our earlier results which were done that way. We then set up a MATLAB for-loop to step the calculations through the list of frequencies. Next we need to define a matrix made up of the coefficients on the left side of our equation set. We then invert this matrix and multiply this inverse by a column vector that selects which of the six inputs is to be applied. From the result of this multiplication, we get *all six* of the transfer functions that involve the chosen input and *all* the unknowns. To select a *new* input, we would define a new column vector and multiply it with the same matrix, getting in this way *all* the transfer functions involving that input and *all* the outputs (unknowns). Repeating this procedure we can get 36 transfer functions in all. We now need to show in detail the program steps that were just briefly sketched out.

```
% this program computes matrix frequency response for the asymmetrical
  damped thrust stand
% first, let's set up the frequencies
format short e
w= linspace(0,60*2*pi,500);
f=w/(2*pi);
% now define s = iw, to be used in the matrix
s=w*1i;
% now set up the for-loop
for i=1:500
% now define the matrix, using s for D
A=[250*s(i)^2+1.3e6*s(i)+1.8e6 +8400 +12000 +26880 -5.64e6 -4.7434;...
8.4e3 250*s(i)^2+1.3e3*s(i)+1.8e6 +12000 +5.4375e6 -4.884e4 -1.71372e5;...
12000 12000 250*s(i)^2+1.8e3*s(i)+3.6125e6 1.0524e5 -6.72e4 2.196e4;...
26880 5.4395e6 1.0457e5 3333*s(i)^2+2.3e4*s(i)+4.5462e7 -5.7335e5
  -1.5194e6;...
-5.6394e6 -4.8838e4 -6.72e4 -5.7335e5 3333*s(i)^2+2.3e4*s(i)+4.7762e7
  7.2616e5;...
```

```
-4.73e4 -1.7137e5 2.196e4 -1.5198e6 7.2615e5 1111*s(i)^2+1.9e4*s(i)+8.
  9291e7];
AI=inv(A);
c1=[0 1 0 0 0 0]';
QO1=AI*c1;
QO12(i)=QO1(2);
amp=abs(QO12);ang=angle(QO12)*360/(2*pi);
end
plot(f,amp);pause;plot(f,ang);
```

We originally wrote the equation set using *D*-operator notation. In the program, we substitute the Laplace variable s, and then set $s = i\omega$ to get the sinusoidal response. The matrix A is entered line by line; you can do it other ways if you prefer. The column vector c1 contains the inputs (driving forces or torques) in the top-to-bottom sequence F_x, F_y, F_z, M_x, M_y, M_z; we want F_y so c1 is as shown in the program. Since we always set our inputs to the number 1.0, the matrix calculation, which really computes the *output* quantities as the vector QO1, also gives the *ratio* output/input, which is the sinusoidal transfer function that we want. The output quantities appear in the vector QO1 in the sequence x, y, z, a, b, c, so to get y/F_y we ask for QO1(2). We can get all the other output/F_y transfer functions by simply asking for QO1(1), QO1(3), QO1(4), etc. If we want to explore other *inputs*, we need to rerun the program with c1 = [1 0 0 0 0 0] if we want F_x as the input, c1 = [0 0 0 0 1 0] if we want M_b as input, etc. If we did a lot of thrust stand design, we could augment the above program with a "front end" that calculated the coefficients in the matrix from the basic thrust stand physical parameters. You would then be able to efficiently explore the effect on frequency response of alternative design concepts.

Running the above program as it stands gives us the y/F_y frequency-response curves of Figure 9.17, which show some slight changes in the peak frequencies and larger changes in the peak heights compared with the symmetric system of Figure 9.16. The phase angle curve is more *conventional* than that of Figure 9.16 and seems to display the expected shape. Note that this transfer function (y/F_y) visually shows only two peaks, whereas we know that the system has six lightly damped natural frequencies. The y force input thus is not effective in exciting y displacement components at the other four frequencies. These other frequencies may, however, show up if we look at other transfer functions such as x/F_y, z/F_y, a/F_y, b/f_y, and c/F_y. Also, there are a total of 36 transfer functions if we consider all the inputs and all the outputs available from our matrix calculations, so the other frequencies may show up clearly in some of these other calculations. When interpreting the words *natural frequency* in linear dynamic systems, recall that we need to be careful to distinguish three *versions* of this concept; the *undamped natural frequency*, the *damped natural frequency*, and the *frequency of peak forced response*. The undamped natural frequency cannot be directly observed in any real system since there must always be *some* damping. When we disturb a system

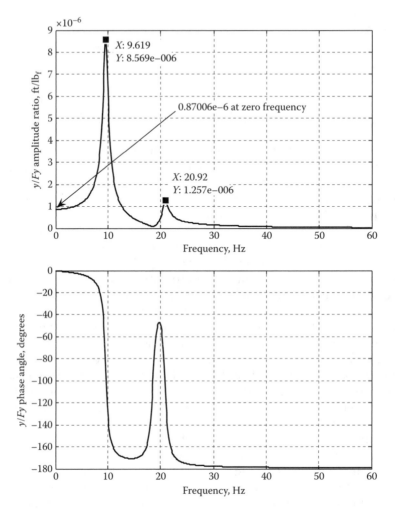

FIGURE 9.17
Frequency response, y/F_y, for the asymmetric system.

and observe its transient oscillations, those frequencies are the damped natural frequencies whose numerical values are obtained from the roots of the system characteristic equation. When we drive a system sinusoidally and scan through a frequency range to find a peak amplitude, that frequency is the frequency of peak forced response. When the damping is zero, all three frequencies are the same. For a simple second-order system, but *not* for more complicated systems, if the undamped natural frequency is called ω_n and the damping ratio ζ, then the damped natural frequency is $\omega_n\sqrt{1-\zeta^2}$ and the frequency of peak forced response is $\omega_n\sqrt{1-2\zeta^2}$.

9.9 Simulation of the Asymmetric System: Use of Simulink *Subsystem* Module

While no new basic ideas are involved, the diagram for this simulation can become quite messy and hard to follow because of the space it takes up on the screen and the many signals and coefficients involved. We can use Simulink's *subsystem* capability to get a much neater and clearer diagram. One way to do this is to first make up a diagram for, say, the x axis equation, using a summer with eight inputs, a gain block for $1/M$, two integrators, and gain blocks for the damping and spring terms (fed back to summer inputs). We provide, as input to one summer port, a clock and lookup table for creating a transient input force, a *to workspace* icon to record x, and an *outport* module connected to the x signal. We can then *duplicate* this five times, to create similar diagrams for the other five equations, inserting, of course, the appropriate inertial, spring, and damping numbers in each case. If you then apply the subsystem procedure to each of these six diagrams, they will each be *condensed* to *subsystem blocks*, which take up much less space. There will also be provision for entering, for each subsystem, the five summer inputs that come from the other five subsystems. These five inputs need to be brought in through gain blocks to allow setting of the system coefficients. Figure 9.18 shows the steps in this procedure. Once we have created the diagram of Figure 9.18a, we *edit>select all*, and then *edit>create subsystem*, which produces the display of Figure 9.18b. We can then duplicate this five times by using *copy* and *paste*, or right-click, drag. If we click on one of the subsystem blocks, as in Figure 9.18b, the display *opens up* to show what is *inside* that subsystem block (Figure 9.18c), which can then be edited in the usual ways (enter numbers into gain blocks, write text labels, move lines, etc). Closing a display as in Figure 9.18c returns one to the *compact* display of Figure 9.18b.

Figure 9.19a shows the block diagram when we *do not* use the subsystem method. I probably could have cleaned this up somewhat, but we see in Figure 9.19b that the subsystem technique does allow a clearer picture. In addition to the use of subsystems, Figure 9.19b also uses a *multiplexer* (mux) and six *demultiplexer* (demux) blocks. The multiplexer *bundles* multiple signal lines into a single "cable," reducing clutter, while the demultiplexer un-bundles the signals when you want to extract them. These features can of course be used whether we use subsystems or not. The simulation results would of course be identical for either block diagram. I applied a ramp z force with the lookup table in that axis and set all the other input forces/moments to zero, with initial conditions of zero for all the axes. Figure 9.20 shows all the system responses to the z-force input. Response to an a-axis moment in the form of a fast ramp to 100 ft-lb$_f$ is shown in Figure 9.21, where we see a strong response in the y axis, as we might guess intuitively from Figure 9.2 or Equations 9.15 and 9.17. While this *cross coupling* of load-cell readings can be

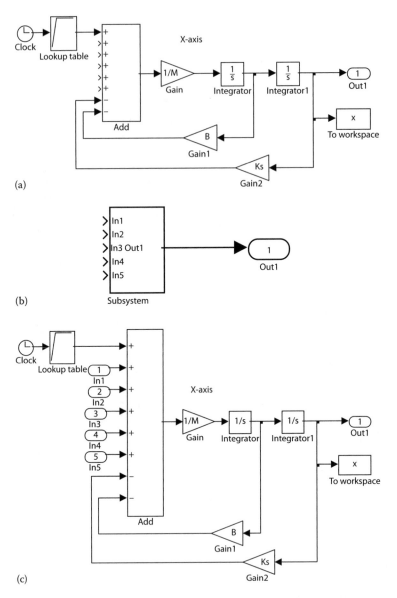

FIGURE 9.18
Steps in creating a subsystem.

taken care of in the calibration equations, thrust stand designs which use an *orthogonal* orientation for the load cells would have this problem only to the extent that the axes were misaligned due to unavoidable errors in machining and assembly. This feature is one of several reasons why the orthogonal configuration is usually used.

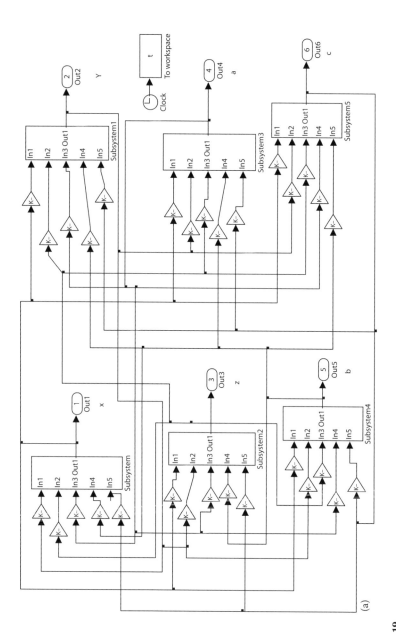

FIGURE 9.19
Comparison of (a) *ordinary and*

(continued)

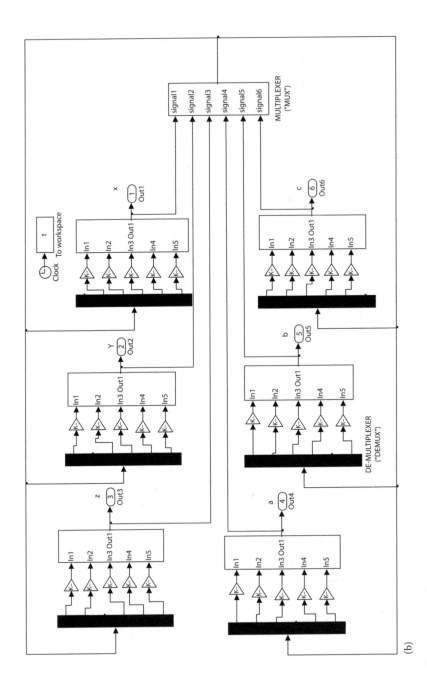

FIGURE 9.19 (continued)
(b) *subsystem* block diagrams.

(b)

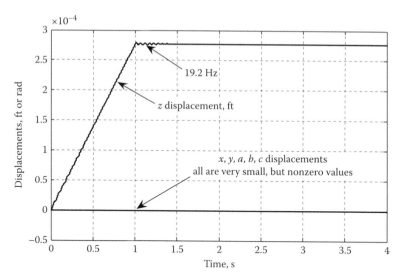

FIGURE 9.20
Response of asymmetric system to Z-axis ramp input force.

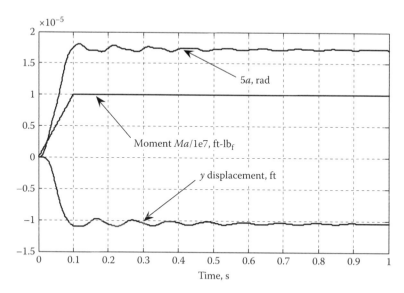

FIGURE 9.21
Response to fast ramp of a-axis moment, 100 ft-lb$_f$.

9.10 Static Calibration of Thrust Stands

As with any other measurement system, thrust stands must be calibrated periodically to verify their accuracy for continued application. Smaller stands may be calibrated using *deadweight calibrators*; for larger stands, the approach is often different. Rather than using dead weights that become cumbersome for large force values, the stand instead includes, for each axis, a precision *calibration load cell*, in addition to the measurement cell. It is then also necessary to provide some kind of *loading device* which will apply loads (which are measured by the calibration load cell) that simulate the loads which will be exerted by the engine. The load-application devices most commonly used include electric-motor-driven screw jacks, or pneumatic or hydraulic motors or cylinders. When the calibration is complete, the calibration load cells are inactivated using a pull-rod mechanism. Thrust stand design thus includes the provision of suitable mechanisms for this type of calibration method and designers have come up with various detailed ways of meeting this need. Figure 9.22a[*] shows one such design while Figure 9.22b shows some details used in both the calibration and measurement load columns. Threaded connections at the ends of the load cells and flexures could in principle be *locked up* using jam nuts, but this has been found to often be inadequate in terms of adjustability combined with low hysteresis. Instead, the *split nut* scheme shown has been found preferable. Here, by using both right- and left-hand threads (in the manner of a turnbuckle), the "slop" in the connection can be taken up, with the split nut unclamped. Then the split nut is clamped to solidify the connections without any relaxation of the slop adjustment. In Figure 9.22a, the loading device appears to be a hydraulic cylinder, which extends under pressure to put the *pull rods* into tension and thus place the calibrate and data load cells into compression, simulating the axial thrust force of the rocket motor. When, after calibration, the hydraulic cylinder is retracted, the pull rods become inactive and any rocket thrust force acts only on the data load cell. Such calibration mechanisms may be provided for each of the thrust stand measurement axes. Figure 9.23 shows a widely used universal flexure originally developed by Ormond Inc. (Ormond Inc. became Force Measurement Systems[†] in 2006) and the original patent drawing of 1960. The full patent document is easily accessed from GOOGLE PATENTS and the text portion gives quite a complete description of its operation.

Figure 9.24 (from Runyan et al., 1988, p. E-7) shows one view of a thrust stand used to test rocket motors mounted in vertical orientation. It shows several of these calibration mechanisms. This stand has all the load

---

[*] Runyan et al., AFAL-TR-88-060, p. 46.
[†] www.forcems.com

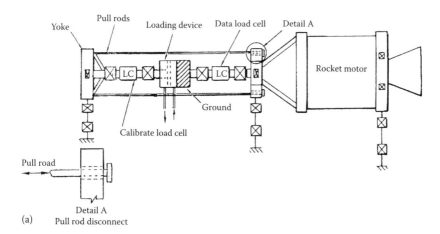

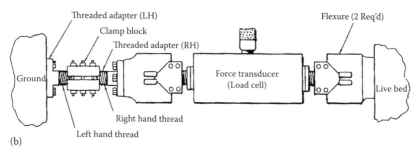

FIGURE 9.22
Hardware details of calibration facility.

columns arranged *orthogonally*; this design is actually more common than the "angled" type of our Figure 9.2, which has been the subject of all of our analysis. We chose this non-orthogonal design only because it is geometrically more general, and would thus give us practice using methods which would apply more easily to any requirements that might arise in the future. All our methods of course apply to both orthogonal and non-orthogonal designs. Arranging all the load columns to align with the *x*, *y*, and *z* axes simplifies the analysis and may also allow easier manufacture and erection. Of course, any actual thrust stand is bound to have some unintentional misalignment which defeats the intended perfect orthogonality, so our general treatment is necessary for studying such imperfections. If one decides to use the orthogonal arrangement whenever possible, there are still some choices to be made. Figure 9.25 (page 24 of Runyan et al. (1988)) shows a number of possible orthogonal configurations. Which of these one chooses would be influenced by the geometry of the particular engine and other such practical considerations.

Returning to the question of static calibration of our example thrust stand, we now want to do a "virtual" calibration using calculations rather than

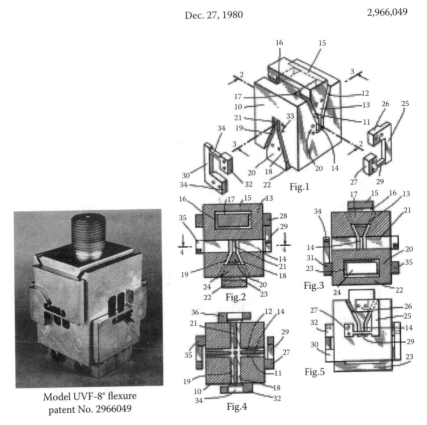

Dec. 27, 1980   2,966,049

Fig.1
Fig.2
Fig.3
Fig.4
Fig.5

Model UVF-8° flexure
patent No. 2966049

FIGURE 9.23
Universal flexure from Ormond (now Force Measurement Systems).

the actual measurements of a *real* calibration. That is, we do not have an *actual* thrust stand, but we still want to explain how such a stand might be calibrated. This can be done by simulating the real calibration with suitable equations and calculations. The first problem we encounter is that the equations we have used so far allow us to find the six *motions* of the thrust stand, but what we now need are ways to compute the *readings* of the measurement load cells, since that is the information that we would gather when calibrating a *real* thrust stand. Our first task then is to use the available motion equations to get new equations for the load-cell readings.

The "reading" of the actual load cell during a real calibration will generally be a *voltage* proportional to the tension or compression load carried by that cell. Because of the excellent linearity of such load cells, we can act as if the load-cell *force* were our *reading* when doing a virtual calibration. Fortunately, the load-cell force is directly related to the six motions for which we already have equations, but it will require some additional analysis to develop that relation. When performing a real calibration the general procedure is to

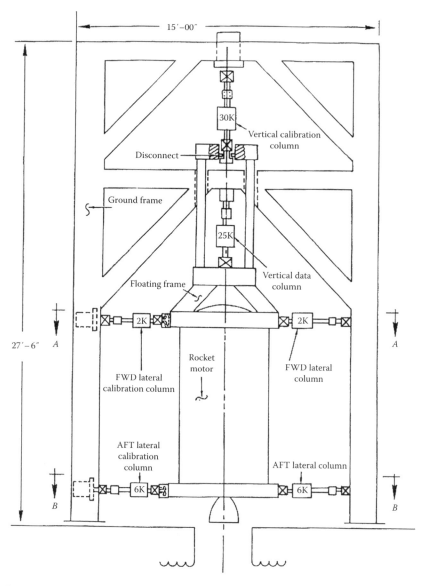

FIGURE 9.24
Rocket motor thrust stand of orthogonal design. (From Runyan et al., AFAL-TR-88-060, p. 46.)

apply, sequentially, a known force or moment to each of the six axes, and get the readings of all six load cells for each applied force or moment. Thus we would apply, say, a 1000 lb$_f$ force to the x-axis and get readings from all six of the load cells. With orthogonal designs, a *perfect* thrust stand would give exactly *zero* readings for all the cells except those aligned with the x-axis. For a *real* thrust stand (orthogonal or not), *every* load cell would give a nonzero

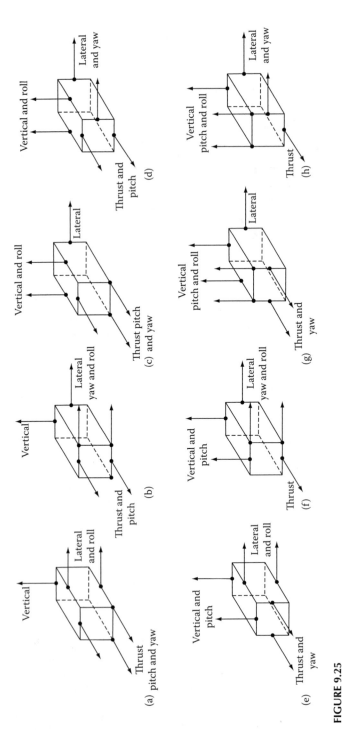

FIGURE 9.25
Various orthogonal-axes thrust stand configurations.

reading, since there is always some misalignment. One purpose of calibration is to take this misalignment into account. For each of the six load-cell readings, we could form the ratio (reading/load) to get six coefficients relating x-axis load to load-cell reading. Repeating this procedure for the y, z, a, b, and c axes, we generate a 6×6 matrix of coefficients relating loads in the six axes to readings of the six load cells. If we assumed perfect *linearity* of all the load cells, this approach would be sufficient for calibration. Since real load cells are very good but *not* perfect, a more comprehensive calibration procedure is used if we want to account for deviations from perfect linearity. Instead of applying a *single* x-axis load, we apply a range of loads from zero to full scale, taking voltage readings at each load and plotting voltage versus load for each load cell. We then fit least-squares straight lines to these data and get the *slope* of these lines, in units of V/lb$_f$. These slope values then become the coefficients in our 6×6 matrix. In our virtual calibration, this line fitting and slope finding will not be possible or necessary since our equations are perfectly linear. The details of how we *use* these 6×6 matrices will of course be the same whether we got the matrix from *single* loads or from the slopes of lines obtained from multiple loads. It should also be clear that for our virtual calibration, due to perfect linearity, the *value* of our single load has no effect on the results, so we choose a convenient value of 1000 lb$_f$ or 1000 ft-lb$_f$ for each force or moment. I should also point out that the use of this 6×6 matrix approach will deal with the most complex arrangements and the highest accuracy levels, but may not be necessary in all applications. Some thrust stand manufacturers invest more time and effort in fabricating and setting up the stand with minimal misalignment, and then reap the benefit of reduced interaction among the various axes, allowing simplified calibration schemes.

The relationship, in the calibration scheme described above, between the applied loads (F's and M's) and the resulting load-cell readings (R's) is given by the following matrix multiplication:

$$\begin{bmatrix} R_1 \\ R_2 \\ R_3 \\ R_4 \\ R_5 \\ R_6 \end{bmatrix} = C \cdot \begin{bmatrix} F_x \\ F_y \\ F_z \\ M_x \\ M_y \\ M_z \end{bmatrix} \tag{9.49}$$

Here C is the 6×6 matrix of coefficients found by the experiment described. Equation 9.49 is really six equations, one for each load-cell reading. For example, R_1 is given by a superposition of the effects of the three forces and three moments, each force or moment multiplied by one of the matrix elements. Once this calibration data has been gathered, we can then use it to compute applied forces and moments from measured load-cell readings. We know from linear algebra that this conversion can be made by getting the *inverse* (C^{inv}) of the matrix C and then computing from the following matrix equation:

$$
\begin{bmatrix} F_x \\ F_y \\ F_z \\ M_x \\ M_y \\ M_z \end{bmatrix} = C^{inv} \cdot \begin{bmatrix} R_1 \\ R_2 \\ R_3 \\ R_4 \\ R_5 \\ R_6 \end{bmatrix}
\tag{9.50}
$$

Again, this is really six equations, one for each applied force or moment. From the load-cell readings and the C matrix's coefficients as numerical values, we are able to compute the acting forces and moments. This completes the discussion of how calibration data and load-cell readings are used to get force and moment values in a *real* calibration.

Let us now see what is involved in getting values for the load-cell readings in our *virtual* calibration. We will assume that the flexures are perfect and thus that the load-cell readings are affected *only* by the forces along the load cell sensitive axis. One approach would be to write a program based on the general equations (such as Equations 9.11 and 9.12) so that we could routinely handle any configuration that might come up. Another would be to just apply basic static force analysis to the specific arrangement of Figure 9.2. For the general approach we could start with Equations 9.43 through 9.48 and eliminate all the derivative terms to get a set of equations relating *static* forces and displacements. In matrix form this would be

$$
\begin{bmatrix} F_x \\ F_y \\ F_z \\ M_x \\ M_y \\ M_z \end{bmatrix} = G \cdot \begin{bmatrix} x \\ y \\ z \\ a \\ b \\ c \end{bmatrix}
\tag{9.51}
$$

where G is a 6×6 matrix of coefficients defined by Equations 9.43 through 9.48. We will instead apply our methods to the simpler *symmetric* thrust stand described by Equations 9.14 through 9.19, thus the G matrix would still be 6×6, but would have a lot of zeros in it. Note that these matrix coefficients all came from computer programs which we have previously set up, so their calculation is already automated, which allows us to get them "painlessly" whenever we want to explore a new configuration. To simulate the act of applying, in turn, x, y, z forces and moments (as we do in a *real* calibration), we need to use Equation 9.51 to compute the motions caused by these forces/moments, and then use these known motions to deduce the *readings* of all the load cells. The first step is easily accomplished by applying the inverse of matrix G as follows:

$$\begin{bmatrix} x \\ y \\ z \\ a \\ b \\ c \end{bmatrix} = G^{\text{inv}} \cdot \begin{bmatrix} F_x \\ F_y \\ F_z \\ M_x \\ M_y \\ M_z \end{bmatrix} \tag{9.52}$$

To use Equation 9.52 to compute, for example, the motions caused by applying a $1000\,\text{lb}_f$ force along the x-axis, we would write

$$\begin{bmatrix} x \\ y \\ z \\ a \\ b \\ c \end{bmatrix} = G^{\text{inv}} \cdot \begin{bmatrix} 1000 \\ 0 \\ 0 \\ 0 \\ 0 \\ 0 \end{bmatrix} \tag{9.53}$$

This would give us six numerical values for the six motions caused by the $1000\,\text{lb}_f$ force along the x-axis. From these known motions, we now need to compute the *readings* (forces) in all six load cells. The various terms in Equation 9.11 are all forces in the x direction, each one *caused* by an x, y, z, a, b, or c motion. If we want the total x force felt by load cell 1, we need to evaluate these terms, *including in the summations only those terms that pertain to load cell 1*. For example, the term involving x uses a summation of

k_{xx1}, k_{xx2}, ..., k_{xx6}, but we only want k_{xx1}, which we then multiply by the x motion computed from Equation 9.53, to get the x force contributed by x motion. Similarly in Equation 9.11 there is a contribution by the y *motion* to the x force, and again we want only the portion that involves load cell 1. We need to do this also for the z, a, b, and c motion terms of Equation 9.11. Finally we algebraically sum up these six components of the x force in load cell 1 to get the total x force due to the combined effects of the six motions. However, in general, load cell 1 could simultaneously have y and z components of force, caused by the same six motions. These can be similarly computed from y and z equations similar to Equation 9.11, but we never stated these other two equations but referred you to the reference. My programs of course include these other equations, so that I can give you the results needed in our various discussions. Once we have computed the x, y, and z components of the force in load cell 1, we can get the *total* force (which is what the strain gages would read) by computing the square root of the sum of the squares of the individual forces. Recall though that this result is the *reading* of load cell 1 due to *only* a 1000 lb$_f$ force in the x direction. This load-cell *reading* is then divided by 1000, to put it on a *per pound* basis. It is, however, exactly what we want in our *virtual* calibration, since it mimics what we physically do in the actual calibration: apply an x force and read all the load cells. Of course, we have only done load cell 1, so we now have to repeat this procedure for the other five load cells. Furthermore, we then have to repeat all this for an applied y force, z force, x moment, y moment, and z moment, just as in the real calibration. For the most general case, this clearly becomes a complex calculation. However, as we stated earlier, if one is doing a lot of thrust stand design, the "dog work" of programming all this has to be done only once, and we can then easily and quickly *plug in numbers* for any special case we might want to study. To get, for our virtual calibration, the matrix C in Equation 9.49, the first column comes from the readings of load cell 1, with each row giving the contribution of one of the six forces or moments, with x force being row 1 and z moment being row 6. The second column is obtained similarly for load cell 2, third column for load cell 3, and so forth. For our example system of Figure 9.2, the C matrix is given by Equation 9.54.

$$C = \begin{bmatrix} -0.57732 & 0.33333 & 0 & 0 & 0 & -0.6666 \\ 0 & -0.66667 & 0 & 0 & 0 & 0.06667 \\ 0.57732 & 0.33333 & 0 & 0 & 0 & -0.06666 \\ -0.25000 & -0.433 & -0.33333 & 0.14334 & -0.08333 & 0 \\ 0.50000 & 0 & -0.33333 & 0 & 0.16667 & 0 \\ -0.25000 & 0.433 & -0.33333 & -0.14333 & -0.08333 & 0 \end{bmatrix}$$

$$(9.54)$$

The inverse is given by Equation 9.55:

$$
C^{inv} =
\begin{bmatrix}
-0.86607 & 0 & 0.86607 & 0 & 0 & 0 \\
-1.50017 & -3.00013 & -1.50017 & 0 & 0 & 0 \\
0 & 0 & 0 & -1 & -1 & -1 \\
-4.5005 & -9.0004 & -4.5005 & 3.4642 & 0 & -3.4642 \\
2.59821 & 0 & -2.59821 & -2 & 4.00001 & -2 \\
-15.00165 & -15.00132 & -15.00165 & 0 & 0 & 0
\end{bmatrix}
$$

$$(9.55)$$

I actually did all this programming in Mathcad, making the usual mistakes in such a long calculation and had to search for and correct various errors that were detected. Of course, for our example thrust stand of Figure 9.2, because of the symmetry, many terms in the general calculation are zero, leading to many zeros in the various matrices mentioned above. While the general approach just explained, once programmed, makes it easy to get results for any configuration of thrust stand, a more direct scheme is available and much quicker for an individual case of simple configuration. This method is no more than a simple *static force analysis*. For, say, a given x force, we can easily find the forces in each of the six load cells, giving the first column of matrix C. We then repeat this for the y and z forces and the three moments, filling in the rest of the 6×6 matrix. Figure 9.26 shows this approach for the case of an applied x force using a *stick figure* representation of Figure 9.2. The results of the force analysis agree with the first column of matrix C in Equation 9.54. We can check the method of Equation 9.50 for using load-cell readings to get the applied forces and moments. One way to do this is to consider simple cases where we *know* what load-cell readings will be produced by certain applied forces or moments. For example, if we apply a pure z force, then we know that load cells 4, 5, and 6 will all have readings equal to 1/3 of the applied z force, and the other three load cells should read zero. Let us use a z force of $3000\,lb_f$ and the matrix values of Equation 9.55 to carry out this check calculation, using matrix multiplication, as in Equation 9.50. A Mathcad calculation gives

$$
R := \begin{bmatrix} 0 \\ 0 \\ 0 \\ 1000 \\ 1000 \\ 1000 \end{bmatrix}
\qquad F := invS \cdot R \qquad
F = \begin{bmatrix} 2.22046 \cdot 10^{-13} \\ -2.56403 \cdot 10^{-13} \\ -3 \cdot 10^{3} \\ -9.09495 \cdot 10^{-13} \\ 0 \\ -8.32759 \cdot 10^{-13} \end{bmatrix}
$$

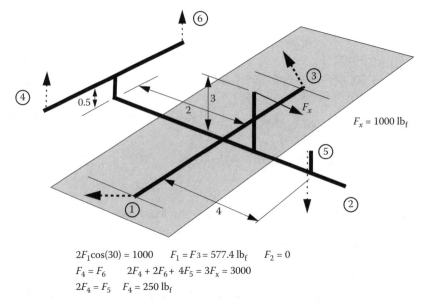

$$2F_1\cos(30) = 1000 \qquad F_1 = F_3 = 577.4 \text{ lb}_f \qquad F_2 = 0$$
$$F_4 = F_6 \qquad 2F_4 + 2F_6 + 4F_5 = 3F_x = 3000$$
$$2F_4 = F_5 \qquad F_4 = 250 \text{ lb}_f$$

To get the *readings* of the load cells one must look at Figure 9.2
and see which are in tension (+reading) and which are in compression
(−reading). For the positive applied x force shown above, the readings
of load cells 1, 4, and 6 are negative (compression) and those
of load cells 3 and 5 are positive (tension).

FIGURE 9.26
Static force analysis verifies matrix calculation for x-force input.

In the Mathcad program I used the name *invS* for C^{inv}. The very small F
values are all really zero, so the method seems to be working properly. The
readings of +1000 lb$_f$ (tension) for load cells 4, 5, and 6 agree with the downward (negative) z force of 3000 lb$_f$.

9.11 Damping of Thrust Stands

The literature that I was able to find on thrust stand design indicates that the
inclusion of intentional viscous dampers, to control the strong resonances of
this kind of equipment, has never been attempted. Since one can never be sure
that *all* the pertinent literature on any engineering topic has been discovered,
I also checked with an active and experienced thrust stand designer.* Scott
indicated that he had never used dampers on any of his stands and also had
not heard of anyone else doing this. The standard approach seems to be to

---

* Scott McFarlane, Ormond Inc. (now Force Measurement Systems, www.forcems.com).

"live with" the low inherent damping and just use electrical low-pass filters to remove the vibration effects from the load-cell voltages. In fact, when I mentioned to Scott the possible use of, say, Butterworth or Bessel filters he said that often simple *first-order* filters were actually sufficient. Since such filters *roll off* very slowly with frequency, it seems that this would work only if one wanted just the *steady* components of thrust, rather than any dynamic features. It is probably a fact that most thrust stands are *not* intended to measure time-varying thrust components, so these simple filters may be adequate. For vehicles whose engines use *thrust-vectoring*, such as the British Harrier or the U.S. Joint Strike Fighter, the dynamic response of the thrust vector angle to swiveling commands from a pilot or autopilot may be vital information for overall control system and vehicle maneuvering performance, and would need to be accurately measured. In such cases, if the frequencies of nozzle swiveling and thrust stand resonances were close to each other, then the sharp cutoff of the more sophisticated filters might be necessary.

The use of electrical filters rather than mechanical dampers is partly based on cost and convenience, but also on the *technical difficulty* of actually designing a damper for such small motions. A similar situation has been encountered in various *space structures*, where again the inherent structural damping is very small and in addition there is *no air* to provide fluid damping when the structure vibrates in the vacuum of space. In these applications, however, the poor damping *cannot* be tolerated and *must* be improved in one way or another. Hundreds, perhaps thousands, of projects and reports have been devoted to various schemes, many of which are complicated feedback control schemes (active damping), for improving the damping of space structures, such as the Hubble telescope and the International Space Station. At least one company, Honeywell, explored the much simpler approach of "ordinary" viscous dampers (passive damping). The problem, of course, just as in the thrust stand, is that the motions to be damped are so small. Conventional *piston/cylinder* dampers must have seals, and these seals will have sufficient *stiction* (static friction) that the tiny motions will never break this loose and allow the relative displacement of piston and cylinder which pumps the damping liquid through an orifice of some sort to dissipate the vibratory energy. Some years ago I obtained, through the help of L. Porter Davis (Technical Director, Isolation and Structural Control), a number of Honeywell reports on their damping device, called the D-Strut. One of these reports* described a D-Strut which could be adjusted by remote control to set the damping (*B* value) to meet the needs of a specific structure and environment.

Figure 9.27 shows a simplified sketch of this *adjustable* D-Strut, which uses metal bellows to allow the needed relative motion without the friction of conventional seals such as O-rings. The fluid resistance element, where the

---

* L. Porter Davis et al., Adaptable passive viscous damper (an adaptable D-Strut), *SPIE North American Conference*, February 1994, Orlando, FL. Honeywell Inc., Satellite Systems Operation, Glendale, AZ.

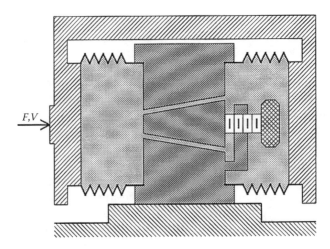

FIGURE 9.27
Adjustable D-strut damper.

desired damping effect is realized, is the annular gap between the conical hole and solid cone. The screw adjustment varies the size of this annular gap and thus the *strength* of the damping effect. In the actual device, this screw adjustment is made with a stepping-motor drive, remotely controlled. (When adjustability of the damping is not required in an application, versions of the D-strut use an ordinary fixed orifice for the fluid resistance element.) While the reports gave detailed descriptions of the actual construction and the results of experimental frequency-response testing, the design formula relating device dimensions and fluid properties to the B values for the device of Figure 9.27 was not provided. Using some simplifying assumptions and basic undergraduate fluid mechanics analysis, I derived the following formula:

$$B \triangleq \frac{F}{V} \triangleq \frac{6\mu L A^2}{\pi h^3 [r_1 - r_s]} \ln\left[\frac{r_1}{r_s}\right] \tag{9.56}$$

where
    μ is the fluid viscosity
    h is the perpendicular distance between the two conical surfaces
    L is the length along the solid cone
    two r's are the radii of the solid cone at the large end (l) and the small end (s)

To relate force F to pressure in the left bellows, and velocity V to volume flow rate q, we need some kind of effective area A for the bellows. This would be most accurately found by experiment for an existing bellows, but is well approximated by the area A of the circular *force pad* welded to its left end. (The pressure/force relation using A neglects the small spring stiffness of the bellows.) Silicone damping fluid (available in a wide range of viscosities) is used and the fluid is pressurized with a preload spring (not shown in Figure 9.27). This preloading is probably an important feature contributing to the ability of the damper to work at very small motions, and it also prevents cavitation. A further design feature contributing to the small-motion capability is the area ratio between the bellows and the annular damping gap. When the bellows area is much larger than the damping gap area, a small velocity of the bellows force pad results in a large fluid velocity in the damping gap. Testing has verified the existence of useable damping for motions as small as 100 nm. Figures 9.28 through 9.30, taken from various Honeywell reports, show some details of actual D-strut devices as used in a number of space structures. Figure 9.29 shows the frequency-response testing of a D-strut using a small electrodynamic vibration shaker. Such testing is vital to check the actual behavior against theoretical predictions, and to suggest changes to theoretical models when measured responses deviate from predicted.

While damping seems to never have been used in thrust stands, the existence of the Honeywell D-strut technology is probably unknown to thrust stand designers, so they likely have not considered it. A survey of load-cell manufacturers shows that here also, damping is generally not attempted. Let us briefly explore the possibility of using the D-strut technology with load cells. Since engine thrust forces are usually thousands of pounds, strain-gage load cells using the direct tension/compression design* are very common. The member that carries the gages might have various solid or hollow cross sections, but all these can be characterized by their cross-sectional area

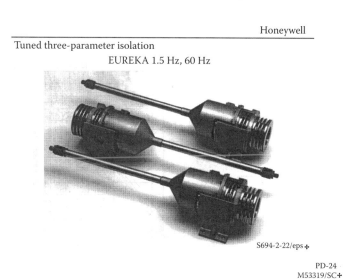

Honeywell

Tuned three-parameter isolation
EUREKA 1.5 Hz, 60 Hz

S694-2-22/eps +

PD-24
M53319/SC+

FIGURE 9.28
Honeywell D-strut damping device.

* E.O. Doebelin, *Measurement Systems*, 5th edn., McGraw-Hill, New York, 2003, pp. 446–452.

FIGURE 9.29
Frequency-response testing using a small electrodynamic vibration shaker. (Courtesy of Honeywell.)

A and length L. Also, the use of steel at a maximum stress level of about 45,000 psi (strain of 1500 με) is typical. For use in a thrust stand, to keep natural frequencies high, we require a certain minimum stiffness in the load cell. When selecting or designing a load cell, a thrust stand designer would specify a certain stiffness and also a full-scale load, at which the 45,000 psi stress would be developed. Using simple strength-of-materials formulas for direct tension/compression loading of steel load cells, we can develop formulas relating column length L and cross-sectional area A to full-scale load P and desired stiffness K_s.

$$K_s = \frac{F}{\delta} = \frac{AE}{L} \qquad \frac{A}{L} = \frac{K_s}{E} = \frac{K_s}{3e7} \tag{9.57}$$

For thrust stand *generic* design, we can use typical engine characteristics to estimate some needed numerical values. As stated earlier, thrust/weight ratios for typical engines might be about 5. Using this value and assuming a particular engine and thrust stand axis required a 10,000 lb$_f$ full-scale force, we can find the load-cell stiffness needed to give a minimum natural frequency of, say, 20.0 Hz, modeling the vibrating system as a simple uniaxial spring/mass system with a mass which weighs 2000 lb$_f$. This stiffness turns out to be 81,821 lb$_f$/in., making $A/L = 0.002727$. For a stress of 45,000 psi, the cross-sectional area must be 10000/45000 = 0.2222 in.^{2}, which makes the maximum allowable length 81.5 in. Since a *shorter* length gives a stiffer load cell, which is good, we can choose our length to suit the space needed for strain gages of convenient large size. If we chose a solid cross section of round or square shape, the diameter would have to be 0.532 in. or the square 0.471 in. on a side. These dimensions are a little cramped

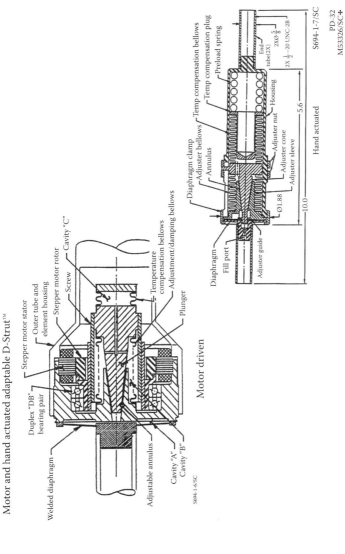

FIGURE 9.30
Remotely adjustable D-strut damper.

for applying large gages, so we consider a *hollow* cross section, square on the outside (to ease gage application) and round on the inside (easily machined). If a is the length of the square's side and d is the hole diameter, and if we stipulate a minimum wall thickness of, say, 0.10 in., we can solve for the dimensions, finding $a = 0.5785$ in. and $d = 0.3785$ in. The a dimension is large enough to provide space for the four strain gages that are used in this type of load cell (see Figure 9.31) if we make the length L long enough. Actually, one likes to place the gages some distance away from the larger diameter sections on each end to allow the stresses to *smooth out* at the gage locations and conform more closely to the theoretically predicted values (St. Venant's principal). There is no *magic number* for the length except we do not want it too long; to avoid potential buckling problems, let us take it as 3 in. This makes the stiffness about 2.22e6 lb$_f$/in., so the full-scale deflection is 0.00450 in.

Neglecting for the moment the need for flexures at each end of the load cell, consider a viscous damper that feels the same relative displacement as the ends of the load cell; that is, the damper is connected *in parallel* with the load cell. Let us find the B value needed to give a system damping ratio of 0.65 for the load cell we just designed. We have here a simple spring/mass/damper second-order dynamic system, so we can use the well-known definition of damping ratio to compute the needed B value.

$$\zeta = 0.65 = \frac{B}{2\sqrt{K_s M}} \qquad B = 1.30\sqrt{\frac{2.22e6 \cdot 2000}{386}} = 4409 \text{ lb}_f/(\text{in./s}) \qquad (9.58)$$

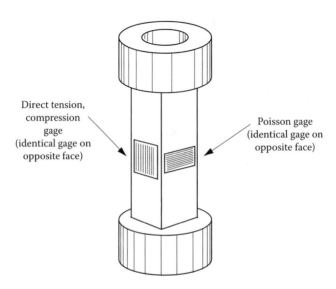

Direct tension, compression gage (identical gage on opposite face)

Poisson gage (identical gage on opposite face)

FIGURE 9.31
Strain-gage load cell.

The load-cell deflection of 0.0045 in. is well above the smallest motion (100 nm) found to be effective in Honeywell's testing of D-struts, so that aspect of feasibility is not in question. It is however *not* obvious that the device of Figure 9.27 can be designed to provide the B value just calculated; one needs to perform design calculations and probably some experimental testing, since this application to *load-cell* damping has, to my knowledge, never been tried. We cannot here pursue such a thorough study but some preliminary checks are possible. To get a rough idea of dimensions, commercial 10,000 lb load cells are about 3 in. in diameter and 4 in. long. Our D-strut probably should not be larger than this. In Equation 9.56 we might thus try $L = 3.0$ in. and area A equal to that of a circle with diameter 3 in., making $A = 7.0$ in.^{2}. We can use a small Mathcad program to explore various combinations of cone radii r_1 and r_s, and gap height h. Silicone damping fluids* are available in a wide range of viscosities (about 1 centistoke to 2.5e6 centistokes) but one cannot always use the very high viscosity fluids in every application because they are more like grease than oil and may not reliably fill small gaps like those used in a D-strut. The Mathcad results below show that extremely high viscosity is not required with h gaps that are large enough to not cause manufacturing, assembly, or thermal expansion problems.

$$A := 7.0 \qquad L := 3.0 \qquad rl := 1.4 \qquad kr := 0.5 \qquad rs := kr \cdot rl$$

$$cstoke := 1000.0 \qquad mu := cstoke \cdot 0.98 \cdot 1.45 \cdot 10^{-7} \qquad mu = 1.421 \cdot 10^{-4} \qquad rs = 0.7$$

$$h := 0.01, \ 0.02.. \ 0.10 \qquad B(h) := \frac{6.0 \cdot mu \cdot L \cdot A^2}{\pi \cdot h^3 (rl - rs)} \cdot \ln\left(\frac{rl}{rs}\right)$$

h =	B(h) =
0.01	$3.95 \cdot 10^4$
0.02	$4.938 \cdot 10^3$
0.03	$1.463 \cdot 10^3$
0.04	617.249
0.05	316.031
0.06	182.889
0.07	115.172
0.08	77.156
0.09	54.189
0.1	39.504

---

* www.dowcorning.com; E.O. Doebelin, *System Dynamics*, pp. 729–731.

While the D-strut *concept* seems, at least initially, to be feasible here, detail design of components such as the metal bellows at each end (which might have to survive high pressure transients) remains unexplored. Also, how will the D-strut actually be *mounted* so as to be *in parallel* with the load cell? Design against mechanical failure requires some knowledge of the *forces* that the D-strut will encounter during thrust stand operation. Let us assume that we *are* able to somehow connect the D-strut in parallel with the load cell and that the maximum force of 10,000 lb is applied as a terminated ramp with rise time of 0.10 s. We need to also include the effect of the flexures at each end of the load cell; let us model them as pure springs with the same stiffness as the load cell. We now have a mechanical system, as in Figure 9.32, which we can analyze and simulate to provide information on the maximum forces to be expected. The three springs all have the same stiffness (2.22e6 lb_f/in.) and the main mass M weighs 2000 lb_f. Universal flexures with a rating of 10,000 lb_f weigh about 1 lb while a 10,000 pound load cell might be about 3 lb. The three Newton's law equations for this system are easily set up and lead to the simulation diagram of Figure 9.33. I have not displayed the system equations; you should be able to *work backward* from the simulation diagram, a skill that is worth cultivating. The two small masses represent the mass effects of the flexures and the load cell, in lumped form. The left flexure contributes half of its mass to the immovable left wall and half to the *joint* between that flexure and the left end of the load cell. The left end of the load cell contributes half of the load-cell mass to this *joint*, making $M_1 = 2/386$. Similarly, M_2 is taken as 2/386 and the other half of the load-cell mass is attached to the main mass, making it 2001.5/386.

The two small masses and the associated stiff springs create some *very* fast-changing dynamics, requiring a very small computing increment (1e−6 s), even though the *main* dynamics were designed to give 20 Hz. That is, this system has both *slow* dynamics and *fast* dynamics, if we would consider *all* the actual natural frequencies, not just the one we intentionally designed.

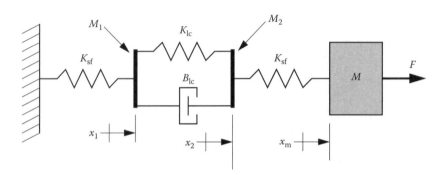

FIGURE 9.32
Damped load cell with end flexures.

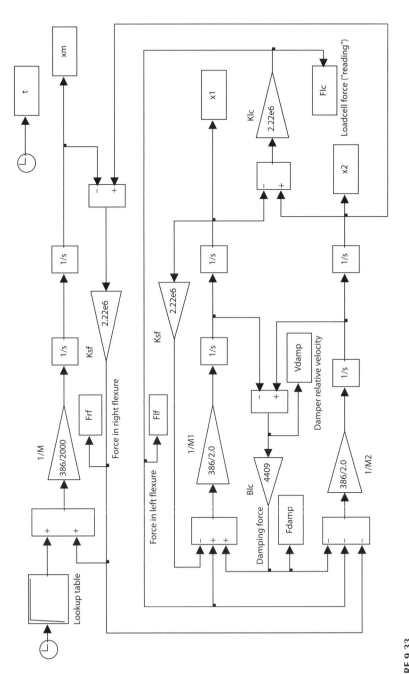

FIGURE 9.33
Simulation of damped load cell with undamped flexures.

Such systems are called *stiff* and can present simulation problems. If we set the computing increment to suit a 20 Hz response, a time step of 0.001 s would be fast enough. If we use this value in our system, the computation is numerically unstable and predicts all variables going toward infinity, and you get an error message. Decreasing the step size finally gets a working simulation, but the computing time is much longer than it would be if the fastest dynamics were 20 Hz. *Much longer* is here strictly *relative*; it still only takes a few seconds, and is thus not a real problem. An alternative approach is to try one of Simulink's other integrators, rather than the Runge-Kutta fixed-step size that I usually start with. In particular, use one of the *stiff* integrators; Simulink offers four of them. These use *self-adjusting* step size, using small time steps when needed and large time steps when allowable. This sound like a good idea and sometimes works well; in our case it was not really needed since our simulation is a simple one and runs quickly enough. Finally, since the two small masses are *so small* relative to the main mass, we are temped to call them zero. If you try this you will get some *algebraic loops*, which Simulink automatically tries to fix, but which may prevent the simulation from running.

Figure 9.34 shows some useful results when the applied force is a terminated ramp with a fast 0.01 s rise time. Recall that we designed our damper to give an *optimum* damping ratio of 0.65 based on a system *without* the flexures. With flexures each as stiff as the load cell, we see that the damping is *not* this good, since the load-cell force (which is proportional to the voltage output) has some fairly strong oscillations. The reason for this is that it would take *infinitely stiff* flexures to make the system behave like the simple second-order model we used to estimate the needed damping B. By increasing the flexure stiffness by a factor of 10, we get in Figure 9.34 a much better damped response, very close to what we designed for. The peak damping forces are 2000–3000 lb depending on the flexure stiffness. These force values would be needed in detailed design of the damper, since they influence stresses and fluid pressures. Note, however that the input force is *very* fast (0.01 s rise time) and that the damping forces would be lower if the actual rise time were not this fast. If we had some previous experience with the rocket motor under test, we might have a good idea of what the fastest rise time might be. The relative velocity of the two damper ends is what causes the damping force, so we show this in the lower graph. Note that the peak velocities are modest (less than 1 in./s), but the large B value causes the large forces.

Based on our cursory analysis, the use of a damper has not been shown to be obviously impossible, so one could pursue the analysis into detail design to see whether a practical device might be built and subjected to experimental testing and development. As mentioned earlier, we choose not to go to this level of detail, however, we *should* explore briefly how such a damper would be attached to the load cell in the *parallel* configuration shown schematically in Figure 9.32. Some sort of *fixture* to achieve this side-by-side mounting could

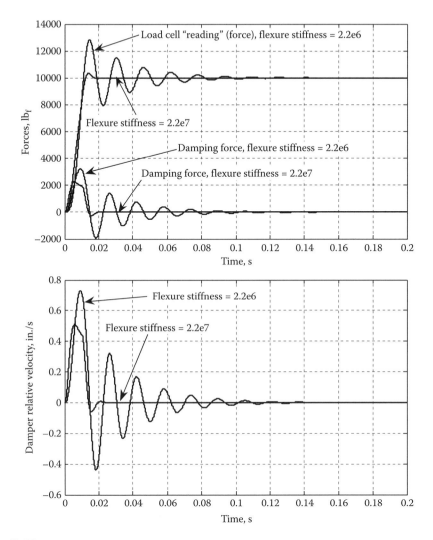

FIGURE 9.34
Response of damped load cell to fast (0.01 s rise time) ramp.

be designed, but because of the large forces and small motions involved, does not appear to be practical due to the stringent alignment requirements of the load cell. It thus seems that one would have to redesign the load cell itself to incorporate the damping means within it. Figure 9.35 shows one of the few damped load cells that are commercially available. It uses the popular *binocular* type of elastic element rather than the direct tension/compression of Figure 9.31. In Figure 9.36, I have "invented" a version of the D-strut concept which might be built into a tension/compression load cell. I have not analyzed this in detail nor built a working model; it is just an idea to suggest how it might be done.

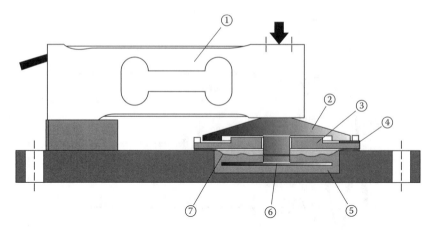

FIGURE 9.35
Damped load cell. (Courtesy of Scaime, Annemasse, France, www.scaime.com.)

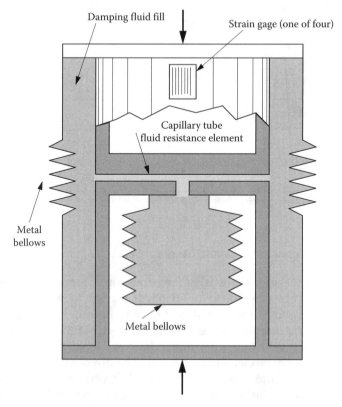

FIGURE 9.36
Possible design for a damped tension/compression load cell.

9.12 Flexure Design

There are a number of different types of flexures that are used for various purposes in thrust stand design and elsewhere. Some of these flexures, such as the Ormond Universal type (Figure 9.23), are quite complex in form and thus difficult to analyze and machine. *Electrical wire discharge machining* is often the only practical machining method for these geometrically compli-cated forms. In this section, we choose to examine the *simplest* type of flexure, which yields to relatively straightforward analysis, using methods and results from basic strength-of-materials approaches which most readers have been exposed to. This is the *plate flexure* exemplified by the Ormond (now Force Measurement Systems) Model EPF shown in Figure 9.37a. A typical applica-tion for such a flexure would be to support the weight of a rocket engine in a thrust stand designed to measure *only* the main thrust component, along the longitudinal axis of the engine when it is fired in the horizontal direction, as

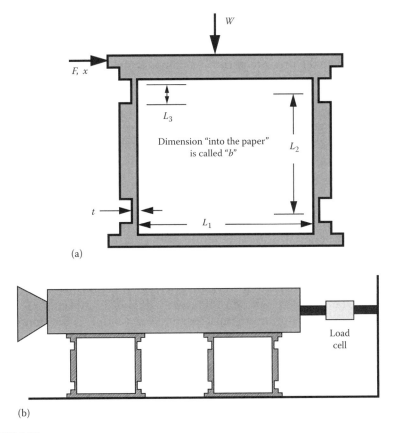

(a)

(b)

FIGURE 9.37
Plate flexure and application.

in Figure 9.37b. (For multiaxial thrust stands the universal flexures of Figure 9.23 would be preferred.) Here we use two such flexures; for a small engine, one might be sufficient. In any case, one requirement on the flexure is that it be sufficiently *soft* in the axial direction so that it absorbs only a small fraction of the thrust, with almost all of it being felt by the load cell. It must also support the total weight of the engine without being overstressed or buckling. Since *softness* in the axial direction requires the elastic "hinges" to be quite thin, and this increases the stress and buckling tendency, the design must strike a compromise between these conflicting goals.

While the flexure type shown in Figure 9.37a is available commercially from Force Measurement Systems (FMS), I was unable to find a published analysis of its behavior, useful for design purposes. I believe we can use an energy approach such as that I developed for a binocular load cell[*] to establish some relations sufficient for analysis and design. We will treat the four thin sections as elastic pin joints in a four-bar linkage with otherwise rigid links, so that the motion of the *platform* (the top link) is a pure translation with horizontal component x, while the two vertical links will rotate through an angle, pivoting about the two lower *pin joints*. We need to compute the horizontal stiffness F/x and the stresses and critical buckling loads caused by forces F and W.

The FMS plate flexures are partly "machined from solid" and partly bolted together. They usually use these flexures in assembled pairs but some of their customers buy the vertical members only, supplying the horizontal pieces themselves. A related form of flexure that is *entirely* machined from solid, and thus avoids the hysteresis effects associated with bolted or welded joints, has been analyzed and the results published.[†] Smith and Chetwynd (1994) consider such flexures mainly as motion stages for micromotion actuators (see our Chapter 5), where the vertical load W is mainly the weight of the "platform," rather than a large load such as the weight of a rocket engine. This type of flexure, called the *notch* type, is shown in Figure 9.38a. Here, the thin sections are formed by drilling accurately spaced holes and then milling away the unwanted material, or by using wire electrical discharge machining. The important dimensions for the notch type of flexure are the same as in the plate flexure, where I have labeled them (Figure 9.37a). In the notch flexure, the thin sections are defined *only* by the thickness t, which is now just the *minimum* thickness. The analyses, whose results we will shortly present, put certain requirements on this minimum thickness and the radius R of the circular profiles formed by the drilled holes. If we conform our design to these restrictions, then t is the only value to be decided by design (there is no L_3); R must simply fall in the required *range* of values.

[*] E.O. Doebelin, *Measurement Systems*, 5th edn., McGraw-Hill, New York, 2003, pp. 443–446.
[†] S.T. Smith and D.G. Chetwynd, *Foundations of Ultraprecision Mechanism Design*, Gordon and Breach, Switzerland, 1994, pp. 104–108.

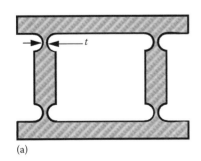
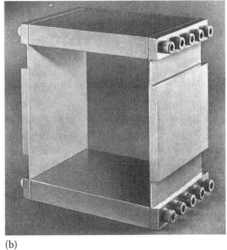

(a)   (b)

FIGURE 9.38

(a) Notch-type flexure and (b) plate flexure pair. (Courtesy of Force Measurement Systems, Fullerton, CA.)

Let us now derive the needed relations to design the plate flexure. I could not find any such analysis, so these results have not been verified by any other analyst or by experimental testing. However, the technique is the same as that used in the binocular load-cell analysis mentioned earlier, and I *did* verify that by both finite-element software methods and also experimental testing. Thus I feel that the equations developed are close enough for designing a device which will be experimentally tested before being used in any critical application. We treat the thin sections of length L_3 as if they were elastic pin joints with a resisting moment proportional to the angular deflection $\theta \approx \tan \theta = x/L_2$. We assume that all four joints have the same torsional stiffness and experience the same angular twist. When an applied force F causes a translation x, the work done is $Fx/2$, which must show up as stored elastic energy in the four pin joints. The energy stored when a moment M acts through an angle theta is $M\theta/2$, so we have $M = FL_2/4$. Knowing the applied moment at each thin section, we now treat these as cantilever beams with an end couple M, and use the result from strength of materials $\theta = (ML_3)/(EI)$, where E is the modulus of elasticity and $I = bt^3/12$. We now can easily get one of our desired results, the *stiffness* of the flexure, $F/x = 4EI/L_3L_2^2$.

Next, we need to consider the stresses tending to cause failure. Each thin section will feel a bending stress due to the moment M, a direct tension or compression stress due to the vertical load W, a horizontal shear stress due to F, and a direct tension/compression stress due to vertical loads caused by the force F. Recall that the motion x will be quite small, since it is just the elastic deflection of the load column under the action of the applied thrust force from the engine. The bending stress, which will be both compression and tension (depending on which surface you consider) is given by the usual $Mc/I = (FL_2/4)(6/bt^2) = 3FL_2/2bt^2$. The direct stress due to W depends on how many flexures support the engine. For a single *flexure pair*, assuming that the load is equally shared between the two vertical members, the stress would be $(W/2)/(bt)$, which could be either tension or compression depending on whether the engine is mounted above the flexure or hangs from it. The horizontal shear stress would be $F/(2bt)$. The horizontal force F causes a tension force in the left vertical member and an equal compression force in the right vertical member. These forces can be found by mentally making a horizontal cut through the lower hinge joints, applying at these locations the known shear forces and bending moments, and then taking moments about the lower left hinge. The vertical forces thus found have a magnitude $(F/2)(L_2/L_1)$.

We next consider buckling, which would only be a potential problem if the engine is mounted *above* the flexure and thus puts the vertical flexure members into compression. For simple situations like a single uniform-section column subject to an axial compression load, handbooks give the critical buckling loads for various types of end fixity of the column. For example, a column of length L, fixed at one end and free at the other has a critical buckling load given by $\pi^2 EI/4L^2$, which is the worst of the end conditions. For a column with pin joints at both ends, the critical buckling load is $\pi^2 EI/L^2$, while for both ends fixed the load is $4\pi^2 EI/L^2$. Note that the range of these buckling loads is 16-to-1, showing how sensitive this mode of failure is to column end conditions, which are always difficult to accurately define in practice. Unfortunately, our flexure system is *not* a simple column of uniform cross section. We have *two* columns connected by a rigid member, and the columns have regions of radically different cross sections. Smith and Chetwynd (1994) suggest a way of dealing with the buckling of our flexure, which is relatively simple and seems to be reasonable. The authors however do *not* quote any verification by experimental testing, so we should proceed with some caution here. Their reasoning goes as follows. The phenomenon of buckling in general refers to a situation where the moment of the applied load exceeds the elastic resisting moment of the structural element. If a column's axial load were always *perfectly* aligned with the center of the cross section, *buckling would never occur*; the column would fail by elastic compressive *yielding* as we gradually increased the load. In reality, such perfect alignment is never possible, so the load will always be at least a little *off center*. In our flexure, $x = 0$ is the case of perfect alignment, and when x is not equal to zero, buckling can occur. It *will* occur when the applied moment Wx exceeds the elastic restoring moment of the four hinges, which gives

$$\frac{4EI\theta}{L_3} = \frac{4EI}{L_3} \cdot \frac{x}{L_2} = W_b x \quad \text{for buckling} \tag{9.59}$$

$$W_b = \frac{4EI}{L_2L_3} \qquad (9.60)$$

For the notch type of flexure, Smith and Chetwynd (1994) offer two alternative analyses relating the angular twist of the elastic hinges to the applied bending moment. The first analysis is quoted from another reference and assumes that the thickness h of the rigid link is related to the circular radius R and the minimum thickness t by

$$\frac{h}{2R+1} \approx 1.0 \quad \text{that is, the notches are nearly semi-circular} \qquad (9.61)$$

If we design to this requirement, then the twist angle θ is related to the applied moment M by

$$\theta = \frac{9\pi R^{0.5}}{2 \cdot Ebt^{2.5}} M \qquad (9.62)$$

The second analysis is due to Smith and Chetwynd (1994) and gives

$$\theta = \frac{2KRM}{EI} = \frac{24KRM}{Ebt^3} \quad \text{with the requirement, } t < R < 5t \qquad (9.63)$$

where

$$K = 0.565\frac{t}{R} + 0.166$$

These results are based on finite-element studies. The maximum allowable moment at the notch based on a design stress σ_{max} is given by

$$M_{max} = \frac{bt^2}{6K_t}\sigma_{max} \quad \text{where stress-concentration factor } K_t = \frac{2.7t + 5.4R}{8R + t} + 0.325 \qquad (9.64)$$

Smith and Chetwynd (1994) do not state a stress equation for the assumptions of Equation 9.62, so the presumption is that one just uses Mc/I, with I computed using the minimum thickness t.

The above two approaches also lead to two alternative expressions for the stiffness, F/x. Using the method of Equation 9.62, the result is

$$\frac{F}{x} = \frac{8Ebt^{2.5}}{9\pi L_2^2 R^{0.5}} \qquad (9.65)$$

while the result for the method of Equation 9.63 is

$$\frac{F}{x} = \frac{Ebt^3}{6RL_2^2\left(0.565\dfrac{t}{R}+0.166\right)} \tag{9.66}$$

We can now also compute buckling loads for these two approaches, using the same concept as was used for Equation 9.60. For the method of Equation 9.62, the buckling load is predicted to be

$$W_b = \frac{8Ebt^{2.5}}{9\pi L_2 R^{0.5}} \tag{9.67}$$

while for the method of Equation 9.63, the result is

$$W_b = \frac{Ebt^3}{6RL_2\left(0.565\dfrac{t}{R}+0.166\right)} \tag{9.68}$$

We now want to apply some of these results to the design of a flexure for a particular application.

The design of a simple plate flexure must meet several requirements. Its stiffness must be sufficiently low (relative to the load column's stiffness) that it does not "absorb" too much of the thrust force. It must not fail by yielding under all its applied loads. It also must not buckle under the applied loads. For our design exercise, let us assume that the load column stiffness is 200,000 lb$_f$/in., that the weight of the rocket motor to be supported is 2,000 lb$_f$, and that the maximum rocket thrust is 10,000 lb$_f$. We require that the flexure absorb no more than 1% of the thrust force, so that 99% is felt by the load cell. (This does *not* mean that we accept a 1% error; calibration takes the flexure stiffness into account.) The frame to which the rocket is bolted is 10 ft long and 2 ft wide (the 2000 lb$_f$ weight quoted above includes this frame). Suppose the flexures must be mounted *below* the frame, thus they are subject to possible buckling. Force Measurement Systems usually uses SAE 4340 steel for flexures, but occasionally employs 6061T6 aluminum, or stainless steel.

One possible arrangement is to consider the vertical members in Figure 9.37a as two *separate pieces*, which would be bolted to the rocket support frame at each end of its 10 ft length, and which could be purchased from suppliers such as FMS, or made in-house. This configuration would be statically *determinate*; that is we can solve for the individual forces using only the equations of statics. Pairs of flexures, bolted together as in Figure 9.37a are also available and could be used as in Figure 9.37b, but now this structure is *statically indeterminate* or *redundant*, and calculating the forces in the members will require application of both statics and *elasticity* relations involving not just the flexures but the frame to which they are mounted. Such calculations are more complicated and less accurate than those for statically determinate structures. With accurate machining and assembly, however, the use of the paired flexures (Figure 9.38b) as *units*, bolted to the support frame at several locations (such as the four corners) can be successful. The indeterminacy of the forces can be slight and will be taken into account in the calibration process, allowing high accuracy force measurements. Plate flexures can also be used *singly*, one each at the four corners of the support frame, as shown in Figure 9.39 from Force Measurement Systems.* Since "3 *points determine a plane*," such arrangements are also statically indeterminate, but feasible in practice, as mentioned above. We will use this arrangement in our example, and analyze it as if it were statically determinate, using the symmetry to calculate the loads in the members.

Assuming that the dead weight of the engine and support frame acts at the geometric center of the 2 ft by 10 ft support frame, and that this weight is equally shared between the two "rear" flexures and the two "front" flexures, we can base our analysis on, say, just the two front flexures, according to Figure 9.40, which we use as our analysis sketch. The full-scale thrust force of 10,000 lb will cause a load-column horizontal deflection of 0.0495 in., deflecting the flexures from their vertical position. Plate flexures are limited in how much angular motion they can withstand,

FIGURE 9.39
Use of plate flexures in single-axis thrust stand. (Courtesy of Force Measurement System, Fullerton, CA.)

---

* Fullerton, CA, www.forcems.com

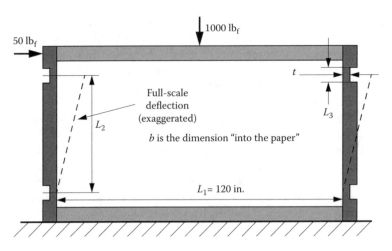

FIGURE 9.40
Front-half of flexure support system.

since the vertical load creates a larger bending moment as the deflection angle increases, tending toward buckling. Force Measurement Systems reduces the vertical load capacity in proportion to the angular deflection, with full rated capacity being available up to 0.75°, and reducing to 75% of allowable full-scale load at an angular deflection at 1.0°. If we limit full-scale angular deflection to 0.75°, we can calculate the minimum value of dimension L_2 as $0.0495/\tan(0.75) = 3.78$ in. If we wish for some reason to use larger L_2's, that is acceptable since it will make the angular deflection less. Assuming equal load sharing, the two front flexures, due to the dead weight load, will each be subject to vertical loads of 500 lb_f, creating compressive stress over the area bt in the thin sections. Additional vertical forces are caused by the horizontal force of 50 lb on the front half of the structure. We earlier stated that this horizontal force caused vertical forces of $FL_2/2L_1 = 25L_2/120 = 0.208L_2$ in the two vertical members, tension in the left and compression in the right. The right member will thus have the largest compressive force, which is $500 + 0.208L_2$. We must choose b and t so as to resist the stress caused by this force, but these two dimensions also influence the bending stiffness of the flexure, which must not be too great, since we allow the horizontal force due to this deflection to be no more than $(10,000)(0.01)/2 = 50\,lb_f$. The maximum allowable stiffness is then $50/0.0495 = 1010$ for the front pair of flexures. We showed earlier that this stiffness is given by $4EI/(L_3L_2^2)$, where $I = bt^3/12$. There are enough adjustable parameters that there will be many acceptable combinations of dimensions that will satisfy all our criteria. Tentatively choosing $L_3 = 0.50$ in., $L_2 = 6.0$ in., $b = 4.0$ in., and exploring a range of t values, a Mathcad program produces the following results.

This program aids in the design of the plate flexures in the last part of this chapter.

Let us call it crc_book_chap9_flexures. First list the parameters.

$$\text{ORIGIN} := 1$$

$$L1 := 120 \qquad L3 := 0.50 \qquad L2 := 6.0 \qquad b := 4.0 \qquad t := 0.01,\ 0.02 \ldots 0.10$$

$$E := 29.7 \cdot 10^{6} \qquad I(t) := \frac{b \cdot t^{3}}{12}$$

$$Ks(t) := \frac{4 \cdot E \cdot I(t)}{L3 \cdot L2^{2}}$$

Ks(t) =

	1
1	2.2
2	17.6
3	59.4
4	140.8
5	275
6	475.2
7	754.6
8	$1.1261 \cdot 10^{3}$
9	$1.604 \cdot 10^{3}$
10	$2.2 \cdot 10^{3}$

I(t) =

	1
1	$3.333 \cdot 10^{-7}$
2	$2.667 \cdot 10^{-6}$
3	$9 \cdot 10^{-6}$
4	$2.133 \cdot 10^{-5}$
5	$4.167 \cdot 10^{-5}$
6	$7.2 \cdot 10^{-5}$
7	$1.143 \cdot 10^{-4}$
8	$1.707 \cdot 10^{-4}$
9	$2.43 \cdot 10^{-4}$
10	$3.333 \cdot 10^{-4}$

t =

	1
1	0.01
2	0.02
3	0.03
4	0.04
5	0.05
6	0.06
7	0.07
8	0.08
9	0.09
10	0.1

Now check for compressive stress in the right member.

$$Sc(t) := \frac{500 + 0.208 \cdot L2}{b \cdot t}$$

Now check for bending stress in each hinge.

$$M := \frac{50 \cdot L2}{4} \qquad Sb(t) := \frac{M \cdot t}{2 \cdot I(t)}$$

Now check for buckling.

$$Wb(t) := \frac{4 \cdot E \cdot I(t)}{L2 \cdot L3}$$

t =			Sb(t) =			Sc(t) =			Wb(t) =		
			1				1				1
0.01		1	$1.125 \cdot 10^6$		1	$1.253 \cdot 10^4$		1	13.2		
0.02		2	$2.812 \cdot 10^5$		2	$6.266 \cdot 10^3$		2	105.6		
0.03		3	$1.25 \cdot 10^5$		3	$4.177 \cdot 10^3$		3	356.4		
0.04		4	$7.031 \cdot 10^4$		4	$3.133 \cdot 10^3$		4	844.8		
0.05		5	$4.5 \cdot 10^4$		5	$2.506 \cdot 10^3$		5	$1.65 \cdot 010^3$		
0.06		6	$3.125 \cdot 10^4$		6	$2.089 \cdot 10^3$		6	$2.851 \cdot 10^3$		
0.07		7	$2.296 \cdot 10^4$		7	$1.79 \cdot 10^3$		7	$4.528 \cdot 10^3$		
0.08		8	$1.758 \cdot 10^4$		8	$1.566 \cdot 10^3$		8	$6.758 \cdot 10^3$		
0.09		9	$1.389 \cdot 10^4$		9	$1.392 \cdot 10^3$		9	$9.623 \cdot 10^3$		
0.1		10	$1.125 \cdot 10^4$		10	$1.253 \cdot 10^3$		10	$1.32 \cdot 10^4$		

We see that a thickness t of about 0.060 in. gives an acceptable value of stiffness K_s and buckling load W_b, with a maximum stress of $31250 + 2089 = 33339$ psi. (The direct shear stress is only 50/2bt psi, which is negligible.) Using AISI 4340 steel, oil quenched at 800°C and tempered at 540°C, the yield strength is about 169,000 psi,[*] so we have a large safety factor for yielding failure. The reference does not quote a value for *fatigue limit*, but for steel a good estimate is about half of the ultimate tensile strength, which is given as 182,000 psi. Without specific application data available, it is difficult to evaluate the design for fatigue failure, since one needs to estimate the amplitude of cyclic stress and the number of cycles anticipated. However, a fatigue limit of about 91,000 psi indicates that fatigue failure is unlikely.

While the plate type of flexure is more common for thrust stand applications, we will now look at the notch type, mainly to see the comparison. As in the plate type, we again use *bolted* construction rather than the *monolithic* type of Figure 9.37, and four *single* flexures, one at each corner. As shown earlier, we have available two alternative analysis/design approaches, and will look at both, using the same values for L_2 and b that we used for the plate flexures. The dimension L_3 does not apply here, but we do need to specify the *minimum thickness t*, which we again allow to vary over the range 0.01–0.10 in. The notch radius R needs to conform to the requirements of Equations 9.61 and 9.63, respectively, for the two calculation methods. From Equation 9.63, we might choose $R = 3t$ as a compromise; this choice also can satisfy Equation 9.61 since it is not hard to make h approximately equal to $7t$, and h does not appear in any of the formulas. Again a Mathcad program eases the calculations.

Let us now design the notch-type flexures, using both the available approaches.

---

[*] www.matweb.com

$R(t) := 3 \cdot t.$ Now compute the stiffness for the two methods.

$$Ks1(t) := \frac{8 \cdot E \cdot b \cdot t^{2.5}}{9 \cdot \pi \cdot L2^2 \cdot R(t)^{0.5}}$$

$$Ks2(t) := \frac{E \cdot b \cdot t^3}{6 \cdot R(t) \cdot L2^2 \cdot \left(0.565 \dfrac{t}{R(t)} + 0.166\right)}$$

t =	Ks (t) =	Ks2 (t) =	Ks1 (t) =
0.01	2.2	51.74	53.908
0.02	17.6	206.961	215.631
0.03	59.4	465.663	485.169
0.04	140.8	827.846	862.523
0.05	275	$1.294 \cdot 10^3$	$1.348 \cdot 10^3$
0.06	475.2	$1.863 \cdot 10^3$	$1.941 \cdot 10^3$
0.07	754.6	$2.535 \cdot 10^3$	$2.641 \cdot 10^3$
0.08	$1.126 \cdot 10^3$	$3.311 \cdot 10^3$	$3.45 \cdot 10^3$
0.09	$1.604 \cdot 10^3$	$4.191 \cdot 10^3$	$4.367 \cdot 10^3$
0.1	$2.2 \cdot 10^3$	$5.174 \cdot 10^3$	$5.391 \cdot 10^3$

Now compute the stress for the two methods.

$$K(t) := \frac{0.565t}{R(t)} + 0.166$$

$$Kt(t) := \frac{2.7 \cdot t + 5.4 \cdot R(t)}{8 \cdot R(t) + t} + 0.325$$

$$M1(t) := \frac{2 \cdot E \cdot b \cdot t^{2.5} \cdot 0.0131}{9 \cdot \pi \cdot R(t)^{0.5}}$$

$$M2(t) := \frac{E \cdot b \cdot t^3 \cdot 0.0131}{24 \cdot K(t) \cdot R(t)}$$

$$s1(t) := \frac{M1(t) \cdot t}{2 \cdot I(t)}$$

$$s2(t) := \frac{M2(t) \cdot 6 \cdot Kt(t)}{b \cdot t^2}$$

s1 (t) =	s2 (t) =	M1 (t) =	M2 (t) =
$9.534 \cdot 10^4$	$9.891 \cdot 10^4$	6.356	6.1
$9.534 \cdot 10^4$	$9.891 \cdot 10^4$	25.423	24.401
$9.534 \cdot 10^4$	$9.891 \cdot 10^4$	57.201	54.902
$9.534 \cdot 10^4$	$9.891 \cdot 10^4$	101.692	97.603
$9.534 \cdot 10^4$	$9.891 \cdot 10^4$	158.893	152.505
$9.534 \cdot 10^4$	$9.891 \cdot 10^4$	228.806	219.607
$9.534 \cdot 10^4$	$9.891 \cdot 10^4$	311.43	298.909
$9.534 \cdot 10^4$	$9.891 \cdot 10^4$	406.766	390.412
$9.534 \cdot 10^4$	$9.891 \cdot 10^4$	514.813	494.115
$9.534 \cdot 10^4$	$9.891 \cdot 10^4$	635.572	610.019

We see that the stiffness of the notch flexures is considerably higher than for the plate design, and that the two calculation methods agree pretty well. The buckling loads also follow this same pattern. The stress results at first appear peculiar, since, while the two methods nearly agree, *the stress seems to be independent of the t value!* A little study shows that this is indeed true and is a consequence of relating R to t with the requirement $R = 3t$. This causes t to cancel out in the stress calculations. We could choose R and t by some independent and arbitrary rule, but making R proportional to t is *not* a foolish or unrealistic choice, so there is probably no good reason to scrap such a requirement, even though this could make stress vary with t.

This completes our study of thrust stand design and analysis. Our goal was to expose some of the basic tools that one can employ in such a project, especially the treatment of dynamic effects pertinent to accurate measurement of unsteady forces and moments. Many practical details were of necessity not explored. The text by Runyan et al. (1988) (AFAL-TR-88-060) explains some of these, mainly relating to steady-state measurements. This is a federal government publication and thus is not constrained by proprietary considerations. Many thrust stands are designed and constructed by private companies, such as Force Measurement Systems, for use by government labs or private engine companies. Designers at these companies are rightly concerned about revealing *trade secrets* to potential competitors, so it is rare to find them publishing articles or books that explain critical details. Hopefully, readers of this chapter will be better prepared to interact with these designers if they ever have to commission the construction of a thrust stand or participate in the operation of an existing facility. For the simpler applications, the material of this chapter might be useful for the *in-house* design of a stand, assuming that it is judiciously augmented by contact with professionals in the field and the usual *learning curve* of experimental development. Engineering society groups such as the Ground Testing Technical Committee of AIAA* can be a useful source of information for this area.

---

* http://www.aiaa.org/tc/gt/gttchome.html

10

Shock Calibrator for Accelerometers

10.1 Introduction

Calibration is an important part of any measurement application. Accelerometers are the most used sensors for all kinds of shock and vibration measurements. Accelerometer calibrations may be sinusoidal tests to determine the frequency response, or shock tests to determine the suitability for shock measurements. PCB Piezotronics* has developed and offered for sale a convenient pneumatic shock calibrator Model 9155C-525. This device makes an interesting study in system dynamics because it includes thermodynamics, fluid mechanics, and mechanical motion aspects. It uses the *back-to-back* method of comparing the response of the accelerometer to be calibrated with that of a standard (reference) accelerometer known to be much more accurate and thus usable as a *transfer standard*. In this method, the two accelerometers are fastened to an "anvil" in a back-to-back fashion. The anvil is subjected to an impact from a "projectile" launched from a pneumatically driven "gun." If the assembly of anvil and two accelerometers behaves essentially as a single rigid mass, then the two accelerometers will feel the same acceleration pulse and we can easily compare the two dynamic response voltages from the instruments, taking the reference accelerometer signal as *perfect*. By adjusting the parameters of the calibrator, shock pulses of various magnitudes and time durations may be selected, thus exercising the tested accelerometer over a range of conditions. Most purchasers and users of this calibrator have little interest in the design or analysis of the device; they just *use* it according to the instrument's instruction manual. Since a more complete understanding of the device develops our capability for the design of instrumentation systems in general, we will pursue a detailed analysis and simulation. Although PCB could not, for proprietary reasons, supply such an analysis, they were very helpful in providing certain details.

---

* www.pcb.com, www.modalshop.com

10.2 Description of the Calibrator

A simplified sketch of the apparatus is shown in Figure 10.1. Two projectiles are provided: one metal and the other plastic. We shall see later that the masses of the projectile and the adapter assembly, together with the effective spring constant of the bumper, largely determine the duration of the acceleration pulse. The provision of the two projectiles with different masses allows the needed adjustability of pulse duration. Further adjustment is provided by having two anvils, one steel and one aluminum, again affecting the masses of the impacting bodies. The solenoid valve is driven by an electronic timer circuit so that one can set the time that the valve is open, thus controlling the time interval during which the projectile is driven by the air pressure. The air supply has a regulator so that the pressure to the valve can be adjusted

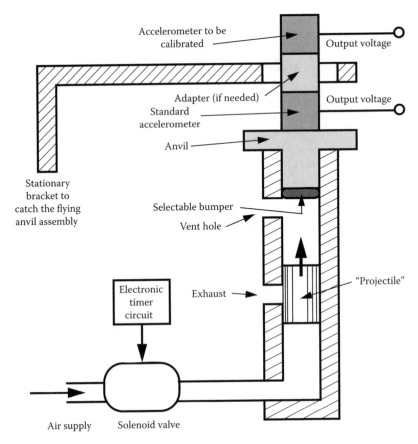

FIGURE 10.1
Schematic diagram of accelerometer shock calibrator.

over a range, up to about 80 psig. These two adjustments together allow one to achieve a useful range of projectile impact velocities. A vent hole is used to prevent the projectile from compressing the air above it and thus slowing the velocity. While the solenoid valve is a *three-way* type (when the valve is *off* the pressure is disconnected *and* the port is open to atmosphere), an exhaust port in the gun barrel dumps the air when the projectile passes above it, providing a more rapid decrease of pressure.

The flying projectile impacts a selectable *bumper* cushion, causing a pulse of upward acceleration, which is felt equally by the two accelerometers (see Figure 10.2). Figure 10.3 shows a typical display from the software associated with the calibrator system. Cushions may be as simple as a piece of paper, tape, or felt, or more "engineered," such as suitably shaped elastomer materials. Just as in instrumented hammers for impact testing, and drop-type shock machines, cushions of some type *must* be used. If we allow two flat metal surfaces to impact, the resulting shock becomes extremely high and its waveform becomes uncontrollable. The selection of cushion material and shape allows some control of the acceleration pulse shape, but it usually approximates a half-sine wave. The apparatus *does* allow good control of the pulse duration and peak value, which are the major goals of the calibrator. Since the anvil assembly slides freely in the *gun barrel* and may reach significant velocities after the impact, a stationary bracket is provided to catch it. The impact of the anvil assembly with the bracket occurs much after the

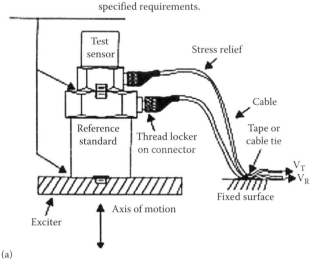

(a)

FIGURE 10.2
(a) Mounting instructions for PCB 301A12 standard accelerometer.

(*continued*)

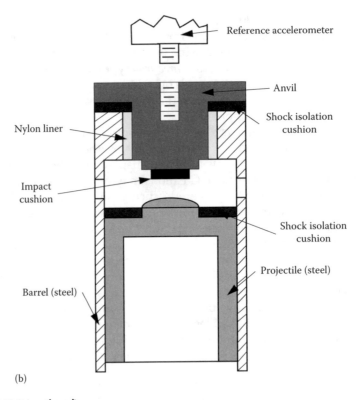

FIGURE 10.2 (continued)
(b) Details of projectile/anvil/barrel assembly.

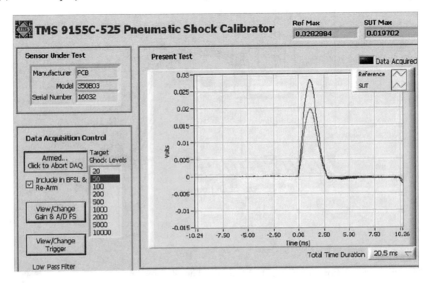

FIGURE 10.3
Sample display from PCB calibrator system.

impact of the projectile with the cushion, so it is easily ignored by proper triggering of the data acquisition system or oscilloscope. While a suitable reference accelerometer from any manufacturer could be used in this calibrator, PCB recommends its own, the Model 301A12. This accelerometer is specially designed for use as a reference accelerometer, and has a tapped hole on the top and an integral threaded stud on the bottom. Thus, the two accelerometers in Figure 10.1 can be *directly* screwed together without the need for an adapter. The calibrator is useful over the range 10*g*–10,000*g*, and pulse durations from 0.1 to 5.0 ms.

10.3 Review of Basic Impact Calculations

The analysis of impacts* is one of the most complex problems in mechanics, but useful approximations are available for preliminary design purposes such as ours. A major difficulty is the short duration of the impact phenomena and the deviation of material properties from the familiar *steady-state* values regularly used in conventional machine design. A common approach, which you may have encountered in undergraduate physics and mechanics courses, ignores the details of the time-varying contact forces and instead focuses on the relation between conditions before and after the impact.† Starting with Newton's law for a rigid body in uniaxial translation, we can derive impulse/momentum relations and a conservation of momentum equation:

$$F = M \cdot \frac{d^2x}{dt^2} = M \cdot \frac{dv}{dt} \qquad \int_{t_1}^{t_2} F dt = M \cdot (v_2 - v_1) \qquad (10.1)$$

When two bodies collide, the impact force acting on each must be the same in magnitude but opposite in direction. The time integral of a force is called its *impulse*. The *momentum* of a mass is defined as the product of mass and velocity. For the two colliding bodies, the impulse acting on each must be the same in magnitude but of opposite sign; thus the total impulse for the system is zero and the change of momentum of the system is zero, that is, the *momentum is conserved. Without the need to consider the detailed nature of the time-varying impact force, we can make useful statements about conditions before and after the collision.* Because we ignore the detailed time variation of forces and motions during the impact, our *before and after* conditions are separated by an indeterminate length of time. While we *lose*

---

* W. Goldsmith, *Impact: The Theory and Physical Behavior of Colliding Solids*, Dover, Mineola, NY, 2001 (original publication was in 1960).
† G.W. Housner and D.E. Hudson, *Applied Mechanics: Dynamics*, Van Nostrand, New York, 1960, pp. 45–48, 79–81.

the information about the impact force time history (which prevents us from calculating things like maximum stresses, etc.), the ability to easily estimate conditions such as velocities of masses is often quite useful.

In addition to using the conservation of momentum relation, we can also apply conservation of *energy* to the before and after conditions. Here we can allow for a *loss* of some energy during the impact, which leads to definition of a useful concept, the *coefficient of restitution*. Note that the conservation of momentum relation holds whether there is an energy loss or not! The coefficient of restitution, C_r, is defined as the ratio of the relative velocities of the two colliding masses before and after the collision:

$$C_r \underset{=}{\Delta} - \frac{(V_{2f} - V_{1f})}{(V_{2i} - V_{1i})} \tag{10.2}$$

where
   The subscript i denotes the initial velocities (before impact)
   The subscript f denotes the final velocities (after impact)

We first want to show that the *relative* velocities before and after a *perfectly elastic* impact are the same magnitude but of opposite signs. Note that this corresponds to a coefficient of restitution equal to +1.0. We can then combine this result with the conservation of momentum to get formulas for each of the final velocities (after impact), assuming the initial velocities and both masses are known. For a perfectly elastic impact (no energy losses), the conservation of momentum and the conservation of energy both apply:

$$\frac{M_1 V_{1i}^2}{2} + \frac{M_2 V_{2i}^2}{2} = \frac{M_1 V_{1f}^2}{2} + \frac{M_2 V_{2f}^2}{2} \tag{10.3}$$

$$\frac{M_1 V_{1i}}{2} + \frac{M_2 V_{2i}}{2} = \frac{M_1 V_{1f}}{2} + \frac{M_2 V_{2f}}{2} \tag{10.4}$$

Solving for M_2/M_1 in each equation and then equating these leads to

$$\frac{M_2}{M_1} = \frac{V_{1f} - V_{1i}}{V_{2i} - V_{2f}} = \frac{(V_{1f} + V_{1i}) \cdot (V_{1f} - V_{1i})}{(V_{2i} + V_{2f}) \cdot (V_{2i} - V_{2f})} \tag{10.5}$$

$$(V_{1i} - V_{2i}) = -(V_{1f} - V_{2f}) \tag{10.6}$$

For a perfectly elastic impact, knowing the masses and the initial velocities, we can find the velocities after impact. If we also have a numerical value for the coefficient of restitution, this can be done for inelastic impacts:

$$M_1 V_{1i} + M_2 V_{2i} = M_1 V_{1f} + M_2 V_{2f} \tag{10.7}$$

$$(M_1) \cdot V_{1f} + (M_2) \cdot V_{2f} = M_1 V_{1i} + M_2 V_{2i} \tag{10.8}$$

$$(1.0) \cdot V_{1f} + (-1.0) \cdot V_{2f} = C_r V_{2i} - C_r V_{1i} \tag{10.9}$$

Equations 10.8 and 10.9 are two linear equations in two unknowns and can be solved using the determinant method (Cramer's rule) for the two final velocities.

$$V_{1f} = \frac{M_2 V_{2i} \cdot (C_r + 1) + (M_1 - C_r M_2) \cdot V_{1i}}{M_1 + M_2}$$

$$V_{2f} = \frac{M_1 V_{1i} \cdot (C_r + 1) + (M_2 - C_r M1) \cdot V_{2i}}{M_1 + M_2} \tag{10.10}$$

Setting $C_r = 1.0$ gives the results for the perfectly elastic impact.

The coefficient of restitution cannot be theoretically computed from dimensions and basic physical properties; it is usually found by a simple *drop test* experiment (Figure 10.4). Here one of the two colliding bodies is placed on a rigid "foundation" and the other body is dropped on it from a known height. The initial and final velocities of the second body are assumed to be zero. Using the definition of C_r and the known behavior of a *free fall* (no air friction) we can show that a numerical value for the coefficient of restitution can be found from two simple measurements. The falling mass has, at impact, kinetic energy $M_1 V_{1i}^2 / 2 = 2gh_1$, while the second mass is motionless at all times. At the top of the rebound, mass one has potential energy $2gh_2$, which means its rebound kinetic energy has to be this same value. Since V_{2i} and V_{2f} are both zero in Equation 10.2, then $V_{1f}/V_{1i} = C_r = (h_2/h_1)^{0.5}$. While the apparent simplicity of the coefficient of restitution is appealing, its application to practical problems is fraught with confusing subtleties. While the drop *test* method of finding a numerical value is simple and does come up with a definite number, that number is really useful *only* for the materials, body shapes, drop height, and environmental conditions (temperature, humidity, etc.) actually used. For example, if we use different drop

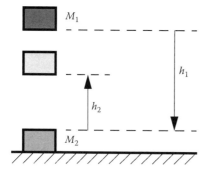

FIGURE 10.4
Coefficient of restitution drop test.

heights, which gives different impact velocities, the coefficient of restitution will very likely be different, perhaps *very* different. Thus doing calculations for different impact velocities using *the same* value for C_r may not be reliable. If you try to find *tables* of C_r numerical values you may have difficulty. The PCB engineers who designed the calibrator were not able to supply values; they did not use this concept in their design. In fact, the design was largely based on past experience and extensive *experimental* development. With all these caveats, one might question the real utility of the concept, but we should remember that for a *given* impact situation, a numerical value for C_r does exist and can be usefully employed in calculations. We just need to realize that these calculations are subject to some uncertainty, to the extent that C_r is uncertain.

Our understanding of the calibrator is enhanced by some further simple calculations. The initial velocity of the anvil assembly is zero and the initial velocity of the projectile is its velocity just before impact with the anvil. Let us first assume a perfectly elastic impact ($C_r = 1.0$). Then Equations 10.10 and 10.11 predict the final velocities of both objects:

$$V_{1f} = V_{1i}\left(\frac{M_1 - M_2}{M_1 + M_2}\right) \qquad V_{2f} = V_{1i}\left(\frac{2M_1}{M_1 + M_2}\right) \qquad (10.11)$$

Note that if the two masses were equal, the final velocity of the projectile would be zero and the final velocity of the anvil would be the same as the initial velocity of the projectile.

10.4 Simulation of the Coefficient of Restitution Drop-Test Experiment

We next want to do a simulation of the drop test often used to get numerical values for the coefficient of restitution for two impacting bodies. To get values of C_r that are less than 1.0, the "bumper" model must include some kind of energy loss effect. Since the actual bumpers are merely "lumps" of rubber, plastic, felt, or some other material, representing their impact behavior is not a simple task. If we decide to limit ourselves to the ordinary-differential-equation models that software such as Simulink® can handle, then we need to make up our bumper models from assemblages of springs, dampers, and masses. We can of course use either linear or nonlinear springs and dampers if we wish. We require that whatever model we adopt, it should be capable of providing C_r values in the range 0–1.0. Also, the shape of the acceleration pulse that is produced at impact should approximate a half-sine wave, since that is the shape that is observed in

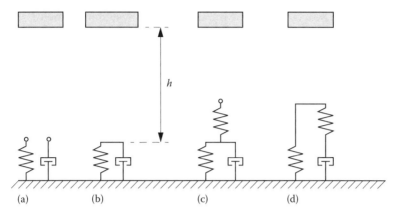

FIGURE 10.5
Some possible models for "lossy" bumpers.

actual experiments. A simple model which first comes to mind is that shown in Figure 10.5a, with a slightly different version in Figure 10.5b. In Figure 10.5a the spring and damper are not tied together, and when the falling mass rebounds, the damper cannot provide a tension force, only a compression force. In Figure 10.5b, as the mass rebounds and the spring expands, the damper is able to exert a *drag* force on the spring, which reduces the net upward force on the mass. In both cases, there can never be a net downward force on the mass due to the nature of the contact between the bumper and the mass.

A little thought shows that both these models might have some problems. For both models, since the falling mass strikes the damper and spring with a considerable velocity V_{1i}, the damper force will *instantly* rise to a value given by $V_{1i}B$. Such an instant rise in force (and thus in acceleration of the mass) is *not* what is seen in actual experiments, where the acceleration pulse is often close to a half-sine wave. This is a valid concern, but a simulation will show that the model might be acceptable if the damper is not too strong. That is, the force will still rise instantly, but if it is small relative to the spring force, the acceleration pulse may still have an acceptable shape. The models of Figure 10.5c and d do not appear to suffer from this defect since the falling mass first encounters a *spring*, whose force builds up *gradually*. Of course, the numerical values of the damper and two spring constants must be properly chosen. We can show by physical reasoning (no detailed analysis is required) that in models of Figure 10.5c and d, both very large and very small B values result in small system damping effects (energy loss), indicating that there exists a maximum energy loss (and thus a minimum value of C_r) for some intermediate B value. This means that arbitrarily low values of C_r cannot be achieved with these two configurations. The reasoning goes as follows. In Figure 10.5c, if B is made very large (stiff), then neither B nor the lower spring is able to move much during the impact, and the system becomes essentially just the upper spring, which is a *totally undamped* system. Similarly in Figure 10.5d, a stiff *B* means that the lower end of the right spring is essentially *fixed*, which gives us a system with two springs *in parallel*, also an undamped system.

To do a simulation, we of course need to have numerical values for the system parameters. Let us take

$$h = 18.0 \text{ in.}$$

$$\text{spring constant} = 15{,}000 \, \text{lb}_\text{f}/\text{in.}$$

$$\text{mass} = 0.00183 \, \text{lb}_\text{f}\text{s}^2/\text{in. (mass weighs } 0.706 \, \text{lb}_\text{f})$$

$$\text{damping } B = \text{try } 1.0 \text{ and } 5.0 \, \text{lb}_\text{f}/(\text{in./s})$$

The system of Figure 10.5a gives the simulation diagram of Figure 10.6. A *relay* module and two *lookup table* modules are used to "turn off" the spring and damper forces when the mass leaves contact. The relay is set so that its output is zero unless its input is greater than zero; then its output switches to 10.0. Figure 10.7 shows the results for *B* = 1.0; the coefficient of restitution appears to be $(13.65/18.0)^{0.5} = 0.87$. The acceleration graph had to be zoomed since, on the time scale of Figure 10.7a, it shows up as a "spike." It is also clear that we used a very small computing increment (1e−5 s) to deal with the very fast changing acceleration. This computing increment is much smaller than is needed during the *free-flight* portions of the response, so computing time might be saved by using a variable-step integrator, but I didn't do that since the total computing time was quite short. Note that the overall shape of the acceleration response is nearly a half-sine wave, but the first part does show the initial sharp change due to the damping force. The peak acceleration is about 3e5/386 = 777 *g*. I "connected" an accelerometer (not shown in the simulation diagram) modeled as a second-order instrument with a natural frequency of 10,000 Hz and an optimal damping of 0.65 to the xdot2 signal in Figure 10.6, with the results shown in Figure 10.8. We see that these accelerometer dynamics tend to *smooth* the response, making it look more like the desired half-sine wave. Piezoelectric accelerometers, often used for shock measurements, have very poor damping (less than 0.05), but their output is usually passed through a low-pass filter, which gives a similar effect, so the nice *half-sine* pulses often observed are partly due to the instrumentation, rather than being the actual acceleration.

Changing the damping value to *B* = 5.0 gives the results of Figure 10.9. The coefficient of restitution is now 0.567, but the pulse shape is now so distorted that the accelerometer dynamics do not smooth it enough to give the

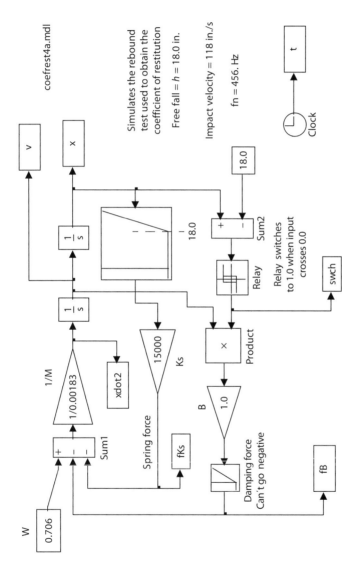

coefrest4a.mdl

Simulates the rebound
test used to obtain the
coefficient of restitution

Free fall = h = 18.0 in.

Impact velocity = 118 in./s

fn = 456. Hz

FIGURE 10.6
Simulation of system of Figure 10.5a.

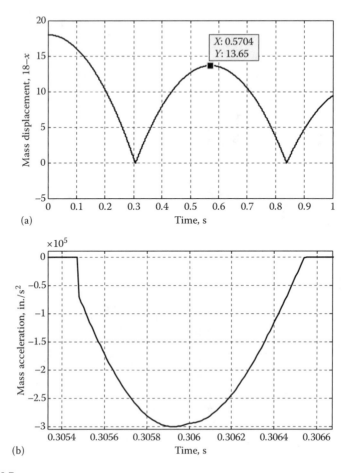

FIGURE 10.7
Displacement and acceleration for system of Figure 10.5a, $B = 1.0$.

shape usually seen in practical testing. Thus the model of Figure 10.5a is limited to larger values of C_r. Going now to the system of Figure 10.5b, we get the upper part of Figure 10.10 as the simulation diagram (note that I now show the accelerometer mentioned earlier). We find that the results for this configuration, while slightly different from those of Figures 10.7 and 10.9, suffer from the same limitation; large B values give acceleration pulse shapes that differ radically from the desired half-sine shape. Moving on to the configuration of Figure 10.5c, we get the simulation diagram seen in the lower part of Figure 10.10. Here we have taken both spring constants to be 30,000 lb$_f$/in.; this was done to make the deflection of the impact point for a static applied force the same as for the previous two models. Note at the lower right an alternative way to simulate this model; it is

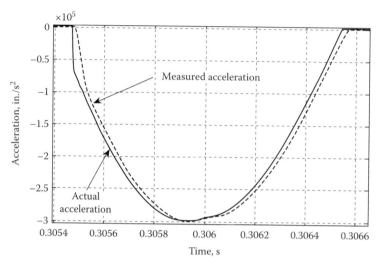

FIGURE 10.8
Accelerometer dynamics makes pulse look more sinusoidal.

more compact but does not show as much detail as the other method. The results are of course identical for each. We find (see Figure 10.11a) that the smallest possible C_r value is found in the range $B = 10$–15; larger or smaller values give *higher* coefficients of restitution. The smallest possible value is $(8.364/18)^{0.5} = 0.68$. Fortunately, $B = 15$ also gives the very nice acceleration pulse shape of Figure 10.11b.

The final configuration, that of Figure 10.5d, leads to the results (simulation diagram not shown) of Figure 10.12. For this system, the two spring constants must each be taken as 15,000, to give the same static deflection. We see that the smallest possible C_r is $(11.45/18)^{0.5} = 0.80$, corresponding to $B = 3$. The acceleration pulse shape for this B value is acceptable, but not as "perfect" as that of Figure 10.11b, which also has a better C_r value. While none of our proposed models is able to provide coefficient of restitution values spanning the entire desired range (0–1.0), the model of Figure 10.5c is the best of the four. When I first started studying this aspect of the problem, I did not expect this much trouble in making up simple models of the calibrator *cushions* that would provide *both* low values of the restitution coefficient and also give the half-sine wave pulse waveform that is often desired in testing. I have tried to find discussion of this problem in the literature but without any real success. I am wondering whether the "nice" waveforms actually measured could, at least in some cases, be explained by the inability (shown earlier in Figure 10.8) of real accelerometers to accurately follow steep wavefronts that we find in some of our simple cushion models. That is, the accelerometers are basically second-order dynamic systems with finite natural

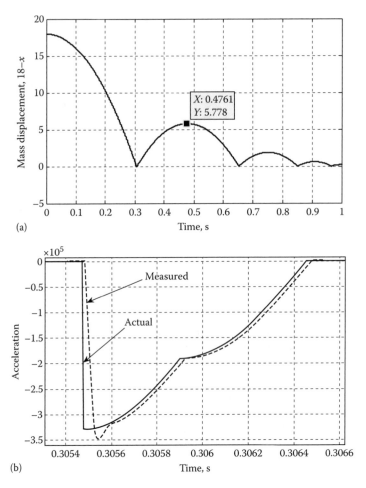

FIGURE 10.9
Large *B* value gives a distorted pulse.

frequencies, so their response to wavefronts with very steep rises cannot be perfect. Maybe cushion models that produce these steep wavefronts are not really wrong; the accelerometers actually used in experiments may not be capable of accurately following them, leading the experimenters to wrongly think that the apparatus is really producing the nice half-sine wave shape. Some simple calculations may give a little insight into this possibility. We might take the PCB reference accelerometer used with their calibrator as typical and check how it responds to very steep waveforms. This accelerometer is quoted as having a natural frequency "greater than" 30 kHz. Let us take the natural frequency as 35 kHz, assuming that if it were "greater than" 40 kHz, PCB would have said so. Since we are dealing with fast inputs, the

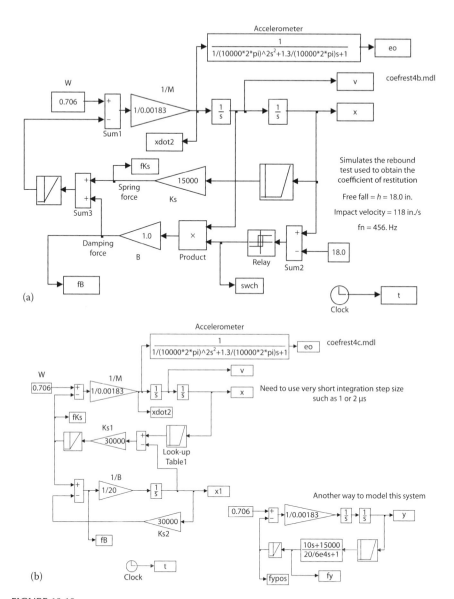

FIGURE 10.10
Simulations for systems of Figure 10.5b and c.

$\tau D/(\tau D + 1)$ *electrical* part of the transfer function can be neglected, leaving only the mechanical part, which is a simple second-order system, for which we can take the damping ratio to be zero. The *fastest* input that one can apply in theory is a perfect impulse, so we might use the impulse response of our second-order system as a benchmark for evaluating the accelerometer

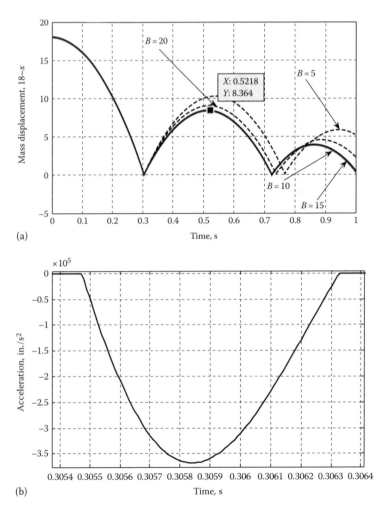

FIGURE 10.11
Response of system of Figure 10.5c.

response to fast-rising waveforms. This impulse response is just a perfect sine wave at the natural frequency, so we see already that the accelerometer output cannot rise faster than this sine wave. For 35 kHz the time to the first peak of the sine wave is 1/4 of the period; about 7.1 μs. The response to a rectangular input pulse is shown in Figure 10.13 where we again see a *rounding off* effect. We should also note that low-pass filters are regularly used in practical testing; these further contribute to the rounding off of steep wavefronts.

Since the *perfect* half-sine acceleration pulse corresponds to a perfectly linear-elastic bumper (pure and ideal spring element), any *real* bumper

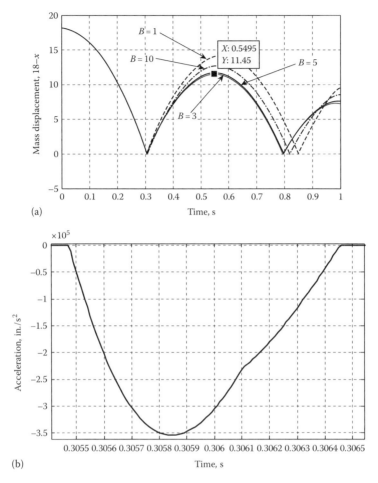

FIGURE 10.12
Response of system of Figure 10.5d.

must produce a pulse which *differs* in some way from that shape, to the extent that the bumper material is not perfectly elastic and linear. A *pure* but *nonlinear* spring element will produce a pulse shape different from the half-sine, but does *not* provide any energy loss and thus its C_r value would be 1.0. To produce energy loss there must be a component of bumper force that *opposes the mass's velocity,* since then we withdraw energy from the system at an instantaneous rate equal to the product of that force and the mass velocity. This energy-dissipating force is one of the forces on the mass and must thus distort the shape of the total force pulse acting on the mass, which of course distorts the acceleration pulse shape, according to Newton's law ($F = ma$). The simplest familiar *device* which

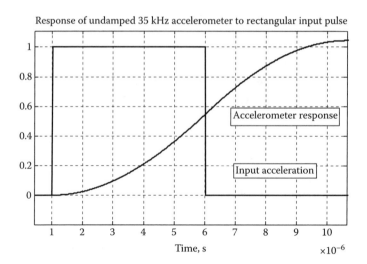

FIGURE 10.13
Accelerometer cannot follow steep wavefront.

does this is the pure/ideal damping element, which we have used in our models. It appears thus that the near-half-sine pulses observed in measurements can be explained by some combination of high C_r values (low bumper energy loss) and/or the *smoothing* effect of accelerometer and low-pass filter dynamics.

A "nonphysical" simulation that *does* allow the full range of C_r values (0–1.00) can be developed, but is of little practical use in our application. We simply model the bumper as a pure linear spring element, with *no* modeling of any physical energy loss component. When the mass rebounds with the same velocity it had just before the impact, we *very quickly* force that rebound velocity to drop to the proper value as given by C_r. This simulation is a little awkward since we must calculate the proper instants when the velocity correction is to be applied, and how large the correction must be to give the proper after-impact velocity. The calculation is not complicated, and we use simulation results to adjust the numbers for the desired behavior, but it must be redone for each successive impact as the mass "bounces down" to its final rest position, and is different for every different C_r value. Figure 10.14a shows the simulation diagram. The analysis is carried out only through the second impact, since, with $C_r = 0.10$, there are not many *bounces* large enough to be of interest. In Figure 10.14b, I show the displacement, velocity, and acceleration of the mass. The velocity during the *free rise/fall* portions of the motion are straight lines, since the acceleration then is just that of gravity, and these sections cover most of the total time of 0.40 s. The impact pulses last only a few milliseconds,

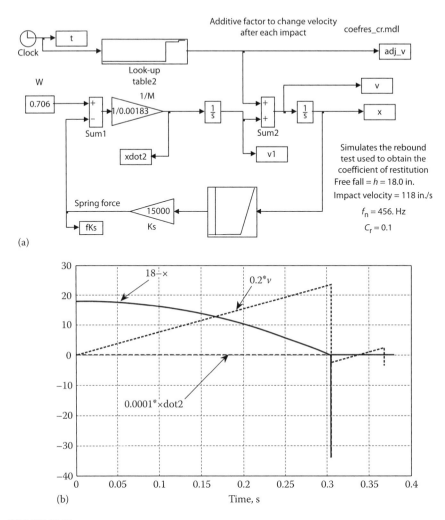

FIGURE 10.14
(a) Simulation usable for C_r in range 0.0–1.0 and (b) results for $C_r = 0.10$.

during which the velocity changes very rapidly, compared to the free fall sections. Since the only force acting (except for the small weight force) is that of the spring, the acceleration pulse is a perfect half-sine waveform. Figure 10.15a shows that the velocity adjustment provided by the lookup table must be carefully timed to occur just an instant *after* the acceleration pulse has caused the impact velocity to reverse and reach its peak upward value. Figure 10.15b shows how the velocity adjustment factor changes with time, to accomplish the correct amount of adjustment at just the right time for each impact.

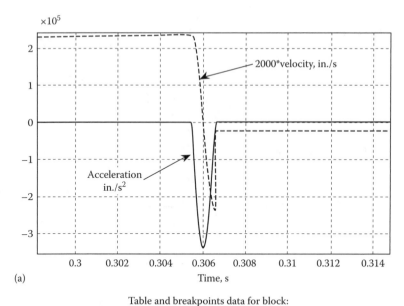

(a)

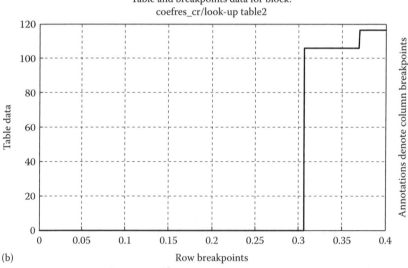

(b)

FIGURE 10.15
Details for simulation of Figure 10.14a: (a) zoomed acceleration and velocity for the first impact and (b) lookup table values for velocity adjustment factor.

10.5 Some Analytical Solutions

While our simulations have been useful, *analytical solutions* of some simple cases will provide *relationships* that are not revealed by simulation, which never provides results in letter form. The simplest model that allows us to

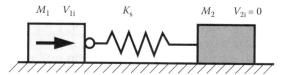

FIGURE 10.16
Analysis model for the analytical solution.

estimate this force models the *bumper* as a perfectly elastic linear spring. We will analyze a system that consists of two rigid masses, one of which has an attached spring (the bumper) and is initially stationary; the other mass impacts the spring with an initial velocity V_{1i}. We will set up and solve the differential equations that give the motion of the two masses up to the point where the spring leaves contact with the projectile upon rebound. If we use the *classical* D-*operator* method of solution, we find that the initial conditions require some care. The Laplace transform method handles the initial conditions more easily, though both methods give exactly the same results. It is acceptable to treat the problem as *horizontal*, as shown in Figure 10.16, even though the actual calibrator operates in a vertical direction; the gravity effects are very negligible because the impact forces in the calibrator correspond to hundreds or thousands of g's of acceleration. That is, the gravity forces on the masses are very small relative to the impact force.

$$(M_2 D^2 + k_s)x_2 + (-K_s)x_1 = 0 \tag{10.12}$$

$$(-K_s)x_2 + (M_1 D^2 + K_s)x_1 = 0 \tag{10.13}$$

$$x_2 = \frac{\begin{vmatrix} 0 & -K_s \\ 0 & M_1 D^2 + K_s \end{vmatrix}}{\begin{vmatrix} M_2 D^2 + K_s & -K_s \\ -K_s & M_1 D^2 + K_s \end{vmatrix}} = \frac{0}{M_1 M_2 D^2 (D^2 + \omega_n^2)}$$

$$D^2 \left(\frac{D^2}{\omega_n^2} + 1 \right) x_2 = 0 \qquad \omega_n^2 \triangleq \frac{K_s}{\dfrac{M_1 M_2}{M_1 + M_2}} \tag{10.14}$$

The differential equation stated in Equation 10.14 is linear with constant coefficients, so it can be solved analytically, using either the *classical* D-*operator* method or the *Laplace transform* method. Since knowledge of both methods allows one to select the one most suitable for a given problem, we will use both.[*] In addition to getting solutions, the two methods are complementary in providing certain alternative viewpoints useful in analysis. A necessary initial step in *both* methods is to find the roots of the system's characteristic equation. Our differential equation is fourth order, so the characteristic

---

[*] E.O. Doebelin, *System Dynamics*, Marcel Dekker, New York, 1998, pp. 337–402.

equation is a fourth degree polynomial set equal to zero, and there are thus four roots. Fourth-degree polynomials *can* be factored analytically (there are formulas analogous to the well-known quadratic formula for second-degree equations) but the equations are so complicated that they are rarely, if ever, used. It has been proven in algebra that the roots of equations *higher* than fourth degree are *impossible* to find in letter form. Thus, in practice, we generally need to find roots by approximate (but very accurate) *numerical* methods if our system differential equation is fourth order or higher. In fact, dealing with even third-order equations in letter form is quite tedious. Thus when we state that *any* linear differential equation with constant coefficients can be analytically solved by following a simple routine, we acknowledge that the vital step of *root finding* often cannot be done analytically but must apply numerical algorithms, which means that *all the system parameters must be known as numbers, not letters*. In any analysis, working with letters is always much preferred, since it allows the formulation of *relations* among the parameters, which can then be invoked to guide our choices of parameter values to meet design specifications.

Fortunately, our present example is a *degenerate* fourth-degree polynomial; that is, two of the four roots are clearly zero, and the remaining two are easily found, giving the complete set of roots as $0, 0, \pm i\omega_n$. Using the classical D-operator method, once one has the roots one immediately writes out the complementary solution, by applying a fixed set of rules. Since our equation has a right side equal to zero, the complementary solution will be the *total* solution; there is no particular solution. Using the mentioned rules, we get

$$x_2 = C_1 + C_2 t + C_3 \sin(\omega_n t + C_4) \tag{10.15}$$

The C's are constants of integration, which must now be found by applying the *initial conditions*, of which there must be four, to find the desired numerical values for the C's. Recall that in this D-operator method the word *initial* must be interpreted as an *infinitesimal* time *after* the system's inputs are applied. This instant is called 0+. Using the Laplace transform method, which also requires values of initial conditions, the word initial has a *different* meaning; it refers to a time *before* any inputs are applied. This interpretation makes the solution simpler, as we will shortly see when we use that method. In the D-operator method, for our example, we will need to find initial numerical values for the unknown x_2 and its first three derivatives. These values will then be inserted into the solution (Equation 10.15 and its derivatives) to give four algebraic equations which are solved for the C values.

$$x_2 = C_1 + C_2 t + C_3 \sin(\omega_n t + C_4) \tag{10.16}$$

$$\dot{x}_2 = C_2 + \omega_n C_3 \cos(\omega_n t + C_4) \tag{10.17}$$

$$\ddot{x}_2 = -\omega_n^2 C_3 \sin(\omega_n t + C_4) \qquad (10.18)$$

$$\dddot{x}_2 = -\omega_n^3 C_3 \cos(\omega_n t + C_4) \qquad (10.19)$$

Since x_2 is *given* to be initially at rest with displacement of zero, $x_2(0+) = 0$ and $dx_2/dt(0+)$. The second derivative is also zero because at $t = 0+$ the spring force is still zero (the spring has not had time to compress a finite amount). The third derivative requires a little more effort. Since the spring force F is the only force acting on mass 2, $F = M_2 D^2 x_2$ and thus $DF = M_2 D^3 x_2 = K_s(Dx_1(0+) - Dx_2(0+)) = K_s Dx_1(0+)$, where $Dx_1(0+)$ is the known initial velocity of the flying mass. We thus see that $D^3 x_2(0+) = V_1(0+)K_s/M_2$. From Equation 10.18 we see that either C_3 or $\sin(C_4)$ must be zero and it must be the latter, since letting C_3 be zero removes that entire term from the solution. The *phase angle* C_4 is thus taken to be zero, and then from Equation 10.16, C_1 is zero. Equation 10.19 then gives $C_3 = -V_1(0+)K_s/M_2\omega_n^3$, which makes $C_2 = V_1(0+)K_s/M_2\omega_n^2$. The final solution is thus

$$x_2 = V_1(0+)\left(\frac{M_1}{M_1 + M_2} \cdot t - \frac{1}{\omega_n}\frac{M_1}{M_1 + M_2} \cdot \sin(\omega_n t) \right) \qquad (10.20)$$

A similar process solves for x_1:

$$x_1 = V_1(0+)\left(\frac{M_1}{M_1 + M_2} t + \frac{1}{\omega_n}\frac{M_2}{M_1 + M_2} \sin(\omega_n t) \right) \qquad (10.21)$$

Turning now to the Laplace transform method, recall that when one transforms the original differential equations from the t *domain* to the s *domain*, the transform of every derivative term includes a certain initial condition; that is, in this method the needed initial conditions are *automatically* introduced. *Initial* now means just *before* the inputs are applied, a time we shall call $t(0-)$. For our example, the highest derivative present in the two original equations is the second derivative, which means that the only initial conditions which show up in the transformed equations are the two initial displacements and the two initial velocities, all of which are zero except $V_1(0-)$, which is of course the same as $V_1(0+)$.

$$(M_2 s^2 + K_s)X_2 - K_s X_1 = 0 \qquad (10.22)$$

$$-K_s X_2 + (M_1 s^2 + K_s)X_1 = M_1 V_1(0-) \qquad (10.23)$$

These Laplace transformed equations *are* now algebraic equations, which can be solved by any convenient algebraic method (determinants are the most systematic) for the functions $X_1(s)$ and $X_2(s)$. One then uses a table of Laplace transforms to obtain their *companion* functions $x_1(t)$ and $x_2(t)$. These solutions of course come out identical to Equations 10.21 and 10.20.

Our main interest is in the acceleration of mass two, since that is where our accelerometers are attached. Inserting the known values of C_3 and C_4 into Equation 10.18 gives

$$\ddot{x}_2 = V_1(0)\frac{M_1}{M_1 + M_2}\,\omega_n \cdot \sin(\omega_n t) \qquad \omega_n^2 \triangleq \frac{K_s}{\dfrac{M_1 M_2}{M_1 + M_2}} \qquad (10.24)$$

This formula of course holds only for the time interval from zero to the point where mass 2 leaves contact with the bumper spring, and shows that the pulse waveform *is* the perfect half-sine wave that we have been using as our *ideal* shape. We see that if the two masses had the same value M

$$\text{peak value of acceleration pulse} = V_1(0)\sqrt{\frac{K_s}{2M}} \qquad (10.25)$$

This relation clearly shows how we can adjust the parameters to get the peak value we desire. Because our model is approximate, we should not expect such predictions to be *exact*, but the relation is certainly useful. It is also easy now to estimate the *duration* of the pulse:

$$\text{pulse duration} = \frac{\pi}{\omega_n} \qquad (10.26)$$

When the two masses are *not* equal, and we know the value of each one, we can of course easily estimate the pulse peak and duration for that situation.

While the above calculations were not very difficult and provide useful guidance in calibrator design, a simulation of the system of Figure 10.16 can also be useful. Let us use some values for the actual calibrator. The reference accelerometer has a mass of 40 g; let us assume the accelerometer under test has the same mass. Of the two available anvils, let us use the one made of steel. Its mass is 80 g. The steel projectile has a mass of 320 g. While Robert Sills of PCB, the inventor/designer of this calibrator graciously supplied me with several actual bumpers made of rubber, felt, and Kapton plastic, he did not know their spring constants since they are selected mainly by trial and error, to get desired results. For our example, let us take the spring constant as 15,000 lb$_f$/in. Let us take the projectile velocity at impact as 1000 in./s. Using these values we get the simulation diagram and results of Figure 10.17. The peak accelerations and pulse durations check with the theoretically predicted values from our equations.

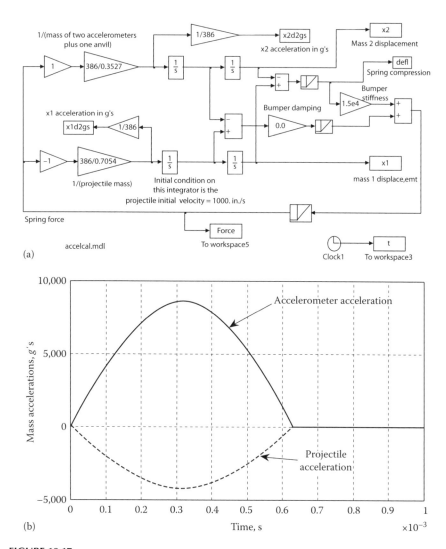

FIGURE 10.17
(a) Simulation and (b) results for system of Figure 10.16.

10.6 Simulation of the Pneumatic Shock Calibrator Apparatus

Now that we are somewhat familiar with some of the basic features of simple impacts, it is now time to return to our main interest, the simulation of the pneumatic calibrator which produces controllable impacts for the purpose of calibrating accelerometers. In addition to the manufacturer's sales brochures, several papers* have been published describing this apparatus, but not, of course, documenting its *design*. Bob Sill is the inventor/designer of this calibrator and provided me with much useful information regarding its construction. Because we have already developed simulations that accept a known projectile impact velocity as input, the new simulation will only model the behavior up to the point where this velocity is known. That is, the new simulation will provide a numerical value for the projectile velocity, but we then use this number as an input to our earlier simulations or calculations. Of course, it is possible to develop a single simulation which models the *entire* process, but we prefer to break it up into these two components for simplicity. Referring to Figure 10.18,

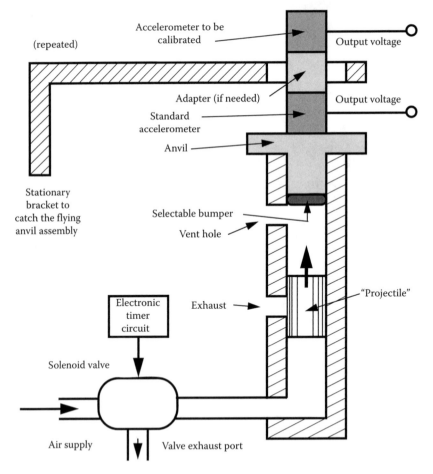

FIGURE 10.18
Figure 10.1 repeated.

---

* R.D. Sill and S.H. Kim, Accelerometer shock sensitivity calibration using a pneumatic exciter (no date given), PCB Piezotronics and The Modal Shop, www.pcb.com and www.modalshop.com; M. Peres and R.D. Sill, Shock and vibration calibration of accelerometers, *Sound and Vibration*, March 2006, pp. 8–12, downloadable at www.sandv.com

we see that we need to model the flow of air from a source of fixed and known pressure (up to about 100 psig) into the chamber of the *gun barrel*, where a pressure will develop and exert an accelerating force on the projectile. This inflow of air will be maintained for a timed period, as set by the electronic timer circuit, and then the solenoid valve is closed. This valve is a so-called three-way type, which means that when it shuts off the air supply it simultaneously opens a port to the atmosphere, dumping the air out of the chamber. The gun barrel exhaust port shown in the figure also aids in dropping the chamber air pressure to atmosphere (0 psig) as quickly as possible; it opens as the projectile passes by. When the chamber pressure has dropped to 0 psig, the projectile essentially *coasts* at the highest velocity it attained while accelerating. We take this velocity as the impact velocity on the anvil assembly and use this number as an input to our earlier simulations (or calculations) of the impact event. To prevent a retarding *back pressure* from developing *ahead* of the projectile, we provide the vent hole shown. This also prevents the anvil from *lifting off* before the impact.

Since the air supply pressure may be as much as 114.7 psia, and the initial chamber pressure is 14.7 psia, the initial flow into the chamber may occur under *choked flow* conditions, since the critical pressure ratio for air is 0.528. As the chamber pressure rises, the valve supply-port pressure drop will decrease, possibly bringing the flow out of the choked condition. Our simulation thus needs to properly sequence the flow model to accommodate these two possible valve flow regimes as they occur during calibrator operation. In typical undergraduate fluid mechanics courses, one learns how to model the flow through simple passages such as orifices and pipes or tubes. When one examines the flow passages in real-world valves, they are not this simple because they must allow for reliable opening and closing, and the shapes must be such as to allow the parts to be easily manufactured. The solenoid valve actually used by PCB in the shock calibrator is the model 8026588.0801 from Norgren-Herion,* shown in cross section in Figure 10.19. The valve shown actually includes *two* valves: a *pilot* valve and a *main* valve. This dual-valve arrangement is used when very rapid action is required and the main valve is fairly large. The electromagnetic solenoids, which transduce a voltage command into a force and then mechanical motion of the valve stem, become quite large when they must apply large forces to rapidly move the mass of the valve stem against friction forces of seals and other valve parts.

To keep the solenoid small, we can instead let the solenoid move a *small and light* pilot valve stem, which opens the pilot valve to supply pressure, and quickly applies this pressure to the upper end of the main-valve piston. This end of the piston can develop large force quickly because of its area and the small volume which has to be pressurized by the pilot valve. This force is applied to the main-valve stem to rapidly open or close it. (Supply pressure is also pushing on the bottom end of the piston, but the area here

---

* www.norgren.com/usa/

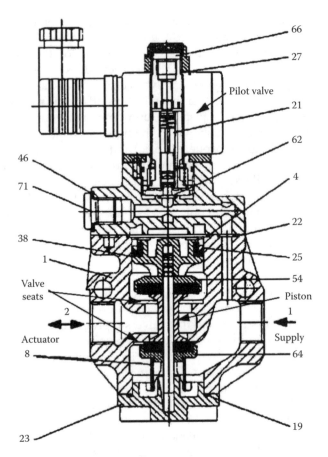

FIGURE 10.19
Norgren-Herion solenoid valve.

is smaller than that at the piston top, so the net force is downward, opening the valve.) As the valve starts to open, pressure now acts on *both* sides of the bottom end of the piston, giving even more net downward force, thus speeding up the opening of the valve. The assembly drawing shown (Figure 10.19) cannot clearly show all these details but you can get the general idea. The main-valve flow passages are the ones we need to model in our simulation; we will not bother to model the action of the pilot valve since the manufacturer supplies a *valve response time* (0.010 s for this valve). This response time tells us how long it takes for the main valve to open or close, which of course includes the response time of the pilot valve.

From the drawing it appears that the main flow passages are somewhat complex annular spaces between the machined metal ports and the elastomer seals on the two ends of the valve stem. Rather than using complicated computational fluid dynamics software to predict (pressure drop)/(flow rate) relations in valves, manufacturers of all kinds of valves have for many years used an *experimental* characterization leading to a single descriptive number called C_v. This *flow factor* C_v is numerically equal to the *measured* flow rate (gal/min) of 68°F water through the valve when the valve pressure drop is 1.0 psi. For our valve the manufacturer gives a C_v value of 4.8 for the flow from supply to *actuator* and 6.1 for the exhaust flow from actuator to atmosphere. These numbers will allow us to model the flows needed in our simulation. This is done using accepted formulas for incompressible flow through *orifices*, as found in fluid mechanics or some system dynamics* texts. In Equation 10.27, C_d is the dimensionless discharge coefficient of the orifice and A_o is the orifice area. We can think of their product as the *effective* flow area and get a number for this product from the given C_v values by substituting known numbers (water density, 1.0 psi pressure drop, 4.8 gal/min).

$$\text{volume flow rate} = C_d A_o \cdot \sqrt{\frac{2\Delta p}{\rho}} \tag{10.27}$$

Using the given parameter values, the numerical value of the effective area $C_d A_o$ is found to be 0.126 in.^{2} for flow into the actuator and 0.161 in.^{2} for exhaust flow out of the actuator. Note that we are using a theoretical formula derived for a simple orifice for the much more complicated shape of the valve's flow passages.

We can get the compressible flow formulas needed to model the various valve flows from any undergraduate fluid mechanics text. Basically, two formulas are needed: one for those time periods when the valve flow is choked, and another for the unchoked periods. For choked flow ($p_d < 0.528 p_u$) we have

$$G = C_d A_o \sqrt{32.17 \rho p_u k \left(\frac{2}{k+1}\right)^{\frac{k+1}{k-1}}} \tag{10.28}$$

whereas for unchoked flow ($p_d > 0.528 p_u$) the relation is

$$G = C_d A_o \sqrt{2 \cdot 32.17 \rho p_u \frac{k}{k-1} \left[\left(\frac{p_d}{p_u}\right)^{\frac{2}{k}} - \left(\frac{p_d}{p_u}\right)^{\frac{k+1}{k}}\right]} \tag{10.29}$$

where
  G is the mass flow rate in lb_m/s
  p_u is the upstream pressure in lb_f/ft^2

---

* E.O. Doebelin, *System Dynamics: Analysis, Modeling, Simulation, Design,* Marcel Dekker, New York, 1998, p. 229.

p_d is the downstream pressure
$C_d A_o$ is the effective flow area in ft^2 as obtained for the valve in Equation 10.27
k is the ratio of specific heats (we take it as 1.4 for air)
ρ is the density at upstream pressure and 530°R temperature

In units consistent with the above formulas, density is found from the relation

$$\rho = \frac{p_u}{53.3 \cdot 530} \qquad (10.30)$$

In computing mass flow rates and also in any later modeling of chamber pressure and volume changes, we choose to treat temperature as constant at "room" temperature of 70°F (530°R). This is often a good approximation for such processes since the *percent* change in absolute temperature is small. For example, a 35° change from the base value of 70 is *not* a 50% change but rather a (35)(100)/530 = 6.6% change. You may be recalling from thermodynamics texts that various expansion/compression processes such as isentropic or polytropic allow quite simple accounting for temperature changes, but these apply directly only to *closed* systems (like a piston/cylinder with a fixed amount of gas inside) whereas our volume will have mass added and subtracted as the valve opens and later exhausts. Also, the temperature is affected by heat transfer and frictional energy losses, both of which are hard to model accurately. If you search most thermodynamics texts you will probably find that problems such as our present one are *not* treated. Ignoring temperature changes, together with the many other assumptions made in our study, can of course ultimately be validated (or refuted) by careful experiments once we build the apparatus.

Before proceeding further, I need to digress for a moment and discuss some questions relative to the modeling of pneumatic valve flow processes. As mentioned earlier, the conventional method used by most valve manufacturers uses the flow coefficient C_v, as discussed above. A perhaps "more scientific" method has been proposed in some standards documents* and adopted by some valve makers, especially outside the United States. The conventional definition and use of C_v suffers from a number of defects when used for compressible fluids, most commonly air. The numerical value is determined using *water*, an essentially *incompressible* fluid, and only at a single pressure-drop value, which is also very low (1 psi). Such a definition and measurement can in no way comprehend the subtle behavior of gas flows through complex flow passages; particularly the phenomena of *choked and unchoked flow*, as modeled for an *isentropic nozzle* in Equations 10.28 and 10.29. Of the various alternative methods of

---

* ISO 6358 *Pneumatic Fluid Power—Components Using Compressible Fluids—Determination of Flow-Rate Characteristics*, ISO, Geneva, Switzerland, 1989.

characterizing pneumatic valve flows, that described in ISO6358 seems to be the most defensible and appears to be gaining acceptance. However when I tried to get the needed flow data for the valve of Figure 10.19, only the C_v value was available, so our analysis will proceed in that way. For valves that have been characterized by the ISO6358 method, we should use that data. Fortunately, while the *numbers* might be significantly different, the *procedure* used in our calibrator study will be essentially the same. The ISO method uses flow rate and pressure-drop measurements covering the unchoked and choked flow regimes to come up with two numerical values: the *sonic conductance C* and the *critical pressure ratio b*. The ISO test procedure requires more instrumentation and takes more time than the C_v procedure; this may explain why it is not universally accepted by valve manufacturers. Two valves that have the *same* C_v value would usually have *different* values for sonic conductance and critical pressure ratio, revealing again the problem with C_v. Also, we find that the critical pressure ratio for actual valves (or other complex flow passages) can be much different from the "textbook" value (0.528) memorized by many engineers, which applies only for an isentropic nozzle.

If the numerical values of the sonic conductance C and critical pressure ratio b are available, these numbers are used in the following formulas* to compute the valve flow rate.

$$\dot{m} = p_1 C \rho_o \sqrt{\frac{T_0}{T_1}} \sqrt{1 - \left(\frac{\frac{p_2}{p_1} - b}{1 - b}\right)^2} \quad \text{for } \frac{p_2}{p_1} > b \quad \text{subsonic flow} \qquad (10.31)$$

$$\dot{m} = p_1 C \rho_o \sqrt{\frac{T_0}{T_1}} \quad \text{for } \frac{p_2}{p_1} \leq b \quad \text{choked flow} \qquad (10.32)$$

The mass flow rate $\dot{m}$ is in kg/s, and the density ρ_o (kg/m³) and temperature T_0 (K) refer to *reference* conditions such as standard temperature and pressure. Pressures p_1 and p_2 are, respectively, the upstream and downstream values in Pascals, while T_1 is the upstream temperature. Note in Equation 10.31 that when p_2/p_1 is equal to b, the mass flow rate given by the two formulas is the same, as it should be, since unchoked flow becomes choked flow when this ratio becomes greater than the critical pressure ratio. Also, in Equation 10.32, when the upstream air temperature is equal to the reference temperature, the mass flow rate is equal to $p_1 \rho_0 C$, which can serve as a definition of C:

---

* P. Beater, *Pneumatic Drives*, Springer, New York, 2007, Chapter 5.

$$C \triangleq \frac{\dot{m}}{p_1 \rho_0} \quad \frac{\text{m}^3}{\text{Pa-s}} \quad \text{volume flow rate/Pa} \tag{10.33}$$

That is, in the ISO experiment, when we keep the upstream pressure fixed and gradually increase the downstream pressure while measuring the flow rate, a graph of flow rate versus pressure ratio will *flatten out* as we reach and then exceed the critical pressure ratio. This transition from a varying flow rate to a constant (choked) flow rate defines the critical pressure ratio and the sonic conductance. Since, with real experimental data, *this transition is not clear cut*, the standard requires a curve-fitting procedure according to specific rules, so that results from different experimenters are comparable. The fact that b values for real valves are routinely well below the theoretical value of 0.528 (b = 0.20 is not unusual), makes clear that our earlier formulas (Equations 10.28 and 10.29) will not lead to highly accurate results. When the valve of our choice has not been subjected to the ISO characterization however, and we have only a C_v value, those equations are the *best we can do*. As we proceed with the simulation, it will be clear that when b and C values *are* available, no new simulation techniques are needed to accommodate the new equations. Since the standard ISO procedure is somewhat expensive and time consuming, researchers* have suggested improved ways of finding b and C that are claimed to produce comparable results.

We now return to the development of the simulation. We earlier stated that our valve is quoted as having a *response time* of 10 ms. In some applications of pneumatic valves the valve opening or closing time is a small fraction of the total system time of interest, so in that case we could simply start our modeling and simulation just *after* the valve opens, ignoring the phenomena that occur *during* the valve's opening. In the calibrator apparatus, we find that we achieve the desired levels of projectile velocity in only a few milliseconds, so we probably need to include as best we can the phenomena that occur *during* valve opening. When the valve first starts to open, if the supply pressure is sufficiently high, the flow will start out as choked. As the chamber volume is pressurized, the flow can go unchoked, all this happening in a few milliseconds. If the valve takes about 10 ms to completely open, the change from choked to unchoked flow can occur *during* the valve opening process. We thus need to model the valve opening not as instantaneous, but as following some kind of motion curve. We *could* make a separate simulation study of the valve opening process itself, but this would involve complex flow forces and frictional effects which are hard to accurately define. We prefer to use the available *measured* total opening time as being more accurate and simply "make up" a valve motion curve

---

* L.H. Yang and C.L Liu, Measuring flow rate characteristics of a discharge valve based on a discharge thermodynamic model, *Meas. Sci. Technol.*, 17, 2006, 3272–3278; K. Kawashima et al., Determination of flow rate characteristics of pneumatic solenoid valves using an iso-thermal chamber, *J. Fluids Eng.*, 126, March 2004, 273–279.

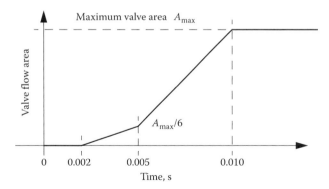

FIGURE 10.20
Assumed area/time characteristic of the valve.

which seems reasonable. Such curves can be modeled easily using a lookup table with time as the input. We will use the curve of Figure 10.20 in our system simulation. The 0.002 s *delay* at the beginning takes into account that the main valve will not start to move until the pilot valve has built up sufficient pressure on the piston to overcome coulomb friction forces and pressure unbalances in the moving parts. The numerical value of 0.002 is of course arbitrary. The overall shape of the "curve" is intended to be roughly *parabolic* as would occur if we had a constant force applied to a given mass. Of course, for closing, the curve starts out at A_{max} and ends up at zero area, but the timing sequence is as shown in Figure 10.20, except "reversed."

The main *output* of this simulation is the numerical value of the projectile velocity when it reaches its *terminal speed* after the air valve is opened and then later shut. (When the pressure in the gun chamber has reduced to 0.0 psig (14.7 psia), no force acts on the projectile and it *coasts* at this terminal velocity.) This projectile velocity would then be used in one of our earlier simulations which requires this value as a given input and produces a time history of the impact acceleration. We start with the supply pressure of, say, 100 psia available at the valve supply port. Starting at time 0.0, the valve supply area changes, as shown in Figure 10.20. Simultaneously, the valve exhaust area changes in the *reverse fashion*; that is it starts out open and then shuts over the 0.010 s time interval. Remember that the exhaust maximum area is *not* the same as the supply maximum area! We will have to configure two separate lookup tables; one for the supply flow and one for the exhaust flow. Note that once the valve is partially open, the supply flow *splits*, part of it going into the gun chamber and the rest being *lost* to the exhaust port (atmosphere). Each of these flows has its own governing pressure difference, which determines whether and when that flow is choked or unchoked. Once the valve is completely open, the exhaust port is completely shut, so now all the supply flow is available to charge up the gun chamber.

Once the valve is completely open, we then need to decide how long to leave it open to achieve a desired projectile terminal velocity and thus a desired peak shock acceleration. This requires trial and error with the simulation, so I will now state the results of that exercise so that we can proceed with the simulation. We will leave the valve open for 0.007 s and then send a *close command* to the solenoid. Due to the 0.002 s *delay* in the curve of Figure 10.20, the valve will start to close at 0.019 s. The lookup table for the exhaust flow will of course again be the *reverse* behavior. The supply area will thus return to zero, and the exhaust area to its maximum value, at time = 0.029 s. Note that the transients will continue for some time *after* t = 0.029 s because the chamber pressure takes a while to decay completely to 0.0 psig, which is when the projectile stops accelerating.

There is a space between the port of the valve and the bottom of the projectile when it is sitting at rest at the bottom of the chamber. The size of this initial volume was not available to me so I arbitrarily took it to be 1.0 in.[3]. When we activate the valve, this initial volume begins to be pressurized, causing the projectile to start to move upward, which increases the volume progressively. We need to properly model all the flows occurring in the valve and then send the *net* flow into this expanding volume. The volume being charged is thus the initial volume plus an additional volume that can be expressed in terms of the projectile cross-sectional area and the projectile displacement x. The displacement can be found from a Newton's law in which the only force acting on the projectile is the time-varying pressure difference across the projectile that exists at any instant in the chamber; we neglect the gravity force as being negligible relative to the pressure force. The projectile is a cylinder with a diameter of 1.25 in. We neglect any friction forces between the projectile and the gun barrel. Let us also assume that the *exhaust port* shown in the gun barrel in Figure 10.18 is not present (we rely solely on the valve exhaust port for discharging the pressure) and that the vent hole keeps the pressure on the top face of the projectile at atmospheric. The relation between valve net flow and chamber pressure can be obtained from the perfect gas law as follows.

$$pV = MRT \qquad \frac{dp}{dt} = \frac{RT}{V} \cdot \frac{dM}{dt} - \frac{MRT}{V^2} \cdot \frac{dV}{dt} \qquad (10.34)$$

Recall that we earlier argued for taking temperature and the gas constant R as *constants*, but in Equation 10.34 the instantaneous mass M in the chamber, and the instantaneous chamber volume V, are *time-varying quantities*. The derivative dM/dt is the net mass flow rate into the chamber, and V can be expressed in terms of projectile displacement x, while projectile velocity dx/dt is used to compute the rate of change of the volume. The instantaneous mass M can be found by integrating dM/dt and p can be similarly found from dp/dt, being careful to use absolute pressure rather than gage pressure and taking atmospheric pressure as 14.7 psia. If we wrote out the complete differential equations, we would find significant nonlinearities, making the equations analytically unsolvable. Simulink of course has no trouble in modeling all these effects. We will find the Simulink module called SWITCH useful in modeling the selection of the proper flow formulas as the flows change between choked and unchoked regimes. The integrator which produces p needs to have a proper initial condition inserted; we can start projectile displacement x at 0.0 at time 0.0.

While any simulation diagram is gradually *built up* piece by piece, I can best explain this particular simulation by displaying the *final* diagram (Figure 10.21) and then explaining the various parts and how they interconnect. The main complications are keeping track of the mass flow rates into and out of the chamber as the pressures change, since we need to change flow formulas when any flow changes from the choked to the unchoked condition. Before time zero, all the pressures start out at atmospheric pressure (14.7 psia). We can pick any supply pressure within the design range of the calibrator; let's arbitrarily choose 50 psia. With this choice, the flow from supply, through the valve, and into the chamber, will start out choked, since the critical pressure ratio for air is 0.528. As the chamber builds up pressure, at some point the supply flow will switch over to unchoked flow. As the supply port of the valve is opened, the exhaust port (which is initially wide open) starts to close. While it is partially open, there will be a flow from the chamber to atmosphere. This flow will be initially unchoked since the chamber pressure starts out at 14.7 psia, making the pressure differential across the exhaust port initially zero. As chamber pressure builds up, if the exhaust port is still partially open, this exhaust flow may become choked. Our simulation needs to provide for all these possibilities, using *switch* modules (with a proper switching signal) to invoke the proper flow equation as conditions change.

Starting at the upper right of Figure 10.21, the lookup table with time t as input provides the mass flow rate (lb_m/s) into the chamber during those times when this flow is choked. In Figure 10.22a, the details of this lookup table show how this flow varies with time as the valve is opened and then later closed. The opening and closing times are fixed by the assumed valve dynamics, but the time spent *wide open* is our choice (0.007 s in Figure 10.22a). This "wide open time" would be determined by trial and error so as to achieve a desired response, such as a maximum projectile velocity of, say, 50 ft/s. The number 0.291 lb_m/s is the computed choked flow rate when the supply pressure is 100 psia and the valve is wide open. The shape of this curve is defined by the varying flow area as a function of time. For choked flow, the shape of the area curve and the mass flow rate curve are identical, since the supply pressure is fixed. The signal *Asup*, divided by 332.6 gives the actual flow area, if that quantity is wanted. The gain block with gain 50/100 allows us to easily change the supply pressure; in this case we have set it at 50 psia. That is, for choked flow, the mass flow rate is directly proportional to supply pressure, and we had set up the lookup table for 100 psia, so if we wanted to set the supply pressure at 100 psia,

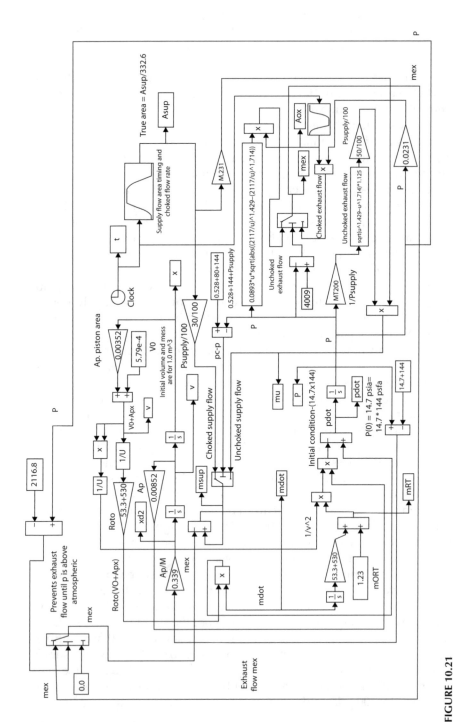

FIGURE 10.21
Accelerometer calibrator simulation diagram.

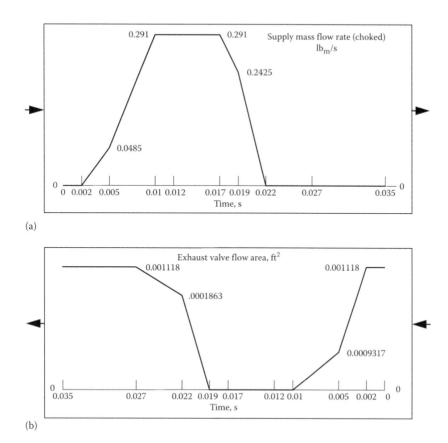

(a)

(b)

FIGURE 10.22
Lookup table details for Figure 10.21.

this gain block would be set at 100/100. The output of this gain block is one of the two alternative inputs to the *switch* module; the other is the unchoked supply flow. The middle input of the switch block is the switching criterion that determines whether the switch module output is the upper input or the lower. This criterion is the difference $pc - p$ between the *critical* pressure for choked flow and the actual chamber pressure p. The critical pressure is 0.528 times the supply pressure (50 psia in our case), expressed in lb_f/ft^2 since our flow equations are written using that pressure unit.

The *unchoked* supply flow must be sensitive to the area variation of the valve supply port, just as the choked flow was. Whether choked or unchoked flow exists, the supply pressure is of course the same, in our case 7200 psfa. The gain block $(1/7200) = (1/p_{supply})$ at the lower middle of the diagram is the input to the formula that calculates the unchoked mass flow rate. Recall that the MATLAB® *function* module always employs symbol u for its input signal, so here $u = p/p_{supply}$, which is what Equation 10.29 requires. The gain block 0.0231 at the lower right is the value of the square root in Equation 10.28 when the upstream pressure (which is also "hidden" in the density) has been taken outside the square root sign. At the upper left, a switch module is used to prevent any exhaust flow from occurring until the chamber pressure exceeds atmospheric pressure.

Equation 10.34 (perfect gas law) has two parts which are implemented separately and then combined to form the rate of change dp/dt of the chamber pressure. The volume V and its derivative dV/dt are needed here; they are formed from the displacement x and velocity v of the projectile. These motion quantities are computed from a Newton's law where the applied force is the pressure difference across the projectile whose mass is M. Volume is simply the initial volume V_0 plus the additional volume created as the projectile moves, while the rate of change is obtained from the projectile velocity v and projectile area Ap. Chamber air mass is computed by adding to the initial mass the mass added by the net mass flow (mass in – mass out) for the chamber. As earlier stated, gas constant R and gas absolute temperature T are taken constant at, respectively, 53.3 and 530. All these signals are combined as in Equation 10.34 to form dp/dt, which is then integrated to get p, as shown at the lower left of the simulation diagram.

To show some typical results we set the supply pressure at 50 psia, as shown in Figure 10.21 This requires setting three gain blocks and one constant at the proper values; their labels include the word *Psupply*. Figure 10.23 displays some selected results. In Figure 10.23a we see that the chamber pressure reaches supply pressure and dwells there until the valve switches to the exhaust position, whereupon pressure drops and actually goes below atmospheric toward the end of the cycle. This drop into the *vacuum* range probably does not occur in the actual calibrator because of the *exhaust port* shown at the bottom right of Figure 10.18, which port we did not include in our simulation. In the real apparatus this port is probably located much closer to the *top* of the tube than the bottom. Thus it would *not* participate in the early part of the activity (as our simulation did not), but at the final moments would probably *prevent* the occurrence of a vacuum since it provides a large passage to the atmosphere. Our simulation allows a vacuum to develop because the rightmost term in Equation 10.34 has a *negative* sign, so if dV/dt is large enough, the expanding volume can reduce the pressure even if all the valve ports were shut.

Our valve cycle is chosen so short that the velocity reaches only about 35 ft/s, with displacement no larger than about 7 in. The actual calibrator has a maximum displacement of about 18 in., so by raising the supply pressure and/or leaving the valve open longer, we can reach the higher projectile velocities needed to get the impact acceleration levels claimed in the calibrator's specifications. Figure 10.23b shows the time variation of the various mass flow rates, as dictated by the valve duty cycle and the choking/unchoking phenomenon.

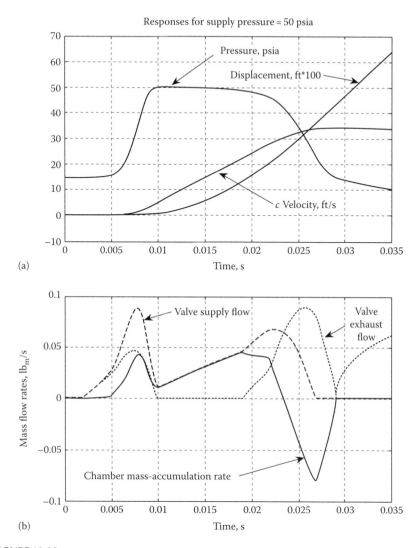

FIGURE 10.23
Simulation results for 50 psia supply pressure.

10.7 Concluding Remarks

Robert D. Sill, the inventor/designer of the calibrator told me that this design was successfully completed without any use of dynamic simulation and was mainly based on his (vast) previous experience with related systems. When we do not personally have such a wealth of background information, we have to rely more on math modeling, theoretical calculations, and simulation efforts. After such efforts have gone as far as is practical, we must finally build and test a prototype, which of course is how expert practitioners such as Mr. Sill built up their design skills over years of work on many projects. A technical paper by Mr. Sill with more details on the back-to-back accelerometer calibration process is available.* When extremely high shock levels are required, the Hopkinson bar† apparatus replaces the pneumatic calibrator.

---

* www.endevco.com, Look for Technical Paper TP-310.
† www.endevco.com, Look for Technical Paper TP-283 (Mr. Sill is the author.), J. Dosch and L. Jing, Hopkinson bar acceptance testing for shock accelerometers, *Sound and Vibration*, February 1999.

11

Shock Testing and the Shock Response Spectrum

11.1 Introduction

Many mechanical failures are caused by shock and vibration, so a design to reduce or prevent such failures is often required when developing new machines or processes. Vibration is usually thought of as a more-or-less continuous motion while shock is a transient occurrence. Fatigue stressing and acoustic noise are two of the common bad effects of vibration, and these may also arise with shock-type loading. Perhaps more often, a *single* excessive shock event will cause an immediate failure due to the plastic deformation or actual breakage of machine parts. Many practical problems involve the protection of delicate products or equipment like electronics "boxes," which might be mounted to a vehicle wall or floor. There is also an entire industry devoted to the engineering of shipping containers to protect the contents from *drops* or *bumps*, which inevitably occur during the loading, transportation, and unloading processes. Another large application area is in military systems where equipment is regularly exposed to blast loading.

For *simple* models of actual machines, the prediction of (and design against failure from) shock inputs can appear quite straightforward. Unfortunately, more realistic treatments reveal significant difficulties. Perhaps the simplest model for a shock-loading situation is shown in Figure 11.1. Here, the "machine" is modeled as a simple spring/mass/damper system with one-degree-of-freedom motion. We assume that the shock input is a prescribed transient motion imposed at the lower end, and described in terms of displacement, x_i, velocity, v_i, or acceleration, a_i, with acceleration inputs being perhaps most common. The mass, of course, responds with its own displacement, velocity, and acceleration, and one needs to choose which of these response quantities is critical with respect to failure. The literature reveals that the most common response quantity is the acceleration of the mass, perhaps because $F = MA$ seems to tell us that *destructive forces* are related to acceleration. This choice is,

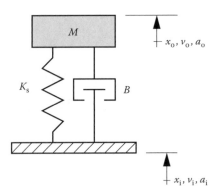

FIGURE 11.1
Simplest model for shock input studies.

however, not espoused by all practitioners, some* of whom argue that other motion quantities predict actual service failures more reliably.

Many types of shock-testing machines have been developed to meet specific needs, and we will be showing some of these as we progress in this chapter. Figure 11.2[†] shows a machine that simulates the shock caused when a dropped object (laptop, cell phone, etc.) impacts the floor. Since such an object

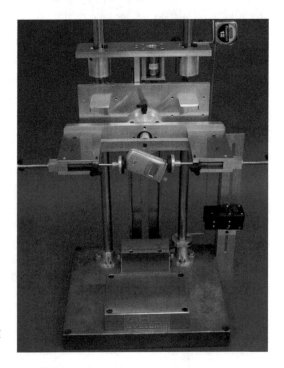

FIGURE 11.2
Drop-test shock machine from GHI systems.

---

* H.A. Gaberson, D. Pal, and R.S. Chapier, Classification of violent environments that cause equipment failure, *Sound and Vibration*, May 2000, pp. 16–23. (downloadable at www.sandv.com)
[†] GHI Systems, www.ghisys.com

might be in an infinite number of angular attitudes at the instant it strikes the floor, and each of these attitudes has its own damage potential, it is useful to be able to *prescribe* this angular orientation and reproduce it from test to test. The test object is held, in any desired attitude, in clamps which can be released just before the impact occurs. These clamps are part of a free-falling slide that can be raised to the desired drop height and then released. Accelerometers can be mounted on the test object to record the shock data, and a velocity transducer gives the object's vertical velocity just before impact. In some cases, high-speed video cameras can be used to study the test object's motion after impact.

In shock analysis and testing, the most commonly chosen input quantity is acceleration, and these acceleration transients can, of course, take many, sometimes complicated, forms. Consider a box of electronic equipment that is bolted to the structure of a naval vessel that experiences a hit from a bomb or torpedo. The structure/box attachment point will experience a transient motion whose acceleration could be measured and this acceleration becomes the input motion to the equipment box. The electronics inside the box is, of course, made up of mechanical objects, which will be dynamically loaded in some way by the shock event. These mechanical parts (circuit boards, wires, connectors, etc.) comprise a very complex mechanical system, which may be almost impossible to accurately analyze, leading to the need for careful dynamic testing to discover weak points and improve the initial designs. This testing takes various forms, one of which involves the *shock response spectrum* (SRS), which we will shortly define and discuss.

Some shock-testing machines, such as the *drop table* type, use simple *standard* input accelerations, rather than trying to duplicate actual recorded shock transients. The goal here is to apply a simple reproducible pulse waveform of acceleration to the device under test (DUT). These machines, in principle, drop a vertically guided mass (to which the DUT is rigidly fastened) onto a table with a cushioning *bumper* (often called a *programmer)* between the falling mass and the machine table. These programmers are carefully designed to produce a desired shape and duration of the acceleration pulse; the drop height is adjusted to control the magnitude of the peak acceleration. An accelerometer fastened to the falling mass records the acceleration (really *deceleration*) to which the DUT is subjected. Programmers can be as simple as small rubber pads, or as complicated as preloaded pneumatic cylinders. Their goal is to produce a controlled pulse shape, usually a *half-sine, sawtooth,* or *trapezoidal* shape.

11.2 Analysis and Simulation of Response to Shock Inputs

Let us now use Simulink® to study the response of the system of Figure 11.1 to idealized versions of these standard pulse inputs. We say *idealized* because the shock machines are not perfect and their pulses are *close to* but not exactly equal to the desired forms. This simulation could be set up in various ways, but we prefer a form where all the possible input and output motion quantities are easily accessible, and the mechanical system is described, not in terms of M, K_s, and B, but rather in terms of the *generic* parameters of *undamped natural frequency* ω_n and *damping ratio* ζ. Using Newton's law, we get the system equation as

$$\ddot{x}_o = \omega_n^2(x_i - x_o) + 2\zeta\omega_n(\dot{x}_i - \dot{x}_o) \tag{11.1}$$

which leads quickly to the simulation diagram of Figure 11.3. This diagram provides for either a sawtooth pulse or a half-sine pulse of acceleration as the input. As shown, by setting the gain block just after the sawtooth pulse lookup table to zero, we have blocked this pulse and allowed the half-sine pulse to be the active one. Rather than inserting numbers for the desired natural frequency and damping ratio, we left these as letters in their gain blocks, so we will need to assign numbers in a MATLAB® statement. At the bottom of the diagram, I copied the sawtooth lookup table and expanded its size to show more detail. Note that the pulse duration is 0.10 s and the "amplitude" is 1.0. These same values are used in the half-sine pulse, but since a lookup table was not used there, we do not have a *picture* of the shape. Of course, once we run the simulation, we can easily plot the input pulse as the variable *halfsine* since we sent it to the workspace.

To get a feel for the shock event, we will run the simulation three times for a range of natural frequencies. In shock response simulations, the damping ratio is often set at about 0.05, since the actual damping of real-world "packages" is usually small but unknown. When the input pulse duration is short relative to the period of the system natural frequency, a pulse of *any* shape begins to act like a perfect impulse function of the same net area, there is no *resonance* effect, and the output acceleration is not excessive (see Figure 11.4a). When the input pulse duration is comparable to the system natural period, because of the light damping, we get an output peak that is larger than the input peak, which we recognize as a sort of resonance. When the system natural period gets very short relative to the input pulse duration (the system is very *stiff*), then the output acceleration begins to look like the input acceleration; the output *follows* the input. In Figure 11.4b, I have zoomed the graph to show both the input and output accelerations when the system has a 100.0 Hz natural frequency (period = 0.01 s).

Figure 11.5 shows the results for the sawtooth pulse. If you compare Figures 11.4 and 11.5, the comparison is not quite fair. A fair comparison requires that two pulses of any shapes must have the same *net area* to make them of the "same strength." The net area of the half-sine pulse is $0.2/\pi = 0.06366$, while that of the sawtooth is 0.05, so we should really have multiplied the sawtooth results by $0.06366/0.05 = 1.2732$. I did this in

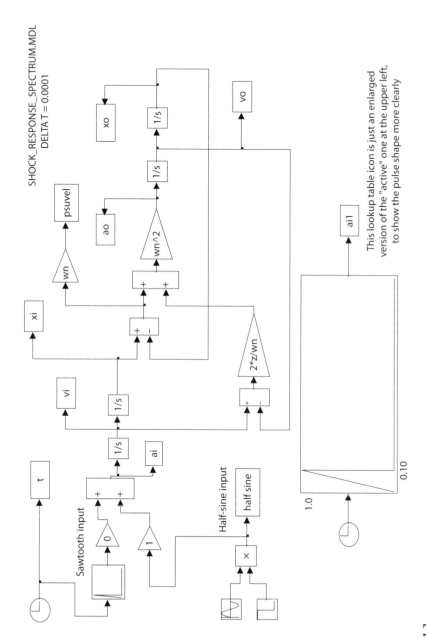

FIGURE 11.3
Simulation diagram for system of Figure 11.1.

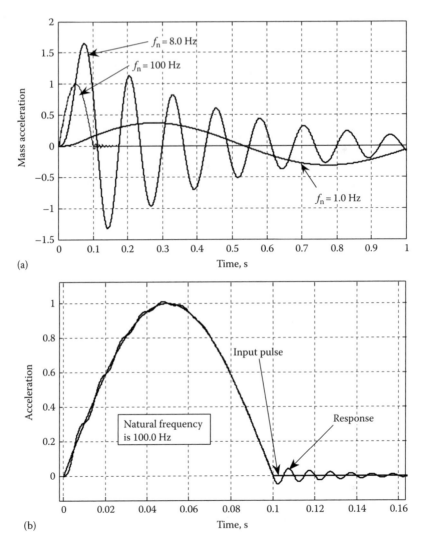

FIGURE 11.4
(a) Response of three systems of different natural frequency to the same half-sine pulse and
(b) zoomed graph for 100 Hz system.

Figure 11.6a. For any shape of pulse, by looking at the response of systems
with a wide range of natural frequencies, we can find which natural frequencies give the *largest peak response accelerations*. In our example, we only
tried three natural frequencies (low, medium, and high) and found that
8.0 Hz gave the largest peak response of the three tried, but we are *not* sure
whether some frequency might be worse, since we tried only three.
The SRS method, which we will shortly discuss in some detail, explores a

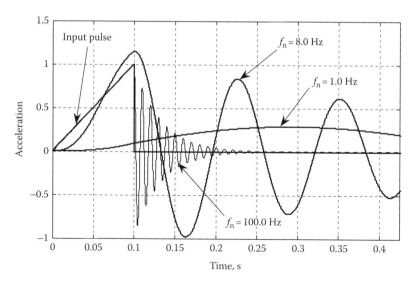

FIGURE 11.5
Response to sawtooth pulse.

very wide range of closely spaced natural frequencies, so that we *cannot* miss the worst one. One use of such an approach, assuming we had the *actual* acceleration input rather than our simple pulses, would be to try to design our system so that its natural frequencies are not close to the one we found to give maximum response.

Before pursuing the SRS further, let us take a look at the *frequency spectra* of the input pulses themselves. For simple pulses, this would be done analytically using the standard formulas for Fourier transforms, while for complex pulses (like many real-world ones) we would have to use a numerical approach with FFT methods available in MATLAB and other software. The frequency spectrum of a pulse tells us how its *strength* is distributed with respect to frequency. When the pulse is applied as the input to some dynamic system, we can get the frequency spectrum of the system output by multiplying the system's frequency response by the frequency spectrum of the pulse (all three quantities are complex numbers with a magnitude and a phase angle). If the pulse has only weak content in a range of frequencies, then the response will also be weak in that frequency range (unless the system has a strong resonance in that range). To get the *time history* of the output signal, we have to do an *inverse* Fourier transform on the frequency function of the output. When designing a system to resist a certain known shock pulse, if we had the transfer functions relating the failure stress of each critical portion of the system to the input acceleration, we could easily compute that stress and decide whether the system is safe or not. As mentioned earlier, practical systems

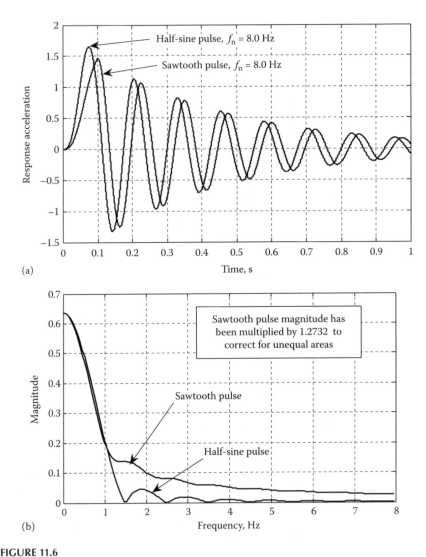

FIGURE 11.6
(a) Response to sawtooth and half-sine pulses of equal area and (b) Fourier transforms of equal-area pulses.

such as *electronics boxes* are much too complex to accurately model and get the needed transfer functions. Thus, knowing the frequency spectrum of the input signal is not directly useful in that way. It *is* useful however in a less quantitative way in that it *does* tell us the frequency range where the pulse is *strong* and where it is *weak*. For example, if a pulse has near-zero frequency content beyond, say, 85 Hz, it will not be able to excite damaging responses beyond 85 Hz.

The MATLAB file below computes the frequency spectra of sawtooth and half-sine pulses from the analytical formulas derived from the definition of the Fourier transform. Both magnitude and phase angle of these complex numbers are computed, but in Figure 11.6b, I show only the magnitudes, which are sufficient to judge the *strength* of the pulses at each frequency. Both the pulses have an amplitude of 1.0 and a duration of 1.0 s, so I have corrected the sawtooth pulse frequency spectrum by multiplying by the pulse area ratio of 1.2732.

```
% pulse_spectra.m computes and plots frequency spectra of shock pulses
% sawtooth pulse
T = 1.0%pulse duration
A = 1.0%pulse amplitude
w = linspace(0.0001,50,200); % define frequency points
f = w/(2*pi);
qr = cos(w.*T)./w.^2 + T.*sin(w.*T)./w -1./w.^2; % compute real part
qim = sin(w.*T)./w.^2 -T.*cos(w.*T)./w; % compute imaginary part
q = (A/T)*(qr-i.*qim); % compute complex value
mag = abs(q); % compute magnitude
ph = angle(q); % compute angle
plot(f,mag); pause
plot(f,ph*57.3);pause
% half-sine pulse, the steps are the same as for the sawtooth pulse
qr1 = -0.5.*(T.*cos(pi/T-w)./(pi/T-w) + T.*cos(pi/T + w)./(pi/T + w)-…
    1./(pi/T-w)-1./(pi/T + w));
qim1 = T.*sin(pi/T-w)./(2.*(pi/T-w))-T.*sin(pi/T + w)./(2.*(pi/T + w));
q1 = A.*(qr1-i.*qim1);
mag1 = abs(q1);
ph1 = angle(q1);
plot(f,mag1);pause
plot(f,ph1*57.3);pause
```

We see that once the pulses are corrected to have equal areas, the low frequency portions of the two spectra appear nearly identical, up to about 1.0 Hz, so in that frequency range, a system *could not tell them apart*. For higher frequencies, important differences appear. First, the half-sine pulse has certain distinct frequencies where it has *no* content, which means that if it were applied to some system, it could excite *no* response at that frequency, even if the system were *capable* of responding there. The sawtooth pulse has no such *zeros* in its frequency spectrum. However, if we were using these pulses as *test* pulses to discover the dynamics of an unknown system, *neither* would be very useful beyond about 1.0 Hz, since the *strength* is so low there that the system under study would not respond enough for accurate measurements. Recall, of course, that the displayed spectra are for a pulse duration of 1.0 s. All we have to do to get spectra that are strong out to higher frequencies is to reduce the pulse duration. The shape of the spectra will be unchanged, but the frequencies will be multiplied in the same ratio as we reduce the pulse duration; a pulse of 0.1 s duration will have the same curves as in Figure 11.6b, but the frequency axis will be labeled 0, 10, 20, ..., 80 Hz.

11.3 Shock Response Spectrum

We now return to our topic of main interest, the SRS. In Figures 11.4 and 11.5, note that the acceleration response curve shows a number of positive and negative *peaks* as time runs on. We are interested only in the *largest* peak since we assume that this value is what might cause mechanical damage or failure. When we apply the *same* pulse over and over to systems with *different* natural frequencies, we find that the size of the largest peak varies as we look at the different systems. An SRS computer program runs the time simulations for a wide range of closely spaced natural frequencies and has some kind of statement which can *pick off* the largest peak value of each time history. If the input acceleration is *all positive* (as in the simulation of Figure 11.3), the largest response peaks will usually be positive rather than negative. For *general* pulses, however, we can get *negative* peaks that are the largest, so to get the largest peak *irrespective of sign*, we can use the absolute value in the statement peak = max(abs(ao)). We imbed this statement within a for-loop which cycles through the time simulations with a fixed damping ratio but with a different natural frequency each time the loop cycles. We accumulate all the peak values and when the for-loop is finished, we plot them against the natural frequency values that correspond to each peak value. The resulting curve is called the *SRS*. There are other, possibly more efficient, ways to calculate the SRS, but this direct method is easy to understand, and with a fast computer it is adequately quick.

When one first encounters the concept of the SRS there seems to be a fundamental defect in the idea. That is, the systems that we deal with in practice are all really *continuous* rather than *lumped* systems, and thus have an infinite number of natural frequencies, yet the SRS gets the response of the simplest possible oscillating system, that with *only one* natural frequency. We regularly model systems that are really continuous (partial differential equations) with approximate lumped models (ordinary differential equations), but even those lumped models often have many significant natural frequencies. The apparent justification for using tools like the SRS seems to rely on the *modal* concept. In simple terms, this states that if the several natural resonances of a system are lightly damped (often true) and *not too close* in frequency, we can treat them (approximately) as individual single-frequency resonances. As with any other engineering tool, acceptance depends on the verification of the predictions by actual applications. The literature, over about 80 years since the method was first introduced, seems to validate SRS as a useful, though not infallible tool. As with most complex concepts, arguments persist about various details; we will examine these shortly.

Let us next show a simple MATLAB/Simulink program which can compute most *versions* of the SRS if it is given the time history of the input motion either as a measured record or as "made up" data. By *versions* we refer to the fact that the input motion (and also the output response) might be chosen to be acceleration, velocity, or displacement. Also, some practitioners have defined response quantities *other* than those just mentioned; we will later explain one of these called *pseudo-velocity,* that claims to be superior to acceleration, the most common response variable. Our program uses the Simulink simulation of Figure 11.3, which was set up to accept an acceleration time history as the input and then computes all the response variables that one might choose for the output quantity. This diagram includes the pseudo-velocity just mentioned, which is *defined* as the product of the relative displacement $(x_i - x_o)$ and the system natural frequency ω_n. We will later give some of the arguments for preferring pseudo-velocity to acceleration, the response variable that seems to be the most used.

```
% shockresponse.m computes shock response spectrum for pulse
   input to second-order system
z = 0.05                    % sets the damping ratio (usually 0.05)
for i = logspace(-2,2,200) % sets the frequency range and number
   of points
fn = i;
wn = fn*2*pi;
[t,x,y] = sim(''shock_response_spectrum'); % runs the Simulink
   simulation
peak = max(abs(ao));                       % picks the peak of
   the response variable (ao in this case)
semilogx(fn,peak,'ko','markersize',3)      % plots the curve,
   one point at a time
hold on
end
```

Since this Simulink simulation includes systems with a wide range of "speeds," we need to be careful to use a proper integration time step. (A 0.01 Hz natural frequency allows a much larger time step than does a 100 Hz.) I took the *easy way out* and used a time step of 0.0001 s for *all* the runs, which makes the low-frequency systems compute slowly, and thus the overall computing time is longer than really necessary. If we wanted to reduce the computing time (it is only a few seconds, so maybe it is not worth it), we could modify the $[t,x,y]$ statement to change the time step each cycle of the for-loop, to suit the current natural frequency.

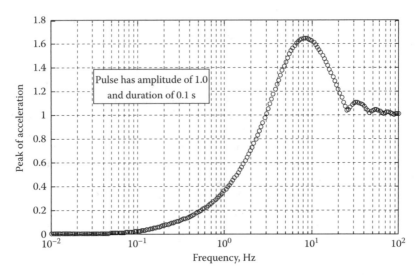

FIGURE 11.7
SRS of half-sine pulse.

The graph of Figure 11.7 shows that, for this pulse, a system with a natural frequency of about 8 Hz will experience the largest peak acceleration, and is thus the most likely to be damaged (this *assumes* that acceleration is the proper criterion for system damage, a view challenged by some workers in the field). This graph uses *symbols* for plotting the points and does *not* allow for the usual "connect the dots" method of getting a continuous curve with *no* symbols. Also, some of the useful features of the MATLAB graphics editor are not available. I found out from the MATLAB tech support people that this is because this program treats each individual point as a *separate plot*, so *connecting the dots* does not make sense to the software. To create a graph without these limitations, we need to *accumulate* all the points *before* we ask for a plot. This can be done by first creating some *empty* vectors and then filling them, one entry at a time, as the for-loop cycles. The program that I finally came up with is displayed below, with the graph as Figure 11.8. Note that if you do not know beforehand the precise shape of any of these curves, you need to be careful to space points closely enough to catch all important details such as peaks and *cusps*.

```
% shockresponse2_log.m  computes shock response spectrum for
                        % pulse input to second-order system
z = 0.05;  % set damping ratio (Simulink needs it), usually 0.05
peak = zeros(1,200);      % create an empty vector for the peak values
f = zeros(1,200);         % create an empty vector for the frequencies
k = logspace(-2,2,200);   % define frequency points
```

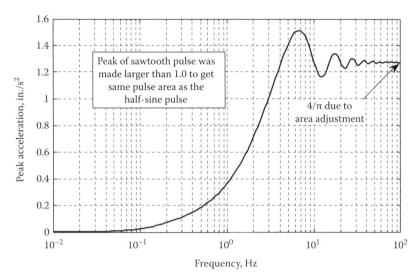

FIGURE 11.8
SRS of sawtooth pulse.

```
for n = 1:200
     fn = k(n);
wn = fn*2*pi;               % Simulink also needs wn values
[t,x,y] = sim('shock_response_spectrum'); % run the Simulink simulation
                            % called 'shock_response_spectrum'
peak(n) = max(abs(ao)); % choose the Simulink variable who's peaks
                        % you want to find
f(n) = fn;
end

semilogx(f,peak)
```

Now that we have a working program, it is easy to experiment with different pulse shapes and different damage criteria. Let us stay with the acceleration damage criterion but take a look at the sawtooth pulse. The only needed change in the program is to modify the Simulink diagram so that the input is now the sawtooth, rather than the half sine. If we want to *compare* this spectrum with that of the half-sine pulse, we also need to multiply the sawtooth input by the ratio of areas for the two pulses, in this case $4/\pi$. This multiplication is suggested *only* if we are *comparing* pulses of various shapes. If we are just looking at the pulse itself, there is no need to do this adjustment.

Figure 11.8 shows the spectrum for the sawtooth pulse with a peak value of $4/\pi$ rather than 1.0, so that we get a fair comparison with the half-sine pulse. This adjustment makes the high-frequency spectrum value go to 1.2732 rather than 1.0 as we saw for the half sine. If we were not *comparing* pulses, we would leave the sawtooth input amplitude at 1.0. Note that after the peak at about 7 Hz, the spectrum becomes nearly flat above about 70 Hz. The reason for this can be seen in Figure 11.5 where the sudden drop in output acceleration after the input reaches its peak (look at the 100 Hz curve) causes rather violent oscillations but none of these *negative peaks* is larger than the "peak" that occurs right at 0.10 s. That is, for high-frequency (stiff) systems, the output just follows the input (except when the input *suddenly* drops and causes the oscillations). With respect to system *damage*, these oscillations can accumulate fatigue stress cycles, but the SRS, as conventionally defined, ignores this feature since it considers only the peak output acceleration. If we want to study the pseudo-velocity damage criterion, we only need to change the `peak = max(abs(ao))` statement to `peak = max(abs(psuvel))` in our program. Also, for those who prefer this criterion, it is conventional to plot the graph using log scales on both axes; it is easily done with `loglog(f,peak)`.

We could do this pseudo-velocity study with the half-sine and sawtooth pulses, but I want to at this point modify the Simulink diagram to accommodate input acceleration pulses of arbitrary shape. Figure 11.9 shows this modification using one of many ways that we could employ to generate *arbitrary* pulse shapes. The input pulse shape and the acceleration response (for a natural frequency of 100 Hz) are shown in Figure 11.10. We see that at this natural frequency, the input acceleration is *magnified* at the output. In Figure 11.11, I compare the response acceleration with the response pseudo-velocity. Except for a scale factor, these two curves look very much alike at this natural frequency. This can be explained by recalling that the pseudo-velocity is nothing but the relative displacement multiplied by the natural frequency. The relative displacement is directly proportional to the spring force and when the damping is so small (zeta = 0.05), the spring force is the *main* force on the mass. Acceleration is proportional to the total force, so the two curves of Figure 11.11 *should* be similar. We next compare the SRS for these two damage criteria using the graphs for each that are conventional (Figures 11.12 and 11.13). I do not show the MATLAB file for this since it is very similar to the one above except for the `peak = max(abs(psuvel))` statement and the fact that I have extended the frequency range out to 1000 Hz, since the input pulse I made up is quite fast.

11.4 Practical Shock Testing and Analysis

Until now I have been using *generic* input shock pulses in order to concentrate on the basic ideas without the possible confusion that might be caused by a number of practical details. We now need to discuss these

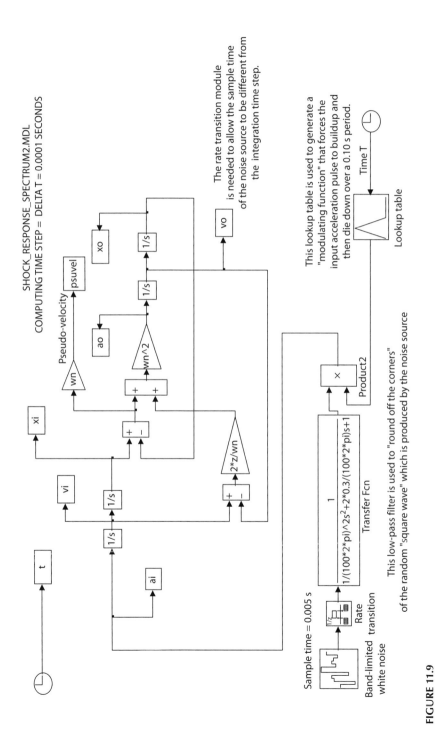

FIGURE 11.9
Simulation that allows *random* pulse shape as input.

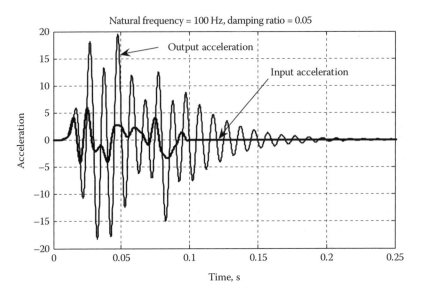

FIGURE 11.10

Response to *random*-shaped input pulse.

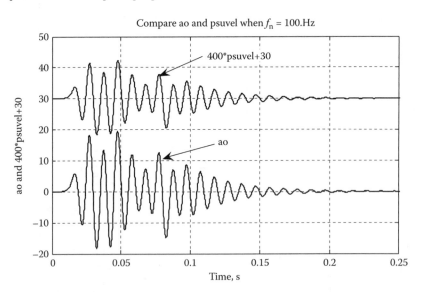

FIGURE 11.11

Comparison of acceleration and pseudo-velocity damage criteria.

details since they are necessary to an understanding of practical shock testing and analysis and the design of shock isolators used to protect sensitive equipment from a given shock environment. A good starting place might be a brief discussion of actual shock-testing machines and the types of shocks they produce. Perhaps most common is the *drop-table* machine,

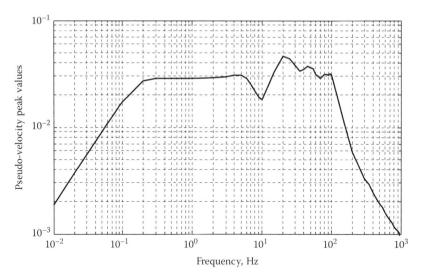

FIGURE 11.12
SRS with pseudo-velocity as damage criterion.

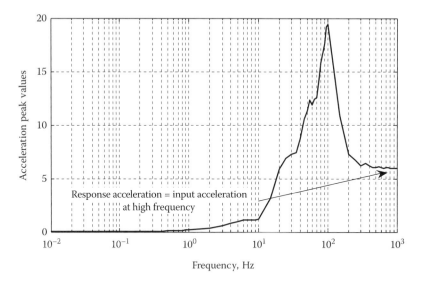

FIGURE 11.13
SRS with acceleration as damage criterion.

as shown in simplified fashion in Figure 11.14. Figure 11.15 shows a real machine. Here the DUT is rigidly fastened to the table, which is then raised to a chosen height and allowed to freely fall onto a *programmer* (such as an elastomer pad), causing a pulse of upward acceleration which is felt by the DUT and recorded by an accelerometer fastened to the table.

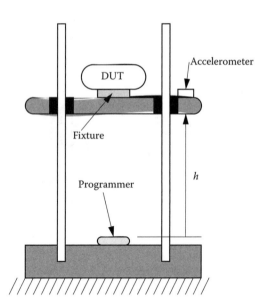

FIGURE 11.14
Basic drop-test shock machine.

Programmers of various types (lead pellets, preloaded air cylinders, etc.) are available to produce the desired shape (half sine, sawtooth, etc.) and duration of pulse. The peak pulse acceleration is set by adjusting the height h of the drop. Since the machines use low-friction bearings for the sliding table, the velocity at impact will be close to that of *free fall*, i.e., $\sqrt{2gh}$, where g is the acceleration of gravity (32.2 ft/s, 386 in./s^2, 9.81 m/s^2). In Figure 11.14, we show a *fixture* used to mount the DUT to the shock machine table. Such fixtures are a necessary feature to connect these two parts and they are carefully designed so that *fixture dynamics* do not unduly affect the test. In fact, there are companies[*] who specialize in the design and manufacture of such fixtures as a service to the vibration/shock testing community. Such fixtures are often made of *magnesium* since it has low inertia and high damping. Also, the location of the accelerometer is critical to accurate measurement and must be carefully chosen to suit each application. The location shown in Figure 11.14 was chosen strictly for convenience in that illustration. The *programmers* usually use an elastomer pad for half-sine pulses, shaped lead pellets for sawtooth pulses, and preloaded air cylinders for square or trapezoidal pulses. With elastomer pads, a soft pad will give a pulse with low peak acceleration and long pulse duration, while a stiff one produces short, high-acceleration pulses (see Equations 10.25 and 10.26). Felt pads or stacked pieces of ordinary paper are also used for half-sine pulses.

Dropping the table leads to a series of physical events, which we need to understand. When the table is released, it falls with nearly 1 g of downward

---

[*] Baughn Engineering, www.baughneng.com

Lansmont's Model 95/115 offers the test engineer a wide range of performance, with its 95 cm × 115 cm (37.4 in. × 45.3 in.) cast aluminum table, 1135 kg. (2500 lb.) payload capacity, and peak half-sine acceleration of 600 g' s. The Model 95/115 can be configured to perform half-sine, trapezoidal, and terminal peak saw-tooth waveforms with minimum set-up times between pulses. The one-piece cast aluminum table, damage-boundary programmers, and integral seismic reaction mass work in harmony to produce extremely clean, repeatable shock pulses. The Model 95/115 is the ideal solution for testing mid-size to larger products where pulse quality and system reliability are of critical importance. The Model 95/115 comes standard with Lansmont's new TouchTest Shock II control system, which allows the operator full control over all shock test parameters and includes advanced features such as shock pulse predictor auto-cycle for consecutive shock pulses, and a wide range of user programmability.

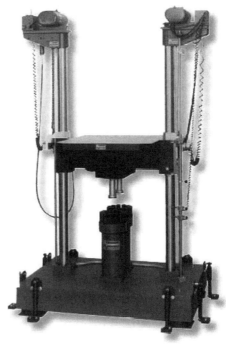

95/115 features:

- Large payload capacity and table surface area make the 95/115 extremely versatile for larger products.

- Wide range of shock pulses are possible with the Model 95/115.

- TouchTest Shock II controls speed up the testing process by simplifying machine set-up and minimizing errors.

- Proven durability and reliability. Dual industrial hoists provide fast, reliable table/payload positioning.

- Redundant safety systems. The Model 95/115 utilizes hydro-pneumatic brakes which automatically engage in the event of power failure. The 95/115 also comes standard with one pressure sensitive safety-mat which returns the machine to a "safe" state if activated.

- Worldwide Customer Service department.

FIGURE 11.15
Drop-test shock machine. (Courtesy of Lansmont Corporation, Monterey, CA, www.lansmont. com.)

acceleration since the bearing friction is low. (Some machines add stretched *bungee cords* (rubber/fabric springs) between the base and the table, to get downward accelerations *higher* than 1 g and thus greater impact velocities.) When the table contacts the programmer (say an elastomer pad, to create a half-sine pulse), an upward force is developed which quickly exceeds the gravity force and produces a large upward acceleration, which is the desired acceleration pulse. When the programmer is maximally compressed, the velocity is zero and the velocity *change* at that instant is equal to the maximum downward velocity (about $\sqrt{2gh}$). If we had been recording the acceleration during the fall, a 1g acceleration would be seen to build up the velocity to its peak; then in a fraction of a second, the huge upward acceleration reduces the velocity to zero. Usually the programmer does not dissipate all of the kinetic energy, and whatever is "left over" causes a *rebound* of the table. After one or more bounces, the table would come to rest (velocity is zero, displacement is the static deflection of the programmer under the weight of the total sliding mass). In an actual shock machine, bounces beyond the first are not a desired part of the test, so an automatic *brake* of some kind catches the rebounding table and holds it fast, to prevent any further impacts.

If we consider what the accelerometer signal would look like for this entire series of events, we can see that this "signature" is more complicated than the simple ideal pulses of Figures 11.4 and 11.5. For example, if we integrate the simple half-sine pulse, at the end of this pulse the upward velocity would be equal to the total pulse area and would continue *forever*, causing the displacement to increase without bound. Real shock machines of the drop-table type include *clamps* which catch the table as it rebounds and hold it fast, preventing a second impact, so we do not need to include provision for this second impact in our adjustments to the input conditions. We *could* include in our input acceleration signal the 1g negative (downward) acceleration which is what actually builds up the impact velocity as the table falls. If this new acceleration signal were integrated, a negative velocity would first be produced and then upon impact, the upward (positive) pulse would provide positive *area*, and at some point the initial negative area would be just canceled by the positive, giving zero velocity at that point in time. Since the peak of the upward acceleration must correspond to the peak upward force created by the programmer (this is assumed to be the point of maximum compression), the zero velocity point must coincide with the *peak* of the acceleration pulse.

Including the initial −1g free fall in our acceleration signal creates a simulation problem in that its time duration is so long (compared with the upward shock pulse) that the computing time for the entire event becomes inconveniently long. A possible way around this is to make use of the *initial conditions* available on any Simulink integrator. That is, in Figure 11.3, vi is the input velocity and it need not start at 0 at time 0.0; we can set this initial condition at any value we want. If we set it at an appropriate negative value, then vi can begin at the velocity value corresponding to the *end* of the free-fall period. That is, we define time = 0.0 to be at the instant of impact, *not* at the beginning of free fall. We can then let time run until the upward (rebound) velocity reaches a value of our choice; this would be related to the *coefficient of restitution* of the elastomer pad. A perfectly elastic rebound would make the final upward velocity exactly equal to the maximum downward velocity.

To find the initial downward (negative) velocity that we should use as the initial condition, we run the simulation with zero initial velocity and note the upward velocity at the instant of peak upward acceleration. We then use the negative of this velocity as our new initial condition, forcing the velocity to be zero at the instant of peak upward acceleration. If this is all that we did, the peak upward velocity would be equal to the initial downward velocity corresponding to a perfectly elastic impact (coefficient of restitution = 1.0). Also, this velocity would continue *forever* and the displacement would go toward infinity. The problem, of course, is that a perfect half-sine pulse of acceleration *cannot possibly* be produced by a real shock machine since it results in velocities and displacements that do not correspond to the *late time* observed behavior. There must be some *energy loss* due to various frictional effects and these energy dissipaters will in some way *deform* the shape of the acceleration pulse. When we measure *actual* shock pulses, all these effects are present and we get an actual pulse shape that results in the proper velocities and displacements. When we *manufacture* pulses for a simulation, we do not get perfectly realistic results unless we try to include some of these features. In Figure 11.16, I have added to the half-sine pulse a downward (negative) acceleration component of about 0.2 at about 0.07 s to crudely simulate a friction effect during the rebound. This feature makes the peak upward velocity less than the initial velocity and also makes the final velocity zero (I got this by forcing the total acceleration to 0.0 at about 0.23 s).

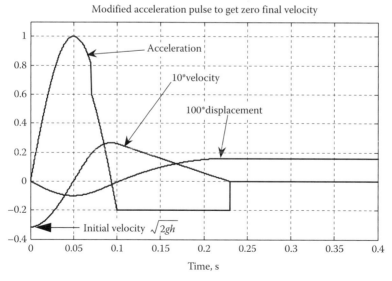

FIGURE 11.16
Input pulse form adjustments to get zero final velocity.

Figure 11.15 shows an actual drop-type machine from one of the large manufacturers. Instead of using elastomer pads (for half-sine pulses) or conical lead pellets (for sawtooth pulses), some machines use so-called *universal programmers*. These can be set up to produce any of the three common pulse shapes using gas cylinders and elastomer pads. Figure 11.17* shows the construction of a universal programmer from MTS Corporation. In Figure 11.17a, to get a half-sine pulse shape, the gas pressure is set high enough so that the gas force on the piston is higher than the peak impact force, thus the piston never moves and the programmer acts as if it were a single rigid piece impacting the elastomer stack. To get a rectangular pulse, the elastomer stack stiffness is set to get the pulse duration wanted, and the gas pressure is set so that the gas force on the piston produces the maximum acceleration wanted. In Figure 11.17b, to get a sawtooth pulse, the elastomer stack is again set to get the pulse duration wanted. The gas pressure is set to give the peak force (acceleration) of the sawtooth. When the impact force exceeds this gas force, the piston suddenly moves upward,

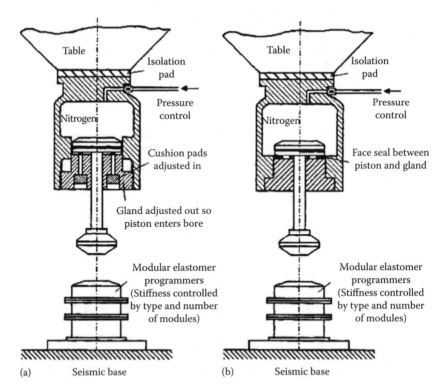

FIGURE 11.17
Universal programmers of MTS Corporation.

---

* C. Lalanne, *Mechanical Shock*, Taylor & Francis, New York, 2002, p. 182.

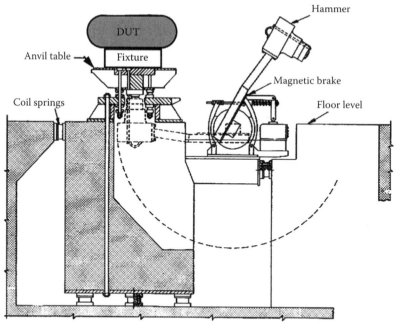

High-impact shock machine for medium weight equipment

FIGURE 11.18
Navy high-impact shock machine.

exposing *both* sides of the piston to the gas pressure, greatly reducing the gas force and thus creating a sudden drop in acceleration.

Figure 11.18 shows a pendulum-type shock machine (U.S. Navy Medium-Weight, High-Impact Shock Machine) with a 3000 lb hammer and 4000 lb anvil used for testing objects (DUTs) weighing up to about 5000 lb. This machine was originally designed in 1942 and is still used today (2009) in one version or another. This machine may not properly simulate some practical shocks of lower-frequency content; the *fixture* shown in Figure 11.18 may be a carefully designed two-degree-of-freedom *mechanical filter* to tailor the shock felt by the DUT to these needs.* Military standard MIL-S-901D specifies the shock test procedures and acceptance criteria for shipboard systems exposed to mechanical shock. The hammer can be raised to a height up to 5.5 ft, and when released, impacts the anvil from the bottom. The anvil and attached DUT are free to move upward (up to 3 in.) before being stopped by a ring of retaining bolts. The two impacts thus produced give a *bidirectional* and complex-waveform acceleration (frequency content up to about 1000 Hz) to the DUT. With this type of machine, it is clear that the DUT velocity and displacement both start and

---

* T.V. Flynn, MSc thesis, U.S. Naval Postgraduate School, Monterey, CA, June 1994.

end at zero. Because of imperfect measurements, the recorded acceleration signal, when integrated once to get velocity and twice to get displacement, often does not give zero final values. To *improve* the raw acceleration
signal, analysts sometimes apply a data-correction called *detrending*. Here
the raw signal is fitted with the best straight line, and then this line is
subtracted from the raw signal. When this adjusted acceleration signal
is integrated, the final velocity and displacement will now be zero. If we
then perform the SRS analysis on this adjusted signal, the result is often
more useful. Commercial SRS software usually includes this detrending
as a selectable option. MATLAB has a simple statement which performs
detrending on any time-varying signal. We will now make up an acceleration signal of this general type, using MATLAB's random number
generator as in Figure 11.9, but we now make the numerical values more
typical of a real application.

Figure 11.19 shows the Simulink diagram for this system. We want acceleration values of several hundred g's, and the equations are written with
units of inches, lb_f, and seconds, so the acceleration values should be in the
neighborhood of $(200g)(386)$, about 80,000 in./s^2. When we run the simulation as shown, we get the signal waveforms as in Figure 11.21a, where it
is clear that the final velocity and displacement are not zero. Since the
signal *ai* (input acceleration) now exists in the MATLAB workspace, we
can use the detrending statement to create a new acceleration signal which
will have final velocity and displacement of zero. We can then substitute
this signal for *ai* in Figure 11.19 and run our SRS analysis based on that
signal. To see the acceleration waveform of Figure 11.21a more clearly,
I zoomed the first part of it in Figure 11.21b. The MATLAB statement for
detrending a signal can be set up to only remove the average value or it
can remove the average value and also remove any linear trend (subtract
the best-fit straight line). If you want *both* the final velocity and the final
displacement to be zero, you have to subtract the best-fit line. This is done
with the statement `aidet = detrend(ai)` where *aidet* is any convenient
name for the detrended acceleration signal. We would then change the
diagram of Figure 11.19 to that of Figure 11.20 and use the Simulink module *from workspace* to bring the detrended *ai* signal from the workspace into
the block diagram. We can then run our SRS program, suitably modified
as shown below.

```
% shockresponse3.m computes shock response spectrum for
% pulse input to second-order system ("peak" means the
% largest value, irrespective of sign)
z = 0.05% sets the damping, usually 0.05
% now create two "empty" vectors to hold the peak and frequency values
peak = zeros(1,200); % empty "peak" vector with 200 points
f = zeros(1,200);      % empty "frequency" vector with 200 points
% k = logspace(-2,2,200);       % use for log-spaced frequencies
k = linspace(0.01,100,200); % use for linearly-spaced frequencies
```

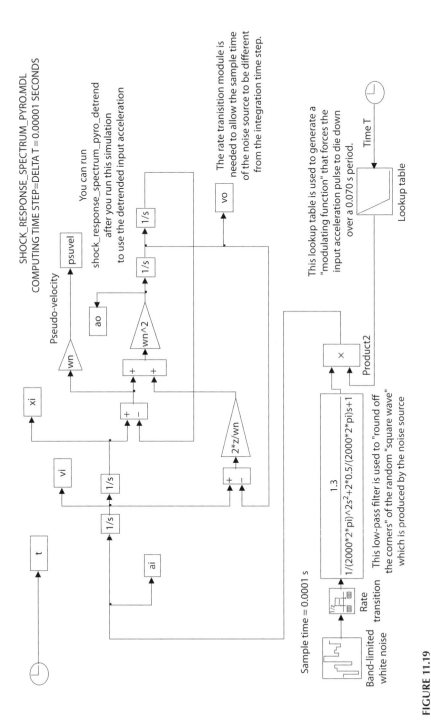

FIGURE 11.19
Simulation for more realistic (random-waveform) shock.

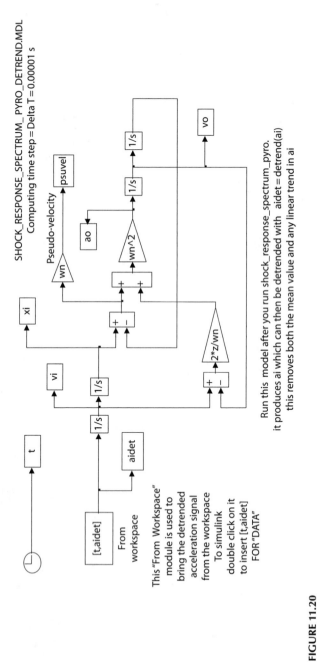

FIGURE 11.20
Simulation to use detrended acceleration input signal.

```
for n = 1:200
    fn = k(n);
    wn = fn*2*pi;
[t,x,y] = sim('shock_response_spectrum'); % use for sawtooth and half
  sine
% [t,x,y] = sim('shock_response_spectrum2'); % use for "general" pulse
% [t,x,y] = sim('shock_response_spectrum_pyro'); % use for pyroshock
% [t,x,y] = sim('shock_response_spectrum_pyro_detrend'); % use for
%                                               detrended pyro
% [t,x,y] = sim('shock_response_spectrum_with_isolater'); % use for
%                                               isolater run
peak(n) = max(abs(ao)); % use for acceleration criterion
% peak(n) = max(abs(psuvel)); % use for pseudo-velocity criterion
% peak(n) = Ei(10000); % use for Tim Edwards' input energy criterion
f(n) = fn; % I tried using fn(n) = fn but that doesn't work!
end
% semilogx(f,peak/386); % use for log-spaced frequencies and accel in
  g's
% semilogx(f,peak);     % use for accelerations in inch/sec^2
% plot(f,peak/386);     % use for linearly-spaced frequencies, g's
plot(f,peak);           % inch/sec^2
% loglog(f,peak/386);   % often used for pseudo-velocity plots
```

11.5 Pyrotechnic Shock

Complex shock signals such as that seen in Figure 11.21 are also produced by explosive devices used to separate sections of space vehicles, (such as disposable fuel tanks) and for other similar functions. These so-called *pyrotechnic** shocks may have frequency content above 100,000 Hz and acceleration levels above 10,000g, depending on the distance away from the explosion site and the severity of the explosion. Bateman (2008) classifies the shock environment into *near-field* (1–6 in.), *mid-field* (1–24 in.), and *far-field* (6–24+ in.). Frequency content beyond about 20,000 Hz may be impossible to measure with even the smallest accelerometers and may be meaningless with regard to damage prediction.[†] Dr. Walter gives four carefully reasoned arguments supporting these conclusions, which have important implications for pyrotechnic shock measurement.

Note that the above program can be used for any type of shock pulse such as the sawtooth and half-sine simulated in the Simulink model *shock response spectrum* (Figure 11.3), the general pulse of *shock response spectrum2* (Figure 11.11), the pendulum machine model *shock response spectrum pyro* (Figure 11.19), and the detrended model which was called *shock response*

---

[*] Vesta I. Bateman, Pyroshock testing update, *Sound and Vibration*, April 2008, pp. 5–6. (downloadable at www.sandv.com)
[†] P.L. Walter, How high in frequency are accelerometer measurements meaningful?, PCB Piezotronics, www.pcb.com, 2008.

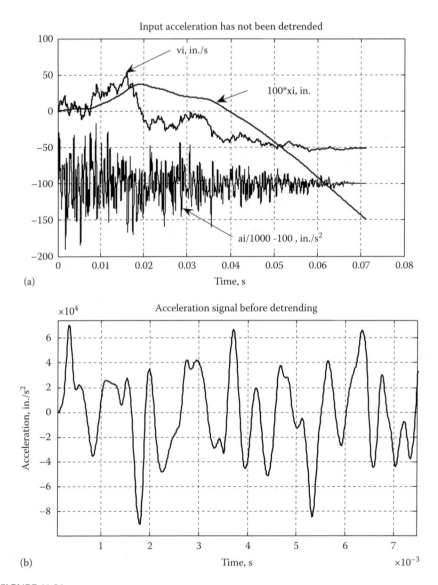

FIGURE 11.21
(a) Responses to random-waveform shock and (b) zoomed acceleration signal.

spectrum pyro detrend (Figure 11.20) by *commenting* or *un-commenting"* certain lines. You could also make up your own Simulink model to suit your needs and then include it in this SRS program. Figure 11.22 shows SRS plots for the acceleration and pseudo-velocity criteria. These two damage criteria are both in use, but there is continued argument as to which is preferred in various applications. We will later return to this important discussion.

When I started to research the area of pyrotechnic shock, I was surprised at the number of such devices actually used in a typical mission. For example, the Space Shuttle, from launch to final landing after a mission's end, uses *several hundred* pyrotechnic devices, each of which produces a shock event with potential damaging consequences. Early space vehicles, such as the Space Shuttle, used explosive bolts/nuts of various types; they are still in use on the Shuttle today. A typical pyrotechnic application is shown in Figure 11.23,

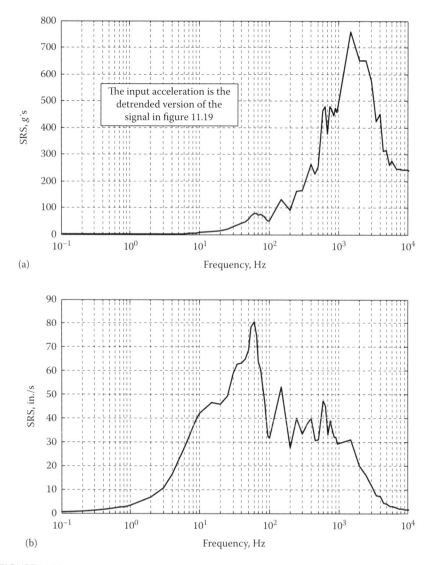

FIGURE 11.22
Comparison of SRS using damage criteria of acceleration and pseudo-velocity.

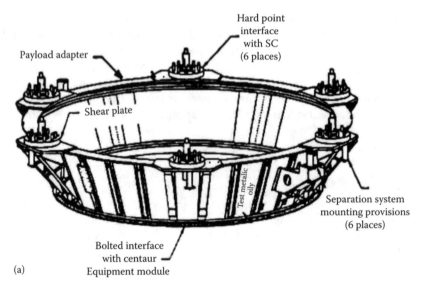

Payload adapter

Hard point
interface
with SC
(6 places)

Shear plate

Test metalic
oily

Separation system
mounting provisions
(6 places)

Bolted interface
with centaur
Equipment module

(a)

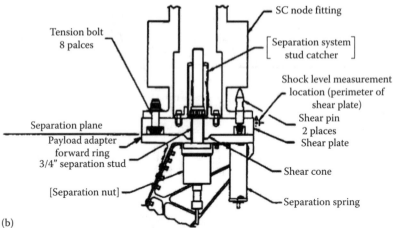

SC node fitting

Tension bolt
8 palces

Separation system
stud catcher

Shock level measurement
location (perimeter of
shear plate)

Separation plane

Shear pin
2 places

Payload adapter
forward ring
3/4" separation stud

Shear plate

Shear cone

[Separation nut]

Separation spring

(b)

FIGURE 11.23
Example of pyroshock from space vehicle separation system. (a) Separation system view of payload adapter and shear plates and (b) section view of separation system. (From Hughes, W.O.
and McNelis, A.M., Statistical analysis of large sample size pyroshock test data set, NASA/
TM—1998-206621, Lewis Research Center, Cleaveland, OH.)

taken from the NASA document referenced there. Unclassified government
documents such as this are a good source of information; private companies
often need to restrict access to internal documents for proprietary reasons.
Figure 11.24 (from a 1973 Langely Research Center report) shows details of
some of these early bolts. Vehicles developed later use other forms of explosive devices with improved behavior. Figure 11.25 shows some alternatives

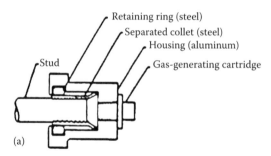

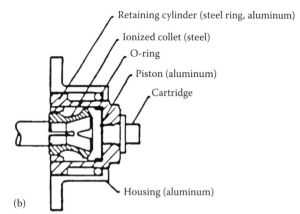

FIGURE 11.24
Early explosive bolt design.
(a) Cross section of noncaptive
separation nut and (b) cross
section of standard design 1.

to explosive bolts assembled from various public domain references. These use *linear shaped charges* to cut through structural members so as to separate parts of the vehicle from other parts. The explosive cords are called *mild detonating fuse* (MDF) and can be configured into various shapes to suit the particular application.

To provide needed separation operations *without* the shocks typical of pyrotechnic devices, engineers have developed strictly mechanical ways, electromechanically actuated, to disengage one portion of a structure from another. One of these called the *Lightband*, from Planetary Systems Corporation,* is shown in Figure 11.26. Such devices still produce shock effects upon disengagement but at a lower level than pyrotechnic devices. However they take up more space and weigh more, so these trade-offs will determine which separation technology is used in a particular application. One drawback of pyrotechnic devices is their *one-shot* nature; they *cannot* be individually tested before their actual use. Mechanical joints like the Lightband can be tested and reused, thus their reliability is more easily verified.

---

* www.planetarysystemscorp.com

In addition to signal transfer, it was also identified that this linear explosive material could be used to perform work. By embedding mild detonating fuse (MDF) into a structural component, the explosive shock and gas output could be used to fracture structural components either bolt fracture or direct fracture of a structural member and allow separation. This approach was used on early strategic missile programs, as well as manned launch programs such as Project Mercury, Gemini, and Apollo and today is widely used on a number of platforms including launch vehicles, strategic missiles, aircraft, and tactical weapons.

Various configurations have been utilized, but typically the assembly consisted of a rigid backing structure (metallic ring or frame) with a locating groove for the MDF, the MDF charge itself, and related explosive transfer hardware (manifolds, transfer line, etc.). This assembly was mounted into the vehicle with the MDF charge precisely located against a specially configured frangible structural member. Upon detonation of the MDF, the explosive shock and gas pressure fractured the structural member in a predetermined stress riser groove, allowing separation of the vehicle.

McDonnell Aircraft Company in 1969 (US 3,486.410, Drexelius et al).

This basic concept was repackaged by Lockheed Missiles and Space Company by placing the expanding tube between two structural couplers, and was sold unde the trade name "Zipcord." An evolution of this configuration used two cords (for redundancy) and is sold under the trade name "Super* Zip." Upon detonation of the MDF, both structural couplers fractured, eliminating the need for the rigid backer. Lockheed patented this system in 1972 (US 3.698.281: Brandt et al.) This approach has been widely used on a number of launch vehicle and spacecraft separation systems.

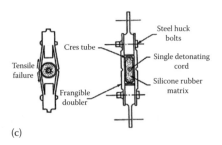

(c)

Super*Zip

Super*Zip, a product of LMSC, is a high load carrying separation system which activates without contamination. This structure-cutting device is commercially used to separate missile stages, payload fairing and spacecraft from their boosters. Super*Zip is a full circumferential ring which joins two shroud structures. Its cross section, as illustrated next, is a flattened tube filled with silicone rubber extrusion with a single strand explosive cord from 9 to 13 gr/ft. Outside the tube, two frangible aluminum doublers with a v-notch in the middle are held together by steel huckbolts. Two detonator blocks are used to actuate the explosive charge.

Shock output from this structure fracturing is quite significant and its shock levels can reach above 10 K g in peak SRS. Shock responses in all three directions are quite pronounced at a distance away. The effects on shock levels of increasing cord sizes are insignificant and can be neglected. Distance attenuation of the shock level in the shelltype structures is extremely slow and can be disregarded over short distances.

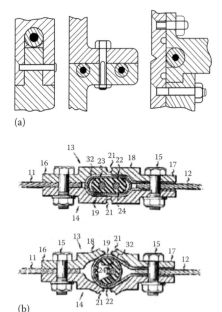

(a)

(b)

FIGURE 11.25

(a) Examples of MDF-based unconfined separation systems. (b) Cross section of Lockheed's "Super*Zip" separation joint before (top) and after function. (c) "Super*Zip" cross section. (Courtesy of McDonnell Aircraft Company, St. Louis, MO and Lockheed Missiles and Space Company, Sunnyvale, CA.)

Test item

•   Lightband is fastened to steel plates with shock accelerometers mounted to the inside center of the steel plates

– The whole test item weights 146 lb

– Each steel plate weights 34 lb

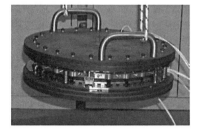

Test setup

Separation shock

Inducing *X*-axis shock

Inducing *Y* and *Z* shock

| The shock the Lightband generates during separation is measured | An external shock is applied to the Lightband in the X-axis by dropping the whole test item on to a circular lead (Pb) target from a certain drop height | An external shock is applied to the Lightband in the Y and axis by hitting the lower plates with a 16 lb sledge hammer |

FIGURE 11.26

LIGHTBAND separation device from Planetary Systems Corporation, Silver Spring, MD.

11.6 Vibration Shakers as Shock Pulse Sources

The SRS of most pyrotechnic shocks are too complex to be accurately simulated by simple shock pulses such as the half-sine or sawtooth forms we earlier studied, thus they cannot be produced in the drop-table type of machine. Pendulum machines which we showed earlier do produce complex, bidirectional accelerations, but the detailed shape of the acceleration waveform cannot be controlled. Faced with this problem, engineers decided to try producing these waveforms on *electrodynamic vibration shakers*, since these systems have an *electrical* input, which can be easily shaped into most any form. The problem, of course, is that the mechanical *output* (actual DUT acceleration) of such shakers may not faithfully follow the electrical command because of various mechanical and electrical limitations of the apparatus. This technique has, however, been successfully developed to the point where it is widely used in practical testing. Sometimes, the shaker is able to reproduce the desired acceleration waveform with sufficient accuracy but in other cases it is only possible to approximate the SRS of the acceleration signal, not its detailed time history. Of course, in some cases it is not possible to produce a difficult waveform with any shaker. Lalanne devotes three chapters* of his book *Mechanical Shock* to this topic indicating its importance in practical shock testing. Commercial software is available which accepts a desired SRS as input and produces a waveform which will give that SRS when applied as a shaker command.

This method is also used with *electrohydraulic shakers* such as the Team Corporation[†] system shown in Figure 11.27. Electrohydraulic shakers cannot reach the high-frequency capability of most electrodynamic shakers, but can accommodate much longer strokes, up to about 10 in. (most electrodynamic

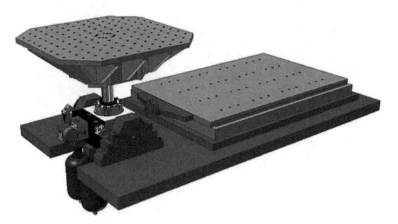

Drawing of Team's SSTS illustrating the trunnion mounted HydraShaker,
T-Film Slip Table, and Vertical Head Expander. All components
are mounted on a steel slab to facilitate installation on the reaction mass,

FIGURE 11.27
Electrohydraulic shaker for shock testing. (Courtesy of Team Corporation, Burlington, WA.)

---

* C. Lalanne, *Mechanical Shock*, Taylor & Francis, New York, 2002, pp. 189–295.
[†] www.teamcorporation.com

vibration shakers are limited to about 2–3 in.). Such long strokes are needed for reproducing earthquake shocks and also have been recently applied to explosive shock effects on equipment mounted on naval vessels. Such shakers are essentially high-performance hydraulic servosystems and require special servovalves and actuators to reach the stroke and speed levels needed. Figure 11.28 shows some of the details of the valves and actuator. Note the unusual configuration of the piston/cylinder actuator, which is used to minimize the volume of oil under compression. Whereas oil is usually treated as incompressible in fluid mechanics courses, in fast servosystems, oil compressibility creates a natural frequency which limits the system's attainable response speed. This natural frequency is raised when we are able to reduce the oil volume under compression in the actuator, improving the high-frequency response.

The shaker systems designed by Team Corporation for defense contractors that supply equipment for U.S. navy vessels had to be able to produce shock acceleration signals that would result in a specific SRS. While the shocks under consideration are the result of explosive events, the shaker system had to deal only with the less severe shocks that would be felt by equipment connected to shock isolators that are already part of the vessel. Team Corporation software was programmed to produce an acceleration input signal to the shaker system that would result in the desired shock spectrum. Several SRS curves are shown at the bottom of Figure 11.29. The *specified* SRS curves (two are shown, for two different values of damping ratio) are plotted as straight-line segments; the *actual* SRS produced by the shaker system is shown as the irregular curves. Note that these spectra are concentrated in rather low-frequency ranges compared to other pyrotechnic shocks which we have displayed earlier. This is because of the mechanical *low-pass filtering* effect of the ship structure and shock isolator hardware. If we were trying to produce the shock as seen at the *input* of the ship structure, it would be more severe in terms of both g level and frequency content.

The acceleration command signal entered into the shaker system is shown at the top of Figure 11.29. Note that the g levels are quite modest. Closer examination shows a definite *symmetry* in this signal; no *real* shock would behave in this way! Note also that the velocity and displacement waveforms obtained by integrating this acceleration signal start and end at zero, and also display symmetry. These features are the result of needing to accommodate the physical limitations of most vibration shakers. We are able to "get away with" this because the customer only required that we meet the SRS criteria; *not* the time history of the acceleration signal. The assumption (that lies behind all applications of the SRS) is that what really matters is the SRS, *not* the detailed time history. Note that the damage criterion used in this SRS is *pseudo-velocity*, not acceleration. This is because the shaker system was to be used to test equipment mounted on naval vessels, and the U.S. Navy has decided that pseudo-velocity is the better criterion for their applications,

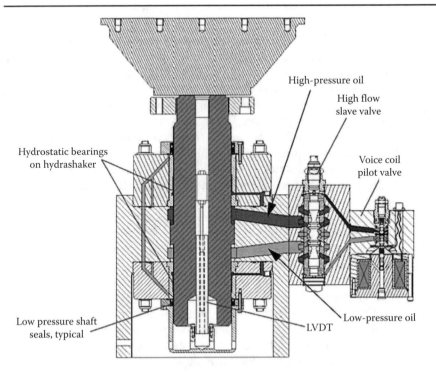

High-pressure oil

High flow
slave valve

Hydrostatic bearings
on hydrashaker

Voice coil
pilot valve

Low pressure shaft
seals, typical

Low-pressure oil

LVDT

FIGURE 11.28

Arrangment of a typical Hydrashaker, showing the internal construction of the voice coil pilot value, the slave valve, and the Hydrashaker piston with hydrostatic rod and piston seals. (Courtesy of Team Corporation, Burlington, WA.)

Servovalve

The key to high frequency response in a servohydraulic vibration system is the ability to control the needed volume of pressurized hydraulic oil to and from the HydraShaker piston. This ability is governed by fluid dynamics and frequency response of the fluid control valve, or servovalve. Team Corporation designs and manufactures the highest performing servovalves in the world, specifically optimized for vibration test system conditions.

A typical servovalve assembly mounted directly to the HydraShaker is composed of two separate valves, the Voice Coil Pilot Valve and the Slave Valve. The Voice Coil Pilot Valve accepts the low level drive signal from the test controller and converts it to a proportional valve spool movement. This results in a proportional flow of hydraulic fluid. The Voice Coil Pilot Valve has a very small spool with low mass and inertia, driven directly by a high power voice coil. Providing a very responsive system with wide dynamic range, the Voice Coil Pilot Valve is able to resolve small differences in drive voltage into correspondingly small differences in spool movement. This high resolution provides the sensitivity to achieve superior test results. For example, the V-20 Voice Coil Pilot Valve can provide full flow to over 600 Hz, rolling off as a second order system but continuing to provide useable, controllable flow to 2 kHz.

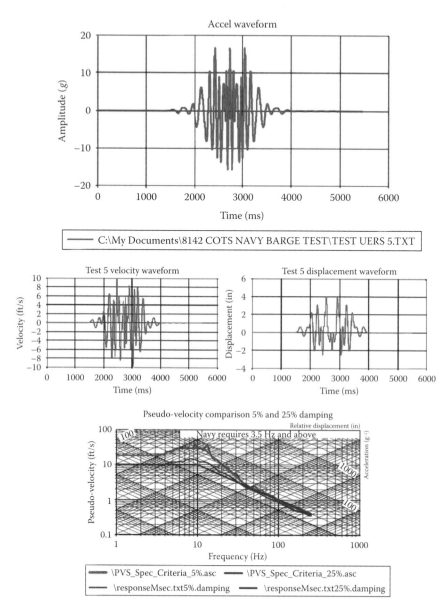

FIGURE 11.29
Results of shock-shaker testing (Team Corporation).

as have the earthquake engineering and nuclear power communities. In the aircraft and spacecraft industries, acceleration is the predominant criterion. In my reading of the literature, and talking to industry practitioners of SRS, I have so far been unable to find any detailed accuracy comparisons of these two approaches, or good arguments for preferring one or the other, which seems peculiar since both methods have been widely used for over 50 years (see Section 11.7 for the one exception). Currently (2007) Tim Edwards and David Smallwood of Sandia National Lab (long time users and developers of SRS methods) are studying a new damage criterion called *Input Energy*. Since Sandia's work is mainly in aerospace applications, they have been using the acceleration damage criterion and continue to do that, until the new method is further developed and evaluated.

While the specific damage criterion associated with the SRS seems to not be a settled issue, there is no question that SRS continues to be the major tool used in dealing with those shock problems where the DUT is too mechanically complicated to allow us to compute or measure the transfer function relating input acceleration to DUT component stress at critical locations in the component. That is, if we had a reasonably accurate transfer function and a measured input acceleration, we could easily use math calculations (or simulation) to get the stress time history at critical points on a component and from this decide if failure would occur, and if so, how to improve the design. For those (many) cases where this is not possible, SRS is the tool of choice. We assume that the acceleration time history of the *attachment point* where our DUT will be connected to the structure has been measured or theoretically predicted. It is also assumed that the DUT is sufficiently small and light (compared to the structure) that its attachment will not significantly modify the assumed input acceleration.

One way that the SRS is used is to specify a *simple* acceleration input whose SRS can serve as a replacement for the *real* input. That is, most real inputs can not be accurately reproduced by any available shock machine, however for many real inputs the *SRS* of the real input *can* be adequately approximated by the SRS of, say, a sawtooth pulse, as easily produced on a commercial drop-table machine. As usual in engineering, we design the pulse test so that its SRS is a *conservative* model of the actual SRS. That is, it should slightly *exceed* the real SRS in the frequency ranges considered significant for the given DUT. This procedure is called *enveloping*, and we see it in the lower graph of Figure 11.29. In that figure, the real acceleration input was strongly *bidirectional*. Such inputs *cannot* be adequately modeled by simple pulses such as half sines or sawtooths, so a *shaker* (rather than a drop-table machine) had to be used there. For shocks that are mainly unidirectional, a half-sine or sawtooth pulse is often a useful model. For an existing DUT, if it survives the shock test using the *model* SRS, it should survive the actual shock. (Figure 11.30 shows an actual shock test using the TEAM electrohydraulic system.) If it does not survive, we need to diagnose the failed components, redesign, and shock test again, until we arrive at a reliable design. Redesign might mean changing the design of internal DUT components, or, we might choose to instead interpose a *shock isolator* between the structure and the DUT. The design of such isolators is facilitated by the SRS concept, as we next show.

An example of a large cabinet being installed on Team's SSTS, prior to testing in the vertical direction.

FIGURE 11.30
Shock testing with the TEAM electrohydraulic shaker.

11.7 Design of a Shock Isolator

In Chapter 2, we treated some methods for protecting sensitive instruments/ machines from vibration inputs. We will now look into those situations where the problem is one of surviving some kind of *shock* input. The general approach is, of course, much the same whether the protected device is an instrument or some other delicate object. If some component in our DUT fails the shock test, it means that the SRS is too high in one or more frequency regions where that component perhaps exhibits a resonance or other excessive response. We can then try to design a shock isolator which will reduce the SRS in those sensitive frequency regions. The most common isolation concept is to simply introduce a "soft" spring between the structure and the DUT. If the spring is soft enough, the natural frequency of the system made up of the DUT total mass and the added spring is very low, which produces a low-pass mechanical filtering action. This filters out the offending frequency content and hopefully may prevent component failure. We, of course, check our isolator design by running the shock test again with the isolator present. We need to warn, of course, that the isolator adds weight, volume, and cost. Also, the required spring may be so soft that its static deflection and/or stability may not be acceptable. In speaking with industry practitioners when preparing this chapter, I found that isolators are often used in naval vessels, but rarely in aerospace applications because of the added weight and space. The aerospace solution to shock environments often consists of "hardening" the components to shock by perhaps encapsulating them in silicone rubber or other media.

When an isolator is feasible, its design might proceed as follows, based on SRS concepts. Usually the needed spring is soft enough that the natural frequency of the DUT mass and added spring is quite low. When this is the case, the DUT really *does* act like a rigid single mass; there is little motion of the inside components relative to the containing *box*. We can then analytically (or with simulation) subject this simple spring/mass system to the given acceleration input and compute the acceleration of the DUT mass. This acceleration of the DUT mass is then taken as the *input* acceleration in our SRS simulation, such as Figure 11.3, and we compute and plot a new SRS for this isolated system. Figures 11.31 through 11.33* show a commercial shock mount which is essentially a damped spring. The spring is made of wire rope bent into a specific three-dimensional curved shape; several manufacturers have produced these for many years in various sizes and shapes. The HERM mount takes one of these wire-rope

FIGURE 11.31
Damped shock isolator of Enidine Corporation.

---

* www.koni-enidine-defense.com/

Enidine Shock and Vibration Isolation Application
HERM offers Naval "Soft Deck" Isolation
By William Shormatz

Application Overview
A major defense contractor and manufacturer of US Navy shipboard electronics cabinets required shock and vibration protection across a wide range of shipboard installations and deck frequencies.

The isolators used to protect these critical control systems would have to isolate both ruggedized and COTS (Commercial Off The Shelf) electronic components within the cabinets. They also needed to isolate shipboard shock inputs of MIL-S-901D Grade A, vibration requirements of MIL-STD-167; and structure borne noise. The isolator needed to be packaged to mount under a 72-inch-high cabinet, facilitating installation into the restricted overheads of shipboard compartments.

Product Solution
Assisted by dynamic modeling, design verification testing and US Navy sponsored barge testing. Enidine developed the High-Energy Rope Mount, or HERM isolator. The HERM isolator incorporates the use of a traditional Enidine helical wire rope isolator encased in a proprietary elastomeric compound. The stainless steel cable of the mount provides for a rugged construction, while the elastomer provides additional damping and stiffness. This unique design results in a fail safe mount with a higher stiffness and energy absorption capacity. The mount is readily scalable and performance easily tuned by varying the wire diameter, loop size, number of loops and elastomeric properties. The HERM isolator has proven particularly strong in low natural frequency "soft deck" applications of 12–16 Hz, reducing output G's to below 15G's. Its sealed nature of construction also provides for easy NBC washdown. Since the mounting size of the HERM isolator is virtually identical to that of standard wire rope isolators used in many shipboard applications, equipment upgrades are both simple and seamless with drop-in replacement capability.

Application Opportunity
The HERM mount is an effective isolator for cabinets, platforms, rafts, cradles, canisters and shipping containers as a shock, vibration and structureborne noise isolator. In addition to electronic cabinets, the HERM has applications on radar systems, weapons and vehicles. Any defense contractor needing shock and vibration protection over a wide deck frequency range could benefit from this application.

FIGURE 11.32
Naval applications of shock isolation. (Courtesy of Enidine Corporation, New York.)

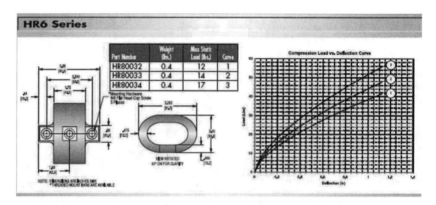

FIGURE 11.33
Characteristics of Enidine shock mount.

mounts and encapsulates it in a special elastomer, which provides the desired damping. Without the elastomer, wire-rope mounts typically can provide a damping ratio of about 0.15 due to the Coulomb friction of the wire-rope strands rubbing against each other. With the added elastomer, this goes up to 0.20–0.25. The nonlinear force/deflection curves of Figure 11.33 show that the mount is a *softening spring*; its stiffness gets less as the load increases. In applications where the mounted object changes mass (like a fuel tank that is not always full) and where we want a fixed natural frequency for the spring-mounted object, we would want a *stiffening spring*. The Enidine folks said that most of their applications involve constant mass, so the softening characteristic is not a problem.

Suppose that we have shock tested an "electronics box" with the shock that produced the acceleration SRS of Figure 11.23a and found that it produced failures of one or several internal components. We might try to diagnose these failures to estimate natural frequencies associated with the failed components. The purpose of such studies would be to identify those frequency regions where the SRS has to somehow be reduced below the damage level. Often such studies are not feasible and we opt to reduce the SRS "across the board," rather than just in certain frequency ranges. In this case, the approach is often to install an isolator with a natural frequency low enough to strongly attenuate the shock over most of the frequency range. Such isolators usually *cannot* use heavy damping since that reduces the attenuation at higher frequencies, so that we will need to accept some *amplification* near the isolator's natural frequency. Usually the shock strength in this range is not excessive, so we can tolerate some amplification there.

Let us assume that our instrument box weighs $20\,\mathrm{lb}_f$, that we want to try an isolator with a natural frequency of 5.0 Hz, and that the shock mount will provide damping of about 0.15. A simple calculation shows that this requires a spring stiffness of $51.1\,\mathrm{lb}_f$ /in., which allows us to select a specific mount, such as in Figure 11.33. (Since the box will probably be supported at its four

corners, we will need four mounts whose individual stiffness is about 51.1/4.)
To build a simulation which will allow us to check out this design, we can
modify that of Figure 11.20 by adding the isolator dynamics as a "pre-filter"
on the acceleration input *aidet,* as in Figure 11.34. The isolator is modeled as
a spring/mass/damping system, which, of course, is the same model that
we use for our SRS calculations, so we just duplicate this model for our pre-
filter, and give it the desired natural frequency and damping ratio. The input
to the isolator system is the detrended acceleration called *aidet,* which we
get in the same way as before. The output is the acceleration *aismass* of the
mass which represents the electronics box. This output is also the *input* to
our usual SRS calculation. In the isolator model, we are also interested in
the absolute displacement *dispmass* of the electronics box (to check how far
it moves under the shock) and the shock mount deflection *reldisp* (because a
particular mount can tolerate only a certain maximum deflection).

Our SRS calculation provides for three possible damage criteria: acceleration, pseudo-velocity, and input energy. While acceleration and pseudo-
velocity are routinely used by the various groups mentioned earlier, *input
energy* is a new concept proposed by Tim Edwards and David Smallwood of
Sandia National Lab, and is presently being studied as a possible addition
to, or replacement for, the two "traditional" criteria. When we run our new
simulation that includes the isolator, using the acceleration damage criterion,
we get the SRS graph of Figure 11.35. Comparing this with Figure 11.23a,
we see that the intensity of the shock event, as felt by the electronics box,
has been greatly reduced. Without the isolator, the SRS shows acceleration of
hundreds of *g*'s over a large range of frequencies, whereas with the isolator,
the highest peak is only about 8*g*. This peak is near the isolator's designed
natural frequency of 5 Hz, and such a peak will always occur with this
type of isolator; it is a price that we pay for the isolator's beneficial effects.
Fortunately, this sort of peak is usually not damaging. Figure 11.36 shows
the time history of the isolator input and output accelerations when the SRS
calculation is doing the 5.0 Hz frequency point. Note that acceleration of the
electronics box now *lasts* much longer than without the isolator, but its level
is now a few *g*'s rather than a few hundred *g*'s. The 8*g* peak value (which is
what the SRS calculation picks off) occurs at about 0.44 s.

We need to check on the motion of the electronics box to see how much
space must be allowed around it, so that it does not bump into adjacent equipment. Getting a plot (not shown) of *dispmass,* we find that the box moves about
6 in. under the action of the shock. Without having a specific application in
mind, we cannot say whether this is acceptable or not. If it were too large, we
would try an isolator with a higher natural frequency, which should reduce
this excursion, but at the price of larger box accelerations in the SRS. Since
our 5 Hz isolator produced very small accelerations, perhaps those produced
by a stiffer isolator would still be acceptable. Figure 11.37 shows the relative displacement between the two ends of the shock mount, which has a
maximum value of about an inch. This would probably be acceptable, and

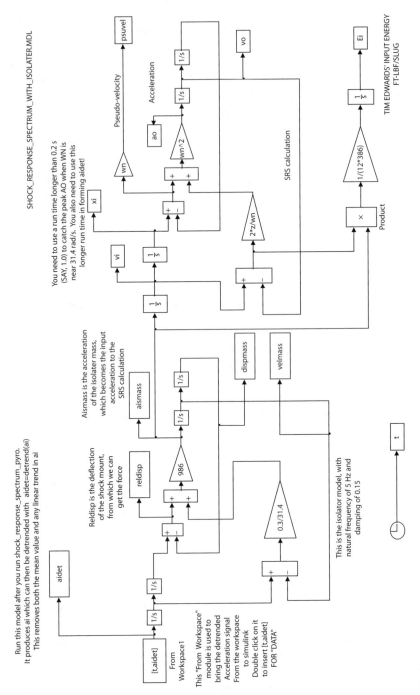

FIGURE 11.34
Simulation for shock isolator study.

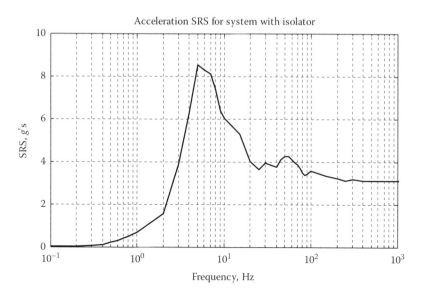

FIGURE 11.35
Isolator improves shock loading of DUT.

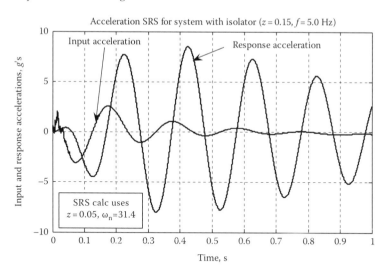

FIGURE 11.36
The SRS calculation at the 5.0 Hz frequency point.

corresponds to a total force (all four of the corner mounts) of about 58.5 lb. We could rerun our simulations using the pseudo-velocity damage criterion but the isolator principle does not really depend on this, so our good results would be essentially duplicated.

At this point, we have covered most of the basic concepts and analysis methods associated with the SRS method. I need to point out that all

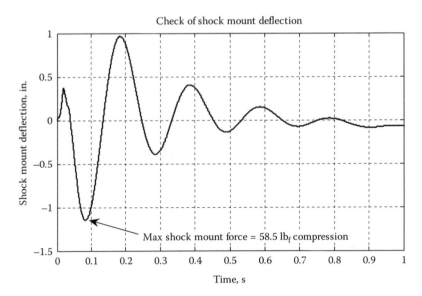

FIGURE 11.37
Extension/compression of the shock mount.

our examples dealt with only the simplest geometrical configuration, a single-degree-of-freedom system. Some practical applications will involve more complicated geometries, that is, shock inputs and responses with several degrees of motion freedom. These complexities are beyond the scope of these notes and have little to do with SRS per se. Rather they address methods used to analyze the motions of multi-degree-of-freedom systems regardless of the damage concept employed. Many books* and papers deal with these areas.

11.8 Relation of SRS to Actual Mechanical Damage

This is the most controversial aspect of the SRS method and in my reading, the least discussed. I have already mentioned several times before the two major "camps" associated with this question: the *acceleration folks* and the *pseudo-velocity folks*. It seems that most users of the SRS method do not concern themselves much with this fundamental question, but rather choose one of the two damage criteria and then proceed to apply that to their own shock problems.

---

* C.M. Harris and C.E. Crede, Eds., *Shock and Vibration Handbook*, McGraw-Hill, New York, 1961, pp. 30-1 to 32-53.

I found one paper* (Gaberson et al. 2000) that deals with this question and reaches a conclusion. Dr. Gaberson worked for many years at one of the Navy labs and has been intimately involved with shock problems over a career of about 40 years. I had an extensive e-mail correspondence with him while developing this chapter, and he graciously supplied a wealth of technical information to me. In addition to the referenced article, he has many other papers and reports on the subject. He has concluded that pseudo-velocity is the proper damage criterion and provides some detailed arguments[†] for that position. Recently I was in touch with Tim Edwards at Sandia National Labs. Together with David Smallwood (a guru of the SRS method with many publications), he is developing a third damage criterion called *input energy*. In early studies[‡] of this method, it seems to be closely related to pseudo-velocity. As an observer of (rather than an active participant in) the SRS scene, I hesitate to endorse any of the commonly used damage criteria. Gaberson et al. (2000) is the only study I found that directly addressed this question and performed actual experiments in an attempt to validate the pseudo-velocity criterion. As a nonexpert in this field, I found it quite convincing. If you are going to use SRS I suggest you read it and reach your own conclusion. The experiments involved subjecting an electric-motor-driven blower (similar to what is in many home heating systems) to one of six different shock events. For each of the shock events, a blower sample was subjected to increasing levels of shock until some kind of failure was observed. Blowers which survived a non-damaging shock were *not* tested again at a higher shock level since there might be some kind of cumulative damage effect.

Shock machines used included a drop-table type, a lightweight Navy pendulum machine, a medium weight Navy pendulum machine, and a Navy heavyweight test using a floating barge subjected to underwater explosive blast. The drop-table tests used three different shocks: a long-duration half-sine pulse, a terminal peak sawtooth pulse, and a short-duration half-sine pulse. These are shown as panels *a, b,* and *c* of Figure 11.38, which is taken from Gaberson et al. (2000). Note that, as we have mentioned before, the *real* shock pulses produced by the various machines are somewhat different from the ideal shape intended. The acceleration traces shown are for the shock level which *just barely* caused some kind of failure. Trace *d* was for the light-weight machine. This shock was the most severe possible with that machine but *did not* cause any failure. Trace *e* was for the medium weight machine, where standard practice requires mounting the DUT on two *somewhat springy* beams. The obvious low-frequency *damped sine wave* seen there is probably the natural frequency of the blower mass on these beams. This trace is for the

---

* H.A. Gaberson, D. Pal, and R.S. Chapier, Classification of violent environments that cause equipment failure, *Sound and Vibration*, May 2000, pp. 16–23 (downloadable at www.sandv.com).
[†] H.A. Gaberson, Pseudo-velocity shock spectrum rules for analysis of mechanical shock, a 36-page unpublished "white paper" available from him at hagaberson@att.net
[‡] T.S. Edwards, Using work and energy to characterize mechanical shock, unpublished "white paper," April, 2007, tsedwar@sandia.gov

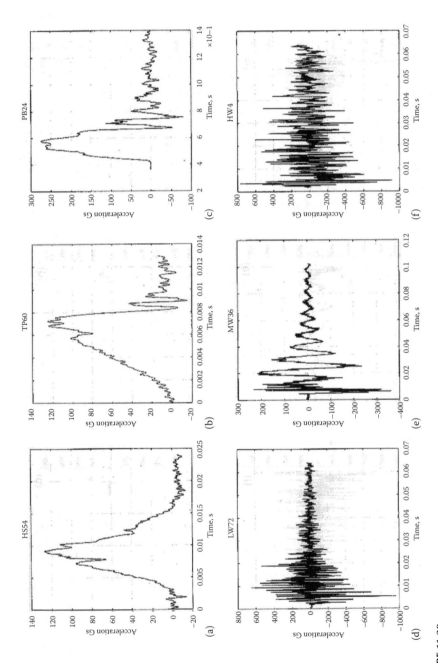

FIGURE 11.38
Measured data from blower shock test.

final test of this type, and it *did* cause failure. Finally, the trace of part *f* is for the barge test, where the blower is mounted on a floating barge. Underwater explosions are produced, moving the explosive closer and closer to the barge until failure occurs; this trace is for the *final* such test, which did produce failure.

All six acceleration inputs were analyzed, using standard SRS methods, for both the output acceleration and the pseudo-velocity damage criteria, and for SRS damping values of 0, 0.05, and 0.20. From these graphical results, it was found that the shock from the lightweight machine (called LW72, panel *d*), which did *not* cause any failure, gave *higher* acceleration SRS values than those of the other five tests, which *did* cause failure, *not* what one would expect. When the pseudo-velocity SRS graphs were compared, and damping of 0.05 or 0.20 was used, the SRS values for the shock which did not cause failure were *lower* (in the frequency range up to about 80 Hz) than those for the five tests which did cause failure. However, at higher frequencies LW72 gave *higher* values. The authors take these results as a validation of the superiority of the pseudo-velocity criterion. As a careful reader of the article, but as a nonexpert in SRS, this evidence is suggestive to me but perhaps not quite compelling. I believe the strongest evidence is the *negative* result for the acceleration criterion; it *falsely* identified the LW72 shock as the worst of the six, irrespective of damping value or frequency ranges. The favorable results for the pseudo-velocity method are only for the lower-frequency range. The authors qualitatively analyze the actual failures (stripped sheet metal screws, deformed bearing-support spider, fractured welds, motor popped out of its mounting bracket, deformed mounting leg) and speculate on the frequency ranges associated with each failure type. Their opinion on these frequency ranges puts them in the low-frequency portion of the spectrum but it seems these conclusions are more in the nature of informed guesses than *scientific* proof. I had hoped that, after publication of this article, these conclusions might be thoughtfully challenged by other experts who favor the acceleration criterion, but I could not find any published evidence of this.

In addition to the above experimental results (the *only* experiments addressing this important question that I have found), the pseudo-velocity proponents offer some *theoretical reasons* why pseudo-velocity should be a good damage criterion. Most of these arguments are discussed and/or referenced in Gaberson (unpublished white paper). Without presenting them in detail here, the essence of the argument goes as follows. Theoretical analysis of steady sinusoidal vibrations of structural elements such as rods (longitudinal vibration) and beams (lateral vibrations) shows that when vibrating at one of the natural frequencies, the *stress at the highest stress point* is proportional to the *velocity* at the highest velocity point. For example, in longitudinal rod vibrations,

$$\sigma_{max} = \sqrt{\rho E v_{max}} = K v_{max} = K\omega x_{max} = K\frac{a_{max}}{\omega} \qquad (11.2)$$

where
   ρ is the material mass density
   E is the modulus of elasticity
   v_{max} is the maximum velocity
   x_{max} is the maximum displacement
   a_{max} is the maximum acceleration
   ω is the frequency

We could thus express the maximum stress in terms of either velocity, displacement, or acceleration, including the frequency, as needed, in Equation 11.2. Pseudo-velocity proponents choose to use the form based on relative displacement x_{max}, giving rise to the definition of pseudo-velocity that we have used earlier in these notes: $(x_i - x_o)\omega_n$. Part of the reason for this choice is that, when defined in this way, SRS plots of pseudo-velocity versus frequency on a special graph paper exhibit low- and high-frequency asymptotes that have a useful physical meaning. Figure 11.39 (taken from Gaberson unpublished white paper) shows a typical pseudo-velocity SRS plot on this paper, called *tripartite* or *four-coordinate* paper. This paper allows plotting of pseudo-velocity values in in./s units on the vertical axis and frequency in Hz on the horizontal axis. One can then read out acceleration values and displacement values on the other two inclined grids. If the input acceleration record includes the initial free fall of a drop-table machine, then the low-frequency asymptote will lie along a constant-displacement grid line with a numerical value equal to the drop height in inches. The high-frequency asymptote will lie along a constant-acceleration grid line equal to the peak input acceleration.

Gaberson (unpublished white paper) also gives some numerical values for pseudovelocities that relate to actual failure. For example, in aluminum alloy 6061-T6, to produce a stress of, say, 35,000 psi, the pseudo-velocity value would be 695 in./s for a rod vibration, 402 in./s for a beam vibration, and 347 in./s for a plate vibration. Quoting from the reference: "For long term and random vibration, fatigue limits as well as stress concentrations, and the actual configuration would make the values much lower."

Figure 11.40 shows another attempt to quantify damage for a particular class of equipment. I got this figure from engineers at NASA Glenn Research Center. It apparently came from the workshop listed on the figure. I tried to get more details about this conference, but only found out that this workshop seems to be a yearly event.

With regard to the ongoing controversy over the relative merits of the acceleration and pseudo-velocity damage criteria, let me offer the following final comment. Figure 11.23a and b shows that the two criteria give rather different indications of the critical frequency regions, at least for the specific

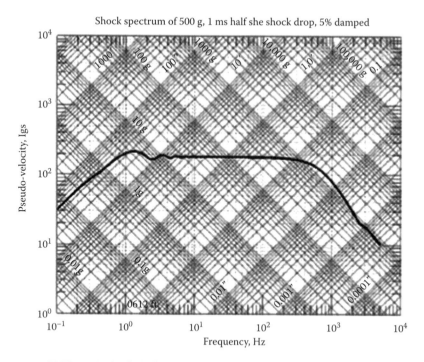

Shock spectrum of 500 g, 1 ms half she shock drop, 5% damped

PVSS on 4CP for the half-sine shock of figure 3. Notice the high frequency asymptote is on the constant 800 g line. That the velocity plateau is at a lilttle under 196 ips. and that the low frequency asymptote is on a constant displacement line of about 50 in.

FIGURE 11.39
Pseudo-velocity SRS plotted on four-coordinate paper.

input acceleration used there. For other input waveforms, this disagreement might be less obvious, but it is always there to some extent. *Yet, both criteria have been used successfully by many engineers for over 40 years!* I tried to find published discussions of this question, but found only the several papers by Dr. Gaberson which directly addressed it. One possible explanation, which I offer for your consideration, goes as follows. For the blower experiment discussed above, the waveform in panel *d* did not cause failure, but its acceleration SRS was *more severe* than that of five other waveforms which *did* cause failure. That is, the acceleration SRS *overestimates* shock severity in this experiment. What if this were always (or often) the case? Then, engineers who used the acceleration criterion would be routinely *overdesigning* their applications, but they would be "successful" because there would not be any (or many) failures. This is pure speculation on my part, and I offer it mainly to stimulate some thought that might produce alternative arguments from readers of this chapter.

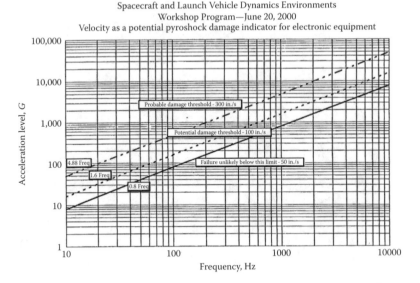

FIGURE 11.40
Some suggested NASA damage criteria.

11.9 Measurement System and Data Acquisition/ Processing Considerations

We have largely *assumed* in this chapter that the measured shock record (usually acceleration) is a proper one for processing through the SRS calculations and applying to our particular application. That is, our focus was on the description and use of the SRS itself, not the instrumentation needed to produce a valid input acceleration record. As usual, the setting up and use of any measurement system deserves critical attention and requires specific knowledge and experience. The intended scope of this chapter did not include such important details but we should at least alert the reader to some sources for this kind of information. Walter (2008) offers good advice on the high-frequency limits inherent in accelerometer shock measurement. Since accelerometers with natural frequencies higher than 20,000 Hz (Walter's suggested limit) are available, the output voltage signal can have components above 20,000, *even if they are really meaningless*. We can apply electrical low-pass filters to the voltage signal, but the accelerometer output voltage may have suffered from overloads that may not be apparent in the filtered signal. An excellent paper explaining this kind of problem and how to deal with it appeared in *Sensors* magazine in 2003.* One possible approach is to

---

* S. Smith, Test data anomalies-when tweaking's OK, *Sensors*, December 2003.

use accelerometers with *mechanical* filters.* A very extensive compilation of papers and reports on various aspects of SRS is available from Tom Irvine[†] but requires a \$40/year subscription fee, or the purchase of a CD. Endevco Corp. offers a wide variety of technical papers[‡] addressing problems in acceleration measurement, in particular, TP 308, 303, 299, 294, 296, 293, 290, 288, 218, and 213. Kistler also has a good selection of technical notes,[§] as does PCB.[¶] Technology Training Inc. offers courses and technical articles** (particularly those by Strether Smith) related to measurement and data acquisition.

---

* E.O. Doebelin, *Measurement Systems*, 5th edn., McGraw-Hill, New York, 2004, pp. 876–878.
† http://www.vibrationdata.com/SRS.htm
‡ http://www.endevco.com/resources/TechPapers.aspx
§ http://www.kistler.com/us_en-us/923_Specials_dyn/app.documentSpecials/Kistler.html
¶ http://www.pcb.com/products/literature.php
** http://www.ttiedu.com/ddaa-articles.html

Appendix A: Basic MATLAB®/ Simulink® Techniques for Dynamic Systems

A.1 Basic Simulink Techniques

A.1.1 A Little History

The earliest engineering application of dynamic system simulation was probably the use of electrical circuits to represent mechanical and acoustical systems. The differential equations that describe many electrical circuits have the same form as those for many mechanical systems. Thus, a mechanical spring and an electrical capacitance play similar roles in their respective equations, while mechanical inertia and electrical inductance, and mechanical viscous friction and electrical resistance are other analogous pairs. One can thus run experiments on electrical circuits and get useful results for mechanical systems. The motivation for actually doing this is that the electrical circuits allow easier changes to their parameters, the *driving forces* in circuits are voltages (which are easily manipulated), and the measurements of results are dynamic voltages, which are easily measured on standard recorders or oscilloscopes. This approach was eventually superseded by more convenient methods and is little used today.

About 1930, the mechanical *differential analyzer* was developed at MIT and was used up to the 1940s for solving differential equations. It used a mechanical integrating device called the *ball-and-disk integrator* together with other mechanical devices. It was not commercially produced and marketed but rather was constructed "as needed" by government agencies or universities. During World War II, it was widely used to compute artillery and bombing ballistics, but is now obsolete.

Next in the development of simulation was the invention (around the 1940s) of the *operational amplifier* (op amp), which first was constructed with vacuum tubes and later used transistors. This device revolutionized simulation since it performed basic mathematical operations, rather than depending on physical analogs. The basic op amp device in the simulation of dynamic systems is the *integrator*, constructed from one op amp, one resistor, and one capacitor. This device is considered basic because the solution of differential equations requires the *integration* of the highest order derivative in an equation to get the next lower one. Having that derivative one just integrates again (and again, and again) until finally we have the unknown quantity itself. *Electronic analog computers* use integrators and other electronic components (devices for setting numerical values of equation coefficients, nonlinear devices such as electronic multipliers, etc.) to simulate entire equations and sets of simultaneous equations. They were the workhorses of dynamic system simulation in the era of the 1940s through the 1960s. One of the largest users of such installations was the aircraft industry. Columbus Ohio had at that time a facility of the North American Aviation Corp. (later North American Rockwell), which designed and built various naval aircraft, including the Vigilante, a Mach 2 aircraft that operated from aircraft carriers. (When I started teaching, I spent several summers working at this facility.) The development of such high-performance aircraft depended on sophisticated simulation capabilities, including one of the world's largest analog computers, which was also interfaced with physical simulations of actual hardware. General-purpose analog computation is now essentially obsolete, but special-purpose computation using op-amp techniques is still a viable approach in many applications.

When the first digital computers appeared, their application to dynamic system simulation soon became practical. At first, general-purpose languages such as FORTRAN had to be used, but soon special-purpose languages were developed, making applications much easier and quicker. The OSU Mechanical Engineering Department for many years used the IBM CSMP (*continuous system modeling program*), which was popular around the world. At first, we had to punch our programs into paper cards and submit them to a large mainframe computer, waiting several hours for results. As minicomputers appeared, the large mainframes used by the entire campus were augmented with departmental minicomputers. Dynamic system simulation languages, such as CSMP, were *command-line* languages where one had to write programs line by line. While the command-line method is still available in most current simulation packages, we usually prefer to use a GUI (graphical user interface)-based approach. Such methods, together with the speed and easy accessibility of the personal computer make dynamic system simulation a powerful tool for engineering analysis and design.

MATLAB/Simulink is one of the most popular simulation packages. While it allows a command-line approach, in this book we will mainly use the GUI. For those who "grew up with" electronic analog computers, the Simulink GUI provided an easy transition. The block diagrams for analog computers and those for digital simulation are essentially the same. However, the digital version is infinitely faster, more versatile, and much more accurate.

A.1.2 Basic Concepts

When Simulink first came out in 1990, it was very easy and quick to learn. As with other software products that have gone through many versions, the latest versions are much harder to learn and use unless some kind of "help" beyond what the company routinely supplies is available. This is due, of course, to the need for software companies to *stay in business* by endlessly producing "new and improved" versions. *Some* added features are really useful and welcomed, but many make the user's task more difficult. This appendix is intended to present a simplified approach, which will still meet the needs of all but the most sophisticated and experienced users.

The basic operation of electronic analog computers (now obsolete) and digital simulation languages for dynamic system study is the *integration* (solving) of ordinary differential equations. These software products can be used to solve both *boundary value* problems and *initial value* problems, but our focus is initial value problems where the independent variable is time *t*. The basic idea is to take the differential equation, as derived from physical principles, and isolate the highest-derivative term on the left-hand side, with all the other terms moved to the right-hand side. We then use a *summing block* to algebraically sum all the terms on the right-hand side of the equation at the input terminals of the summing block. We do not yet actually have "signals" for each of these inputs, but they will become available shortly. If these signals are the inputs to the summer, the output *must* be the highest derivative term. We follow the summer with a *gain* block, which multiples the output of the summer by a constant of our choice. We choose this constant to be the reciprocal of the *coefficient* of the highest derivative term; thus the output of the gain block will be that highest derivative. The heart of the simulation method is the *integrator block*, which integrates with respect to time, whatever its input signal might be. Each integrator has provision for entering a numerical value for its *initial condition*, that is, the value of the integrator *output* signal at time zero. The first integrator produces the second-highest derivative signal and successive integrators produce all the lower derivatives and finally the unknown itself, thus "solving" the equation. Many physical problems are modeled with *simultaneous sets* of differential equations, rather than a single equation. The solution procedure just outlined is followed for *each* of the equations in the set, allowing solution for all the unknowns (say, three equations in three unknowns).

Further explanation of the general approach is best implemented by showing some simple examples. The familiar mass/spring/damper system of Figure A.1 has the equation

$$\Sigma forces = MA = M\frac{d^2x_o}{dt^2} = -K_sx_o - B\frac{dx_o}{dt} + F_i$$

(A.1)

Applying the above instructions gives the Simulink block diagram of Figure A.2. We see three function blocks not mentioned before.

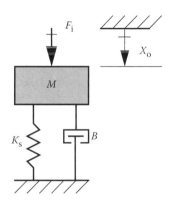

FIGURE A.1
Simple vibrating system.

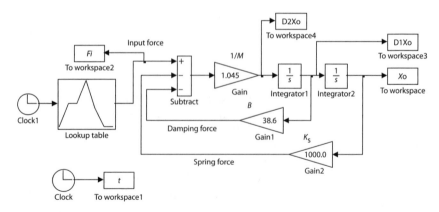

FIGURE A.2
Simulink diagram for vibrating system.

The *clock block* gives us access to the independent variable time t. The *to workspace block* allows us to send the values of any signal in the diagram to the MATLAB workspace, where any additional computations with that variable could be implemented. Often, we use it only to allow plotting of graphs of that variable versus time (or other variables) using standard MATLAB plotting methods. The *lookup table* is one of the most useful modules. It allows us to generate arbitrary functions of the input variable by entering sets of "x, y" data points. Here we use it to generate the input force as a function of time. The input variable can be *any* signal, not just time. For example, if we let the input be spring deflection x_o, then the output could be a nonlinear spring force of any shape that we wish!

Note that such block diagrams give us a strong visual "picture" of system behavior; we can *see* every force and motion and their interactions, giving us a much better understanding of the differential equation than one would get from a *command-line* type of simulation.

A.1.3 Graphing Methods

To get graphs of various signals, we use the MATLAB plot statement. For example, to plot the mass displacement versus time we would enter on the command line *plot(t, Xo)* This brings up a graphing screen which allows us to manipulate the graph in many useful ways. This screen (Figure A.3) was not available in earlier versions of Simulink and is one of the "improvements" which *is* worthwhile. To add many useful features to our graphs, we click on *view* and then select *property editor*. To add grid lines, a title, and axis labels, click on the numerical scales (either x or y) along the graph edges and a screen will appear that allows you to add these features. The default curve line color is blue; usually we want black and a thicker line than the default, which is quite thin when printed. Click anywhere on the curve and a screen will appear allowing choice of line width, color, and type (solid, dashed, etc.)

View   Insert

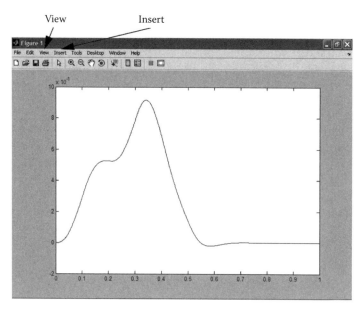

FIGURE A.3
The MATLAB graphics panel.

If you click on *insert*, you can then add text boxes, text arrows, lines, etc. to your graph. As an example, let us enter the plotting command *plot(t,10*Xo,t,D1Xo)*. Figure A.4 shows the type of graph that we usually want. It has faint grid lines (the MATLAB default) since we generally want *working*

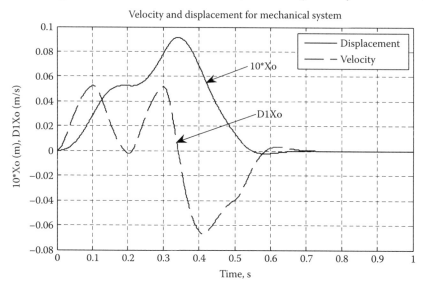

FIGURE A.4
Typical MATLAB graph.

graphs rather than *presentation graphs*. Working graphs need *grid lines* since we use them to pick off numerical values and/or compare several curves on the same graph. Presentation graphs are intended to be projected for a large audience and should have a minimum of *clutter* since they will only be visible to the viewers for a short time and thus their content must be rapidly absorbed. The default *line width* (0.5) results in a rather *faint* curve, so I usually change it to 2.0, as in Figure A.4. The curves originally come up in color, but reports are usually reproduced in black/white, so I change all the curves to black, using the available *line types* (solid, dashed, dotted, etc.) to distinguish one curve from another. The toolbar in Figure A.3 offers many other graphical modifications, but I have explained the basic ones which usually suffice for most graphs. When you are actually using Simulink graphing, it is easy to explore these other options to see how they work. Note that in Figure A.4, to show each curve to a large scale on the same graph, we have plotted 10*Xo rather than Xo itself. This scale factor was selected *after* I had plotted the two curves *without* this factor and saw that, for clarity, an adjustment was needed.

A.1.4 The Simulink Functional Blocks

As mentioned earlier, recent versions of Simulink have added many features which are rarely used, making it more complicated and requiring a longer learning time. This has happened to the *library* of functional blocks, so we want to simplify this aspect of the simulation process. To access Simulink, you must first access MATLAB, which initially displays a screen with a horizontal toolbar across the top. The 8th item from the left selects Simulink, and when you click on it, you get a screen (called the *simulink library browser*) which displays the library of functional blocks broken down into a long list of categories. To open a blank worksheet for building a new model you click on FILE > NEW > MODEL. You then see a screen with a large empty space. By selecting the desired blocks from the library browser, you can drag them, one at a time onto your worksheet, where they will later be properly connected to simulate the equations you are working with. The selection process can be quite time consuming because there are so many choices, and the beginning user does not really know where to look. To avoid this problem, one can *once and for all* create a *template* worksheet which has on it only the most commonly used blocks and save this template with an appropriate name. It shows only the *isolated* blocks; none of them are connected. Having once made this template, whenever you begin a new simulation, you simple call up this file, use SAVE AS to give it a new name, and then *delete* from the worksheet those blocks not needed for the current simulation. (The template needs to have *only one* of each block type; on your new worksheet you can *duplicate* any block by right-clicking and dragging on it.) To set numerical parameters for any block, double-click on it to open a dialog box.

I will now show the template that I personally use and that should serve you well for most of the simulations that you might need to do (Figure A.5). Obviously, there will, at times, be a need to access some blocks that are *not* on this particular template, but this will be rare and probably will not happen until you have become more expert at using Simulink. The easiest way for you to access this template would be for me to attach its file to an e-mail sent to you. If this is not practical, you need to make up the template yourself using the library browser to select the items you want for your own template.

Figure A.6 shows the library browser, and I will now give some directions so that you can find the items that are shown on Figure A.5. The following blocks will be found in the list item called *sources*: **clock, digital clock, step, band-limited white noise, chirp signal, sine wave, constant, signal builder, repeating sequence, pulse generator.** The following blocks will be found in *continuous*: **integrator, derivative, transfer function, transport delay.** The following blocks will be found in *discontinuities*: **backlash, coulomb and viscous friction, dead zone, quantizer, relay, saturation, rate limiter.** The following blocks will be found in *discrete*: **first-order hold, zero-order hold.** The following block will be found in *lookup tables*: **lookup table.** The following blocks will be found in *math operations*: **sign, math function, gain, subtract, multipy, absolute value.** The following block will be found in *signal attributes*: **rate transition.** The following block will be found in *signal routing*: **switch.** The following blocks will be found in *sinks*: **stop, to workspace.** The following block will be found in *user-defined functions*: **Fcn.**

A.1.5 Running a Simulation

Once one has a block diagram that represents the equations of the system to be simulated, some other vital decisions need to be made and implemented. These have to do with things such as which integrator to use, how long will the simulation run, what time step will be used by the integrator, etc. If the window showing the block diagram is open, one needs to click on the item called *simulation* in the horizontal toolbar at the top of this window. Then click on *configuration parameters,* which opens a new window shown as Figure A.7. This window has many *confusing* options so we need to explain a simplified standard procedure. The main choices to be made are the start and stop times for the simulation, the type of integrator to use, and the time-step size; other items appearing in this window can be left to default. The start time of most simulations is taken as 0.0, so this setting need not be changed from the default value. In most simulations, we have enough understanding of the physical problem that we can easily choose a stop time. When we make the first run, we will clearly see whether this initial choice should be adjusted. Simulink offers several different integrators, the default being a variable-step-size type called ode45 (Dormand-Prince). Variable-step-size integrators are designed to speed up computation by using a small step size when things are changing rapidly and large step sizes when they change more slowly, thus *optimizing*

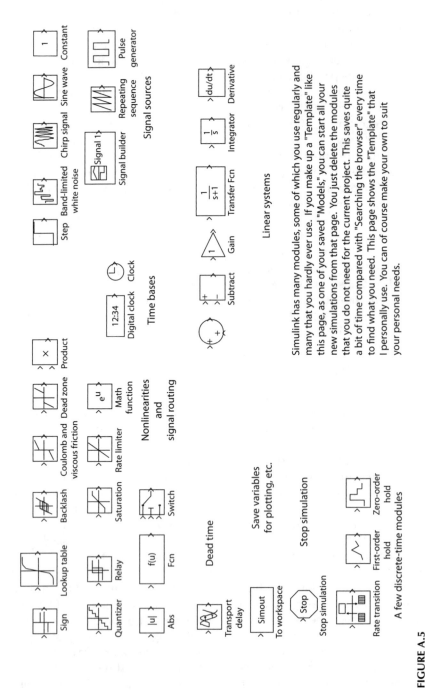

FIGURE A.5
A *Template* of commonly-used Simulink modules.

FIGURE A.6
The Simulink library browser.

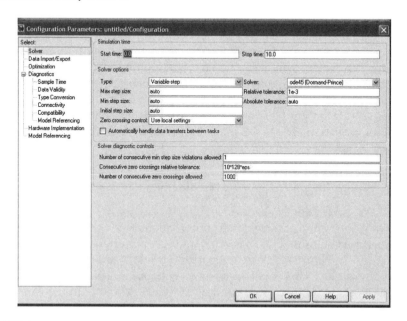

FIGURE A.7
The configuration parameters panel.

the step size. I actually prefer *fixed* step-size integrators for two reasons. First, modern computers are so fast that the speed advantage of variable step size is not really noticeable unless the simulation is *very* complex. Second, variable step size *often* makes the time interval between plotted points so large that curves that are really smooth appear *jumpy*, giving poor printed graphs. Thus I always start a new simulation with a fixed-step-size integrator, going to variable step size only in those rare cases where an advantage is obvious. My choice is the ode4 (Runge–Kutta), which you select by clicking on the check mark next to *solver*, selecting *fixed step size*, clicking on the check mark to the right of *ode45 (Dormand-Prince)*, and then selecting *ode4 (Runge–Kutta)*.

A disadvantage of fixed step size is that *you* need to choose the step size (it is automatic with variable step size). I do not consider this as a *real* disadvantage because it *forces you to think* about this aspect of the simulation. If you choose too large a step size, the calculations may be inaccurate or even numerically unstable, and even if the calculations are correct, the plotted graphs will not be smooth. Too small a step size uses more computing time and could lead to inaccurate calculation due to accumulation of round-off errors. Fortunately neither of these problems is serious, so one should err on the side of choosing small step sizes. How small, of course, depends on the *speed* of the simulated system; fast systems need small step sizes. In practical applications, we often are familiar enough with general system behavior that we can choose trial values for both the total run time and the time step. Once we make the first run, those results are most helpful in adjusting both these values if needed. Fortunately, most simulations will run accurately for a wide range of step sizes; there is no *magic number*. We should always verify the accuracy of calculations by trying a range of step sizes; if the results are essentially the same for a range of step sizes, it is highly likely that we have found proper step-size values. If you have nothing else to go on, many simulations run well when the total number of time points is in the range 1,000 to 10,000. The system of Figure A.2 (which gave the graphs of Figures A.3 and A.4), had a run time of 1.0 s and a step size of 0.001 s.

When all the proper settings have been made, to actually run the simulation enter SIMULATION > START on the toolbar at the top of the block diagram window.

A.1.6 Configuring the *To Workspace* Blocks

The *To Workspace* blocks are used mainly to save any signals that we will later want to graph (or perhaps do other calculations with). However, one of the settings on these blocks is critical for proper operation, so we need to specify this. In Figure A.8, the topmost field is just the name which you wish to give this signal. This is of course up to you, but in general we like to give variables names that remind us of their physical meaning. In Figure A.8 the signal is the input force on a mechanical system, so Fi might be a good name. The second field down can usually be left at the default value (*inf*), since this will save *all* the data points generated in your simulation run. (If you for

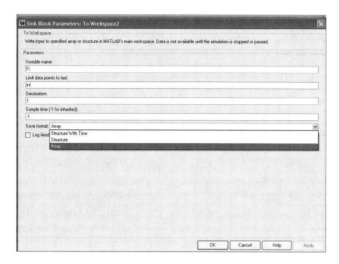

FIGURE A.8
Setting up the *To Workspace* block.

some reason wanted to save only the last 1000 points you would replace *inf* by 1000.) The third field down (decimation) can also be left at the default value; it saves every computed point. If you, instead, wanted to save only, say, every 5th point that was computed, you would enter 5. **The most critical entry in this box is the bottom field; it should be selected as "array," rather than the default value of "structure" or the alternative "structure with time."** If you should go to the online help to learn what this choice is about, you will probably leave confused. They talk about the choices but give you no clue as to why you would choose one or the other. I found out that I wanted "array" simply by accepting the default and finding it "did not work" for my simulations. One could phone their tech support, which is usually pretty good, and see what they have to say; I have not bothered to do this yet. When you are configuring the *To Workspace* blocks, do just one of these and then *duplicate* it (except of course for the variable name) by right-clicking/dragging to set all the other such blocks. This saves time and effort and reduces the chance of forgetting and having the simulation crash because one block was set at *structure*.

A.1.7 "Wiring" the Block Diagram (Connecting and Manipulating the Blocks)

If you have the blocks you need on your worksheet, you next need to connect them in the proper arrangement, such as in our (Figure A.2) example. Blocks can be connected *left-to-right* or *right-to-left*, so you sometimes need to *flip* a block, using FORMAT > FLIP. Blocks can also be oriented vertically, rather than horizontally, using FORMAT > ROTATE. To connect the output of one block to the input of another, just left click on the output *terminal* of the first block and drag the line to the input of the next. You can move both vertically and horizontally. If two blocks are already connected with a line, you can *branch off* of that line by right-clicking on it at any point and dragging to the desired input terminal. Once connected, blocks can be moved around by clicking and dragging. Horizontal and vertical "wires" can be shifted by clicking on them (a *crossed arrows* symbol appears) and then dragging to the new position. The *size* of the diagram can be manipulated by clicking on *view* in the toolbar and then using *zoom in, zoom out, fit system to view,* or *normal (100%)*. You will have to try these to see exactly how they work.

Block diagrams are much more useful to you and others who might read them if you add well-chosen text notations to them. I usually label every signal and every gain block, displaying the parameter symbol whose numeric value is shown in the gain block. Any other blocks that have numeric parameters or other features which should be documented can be similarly labeled. If you double click in any *white space* in a diagram, you can then type text which can be dragged to where you want it. This text can be given the usual size, font, color, etc., formats. The *size* of blocks can also be changed. Click on a block and *handles* appear at its corners. You can drag these handles to change the size of the block. One reason for doing this is to make the block large enough that the text inside it is clearly visible. For example, if a gain block is too small to show the numeric value of that gain (say 1.254e−4), the block will only show a "K," rather than the value of K, making the diagram less useful. When the diagram looks satisfactory to you, you can print it with FILE > PRINT. **One of the biggest problems in practical engineering is inadequate technical documentation, since information must be shared within teams and with other company groups. You need to practice the skill of making your documents (including diagrams such as those for Simulink) understandable, not just to yourself, but to others who did not originate them. This can be justified also by simple self interest. You may understand a poorly documented diagram when you first make it, but just wait a few days or weeks. Even you will have trouble understanding it!**

A.1.8 Setting Numerical Values of System Parameters: MATLAB Scripts/Files

In the block diagram of Figure A.2, I have entered numerical values of all the system parameters directly from that diagram. An alternative method would be to specify the various gains as letter symbols (M, Ks, B) and give them numerical values with MATLAB statements or files. Which method is best depends on the situation. The advantage of entering the numbers directly on the diagram is that one does not need to *look in two places* to interpret the diagram. Also, changing the numerical values directly on the diagram is faster. However, using MATLAB m-files is advantageous, for example, when we want to make multiple runs where we change the numerical values of parameters from run to run. Figure A.9 shows the model of Figure A.2 with the parameters M, Ks, and B given as letters; Figure A.10 shows graphical results. To vary B over three values we write an m-file:

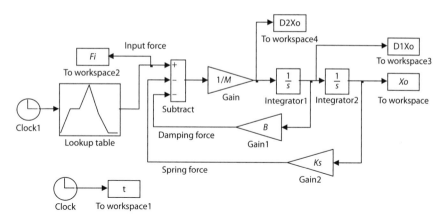

FIGURE A.9
Diagram designed to convey system physical behavior.

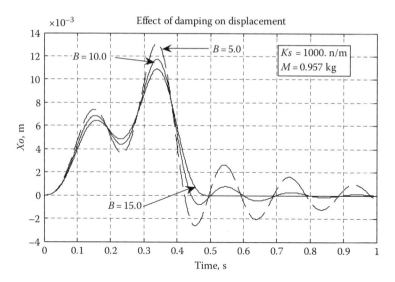

FIGURE A.10
Displaying effects of a multi-valued parameter.

```
% use with Simulink me770newmanual2.mdl
% to demonstrate multiple runs
M = 1/1.045; Ks = 1000.0;
for i = 1:3
    B = 5*i;
    [t,x,y] = sim('me770newmanual2');
    plot(t,Xo);
    hold on
end
```

The [t,x,y] = sim statement actually runs the simulation; it requires only the name of the model file (in single quotes). (There is a more detailed version of this statement which allows setting of initial conditions, start and stop times, and other simulation parameters, but it is usually easier to set these in the usual way *on the model block diagram* and then use the [t,x,y] statement in an m-file which does the manipulations that you want.)

A.1.9 Making Tables of Numerical Values of Variables

We usually are satisfied with *graphs* of our variables but now and then we want *tables* since they show accurate numerical values. MATLAB has a convenient way of displaying such tables *except* that there is no really easy way to head each of the table columns with the *name* of that variable. It seems ridiculous that such a powerful software would not bother to provide such a basic feature. I have called their tech support over the years to complain about this but, so far, to no avail. If you are satisfied with a table *without* headings, this is easily obtained. For example, if we wanted to display a table of the first 10 values of t, Xo, D1Xo, and D2Xo we would simply enter the MATLAB command [t(1:10) Xo(1:10) D1Xo(1:10) D2Xo(1:10)] (leave one or more spaces between the variable names.) If you want to get a table with headings, you can write a very short program using the statement *fprintf* as follows:

```
% prints table of numerical values
fprintf('t Xo D1Xo D2Xo');
[t(1:10) Xo(1:10) D1Xo(1:10) D2Xo(1:10)]
```

The spacing between the variable names in the fprintf statement *does* matter but there seems to be no way to get things to *line up* except by trial and error. However, if you produce the table *without* headings and then copy it to the clipboard with the intention of inserting the table into a word-processor (or other) document, you can easily add the heading (with proper alignment) in the word processor. In the table below I used the program, which gave misaligned headings, and then aligned them in the word processor. Of course, if you know that you are *only* going to copy the table to a word processor document, there is no point in using the program, just copy the table *without* headings and add them with the other software. Note below that the program produces the text ans =, which is also not wanted, but could easily be removed in the word processor.

t	Xo	D1Xo	D2Xo
ans =			
0	0	0	0
1.0000e-003	8.6205e-009	2.5775e-005	5.1201e-002
2.0000e-003	6.8269e-008	1.0171e-004	1.0033e-001
3.0000e-003	2.2807e-007	2.2575e-004	1.4741e-001

```
4.0000e-003  5.3512e-007  3.9585e-004  1.9247e-001
5.0000e-003  1.0345e-006  6.1003e-004  2.3556e-001
6.0000e-003  1.7692e-006  8.6633e-004  2.7671e-001
7.0000e-003  2.7805e-006  1.1628e-003  3.1594e-001
8.0000e-003  4.1076e-006  1.4976e-003  3.5330e-001
9.0000e-003  5.7878e-006  1.8688e-003  3.8882e-001
```

A.1.10 "Detailed" versus "Compact" Simulation Formats (Basic Equations versus Transfer Functions)

The block diagrams of Figures A.2 and A.9 could be called "detailed" simulations since they were derived directly from the basic physical equations that model the system. We often prefer this form since it shows all the basic parameters and all the physical signals (forces and motions in this case) that might be of interest. Sometimes another format using *transfer functions* is useful. If we were only interested in, say, the response of the displacement *Xo* to the force *Fi*, we might manipulate the physical equation into a transfer function *before* doing any simulation.

$$\frac{x_o}{F_i}(D) = \frac{1}{MD^2 + BD + K_s} = \frac{K}{\dfrac{D^2}{\omega_n^2} + \dfrac{2\zeta D}{\omega_n} + 1} \tag{A.2}$$

This form can be easily and compactly simulated using the Simulink *transfer function block,* as shown in Figure A.11. If we wanted the block diagram to directly display the physical parameters M, K_s, and B we would use the upper diagram on Figure A.11. If instead we wanted to put the transfer function into a useful *standard form,* we would use the lower diagram, which uses the standard parameters *steady-state gain K, undamped natural frequency ω_n, and damping ratio ζ.* The transfer function form is particularly useful when we are dealing with a complex system made up of several components, each of which has a known transfer function. The complete system diagram could become quite unwieldy if we did not use the *compact* form of simulation. To enter numerical values into a transfer function block just double-click on the block; this opens a dialog box where you enter the numerator polynomial and the denominator polynomial coefficients in descending powers of s (or D). (MATLAB uses s, not D.)

A.1.11 Simulating Sets of Simultaneous Equations

A physical system need not be very complicated for it to be described by a *set* of equations rather than a single equation. The simulation of such systems is really not any more difficult than for a single equation; however, analysts sometimes choose to *manipulate* the set of equations *before* going to the simulation. Unfortunately this is not only needless work but also results in a

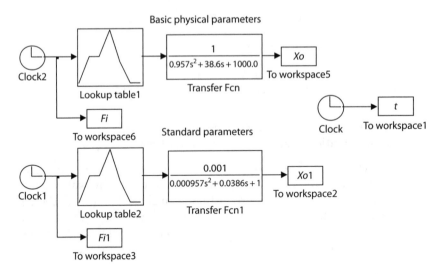

FIGURE A.11
Transfer function block gives a compact diagram.

simulation that is less useful, so we do not recommend it. The *reason* that this path is sometimes pursued is probably that we have learned the *analytical* solution method (which *does* require some manipulation before solving) and thus feel that this must also be a part of the simulation approach. Let us use a simple example to explain these concepts.

Figure A.12 shows a mechanical system with two driving inputs and two response variables. Using Newton's law twice (once for each mass), we can derive the two simultaneous equations which describe the system.

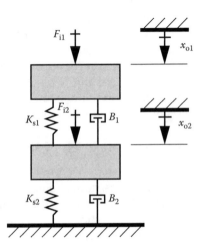

FIGURE A.12
Mechanical system with two inputs and two outputs.

$$M_1\ddot{x}_{o1} = F_{i1} - K_{s1}(x_{o1} - x_{o2}) - B_1(\dot{x}_{o1} - \dot{x}_{o2}) \tag{A.3}$$

$$M_2\ddot{x}_{o2} = F_{i2} - K_{s1}(x_{o2} - x_{o1}) - B_1(\dot{x}_{o2} - \dot{x}_{o1}) - K_{s2}x_{o2} - B_2\dot{x}_{o2} \tag{A.4}$$

Note that the two weights (gravity forces on the masses) do *not* appear in these equations because we have chosen the origin of the two displacement coordinates at their respective *static equilibrium positions*. In most cases, we should develop the simulation block diagram *directly* from these equations; however, we will now act as if we were going to do an *analytical* solution to discuss the differences.

To solve any set of n linear equations in n unknowns, one must first reduce the set of equations to *single* equations, one for each unknown. One then solves those single equations in the usual way. The first step is to either Laplace transform each equation or else represent it in the D-operator form. This changes the equations into *algebraic* equations rather than *differential* equations. (If we use the D-operator method the equations are not *really* algebraic, we just treat them *as if they were*. In the Laplace transform method, the equations really are algebraic.)

Once we are dealing with algebraic equations, any method that you learned in algebra for solving n equations in n unknowns may be applied. For simple equations and when n is small, substitution and elimination might be a practical method. In general, however, we will use a systematic approach which works for any value of n and also is available in computerized form in various software products. This method is called the method of determinants or *Cramer's rule*. The first step is to organize the equation set into rows and columns, with unknowns on the left and given inputs on the right. Once the equations are in this form, we can immediately express any of the unknowns as the ratio of two determinants which are apparent from the row/column form of the equation set. The numerator determinant will be different for each unknown; the denominator determinant will be the same for each unknown. One then needs to expand these determinants to get the single differential equation for each of the unknowns. If you are interested only in, say, one of the unknowns, you need to expand only two determinants. If you are interested in n of the unknowns you need to expand $n + 1$ determinants. If you have available software that provides a *symbolic processor (such as MAPLE*)*, the determinant expansions can be done by that software. A *symbolic* processor is necessary because the determinant elements are *not* numbers but rather letter expressions such as $MD^2 + BD + K_s$.

$$(M_1D^2 + B_1D + K_{s1})x_{o1} + (-B_1D - K_{s1})x_{o2} = F_{i1} \tag{A.5}$$

$$(-B_1D - K_{s1})x_{o1} + (M_2D^2 + (B_1 + B_2)D + (K_{s1} + K_{s2}))x_{o2} = F_{i2} \tag{A.6}$$

---

* www.maplesoft.com

(The Laplace transform versions of Equations A.5 and A.6 would include *initial conditions,* which are treated as *inputs.*)

We can now solve for both x_{o1} and x_{o2} (one at a time) using the determinant method.

$$x_{o1} = \frac{\begin{vmatrix} F_{i1} & -B_1 D - K_{s1} \\ F_{i2} & M_2 D^2 + (B_1 + B_2)D + (K_{s1} + K_{s2}) \end{vmatrix}}{\begin{vmatrix} M_1 D^2 + B_1 D + K_{s1} & -B_1 D - K_{s1} \\ -B_1 D - K_{s1} & M_2 D^2 + (B_1 + B_2)D + (K_{s1} + K_{s2}) \end{vmatrix}} \qquad (A.7)$$

The rule for forming an equation like (A.7) to solve for *any one* of the unknowns is as follows. The denominator determinant is formed from the coefficients of unknowns on the left-hand side of Equations A.5 and A.6. This will be the denominator determinant for *all* the unknowns. The numerator determinant for any unknown is formed by replacing the column of that unknown's coefficients by the column of inputs on the right-hand side. Thus for x_{o2} we would have

$$x_{o2} = \frac{\begin{vmatrix} M_1 D^2 + B_1 D + K_{s1} & F_{i1} \\ -B_1 D - K_{s1} & F_{i2} \end{vmatrix}}{\begin{vmatrix} M_1 D^2 + B_1 D + K_{s1} & -B_1 D - K_{s1} \\ -B_1 D - K_{s1} & M_2 D^2 + (B_1 + B_2)D + (K_{s1} + K_{s2}) \end{vmatrix}} \qquad (A.8)$$

To get the single differential equation for unknown x_{o1}, we need to expand the determinants, *cross-multiply,* and then reinterpret the D operators as time derivatives. Most readers will know how to expand 2×2 determinants; the procedure for larger determinants is routine but tedious, and the computerized approach using a symbolic processor is very welcome. From Equation A.7 we get

$$x_{o1} = \frac{(M_2 D^2 + (B_1 + B_2)D + (K_{s1} + K_{s2}))F_{i1} + (B_1 D + K_{s1})F_{i2}}{M_1 M_2 D^4 + (M_1 B_1 + M_1 B_2 + M_2 B_{11})D^3 + (B_1 B_2 + M_1 K_{s1} + M_1 K_{s2} + M_2 K_{s1})D^2}$$
$$+ (B_1 K_{s2} + B_2 K_{s1})D + K_{s1}K_{s2} \qquad (A.9)$$

Note that this gives the response of x_{o1} to *both* forces acting simultaneously:

$$(M_1 M_2 D^4 + (M_1 B_1 + M_1 B_2 + M_2 B_{11})D^3 + (B_1 B_2 + M_1 K_{s1} + M_1 K_{s2} + M_2 K_{s1})D^2$$
$$+ (B_1 K_{s2} + B_2 K_{s1})D + K_{s1}K_{s2})x_{o1}$$
$$= (M_2 D^2 + (B_1 + B_2)D + (K_{s1} + K_{s2}))F_{i1} + (B_1 D + K_{s1})F_{i2} \qquad (A.10)$$

If we interpret the D operators as time derivatives, Equation A.10 is the single differential equation in the unknown x_{o1} that we wanted. One could solve this differential equation using either the classical D-operator method or the Laplace transform method once the time-varying functions for the two input forces were given. If you use the Laplace transform method the D's would be replaced with s's and the inputs would include any nonzero initial conditions. Note that from Equations A.9 or A.10 it is easy, using the principle of superposition, to define two *transfer functions*, x_{o1}/F_{i1} and x_{o1}/F_{i2}. If we wanted these transfer functions, all the above work *would* be necessary. However, if we wanted to get the responses of this system to any kind of input forces and we had numerical values for all the system parameters, all this work can be *avoided* by simply simulating the entire system from the original equations (A.3 and A.4). Also, this simulation will provide much more information (such as all the velocities and accelerations, and all spring and damper forces). We could, of course, run a simulation using the derived transfer functions but that would *not* provide anything but x_{o1}. Thus if you are going to do simulation of any physical system it is easiest and best to do it *from the original physical equations*. This provides the most information with the least work. There are, of course, often good reasons for doing the analytical manipulations which get us the transfer functions in letter form, but if we are going to do *simulation*, we do it from the original equations.

To simulate any physical system from its original set of equations, we proceed just as for a single equation, but we repeat the process for each equation. That is, we isolate the highest derivative term of each unknown on the left-hand side of its equation and then use a summing block to *add up* all the terms on the right-hand side. The output of this summer is sent to a gain block set at the reciprocal of the coefficient of the highest derivative term. The output of the summer is then integrated as many times as necessary to get the lower derivatives and finally the unknown itself. In the act of doing this for all the equations, all the terms at the inputs of the summers will become available. Figure A.13 shows this for our present example. Note that we have provided several different types of input forces, which could be applied to one or both of the masses, individually or simultaneously. If a block is intentionally left unconnected, you will get an error message, but the simulation will run properly. In Figure A.13, only the *chirp* force is applied, and only to M_2. The chirp signal is sometimes used in vibration testing to quickly get an approximate frequency response of a system. In Simulink, the chirp signal is a *swept-sine* function which has a fixed amplitude of 1.0, but a frequency which varies linearly with time, starting at time zero with a frequency of your choice and ending at a target time with a higher frequency of your choice. In Figure A.13, the frequency sweeps from 5 to 9 Hz in 5 s. The response of X_{o1} to this driving force is shown in Figure A.14, where we see two resonant peaks related to the two natural frequencies of such a system.

Many practical vibration problems involve the response to nonsinusoidal periodic forces because in a *thoroughly warmed-up* machine running at

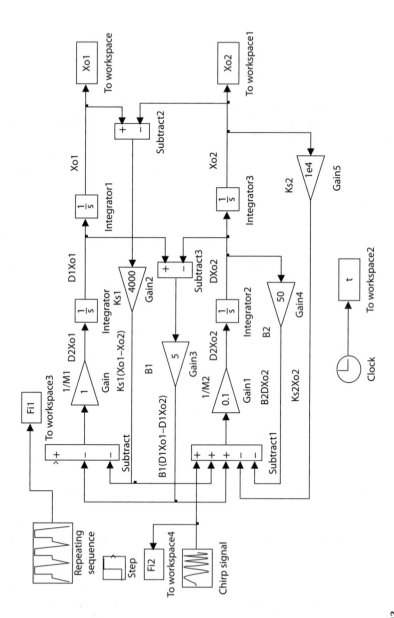

FIGURE A.13
Simulation diagram for MIMO (multiple-input, multiple-output) mechanical system.

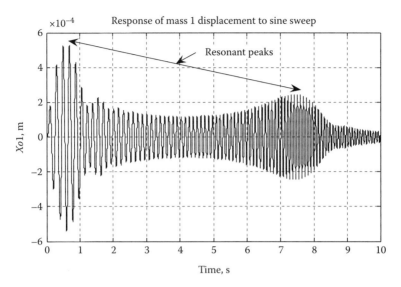

FIGURE A.14
Use of chirp signal to study system resonance.

constant speed, all the forces, pressures, temperatures, etc. will be essentially periodic. The Simulink *repeating sequence* block allows us to generate such force signals with any waveform and repetition rate. One cycle of the waveform is specified by giving a set of "t, x" points which define the shape, and the block then repeats that shape over and over. Figure A.15 shows the force

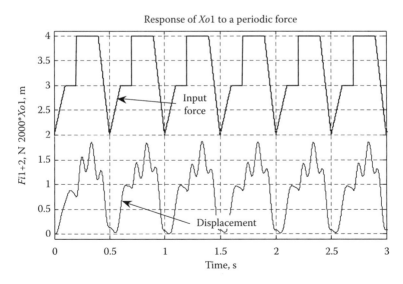

FIGURE A.15
Application of the repeated-sequence module.

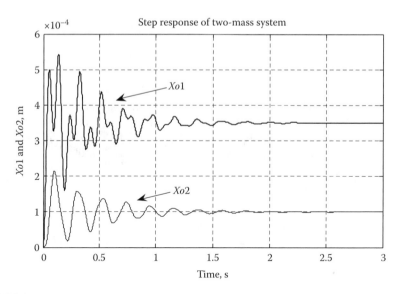

FIGURE A.16
Response of two-mass system to step-input force.

applied to mass 1, and the response of displacement $Xo1$ to that force. Finally, we apply a step input to mass 1 and graph the displacement response of each mass in Figure A.16.

It is easy to apply any combination of the three different types of forces we have used here to either or both of the masses. We also can graph any of the accelerations, velocities, and displacements and the various spring and damping forces. Returning to Equations A.3 and A.4 and Figure A.13 for a moment, note that I have arranged the diagram so that we can explicitly see the spring and damper forces. Some analysts will often use another approach, which is not incorrect, but which gives a less *physical* feel to the diagram and is thus less desirable in general. In this other approach, in Equation A.3 one *multiplies out* the spring and damping terms and forms (in the diagram) terms like $K_{s1}X_{o1}$, and immediately draws a gain block with input X_{o1} and gain K_{s1}. The output of this block has the dimensions of a force but it is a force *that does not physically exist!* This is not mathematically wrong, and the simulation would run properly, but it is physically confusing, and needlessly so.

This situation gets even worse in Equation A.4, where one might gather up all the terms in X_{o1} to give a term $X_{o1}(K_{s1} + K_{s2})$, and then draw a gain block with a gain of $(K_{s1} + K_{s2})$ and input X_{o1}. This is mathematically right, but again creates a *force* signal which does not physically exist. The probable reason for some analysts following this method is that it is what one was *taught* to do when pursuing an *analytical* solution. Thus we need to break this habit when we solve problems by simulation rather than analysis.

With regard to the sine-sweep test of Figure A.14, note that this is not an accurate or efficient way of analytically finding the *frequency response* of any linear system, even though it is a standard tool of vibration testing. The basic problem is that frequency response refers to the *sinusoidal steady state* that occurs when a sine wave of *fixed* frequency is input, and we *wait* for transients to die out, before making any measurements of amplitude and phase angle. Our chirp test sweep may not have been slow enough to allow essentially steady-state conditions as we pass through the range of frequencies. In an experimental test, the *speed* of the chirp is carefully adjusted to satisfy accuracy needs. (A *slow* chirp is perhaps more correctly called a *sweep*.) Also, both input (force) and output (motion) signals are simultaneously measured and processed to obtain the desired amplitude ratio and phase angle data. When we want to compute (rather than measure) the frequency response of a linear system model, we *do* need to get the transfer function, as in Equation A.9, and then use, say, MATLAB tools to get the amplitude ratio and phase. (Actually, we will shortly show a *matrix frequency response method* which does *not* require transfer functions and is sometimes a preferred method.)

If our system model is *nonlinear*, the concept of frequency response does not strictly apply because the response signals (displacements, velocities, and accelerations) will *not* be sinusoidal when the input forces *are* perfect sine waves. When comparing a perfect input sine wave with some periodic (but *not* sinusoidal) output waveform, the concepts of amplitude ratio and phase angle become confused. Simulation of such a system with a chirp (sine-sweep) input force *does* give useful information about the system, even if we cannot get the usual amplitude-ratio and phase angle curves. If our chirp study shows a dangerous *resonance* effect, the fact that the response waveform is not a sine wave is of little practical significance. It is now, however, important to run the chirp test with a *range* of force amplitudes because the response of nonlinear systems is sensitive to the level of the forcing inputs. The behavior for a low amplitude and a high one can be quite different for nonlinear systems, while for linear ones the input amplitude has no effect on the amplitude ratio and phase angle. In fact, a widely used experimental test for nonlinearity is to run tests with *different-size* step inputs, sinusoidal inputs, etc. If the nature of system behavior changes when the input size is changed, that is proof of some kind of nonlinear behavior.

A.1.12 Frequency-Selective Filters

Frequency-selective filters (low-pass, high-pass, band-pass, and band-reject) are useful in many simulations. In earlier versions of Simulink, the block library included a selection of *ready-made filters*, but the version used in this book requires that one make suitable MATLAB statements to generate the filter transfer functions, which can then be loaded into an ordinary Simulink *transfer function block*. Of the various filter types available, we will discuss the three that will meet most simulation requirements: Bessel, Butterworth, and elliptical. Bessel filters are available only as low-pass filters. They do not have as sharp a cutoff as the others, but they have a very linear phase angle curve (nearly constant time delay) over a large frequency range where the amplitude ratio is nearly flat. For signals with frequency content in that range, these filters will accurately preserve the signal waveform. They also have no overshoot for a step input. If a sharper cutoff and a flat amplitude ratio are needed, and if linear phase is not as critical, then a Butterworth type will usually be preferred. Where a very sharp cutoff is required (usually as an antialiasing filter in front of a digital data acquisition system) an elliptical filter will often be chosen.

All three types are available with a selection of *order*. Order is the order of the polynomial in the denominator of the filter's transfer function. Higher order filters will have a sharper cutoff, but will give more time delay (phase shift). High-order filters are quite common; we will show eighth-order Bessel and Butterworth filters and a seventh-order elliptical filter (since elliptical filters have a very sharp cutoff). To place a filter on your Simulink worksheet, you first place a transfer function block there and specify the numerator and denominator polynomials in a certain *letter form*. You then go to MATLAB and enter certain statements that use that same letter form. MATLAB then computes the required polynomials and they are automatically inserted into the Simulink transfer function block. Let us compare low-pass filters of the above three types; this will also show how to set up filters in general. We choose a cutoff frequency of 100 Hz. *Cutoff frequency* generally means the frequency at which the amplitude ratio has dropped to a value of 0.707 (−3 dB), however in MATLAB, when you set up a Bessel filter, the frequency number that you request must be 2 *times* the desired −3 dB frequency. This is because the MATLAB folks decided that it would be more useful to make the entered frequency the end of a frequency range *where the filter time delay was essentially constant* (this is the range over which the phase shift is nearly linear with frequency). Thus for a fair comparison of the three filters we should ask for the *same* −3 dB frequency for each. In addition to the filter *order*, elliptic filters also require that we specify a *maximum passband ripple in dB* (called Rp) and a *minimum stop band attenuation in dB* (called Rs). For our example, we choose these as 0.5 and 40 dB.

To specify a seventh-order low-pass elliptic filter with the above characteristics, we would enter the MATLAB statement: [b,a] = ellip(7,0.5,40,100*2*pi,'s'). The term [b,a] refers to the numerator (b) and denominator (a) of our Simulink transfer function block. We need to set up the numerator as (b/a(8)) and the denominator term as (a/a(8)). The division by a(8) (the last term in the denominator polynomial) puts the numerical values in a more convenient form (the trailing terms in both numerator and denominator are forced to be equal to 1.0). Note that the terms (b/a(8)) and (a/a(8)) use ( ) and not the usual [ ]; using ( ) will display the actual numerical polynomials, rather then just displaying a and b as letters, which is what we get when we use [ ].

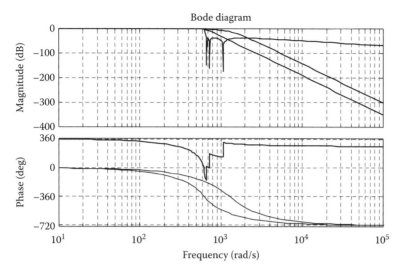

FIGURE A.17
Standard Bode plots may not give the best display.

The filter frequency response is of major importance so we want to display these curves for the three filters we are comparing. Most readers who have some experience with MATLAB would use the bode(num,den) command to produce these graphs (Figure A.17). While Bode plots are useful for some applications, there are often better ways to display the data. Note in the phase plots that the upper curve (the elliptic filter) does not start at zero, *while we know theoretically that it must*! Of course 360° is the same as 0°, but the curve is difficult to compare with the other two, which fortunately are plotted correctly. We can get better phase plots, as explained below.

If we arrange to plot frequency and amplitude ratio on *linear* (rather than log) scales, we get Figure A.18, which is much more useful. We next show a program which implements our desired calculations and plotting. These commands could of course be used *without* writing a program, if that was desired.

```
% me770newmanual6 forms, computes, and plots frequency
% response for elliptic, Butterworth and Bessel filters
[b,a] = ellip(7,0.5,40,100*2*pi,'s');  These 3 commands show
[d,c] = besself(8,200*2*pi);           how to set up any filters
[f,e] = butter(8,100*2*pi,'s');        of these types.
fr = 0:0.1:200;
w = fr*2*pi;
[h1] = freqs(b,a,w);
[h2] = freqs(d,c,w);
[h3] = freqs(f,e,w);
ar1 = abs(h1);
ar2 = abs(h2);
```

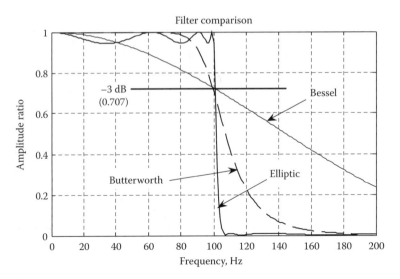

FIGURE A.18
Linear (rather than logarithmic) scales give a clearer display.

```
ar3 = abs(h3);
k = 360/(2*pi);
phi1 = angle(h1)*k;
phi2 = angle(h2)*k;
phi3 = angle(h3)*k;
bode(b,a,'k');hold on;
bode(d,c,'k');hold on;
bode(f,e,'k');hold on;pause
hold off
plot(fr,ar1,'k',fr,ar2,'k',fr,ar3,'k');pause   produces Figure A.18
hold off
plot(fr,unwrap(phi1),'k',fr,unwrap(phi2),'k',fr,unwrap(phi3),'k')
  produces Figure A.19
```

The *fr* = and *w* = statements set up the frequency range from 0 to 100 Hz
(Bode plots, being logarithmic, can **never** go to 0 Hz). The very fine spacing
of these points was found to be necessary to avoid false plotting of the phase
angle curves. Most engineering software has trouble with angles when they
go outside the range ±180°. MATLAB provides the statement *unwrap*, which
usually corrects the error, but sometimes also requires fine spacing of the frequency points, so that the program does not "lose track" of the curve. When
we complained about the elliptic filter phase angle curve in Figure A.17, it was
based on some knowledge of what the *correct* curve should look like. If you
had no idea as to what the correct shape was, you might very well accept the
incorrect curve (after all, it was produced by a computer, and computers are
always right!). We will shortly discuss the correct shape of these curves.

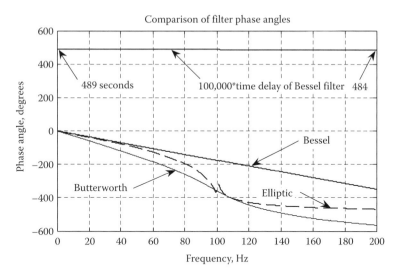

FIGURE A.19
Use of *Unwrap* to correct phase angle curves.

The [*h*] = statements are a very convenient way to compute frequency response data when we do not want Bode plots (logarithmic scales). This statement computes, as a complex number, the sinusoidal transfer function for the system with numerator and denominator polynomials called *b* and *a*. We then use *abs* to get the amplitude ratio and *angle* to get the phase angle in radians. The *bode* statements include 'k' to plot the curves as black lines. When I tried to do *bode* without the 'k,' I got three different colors and was *not* able to change them to black using the *property editor* of the graph screen, which usually does provide this service.

The Simulink block diagram showing our three filters as transfer function blocks is Figure A.20. We see there the polynomials which describe the filter transfer functions. When we set up the transfer function blocks, we gave for the numerator (*d/c(9)*) and for the denominator (*c/c(9)*) in the Bessel filter (and similar statements for the other two filters). This displays the polynomials in the most useful form, with the trailing terms being 1.0. Note that the Bessel and Butterworth filters have only a 1.0 in their numerator, eighth-degree polynomials in the denominator, and all three filters have a steady-state gain of 1.0. For the Bessel and Butterworth filters, we can easily tell what the correct phase angle curve should look like, since the eighth-degree polynomials could be factored into various combinations of first-order and second-order terms. In fact, these filters are designed as four second-order systems in "cascade," thus the phase angle must start at 0 and go monotonically toward −720 (four times −180). This theoretical knowledge helps us detect *incorrect* computer-plotted curves and then try to fix them, as explained above (see Figure A.19).

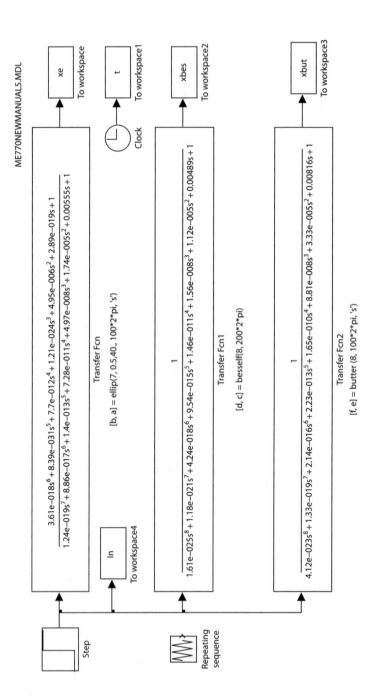

FIGURE A.20
Simulation of elliptic, Bessel, and Butterworth low-pass filters.

The elliptic filter transfer function is more complicated (it *must* be, to get the extremely sharp cutoff seen in Figure A.18). We *can* tell however that the phase must start at 0° and end at −90° because the denominator is one degree higher than the numerator. If we wanted to get more details on the factors of the numerator and denominator, the MATLAB *roots* statement can be applied to both the numerator (b) and denominator (a), giving the results below:

```
roots(b)                          roots(a)
ans =                             ans =
-1.1107e-027 +1.0706e+003i        -1.0098e+001 +6.3043e+002i
-1.1107e-027 -1.0706e+003i        -1.0098e+001 -6.3043e+002i
-2.8422e-013 +7.3212e+002i        -5.0453e+001 +5.9149e+002i
-2.8422e-013 -7.3212e+002i        -5.0453e+001 -5.9149e+002i
 2.0717e-013 +6.7177e+002i        -1.6099e+002 +4.3207e+002i
 2.0717e-013 -6.7177e+002i        -1.6099e+002 -4.3207e+002i
                                  -2.7032e+002
```

We see that the numerator has three pairs of pure imaginary roots (the *computed* real parts are *displayed* as very small numerical values, but they are actually zero) and the denominator has three pairs of complex conjugate roots and one real negative root. One could use these roots to figure out what the phase angle curve should look like, but we will not here bother to pursue those details.

The step-function response of all three filters is easily obtained from the Simulink model and is shown in Figure A.21. While it clearly reveals the *underdamped* nature of the elliptic and Butterworth filters, a perfect step input is perhaps a little *unfair* in practical terms since perfect step changes of real physical signals are impossible. In Figure A.22 I show the response to a periodic triangular wave of 10 Hz frequency, zoomed to show some subtle

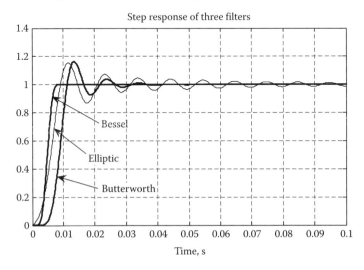

FIGURE A.21
Step response of elliptic, Bessel, and Butterworth low-pass filters.

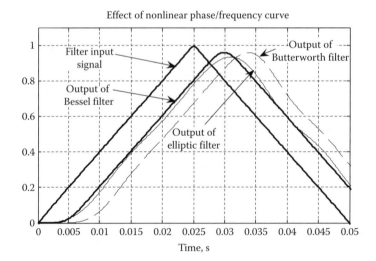

Effect of nonlinear phase/frequency curve

FIGURE A.22
Bessel filter best preserves signal waveform.

detail. All three filters of course exhibit time delay, but because the elliptic and Butterworth filters have a time delay that is not constant with frequency, the shape of their output waveforms shows a small but noticeable curvature, whereas the Bessel output duplicates the straight lines of the input signal. This is because the Bessel filter (see Figure A.19) has a nearly constant time delay of 0.00489 seconds over its pass band.

The other types of filters that we might need are the *band-pass,* the *high-pass,* and the *band-stop (band reject).* These are all available as Butterworth filters and we give below a program which shows how to set them up and then graph their frequency response. Again, we have to look at their transfer functions to see whether the phase angle curves which the computer produces are correct, and then modify our program, if needed, to get curves that agree with the theoretical predictions.

```
% Butterworth bandpass filter design
[wn] = [80*2*pi 120*2*pi];
[b,a] = butter(4,wn,'s');
fr = 0:0.1:200;
w = fr*2*pi;
[h] = freqs(b,a,w);
ar = abs(h);
plot(fr,ar);pause
phi = angle(h)*57.3;
plot(fr,unwrap(phi) + 360);pause
% now do the bandstop filter
[b,a] = butter(4,wn,'stop','s')
[h] = freqs(b,a,w);
```

```
ar = abs(h);
plot(fr,ar);pause
phi = angle(h)'57.3;
plot(fr,phi);pause
% now do the highpass filter
[b,a] = butter(8,100*2*pi,'high','s')
[h] = freqs(b,a,w);
ar = abs(h);
plot(fr,ar);pause
phi = angle(h)*57.3;
plot(fr,phi + 360);pause
```

For the band-pass and band-stop filters, you need to use the *[wn]* = statement to set the low and high frequency ends of the pass band or stop band. The phase angle curve for the band-pass filter needed the *unwrap* statement and also the addition of 360° to agree with the theoretical values. The high-pass filter needed the 360° addition also. Figures A.24 through A.26 show some of the results from this program, including a Simulink model in Figure A.23. Figure A.24 shows how a band-pass filter can extract a *desired* signal component from a *contaminated* signal, while Figure A.25 shows how a band-stop filter can remove an undesired 100 Hz signal component. Finally, Figure A.26 shows a high-pass filter being used to remove a slow zero drift from a signal. Figure A.27 shows the amplitude ratio curves for these three types of filters.

A.1.13 Working with Random Signals

In studying dynamic system of various types, *random signals* are sometimes involved. All measurement systems suffer from some sort of unwanted signals, and often these are not mathematically predictable, but of a random nature. Also, physical systems may be driven by inputs that are inherently random, such as atmospheric turbulence acting on an aircraft wing, or an off-road vehicle going over rough terrain. Analytical treatment of some aspects of random signals and the response of systems to them is sometimes possible, but is limited to *statistical aspects* of the signals; we are *not* able to calculate time histories. If we have a simulation of a random signal source, this can be used in a simulation of any dynamic system to compute time history responses to that input signal. We can also easily compute various useful statistical measures.

MATLAB/Simulink offers a number of statements/blocks that produce random numbers or signals. The most useful random signal source is the one called *band-limited white noise* in Simulink and it meets most needs, so we will discuss no others. Unfortunately, the description of this module found in the software manual and online help does not give a clear picture of how it works or how to use it. The discussion below is an attempt to fill in these gaps, so that a simulation using this signal source can be set up to give the results really desired.

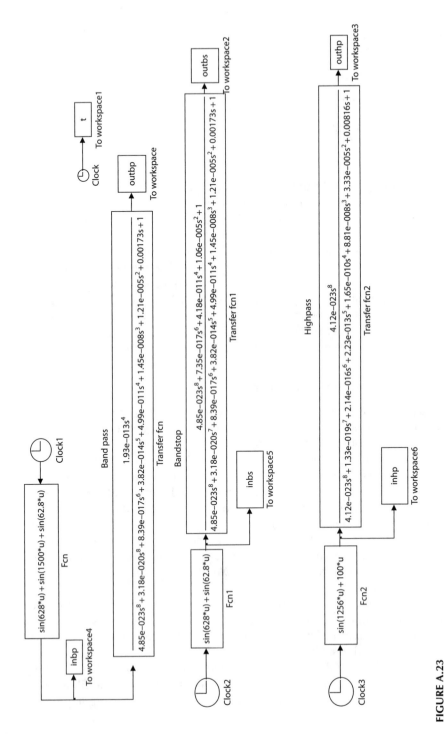

FIGURE A.23
Simulation of Butterworth band-pass, band-stop, and high-pass filters.

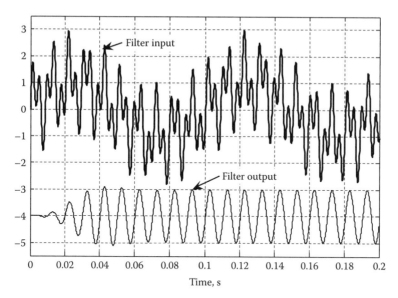

FIGURE A.24
Band-pass filter extracts 100 Hz signal from *"noisy"* signal.

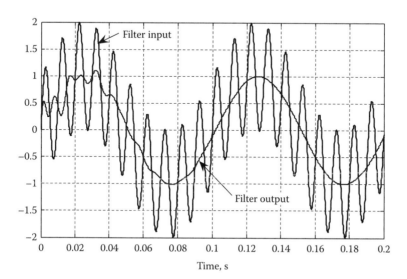

FIGURE A.25
Band-stop filter removes 100 Hz undesired signal.

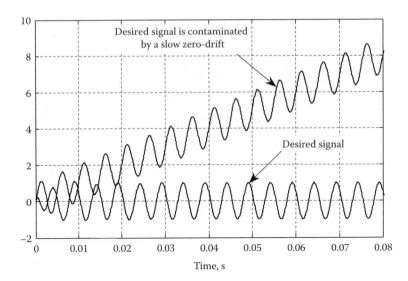

FIGURE A.26
High-pass filter removes slow drift from desired signal.

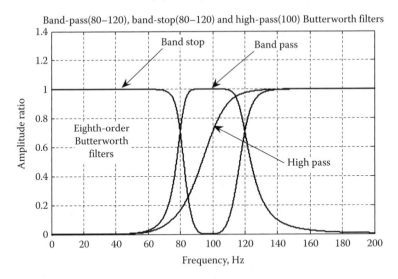

FIGURE A.27
Frequency response of band-stop, band-pass, and high-pass filters.

When trying to set up a signal source so as to mimic some real-world random effect, we often know at least roughly the *frequency spectrum* (power spectral density, PSD) that we want to produce. The usual approach is to start with a *white noise* source (*flat* PSD) and then pass this signal through a properly designed filter to shape the spectrum to our needs. Simulink has spectrum analyzer modules that are intended to compute the frequency spectrum of time-varying signals, but in my experience they do not *automatically* produce proper results. It seems that one must always apply some kind of scaling or other manipulations which are not at all obvious, particularly to a user who is not an *expert*. One often has to perform some kind of *checking* calculation in order to confirm that you are getting what you really want. For example, the integral over all frequencies of a true PSD curve *must* be numerically equal to the total mean square value of the time signal. Recall that the term *PSD* is really somewhat of a misnomer in that it has *nothing* to do with power (mechanical, electrical, fluid, or thermal) and that a more descriptive name would be *mean-square spectral density*. That is, the PSD value at a particular frequency is in principle obtained by narrow-band-pass filtering the time signal, computing the mean-square value of this filtered signal, and then dividing by the filter bandwidth. Thus, the PSD shows how the total mean-square value is distributed over the frequency range, and therefore the integral of this curve must be the *total* mean-square value. In a simulation, it is easy to compute the total mean-square value by squaring the signal, integrating it over time, and dividing by the integration time. This operation must of course be continued until we see the output become more or less constant, at which time we have an estimate of the total mean-square value. To see whether a signal we have "manufactured" has the PSD we want, one check is to compare the total mean-square values obtained in the two ways (*time domain* and *frequency domain*) just described.

When PSDs are computed using fast Fourier transform (FFT) software routines, the PSD is often taken to be equal to the square of the direct Fourier transform value, divided by the frequency resolution of the transform. (If the time signal analyzed lasts for T seconds, the frequency resolution is $1/T$ Hz.) Even this apparently simple relationship can be a source of trouble. In MATLAB, the FFT command does *not* directly produce correct numerical values when you let it operate on your time-varying signal; there *always* is a need to do some *scaling*, which the manual does not explain. The reader is expected to provide this expertise. Also, the *theoretical* definition of quantities such as the PSD may differ from one textbook to another, and the authors usually do not explain how to reconcile one definition with another. For example, while real-world signals are usually thought of as containing only *positive* frequencies, theoretical definitions are often given in terms of a frequency range from minus infinity to plus infinity. Such a definition is not *wrong*, but we do need to know how values computed this way are related to those which consider only positive frequencies. Also, using frequencies in rad/s versus Hz can give different PSD values.

Simulink's *band-limited white noise* generator produces a square wave whose amplitude changes every T seconds, where T is the *sample time* parameter, whose value you may choose. The probability distribution of the amplitudes is Gaussian (normal). You can control the "size" of the random signal with the parameter called *variance* or in some Simulink versions *power*. You may also choose a number for the *seed*; its default value is usually 23341, but you may choose another number as you please. The manual does not give any restriction on the seed value but I would suggest any five-digit odd integer. If you want two signals in a given simulation to be *independent* random variables, you must give them two different seed values. If you want to rerun a simulation at a later date, and reproduce your earlier results, you must use *the same* seed value each time. One purpose of this discussion is to give some guidance on how one chooses numerical values for these three parameters (sample time, variance, seed).

The choice of numerical value for *variance* (power) can be guided by the fact that if you want to, say, double the amplitude of your signal, you need to raise the variance value by a factor of 4. That is, the amplitude changes as the *square root* of the variance value. Usually, when we want to generate a random signal in a simulation, we want it to have a certain frequency spectrum and then *roll off to zero* above the highest frequency of interest. The conventional way of doing this has been to generate a *flat* frequency spectrum (*white noise*) out to a frequency a little higher than we want and then pass the *white noise* signal through a filter of our choice, using the filter to shape the frequency spectrum as we wish. While the Simulink manuals call this module *band-limited white noise*, its raw output is definitely *not* white. I am sure the Simulink software engineers know this but it is not made clear in the manuals. The *actual* frequency spectrum of the random-amplitude square wave is not discussed in any books that I have consulted, except for *Control System Synthesis*,* which is where I got the following information. If you are savvy about random signals you might be able to figure it out for yourself, maybe you already have. For those who do not have this information, I will now present it, using the reference just quoted as my source. The formula is for the PSD of the signal. Since the definition of PSD is itself subject to some interpretation (just look at several different books or papers on the subject), I had to do some *adjusting* to get a result that actually produced the signal desired. The way I checked to see if I was getting what I wanted was to require that the integral of the PSD curve over all frequencies from 0 to infinity be equal to the total mean-square value of the time signal. A reliable calculation of this mean-square value is easily implemented in Simulink using the basic definition. It simply requires sending the time signal into a squaring device and then averaging this squared signal over time, taking a long enough integrating time to see reasonable *convergence* to a steady value. (This mean-square value will of course *never* become absolutely steady, but it does *level off* enough to allow one to stop the calculation and get a good estimate.) The formula for the PSD is

$$\text{PSD} = \frac{\pi}{2} \cdot \text{msq} \cdot T \cdot \frac{\left(\sin \dfrac{\omega T}{2}\right)^2}{\left(\dfrac{\omega T}{2}\right)^2} \tag{A.11}$$

---

* J.G. Truxal, *Control System Synthesis*, McGraw-Hill, New York, 1955, pp. 433–444.

Here

T is the sample time (which you can choose) in seconds
msq is the total mean-square value which you want for your signal
ω is the frequency in rad/s

As an example, if we take, say, $T = 0.01$ s, and a mean-square value of 0.33, we can get the MATLAB graph of Figure A.28. It is clear that this frequency spectrum is *not* flat as a white noise should be. However the signal *can* be used to get a *flat* spectrum by restricting the frequency range to $0 \le f \le \dfrac{0.1}{T}$, (where $\omega = 2\pi f$), in our case from 0 to 10 Hz. That is, from 0 to about 10 Hz, the PSD is *close to* flat at the value given by Equation A.11 for $f = 0$. The value of PSD at zero frequency is an indeterminate form (0/0), but it can be resolved by use of L'Hospital's rule, giving the result $\dfrac{\pi}{2} \cdot$ msq $\cdot T$ which in our case is 0.00518. Thus, if you know what PSD value you want, and the frequency range over which you want the PSD to be reasonably flat, you could choose the sample-time parameter to get the desired frequency range, and the mean-square value to get the PSD value.

Unfortunately, while the choice of the *variance* or *noise power* for Simulink's band-limited white noise generator *influences* the mean squared value, it is not *equal* to the mean-squared value. I tried to find the relation between the two (the manual is of no help) and believe I have a valid relation. The mean-squared value seems to be given by the ratio obtained by dividing the noise variance (also called *power*) by the sample time T. This can be numerically checked by running simulations with several different combinations of T and variance and plotting the mean-squared value. The Simulink diagram (Figure A.29) shows such a calculation. The T and variance values were chosen so that the mean-squared value for each of the four examples would be

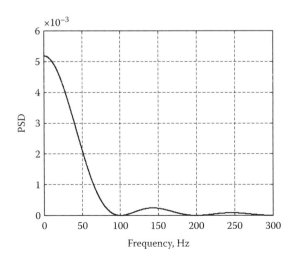

FIGURE A.28
Frequency spectrum of random square-wave.

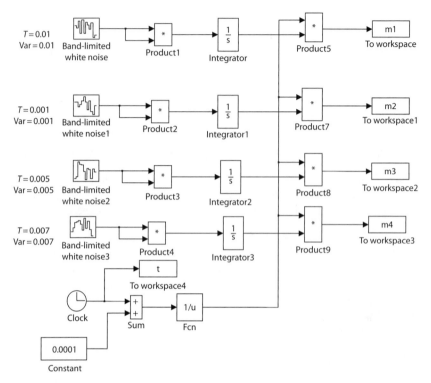

FIGURE A.29
Simulation to compute mean-square values.

the same, and numerically equal to 1.0. I also took the seed values different for all four random signals. Since the calculation for the mean-squared value, as a running variable, requires division by time t, there would be a division by zero at the first calculation. This is avoided by adding a small number (0.0001, for example) to t before the division occurs. The graph of Figure A.30 shows that all four signals seem to have the predicted mean-square value of 1.0, so I think it is safe to use this relation when setting up any simulations which employ the white noise generator.

In setting up the white-noise generator, the following steps can be followed. We assume that the user wants a flat PSD of a certain magnitude, over a range of frequencies from 0 to some upper limiting value. Once such a spectrum has been achieved, one can process the signal through a filter transfer function block which will shape the frequency variation to suit the needs of each application. (We will address this final step later, since it has no effect on the basic setup.) First, one might set the sample time T using the rule that $1/T$ should be equal to 10 times the highest frequency of interest. Thus, if you want a flat spectrum out to, say, 50 Hz, you would choose $T = 0.002$ s. Next, if you want a flat PSD of, say, 1.34, then the noise power (also called variance in some Simulink versions) should be set at a value given by

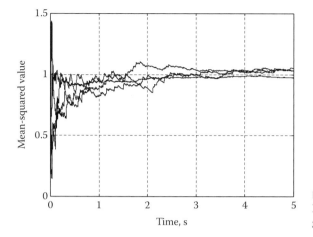

FIGURE A.30
Mean-square value check graphs.

$$\text{variance} = \frac{2}{\pi} \cdot \text{PSD} = \frac{2 \cdot 1.34}{3.1416} = 0.8531 \qquad (A.12)$$

Once the generator has been set up this way, it would be nice to send the *final* signal into a spectrum analyzer to verify the actual spectrum. MATLAB/Simulink offers a number of spectrum analyzers, but I have never been able to get verifiable results from them, though I have tried several times, using several versions of Simulink. If any reader has had a *good experience* with these spectrum analyzers, please let me know. One way to get a check is to use Simulink's narrow-band-pass filters to implement the old *analog* way of measuring PSD. That is, send the signal into a narrow band-pass filter, centered on a frequency of your choice, and with a passband of your choice. Send the filter output to a squaring/averaging operation to compute the mean-squared value of the filtered signal. Divide this mean-squared value by the filter bandwidth to get the PSD value at the filter's center frequency. You could then repeat this at several different center frequencies to check how the PSD varied with frequency. This is tedious but seems to work when I have tried it. I think this *verification* is not really necessary; the relations I have given seem to be correct. If you do try the band-pass filter idea, for the best accuracy you should use for the filter bandwidth *not* the bandwidth stated in the MATLAB function for that filter, but rather the *effective noise bandwidth*, which corrects for the deviation from perfection found in any real filter including the MATLAB filters. I have studied the effective bandwidth of MATLAB's Butterworth band-pass filters and found that the 10th order filter has an effective bandwidth 1.016 times the *stated* bandwidth, so these high-order filters are *almost perfect* and do not really require any correction to the bandwidth number used to calculate the PSD.

I have used DASYLAB's* spectrum analyzers, which *do* work properly (and easily) to verify some of the relations stated above, though it is not possible to verify the relation between variance and mean-squared value, since that is peculiar to Simulink's definitions, which are not given explicitly in the manual. (DASYLAB is a very convenient data acquisition and processing software product; it is *not* a dynamic system simulation software.) When I checked the shape of the spectrum given by Equation A.11 and Figure A.28, and the value of PSD at zero frequency, the results did agree with the relations stated earlier. An interesting, and potentially useful, feature of this study is that DASYLAB's random signal has a *uniform* probability density function, whereas Simulink's is Gaussian. The derivations found in Truxal (1955) show that the spectrum shape given by Equation A.11 *does not depend* on the probability density function; it is the same for all, such as the uniform and the Gaussian. Thus DASYLAB gives the same result as Simulink, even though each uses a different distribution. I have suggested to DASYLAB that they might provide a Gaussian signal in addition to their present uniform one. Until they make this change, one can, within DASYLAB, easily convert the uniform signal to Gaussian. If any reader should need this conversion, contact me for the details.

Let us next implement in Simulink the spectrum defined in Equation A.13 to see how all this actually works. Figure A.31 shows how we can generate the *square wave* signal and then low-pass filter it to get a random signal with a relatively flat PSD curve out to about 50 Hz. This model was actually imported from an earlier version of Simulink (I did this example some years ago) and uses a Butterworth low-pass filter that was a *block* available in Simulink itself. In current versions, such filters must be formed in MATLAB and then *installed* in a Simulink transfer function block, as we saw earlier (Figure A.20). The current version *did* accept this "old" filter, so this model is sort of a *hybrid* of the two versions. Since you will likely be using a *current* version, you would have to use the MATLAB filter (I show it at the bottom of Figure A.31, but do not actually use it in the calculations). The computing time increment is 0.0002 s. In the diagram, I used an eighth order Butterworth low-pass filter, set to *cutoff* at 50 Hz. The usual definition of *cutoff frequency* is the point at which the amplitude ratio has dropped by −3 dB. When we check some of the numerical values, we must always keep in mind that these real filters do not behave exactly like the ideal versions. Thus, frequencies below 50 Hz are not all *treated equally* (the amplitude ratio at 50 Hz is about 0.71 rather than the ideal 1.0) and frequencies above 50 Hz are not *totally wiped out* (the amplitude ratio goes to zero *gradually*, not suddenly).

Figure A.32 shows the time histories of the *square wave* signal which comes out of the *band-limited white noise* generator, and also the low-pass-filtered signal, which should have a relatively flat PSD of 1.34 V²/Hz. (I have arbitrarily given the output signal of the band-limited white noise generator the

---

* E.O. Doebelin, *Measurement Systems*, 5th edn., pp. 981–1014., www.dasylab.com

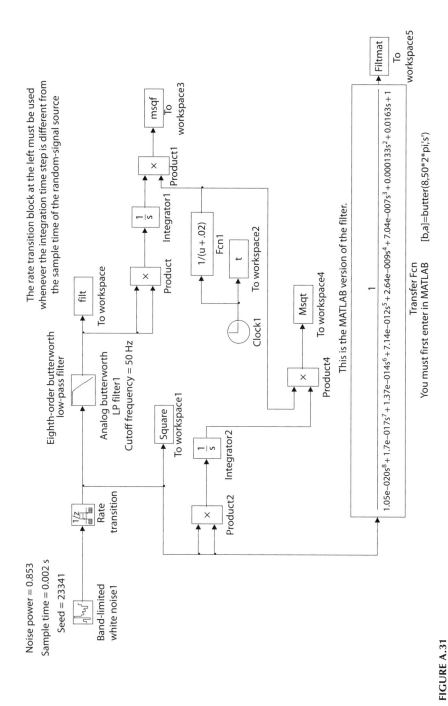

FIGURE A.31

Generation of random signal with desired frequency spectrum.

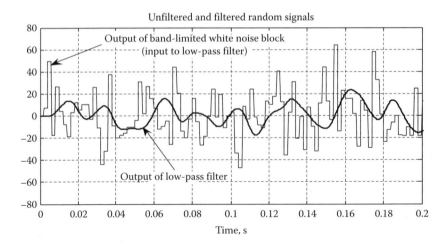

FIGURE A.32
Time histories of raw and filtered random signals.

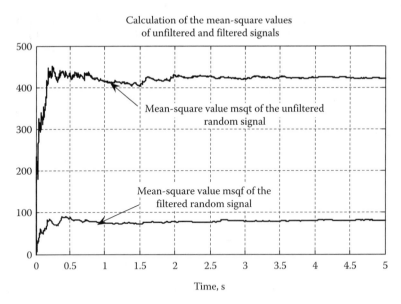

FIGURE A.33
Mean-square values of raw and filtered signals.

units "volts," so that we have some definite units to work with.) Figure A.33 shows the measurement of the total mean-squared values for the two signals. Our theoretical prediction for the square-wave signal was $0.853/0.002 = 427\,V^2$, which is close to what we see on the graph (about 430). For the filtered signal, the total mean-squared value would be given by the total area under its PSD curve. If Simulink had reliable spectrum analyzers, we might use one to get the PSD curve of this signal and then somehow integrate it to get the total area. We can approximate this by simply multiplying the PSD value at 0 Hz by the filter passband of 50 Hz. This operation gives (1.34)(50) = 67, while the graph shows a value of about 70. (As a matter of curiosity, the *old* and *new* Butterworth filters do *not* produce identical signals, even though they are supposed to model the very same transfer functions. Butterworth filters have been around for quite a long time and you would think that their characteristics were pretty well settled. Why the MATLAB software engineers produced this inconsistency is an open question, though in a product as complex as MATLAB, and over a span of years, it is not hard to imagine that errors can creep in. The old filters were only available as *blocks*, and one could not actually see what the transfer function was, so a direct comparison is not possible. Also, the difference in behavior between *old* and *new* is not radical. In any case, we see again that one must be careful in blindly relying on any commercial software product.) Finally, let us sum up how one sets up the band-limited white noise signal source to model a random signal that approximates a white noise from zero frequency to a frequency called f_{max}.

1. Decide on the highest frequency f_{max} for which you want the random signal to be "white." Then set the block's sample time to $0.10/f_{max}$. Thus if f_{max} is 100 Hz, the sample time is 0.001 s.
2. Send the output of the white-noise block to a *rate-transition* block. This allows you to choose the integration time step, and the random signal sample time independently.
3. Follow the rate-transition block with a low-pass filter (an eighth-order Butterworth is usually OK) with the filter cutoff frequency set to f_{max}.
4. If the PSD (more correctly called the mean-square spectral density) that you want is numerically equal to PSD, then set the *variance (sometimes called power)* in the block to $\frac{2}{\pi}$ PSD. The units of PSD are the square of the units of the physical variable divided by the frequency (could be either rad/s or Hz). For example, if the physical variable is pressure in psi, the units of PSD would be $(psi)^2/(rad/s)$. The total mean-square value of the signal is then given by *variance/ (sample time)*. Recall that the *root-mean-square (RMS)* value is just the square root of the mean-square value, and has dimensions, for our example, of $psi/\sqrt{(rad)/(s)}$

While a *white* random signal (flat PSD) is often what is wanted, sometimes we need to provide a random signal that has a *non-flat* PSD. A common way of providing this is to first create a random signal that is white up to the highest frequency needed in our non-white spectrum. We then pass this signal through a suitable frequency-sensitive filter, chosen so as to *shape* the spectrum to our needs. The relation that allows us to choose this filter is

$$\text{output PSD} = (\text{input PSD}) \cdot \left| \frac{q_o}{q_i}(i\omega) \right|^2 \tag{A.13}$$

Since the input PSD is constant (flat), if we know the desired shape of the output PSD, the filter's amplitude ratio (phase does not matter, according to Equation A.13) should follow the square root of the output PSD shape. As a simple contrived example, if our desired PSD shape were $1/(0.01\omega^2 + 1)$, we could use as our *shaping filter* a first order system $1/(0.1s + 1)$, since its amplitude ratio is $1/\sqrt{0.01\omega^2 + 1}$. Figures A.34 and A.35 show this example. While the desired spectrum in our example was chosen for simplicity, one can usually make up transfer functions whose frequency response curve closely matches almost any desired shape.

A.1.14 Generating Random Signals, with Prescribed PSD, for Vibration Testing

Random vibration testing is quite common in practice, so the software which controls the vibration shaker/amplifier needs to be able to generate voltage signals with a random waveform and a prescribed PSD. Since this appendix is addressed to Simulink applications rather than vibration testing, we will not go into complete detail, however the methods used are quite interesting and not easily found in the open literature, so I want to at least show the basic ideas.

I got the basic ideas from an engineer with one of the vibration test equipment companies; I cannot recall his name, it was some time ago. For proprietary reasons he was not able to provide complete detail, but gave enough that I was able to check it out in a general way. The idea is to create a large number of unit-amplitude sine waves whose frequencies cover the range of interest. Each of these sine waves is then given a *random* phase angle chosen from a Gaussian random number generator (MATLAB has one). You then do an *inverse* FFT to convert the frequency-domain values into a time function. The procedure just given would generate a *white* noise; if you wanted a non-white signal you would create the family of sine waves with amplitudes that varied with frequency the way you want.

A MATLAB file that implements this concept follows below. Figure A.36 shows the resulting time signal. The last program statement plots a histogram of the random time signal, to see if it looks Gaussian (it does). It also computes and plots the PSD graph of the random signal using FFT methods. If

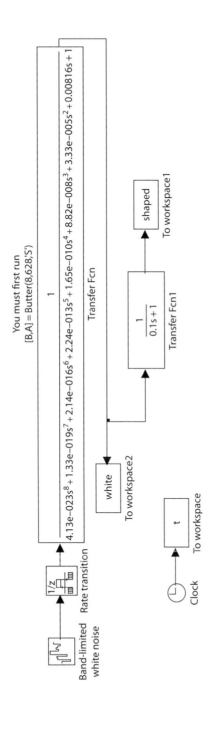

FIGURE A.34

Use of the transfer function block as a shaping filter.

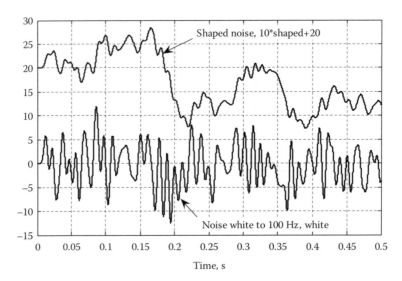

FIGURE A.35
Shaping filter produces nonwhite signal.

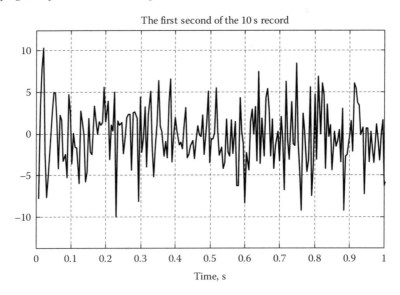

FIGURE A.36
Use of inverse FFT to generate random signal.

you run this program, be prepared for this PSD plot to *not* be a nice horizontal line as expected for a *textbook* white noise; the graph will be *very* uneven. Nothing is wrong. Remember that even if we did have a perfect signal, it takes *averaging* of *many* time records to get a reliable PSD plot; a single record *always* is very rough.

```
% randvib2.m
% program to compute the time history of a Gaussian random signal
% with a specified PSD and a specified record length of T seconds
% the frequency spacing must be 2*pi/T and the highest frequency
% computed should be above the highest frequency specified in the
% PSD curve, such that the total number of frequency points is a
% power of two. That is, if we want a record length of 10 seconds
% then the frequency spacing must be 2*pi/10, and if our highest
% PSD frequency is 100 Hz, then it would take
% 100*2*pi/(2*pi/10) = 1000 points,
% but we want a power of two, so we use 1024 points and thus go
% to a slightly higher frequency than we originally wanted.
% the shape of the PSD curve is most conveniently specified by
% giving a transfer function with the desired shape and using
% it to compute the values which will be inverse FFT'd. If we
% want a "white" noise flat to a given frequency, we might use an
% 8th-order Butterworth filter to compute these values.
% We will use this as the example below.
% suppose we want a white noise "flat" from 0 to 100 Hz,
% and a time record length of 10 seconds.
% compute the magnitudes from the Butterworth filter
[b,a] = butter(8,100*2*pi,'s');
w = 0:2*pi/10:2047*2*pi/10; % highest frequency is 204 Hz
[mag,phase] = bode(b,a,w);
plot(w/(2*pi),mag);
pause
% define random phase angles randn has std.dev of 1.0, so its "range"
% is about plus and minus 3 I use pi/3, below to make the phase
% angle range about plus and minus pi (angles beyond that range are
% not really "different", so why go there)
for i = 1;% for i = 1:5;% i = 1 is almost as good as i = 5;
n = 19731 + 10*i;
randn('seed',n);
ph = [randn(size(w))*pi/3]';
% give the mag values the random phase angles
magr = mag.*exp(1i*ph);
% use inverse FFT to compute the time history. I delete
% the first time point since it seems to always be
% very large
wmax = 2047*2*pi/10;
ft = real(ifft(magr))*wmax/pi;ft = ft(2:2047);
tt = 0:2*pi/wmax:2*pi/(wmax/2047);tt = tt(2:2047);
plot(tt,ft);pause;axis([0 1min(ft) max(ft)]);grid;
pause
% now "scale" the magr values to get true PSD values
% using the relation FFT = sqrt(PSD*delta w)
% let's make the PSD value 0.5 in the "flat" range
delw = 2*pi/10;
magrpsd = magr*sqrt(0.5*delw);
```

```
ftpsd = real(ifft(magrpsd))*wmax/pi;ftpsd = ftpsd(2:2047);
plot(tt,ftpsd);pause;axis([0 1min(ftpsd) max(ftpsd)]);
pause
% now FFT the time function to see what the PSD looks like
delt = 2*pi/wmax;
psd = (delt*abs(fft(ftpsd,2048))).^2/delw;
plot(w/(2*pi),psd)
end
hist(ft,20); h = findobj(gca,'Type','patch'); set(h,'FaceColor',
'w','EdgeColor','k')
```

A.1.15 Example Applications of Some Functional Blocks

The Simulink functional blocks can be used in very versatile ways to model physical systems. We now show in Figure A.37 a sampling of examples which have general application, with graphical results shown in Figures A.38 through A.42. The *step input* block is usually used to apply a step input (force, voltage, etc.) to a model of some physical system. It can, however, be used in other ways. These are based on the fact that the step input can be applied at times of our choice. This capability allows us to *turn on* and *turn off* signals at will. The top part of Figure A.37 shows an example of this that uses also the *multiplier* block with results in Figure A.38.

The time delay (dead time) block (called *transport delay* in Figure A.37) can be used to model physical events such as the propagation of a pressure signal in a pneumatic transmission line. When a piece of pipe or tubing is terminated in a volume, the response can sometimes be modeled as a simple first-order dynamic system. If the tubing is quite long, the finite speed of sound in air (which is also the speed of propagation of pressure signals) requires that the first-order model be augmented with a *dead time* to give a more realistic behavior. Similar effects are also found in a steel rolling mill control system, where the thickness sensor is located downstream from the rolls which are adjusted to control the thickness of the steel strip. If the sensor is 10 ft downstream and the strip travels at 10 ft/s, the time delay between the actual change in thickness and its measurement is 1.0 s. Such time delays can radically affect the control system stability,* so cannot be ignored. In another example, the remote steering of a robotic vehicle on the moon involves sending radio control signals to the vehicle and receiving video signals from a camera on the vehicle. The human operator on earth finds the manual control task quite challenging because of these two delays, which are each about 1.3 s, since the moon is 240,000 miles from earth, and radio signals travel at about 186,000 miles/s. Figure A.39 shows the pneumatic transmission line example. When time delays are imbedded in a feedback loop, the system differential equation

---

* E.O. Doebelin, *Control System Principles and Design*, Wiley, New York, 1985, pp. 274–286.

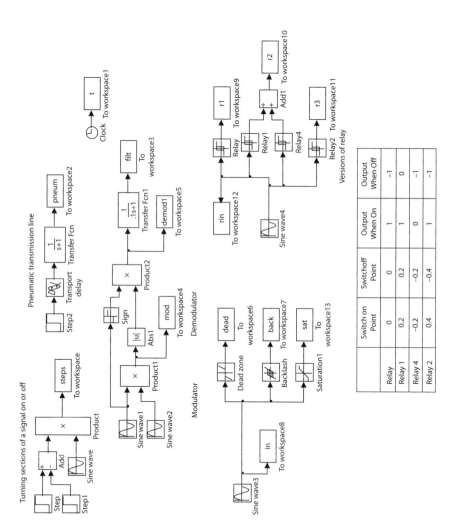

FIGURE A.37
Example applications of some Simulink functional blocks.

	Switch on Point	Switchoff Point	Output When On	Output When Off
Relay	0	0	1	−1
Relay 1	0.2	0.2	1	0
Relay 4	−0.2	−0.2	0	−1
Relay 2	0.4	−0.4	1	−1

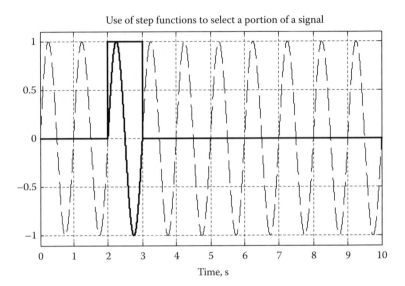

FIGURE A.38

Use of step inputs and multiplier block to select a portion of a waveform.

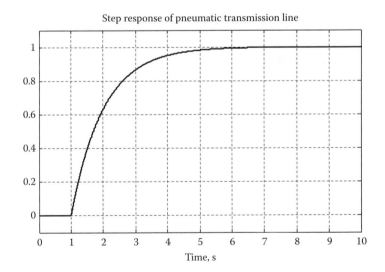

FIGURE A.39

Effect of dead-time on pneumatic tubing/volume step response.

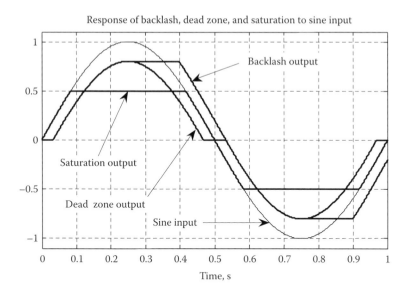

FIGURE A.40
Effects of saturation, dead zone, and backlash on a sinusoidal input.

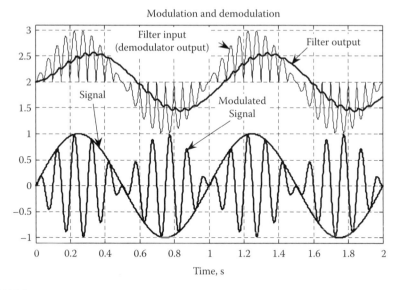

FIGURE A.41
Amplitude modulation and demodulation.

is no longer the analytically tractable *linear with constant coefficients* type, but becomes analytically unsolvable, so simulation is usually *required*.

The *multiplier, sign, sine wave* and *absolute value* blocks are used in Figure A.37 to model a *phase-sensitive demodulator*, a widely used component of measurement systems. Sine wave 1 is a 1 Hz sine wave signal which modulates

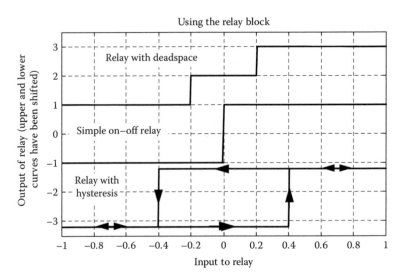

FIGURE A.42
Three types of relay actions.

(multiplies) Sine wave 2, a 10 Hz *carrier.* For example, mechanical displacement signals in an LVDT motion sensor* modulate the high-frequency AC output of the sensor. The demodulator, together with a final low-pass filter, are used to produce a sensor output signal that "looks like" the mechanical displacement. In our example, I have chosen the numbers such that the filter is not really acceptable; it leaves too much *ripple* in its output signal (Figure A.41). In a practical system, we are able to make this ripple so small as to not be noticeable.

At the lower left of Figure A.37, we show three nonlinear blocks useful in many practical applications. The *dead zone* block could, for example, be used to model the flow area variation of a hydraulic servovalve of the *overlapped* type.† Here the valve has a small neutral zone where moving the valve spool does not actually open the valve. The *backlash* block models *lost motion* in a mechanism, such as two meshing gears where the tooth width of one gear is smaller than the tooth gap of the mating gear. Many mechanical and electrical components exhibit *limiting,* which is modeled by the *saturation* block. *On-off* relays‡ are used in many control systems; their nonlinearity makes analytical treatment difficult. They are easily simulated using the *relay* block. The lower right part of Figure A.37 shows several practical relay actions simulated by this block. Using two step functions, a summer, and a multiplier, set up to produce a rectangular pulse of height 1.0, and timed to occur at a moment of our choice, we can *select a portion* of any waveform (Figure A.38). One use of this facility is to create a *transient* waveform from an available *periodic* waveform. The

---

* E.O. Doebelin, *Measurement Systems,* 5th edn., pp. 258–261.
† E.O. Doebelin, *Control System Principles and Design,* pp. 102–113.
‡ *Control System Principles and Design,* pp. 221–257.

transport delay (dead time) block accepts any signal as its input and produces exactly the same waveform, but with the prescribed delay, at its output. The delay *must* be a positive number; you *cannot* produce a time *advance*. *Backlash* and *dead zone* both reduce the *amplitude* of the output signal but backlash also gives a hysteresis effect. The *saturation* block models the limiting effect present in much practical hardware that is nominally linear. An electronic amplifier can have 0.01% linearity in its design range, but will saturate at, say, ±10 V when the input signal becomes too large. Many commercial sensors and signal conditioners allow or require AC excitation (say 5 V at 3000 Hz). This produces an amplitude-modulated signal which has the correct information in it, but *does not look like* the physical variable being measured. A phase-sensitive demodulator and low-pass filter are used to recover the proper waveform. The *relay* block has many uses in addition to the obvious one of modeling electromechanical or solid-state switches. Used with a multiplier, it can turn signals on and off by supplying a 1 or a 0 to one input of the multiplier. The relay actions shown in the table at the bottom of Figure A.37 are implemented in the simulation diagram just above it. The relay block requires that we choose values for four settings, as shown in the table for our examples. (In the graph of Figure A.42, some of the curves have been vertically shifted to display them more clearly.)

We next show some ways to model nonlinear math functions using the *table lookup* block, the *fcn* block, and the MATLAB *spline* command. When two variables in a physical system are related by some nonlinear function, this usually makes the system differential equation nonlinear and thus analytically unsolvable, and simulation becomes necessary. If the nonlinear relation is specified in terms of a table of experimental data, then the use of the *lookup table* seems natural since the *input vector* and *output vector* required in this block are already available. As an alternative that might in some cases be preferable, we could use *curve-fitting software* to fit an *analytical* function to the experimental data and then enter this function into Simulink. If the nonlinear effect is already specified by a known math function, then we would use the *fcn* block directly. Figure A.43 shows the implementation of some of these ideas. We use a mass/spring/damper system for our example, but the damper and spring are both nonlinear. Many fluid dampers or shock absorbers have a nonlinear relation between damping force and velocity because of turbulent flow in the device passages and valves. This relation is often modeled as a *square law* or second-power curve. If we use a *function* block with a squaring relation *directly* we unfortunately get an unsymmetrical curve with *negative* damping when the velocity goes negative. To get the desired positive-damping curve, we need to get the algebraic sign correct, which can be done in several ways, one of which is shown. Our main interest is in the nonlinear spring force, which is given by the relation $F_{spring} = 10^*x + 0.01^*x^\wedge5$. This relation is easily modeled, as shown with the *function* block. Note that when using this block, the input signal *must always* be given the symbol u, no matter what you might have named it on the block diagram (x in our case).

If the nonlinear spring had been experimentally calibrated to determine its force/deflection curve, we might use that calibration data directly in a

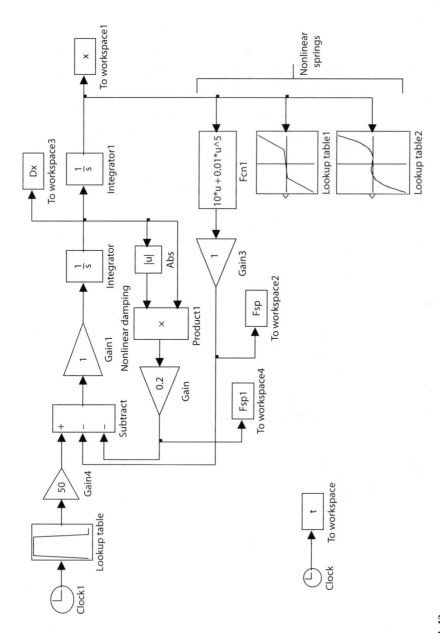

FIGURE A.43
Modeling nonlinear springs and damping.

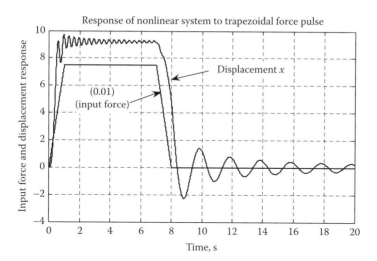

FIGURE A.44
Natural frequency varies with position in a nonlinear system.

lookup table since we would not have a math function to put into a *function* block. In *lookup table* 1, I have used this approach but with unrealistically sparse data, only 5 points, which are however exactly correct for the function given. The sparse data is used to make some effects more obvious to the eye. In *lookup table* 2, I have taken *the same* 5 points but used MATLAB's *spline* function to generate a set of interpolated points which are inserted into the lookup table. Figure A.44 shows the response to an input force pulse (generated in a lookup table) when we use the function block to model the spring exactly. Note that the natural frequency is much higher when displacement is large (near 9) than when it is small (near 0). Linear systems of course have exactly the same natural frequencies *everywhere*.

The lookup table that uses only 5 points and linear interpolation would of course not be an accurate model. To show how the *spline* operation can interpolate a smooth curve through the 5 points we would use the following MATLAB commands.

```
xx=-10:5:10;                define the displacement values of
                            the 5 points
yy=-10.*xx+0.01.*xx.^5;     define the force values of the 5
                            points
xint=-10:0.1:10;            define the x values for the
                            interpolation (any finely-spaced
                            set of points is OK)
yint=spline(xx,yy,xint);    performs a cubic spline
                            interpolation for the 5 given points
ya=-10.*xint+0.01.*xint.^5; computes accurate values of force
                            for the interpolation points
```

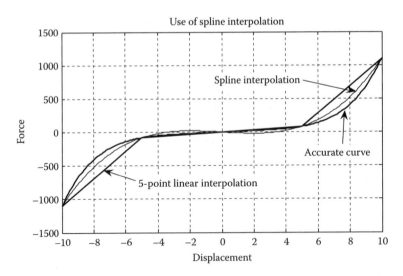

FIGURE A.45
Use of MATLAB spline operation.

We can generate a composite plot of the force/displacement curves for: the 5-point linear interpolation, the 201-point spline-interpolated curve, and the *perfectly accurate* curve computed from the defining formula. We see (Figure A.45) that, even for *very* sparse initial points, the spline operation gets us quite close to the exact curve. In practice we would of course use more than 5 points, but it does not take many more than 5 to get acceptably close. One defect in the spline interpolation using 5 points is just barely visible in the graph. From about −4 to 4 displacement, we see a slight *negative* spring constant (slope). Negative spring constants can cause *instability,* so we need to avoid this; using more than 5 points should fix this. Spline interpolation is easy in MATLAB and has many practical applications. One desirable feature is that it creates curves with continuous derivatives; linear interpolation has slope discontinuities at every original point.

The *switch* block has many useful applications. Basically it allows us to alternate the *path* of a signal between two selections, depending on the value of another signal in the simulation. Figure A.46 shows a tank-filling system which uses a *coarse* pump to rapidly fill the tank until the level h nears the desired value of 10 and then switches to a *fine* pump to accurately reach the final level. We use the *switch* block both to sequence the two pumps and also to model the tank cross-section area where it transitions linearly from the value 5 to the value 8 over the height range 3–5. The switch block has three input terminals and one output terminal (Figure A.47). The two *upper and lower* inputs are used for the two signal paths which we wish to switch between. The *middle* terminal is the signal which causes the switching to occur. The switching point of the switch block is set with a parameter called *threshold.* When the middle signal is *below* the threshold value, the lower

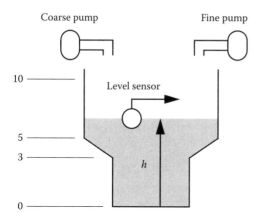

Coarse pump   Fine pump

10

Level sensor

5

3

h

0

FIGURE A.46
Tank level control system.

signal is passed through to the output. When the middle signal reaches and exceeds the threshold value the upper signal is passed to the output. For example, for switch 1, the output is 10 until the liquid level reaches 9.5. Then the output becomes the input to the lower terminal, which comes from the output of switch 2. This output is 1.0 until *h* reaches the value 10, when it goes to 0. Note that this output is at the value 1 *all the time* until *h* reaches 10, but it is *ineffective* until *h* reaches 9.5 because switch 1 blocks it. In creating the variable tank area when h is between 3 and 5, note that Figure A.46 is perhaps misleading. It seems to show a *conical* transition piece, but this would *not* give a linear area change, since area varies as the *square* of the radius. That is, to be geometrically correct, the transition piece should have a *curved* (not straight) profile. Figure A.48 shows the tank-filling process.

This completes our treatment of basic Simulink simulation methods for *analog* systems. It should be adequate for a wide variety of applications. For those readers who will go on to more advanced simulations, with the background of these notes you should be able to find and learn the more sophisticated techniques with the aid of the online help, phone calls to tech support, discussions with colleagues, reference to published papers, short courses offered by MATLAB, etc. We next give a short treatment of simulation methods for *computer-aided systems*, where the application of the *zero-order hold* and *quantizer* blocks are demonstrated. A common application of the *quantizer, zero-order hold,* and *time-delay* blocks is in the modeling and simulation of computer-aided systems, sometimes called *mechatronic* systems. When modeling a complete system, the main effects of the computer portion are *sampling, quantization,* and *computational delay.* Most of the sensors used in such systems are *analog* in nature, producing usually a voltage which varies smoothly with the measured variable. The digital computer will not accept such a voltage; it computes with one specific numerical value at a time, thus the sensor output must be *sampled* periodically, and this sampled value *held constant* until the next sampling instant.

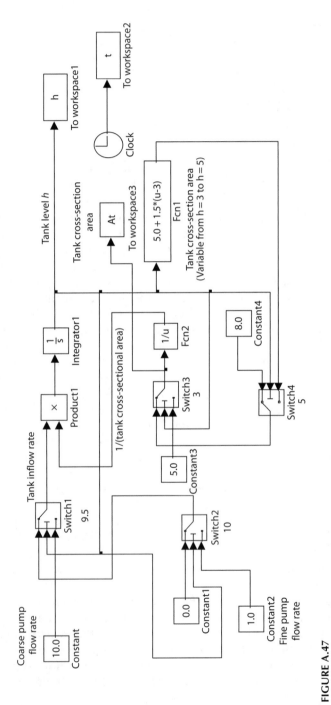

FIGURE A.47
Use of *Switch* module in simulation of tank level control system.

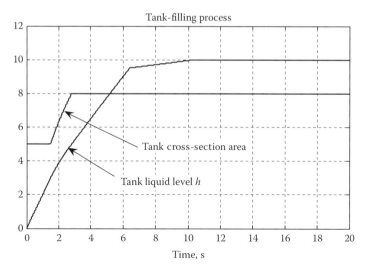

FIGURE A.48
Response of tank level control system.

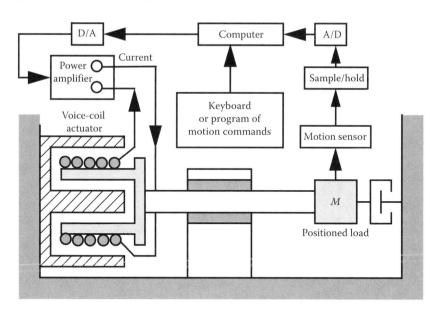

FIGURE A.49
Computer-aided motion control system.

The simple motion control system of Figure A.49 uses this scheme. An analog-to-digital converter operates on the sampled value to produce a true digital signal with a certain number of *bits*, usually 10 or more. In the simulation of Figure A.50, I used a 7-bit converter to make some of the observed effects more obvious. The *zero-order hold* block allows one to set the sampling time

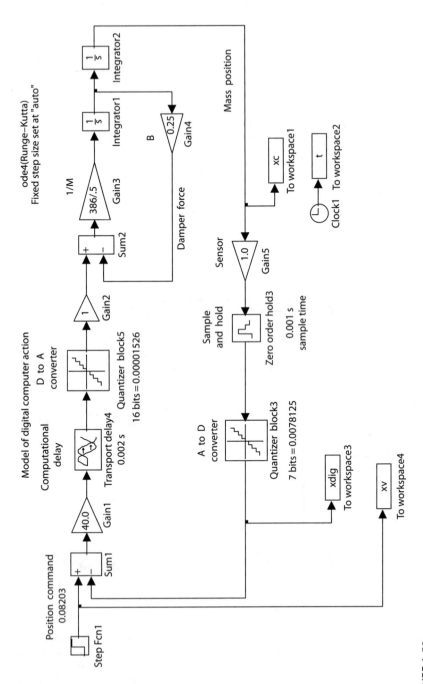

FIGURE A.50
Computer-aided motion control system simulation.

interval when we click on it; I used 0.001 s for this example. The *quantization* block allows us, upon clicking, to set its *resolution*, which for 7 bits is $1/2^7 = 0.0078125$. The *computational delay* of the computer depends on its speed and the complexity of the calculation; I arbitrarily set it at 0.002 s. Most digital computers carry *many* bits, so they do not contribute much to the quantization error; I neglected it completely. Since most actuators (such as the translational dc motor called a *voice-coil actuator* of our example) are analog devices, the digital output of the computer must be converted to an analog voltage in a D/A converter, where again we need to choose the number of bits; I here chose a 16 bit device. (Our simulation allows quick and easy study of the effects of such choices. I made the sensor in this example the focus of resolution questions by choosing a too-coarse resolution. I let the D/A converter have a more typical (*fine*) value so that there would be no confusion as to the source of resolution problems.)

We test the system with a step input command of size 0.08203, as shown in Figure A.51. Using a 7 bit A/D converter on our sensor, the only values available in the neighborhood of our 0.08203 command are 0.078125 (10 quantization units) and 0.0859375 (11 units). That is, the system is *looking for* 0.08203 but *cannot find it*, so it will "hunt" or oscillate between the two values it *can* find. This hunting is usually unacceptable but *must* occur when quantized (rather than smoothly varying) signals are used. Of course one can (and does) set the sensor quantization resolution fine enough to make the hunting unnoticeable. (Real effects such as Coulomb friction in the load bearings (which was ignored in our study) may also stop the hunting.)

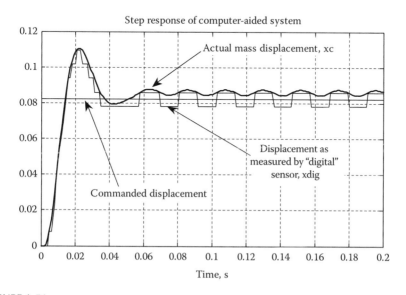

FIGURE A.51
Signal quantization causes oscillations.

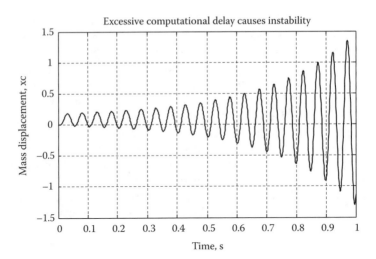

FIGURE A.52
Feedback system instability.

In Figure A.52 the computational delay has been increased to 0.008 s, causing the system to become absolutely *unstable*. We see from these few example test runs how easily one can study the effect of system parameters on the behavior of computer-aided systems of all kinds. My text *System Dynamics: Modeling, Analysis, Simulation, Design*, Marcel Dekker, New York, 1998, studies this system in more detail and also includes many Simulink simulations of mechanical, electrical, fluid, thermal, and mixed dynamic systems.

A.2 Basic Frequency-Domain Techniques in MATLAB

A.2.1 Introduction

Frequency-domain methods refer to the representation of both *signals* and *systems* in terms of sine waves and the response to sine waves. When applied to *signals*, we speak of *converting* the signal time history into a *frequency spectrum*. When applied to *systems*, we are interested in the *frequency response* and limit ourselves to consideration of *linear systems with constant coefficients*. When this class of systems is subjected to a perfect sine wave input, the *steady-state* response will also be a sine wave of exactly the same frequency as the input. The term *steady state* here means the *sinusoidal steady state*. That is, we completely ignore the *starting transients* and wait for these to die out before measuring the output sine wave amplitude and phase angle.

One can, of course, subject (mathematically or experimentally) *nonlinear* or *time-variant linear* systems to sinusoidal inputs, but the response in general is *not* sinusoidal, nor necessarily even *periodic*. If it *does* appear periodic, one could extract (in the sense of *Fourier series*) the *fundamental frequency* component (also called the first harmonic) which would be a perfect sine wave, and find an amplitude ratio and phase angle relative to the input sine wave. A complication is that the response might be different for different input amplitudes. A practically useful analysis technique for feedback control systems which takes this viewpoint is called the *describing-function method.** Since all *real* systems have at least a little nonlinearity and/or time-varying parameter values, when we do *experimental testing* we are always faced with an output signal that is not quite a perfect sine wave. Commercial frequency-response analyzers are designed to take this fact into account, usually by some kind of filtering to extract the fundamental frequency component before computing amplitude ratio and phase angle.

A.2.2 Choosing the Frequency Values to Be Used in a Calculation

Because of the requirements of calculus, frequencies used in theoretical calculations must always be given in *radians/time*, often rad/s. However, when we make experimental measurements, the electronic timer/counters that are used read in Hz (cycles/s). Also, cycles/s relate more directly to machine operating conditions such as RPM. In doing calculations, a convenient scheme is to set up the range of frequencies in Hz and call that variable f; then define another variable w, which converts the f values to rad/s. In addition to the *range* of frequencies that is of interest, the *spacing* of the points is also important. Since too coarse a spacing leads to *rough* plotted curves or *missed peaks/valleys*, and because our computers/software are quite fast, we usually opt to err on the side of *fine* spacing of the points. In MATLAB, there are several ways to actually set up the frequency values.

The most basic way is to use the standard MATLAB *range variable* statement, such as $f = 0{:}1{:}200$; which gives us 201 points from 0 to 200, spaced at intervals of 1.0 Hz. The *companion* statement to get the corresponding w values would be $w = 2*pi*f$; another way is to use the *linspace* or *logspace* statements:

```
f = linspace(0,1000,1001);   gives 1001 points linearly-spaced
                             between 0 and 1000
f = logspace(-1,2,500);      gives 500 logarithmically-spaced
                             points between 10⁻¹ and 10²
```

A.2.3 Units for Amplitude Ratios and Phase Angles

A sinusoidal transfer function (also called frequency-response function, FRF) will always have the units of the output quantity divided by the input quantity. If the input is force in Newtons and the output is displacement in

---

* E.O. Doebelin, *Control System Principles and Design*, Wiley, New York, 1985, Chapter 8.

meters the transfer function must have the unit m/N, and the *true* amplitude ratio must have these same units. However, we sometimes want to use *decibels* (dB) for the amplitude ratio, using the definition

$$dB \triangleq 20 \log 10 \text{ (amplitude ratio)}$$

When we use amplitude ratio in decibels, we usually also use a logarithmic scale for the frequency. When we use frequency in rad/s, the phase angles will always come out in radians, but we usually prefer degrees, so if we had called the phase angle in radians, say, *phir*, we could easily change over to degrees by defining *phid= (360/2pi)∗phir* for plotting or tabulating values. When we are designing a dynamic system to accurately preserve the wave-form of the input signal (measurement systems are a prime example), we want a flat amplitude ratio and a phase angle curve which is linear with fre-quency. To check the phase angle requirement, you *cannot* use standard *Bode plots*, since they have a logarithmic frequency scale and a line which is truly straight will appear curved on the log-scale plot. Linearity of phase angle can however be better checked by computing and plotting the system *time delay*, which is obtained from the phase angle data by using the formula:

time delay in seconds = (phase angle in radians)/(frequency in rad/s) (A.14)

A graph of time delay versus frequency *will* be a horizontal straight line on *either* linear or logarithmic frequency scales if the time delay is constant (phase angle is linear with frequency).

A.2.4 Computing and Graphing the Frequency-Response Curves

To compute the sinusoidal transfer function, we usually start with the *opera-tional (D-operator)* transfer function or the *Laplace (s-domain)* transfer func-tion. When we substitute $i\omega$ for either s or D, the transfer function becomes a *complex number*, whose magnitude is the *amplitude ratio* and whose angle is the *phase angle* that we desire. The quickest way (not always the best way) to plot the frequency response curves for a system with a transfer function whose numerator polynomial is $b(s)$ and denominator polynomial is $a(s)$ is to use the MATLAB statement *bode (b,a)* which computes the numerical values and plots the curves on logarithmic coordinates (dB for amplitude ratio, radians for phase angle, frequency in rad/s). For example, if $b = 3s + 1$ and $a = 2s^3 + 1.23s^2 + 5s + 1$, we enter $b = [3\ 1]$ and $a = [2\ 1.23\ 5\ 1]$, then enter the command *bode(b,a 'k')* we get the curves of Figure A.53. Note that we did *not* have to specify the frequency range, it was set automatically. If you enter *bode(b,a)* you get the same curves but they are plotted in *blue* rather than the black we usually prefer (the "k" gets us black curves). When the plot appears, you *can* invoke the *property editor* as we usually do for our graphs, but it appears that this does *not* allow the usual easy changing of colors, changing line width, etc. (The MATLAB software engineers probably consider the Bode plots as *ready-made* graphs which people will not want to modify.) One probably *can*

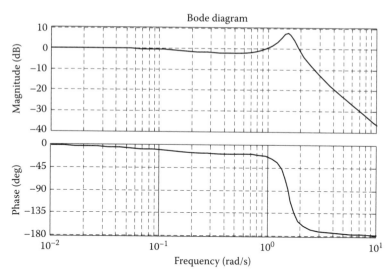

FIGURE A.53
Standard MATLAB Bode (frequency response) graph.

modify the Bode plots if one goes to the basic plot commands but this gets a little complicated. Also, the *numerical values* for amplitude ratio and phase angle are *not* readily available when you use the Bode command.

If you want better control of your graphs, you can get it without too much trouble by generating them "from scratch," using a different version of the Bode command. This will also give the numerical values of frequency, amplitude ratio, and phase angle, which can be displayed as tables if you wish. The command sequence goes as follows:

```
f = logspace(-2,1,500);        500 log-spaced point between 0.01
                                and 10.0 Hz
w = f*2*pi;                     converts to rad/sec
b = [3 1];                      define numerator
a = [2 1.23 5 1]:              define denominator
[mag,phase] = bode(b,a,w);     computes true (not dB) amplitude
                                ratio (mag) and phase (degrees)
                                but doesn't plot anything
semilogx(f,mag);               plots true amplitude ratio (not dB)
                                versus frequency in Hz
semilogx(f,phase);             plots phase angle (degrees) versus
                                frequency in Hz
```

These commands essentially duplicate the graphs of Figure A.53 except we plot the *true* amplitude ratio rather than dB. We also get graphs that allow all the usual adjustments provided by the *property editor* and all the numerical values are available for further manipulation or printing as tables. If you want the amplitude ratio in dB, use

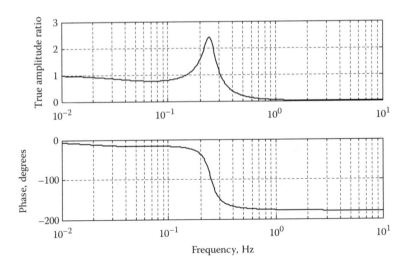

FIGURE A.54
"Non-dB" amplitude ratio gives clearer display.

```
magdB = 20*log10(mag);
semilog(f,magdB);
```

If we want to display both amplitude ratio and phase on a *single graph* (as is done with most Bode plots), use

```
subplot(2,1,1);semilogx(f,mag);
subplot(2,1,2);semilogx(f,phase);
```

This produces Figure A.54, where I was able to easily get a line width of 2, black curves, and labels of my choice, using the property editor in the usual way. Graphs with linear rather than logarithmic frequency scales allow us to go to zero frequency (often desirable) and read frequencies from the graph more accurately. The only change needed in the above commands is to use

```
f = linspace(0,1,500);
w = f*2*pi;
```

With these changes we can plot the amplitude ratio curve as in Figure A.55.

Physical systems sometimes have dead time (time delay) effects in their transfer functions. The frequency response of a dead time has an amplitude ratio of exactly 1.0 for all frequencies, so the inclusion of a time delay has *no* effect on the system amplitude ratio. However, dead time has a negative phase angle which increases linearly with frequency and *without any bound*. That is, it just keeps getting more and more negative as the frequency gets

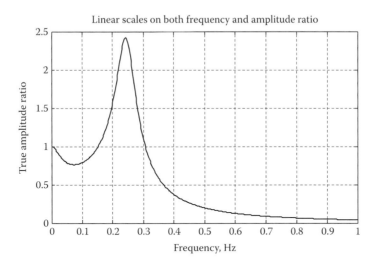

FIGURE A.55
Linear frequency scale shows behavior at zero frequency.

higher. Transfer functions which are the usual ratio of polynomials *always* are bounded in phase angle at some positive or negative multiple of 90°. An ever-increasing negative phase angle in the open-loop transfer function of a feedback system has a very *bad* effect on system stability, so dead time effects are very important when present.

MATLAB's various Bode type statements do not provide directly for inclusion of dead times in the transfer function, so you have to provide the effect separately, which fortunately is easy to do. MATLAB *does* provide a statement to *approximate* the effect of a dead time with various transfer functions which *are* ratios of polynomials allowing approximate analysis of feedback systems that have dead times. These approximations are called the *Pade approximants*; you can get details by typing *help pade*, but we will here not pursue this, since our interest is in the frequency response, which need *not* be approximated. This is one advantage of the frequency-response methods of control system design/analysis; they deal *exactly* with dead-time effects. The phase angle of any dead time is given by $\phi = -\omega\tau_{dt}$ where τ_{dt} is the dead time in seconds, ω is the frequency in rad/s, and ϕ is the phase angle in radians. If our example system above had in addition a dead time of 0.5s, we could take this into account by computing the angle given by the dead time and simply adding it to the angle (called *phase* above) of the original system. This gives the graph of Figure A.56.

Systems are often built up from cascaded *components*, each of which has its own transfer function. If we want the *overall* transfer function of these *series connected* devices, MATLAB provides a convenient command. Taking as an example three identical first-order systems with time constants of 1.0s, we would enter the following commands.

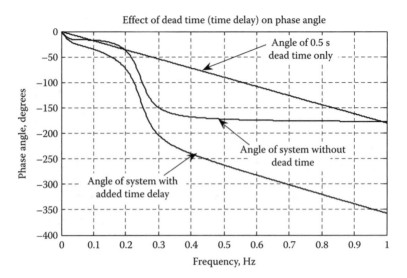

FIGURE A.56
System with dead-time requires manual correction of phase angle.

```
>> num1 = [1];
>> den1 = [1 1];
>> num2 = [1];
>> den2 = [1 1];
>> num3 = [1];
>> den3 = [1 1];
>> sys1 = tf(num1,den1);
>> sys2 = tf(num2,den2);
>> sys3 = tf(numd,den3);
>> sys = series(series(sys1,sys2),sys3)
```

This would produce the following display on your computer screen:

```
Transfer function:
        1
---------------------
s^3 + 3s^2 + 3s + 1
```

A.2.5 Fitting Analytical Transfer Functions to Experimental Frequency-Response Data

The frequency response of physical systems is often measured experimentally to validate and improve theoretical models or to be used directly when theoretical models are not available. Some frequency-response applications (feedback control design for example) can use the experimental data as such. Many times, however, we would like to get *analytical* transfer functions which fit the experimental data. A number of methods* are available for this; we show only an easy-to-use MATLAB routine. If you have used curve-fitting software to fit, say, polynomials to graphical data, you may think that you already know how to deal with the present application. Actually, such routine curve-fitting software will *not* handle this situation. The main problem is that one must fit *two* curves (amplitude ratio and phase angle) at once. The MATLAB routine that provides this service is called *invfreqs(h,w,nb,na)*. The quantity h is a list of the experimental data, in the form of a complex number giving the amplitude ratio and phase angle measured at frequencies w. Parameter nb is the *order* of the numerator polynomial of the transfer function that you wish to fit to the measured data; na is the order of the denominator polynomial. If the numerator polynomial were, say, $b_1s + b_0$ and the denominator $a_3s + a_2s + a_1s + a_0$, then $na = 3$ and $nb = 1$. An example goes as follows.

```
% program to demonstrate the frequency-response
% curve-fitting routine invfreqs
% use a given transfer function to generate the "experimental points"
b = [0.53 5.3]; a = [1 1];        define the transfer function
d = [1]; c = [1/2500 0.8/50 1]; using the "series" command
[num,den] = series([b],[a],[d],[c])
% check if the method gives perfect results for perfect data
w = linspace(0,200,500);    set up the frequencies
ha = freqs(num,den,w);      compute the "experimental" data
[g,h] = invfreqs(ha,w,1,3); invoke the curve-fitting procedure
gs = g/h(4)                 put results in "standard" form
hs = h/h(4)
bode(num,den,'k'); hold on; bode(gs,hs,'k'); pause; hold off
% make data noisy with uniform pdf noise
noiz = rand(size(w))-0.5;     generate the noise
hnoiz = ha + 0.5.*ha.*noiz;   scale the noise and add to
                              "perfect"
% compare perfect and noisy data
plot(w,abs(ha)); hold on; plot(w,abs(hnoiz)); pause; hold off
[gn,hn] = invfreqs(hnoiz,w,1,3); invoke curve fitting for
noisy data
gns = gn/hn(4)
hns = hn/hn(4)
bode(num,den,'k'); hold on; bode(gns,hns,'k'); pause
hold off
```

To check how well the *invfreqs* routine works, we can give it the easiest task; generate the "experimental" data from a *known* analytical function and use exactly the correct na and nb to match this function. Of course, the program is *not* told the transfer function *coefficients*, so it still has to discover these, and we can see how well it performs under these ideal conditions.

---

* E.O. Doebelin, *System Modeling and Response: Theoretical and Experimental Approaches*, Wiley, New York, 1980, pp. 244–251.

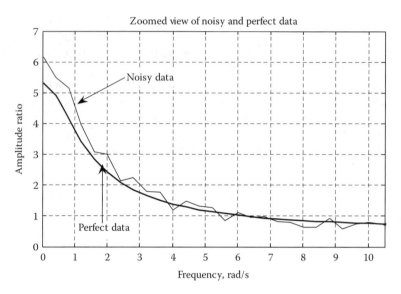

FIGURE A.57
Adding random "noise" to "manufactured" data.

In the MATLAB program above, when perfect data is used and the procedure is given the correct *na* and *nb*, it extracts the proper coefficients almost perfectly and the Bode plots of *data* and *fitted function* lie on top of each other. When we add some random noise to the *data* to make it more realistic, the process still works very well. Figure A.57 shows that the noise in the amplitude ratio data is a significant percentage of the perfect value. (A plot of the noisy phase angle would show similar behavior.) Figure A.58 shows superimposed

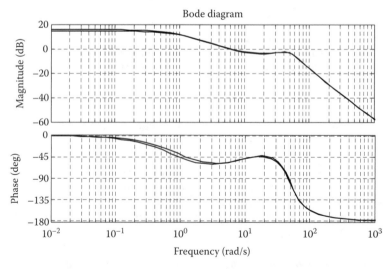

FIGURE A.58
Curve-fitting routine works well with both "perfect" and noisy data.

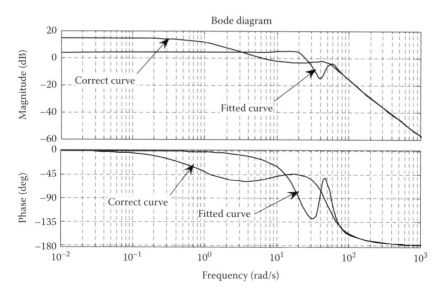

FIGURE A.59
Bad choice of transfer function form causes trouble.

Bode plots of the perfect transfer function and the fitted function for the case of the noisy data. Both the amplitude ratio and phase angle curves for the fitted function are very close to the correct values.

A further check on the capabilities of this process would be to give it noisy data and also *wrong* values for *na* and *nb*, since in practice we would usually be *guessing* these. If we set *na* and *nb* too small, there is a good chance that the fit may not be good. If we set them too large, one might hope that it would make the coefficients of the "excess" terms close to zero and thus still give a good fit. Figure A.59, where *na* = 4 and *nb* = 2 were used with the noisy data, shows that our hopes are *not* justified; the fit is very bad. Of course we always *check* our fits, so there is no reason to ever accept a bad fit.

In choosing initial values for *na* and *nb*, we are guided by our knowledge of how the shapes of the curves are affected by *na* and *nb*. This is another place where the skill of rapidly sketching approximate Bode plots is useful. In the poor fit of Figure A.59, note that the fit is actually very good *at high frequencies*. There is a version (see *help invfreqs*) of this procedure that allows *weighting* of the data points. For a result like Figure A.59, one might try weighting the low frequency points more than the high frequencies and trying the fit again. We will not pursue this here, but it can result in improved fits.

A.2.6 Using *sys* and *tf* Statements to Form Transfer Functions

We used the *sys* and *tf* statements earlier, but did not give a complete discussion there. These statements are quite useful when we want to form more complicated transfer functions from *component parts*. For example:

```
s = tf('s'); h = (s + 0.5)/        the s = tf('s') allows the
((s + .1)*(s + .2))                writing of transfer functions
                                   "directly" as shown. You must
                                   use the standard form where the
                                   highest power of s has the
                                   coefficient 1.0. You can use any
                                   name where
                                   I have used "h"
```

```
Transfer function:
    s + 0.5
-----------------                  The computer responds with this
s^2 + 0.3s + 0.02                  display.
```

```
» h2 = (s + 4)/(s + 10)            Define another transfer function.
Transfer function:
s + 4
------                             Computer responds.
s + 10
» sys = series(h,h2)               Cascade (series connection) the
                                   two transfer functions.
```

```
Transfer function:
      s^2 + 4.5s + 2
-----------------------------      Computer responds.
s^3 + 10.3s^2 + 3.02s + 0.2
bode(sys)                          Ask for a bode plot.
```

Since we often prefer the *other* standard form (where the *trailing* terms have the coefficient 1.0), we can "fool" the scheme by writing our transfer functions as follows:

```
s = tf('s');h = 2*(s + 0.5)/((10*(s + .1)*5*(s + .2)))
```

```
Transfer function:
    2s + 1
-----------------                  Computer responds
50s^2 + 15s + 1
```

Note that this transfer function has a steady-state gain (static sensitivity) of 1.0, whereas the one which used the *other* standard form has a steady-state gain of 0.5/0.02 = 25.0, so these two versions have the same dynamics but different steady-state gain. If I were writing this MATLAB software I would set things up so that transfer functions would *explicitly* show the steady-state gain as a *separate multiplier* (example: 3.67(5s + 1)). Such a format clearly separates the steady-state gain from the dynamics, which in my opinion is a useful feature. Since MATLAB *does not* do this directly, if we really prefer this format, we can easily include the steady-state gain as a separate operation when we decide to compute and graph frequency-response curves. That is, if we want the gain of 25.0 that is implied in the original definition of the above example transfer function, we could easily modify the numerator in a *num, den* statement as follows: num = [50 25] den = [50 15 1].

A.2.7 Matrix Frequency Response Calculation

When a system has *several* inputs and outputs, we can define several transfer functions relating a certain output to a certain input. If we want to see the actual analytical form of these transfer functions, we must go through the *determinants* method discussed earlier, which can be quite tedious for complicated systems. However, if our only interest is in the frequency-response (sinusoidal transfer functions) *for given numerical values* of all the system parameters, matrix methods provide an efficient calculation which gets us *all* the possible pairs of output/input transfer functions *in one fell swoop*. To show how this is done we apply it to the mechanical system of Figure A.12, which has two inputs and two outputs. The example program is easily extended to deal with any number of inputs and outputs giving a general method. The program is displayed below. It is based on conventional linear algebra methods, the details of which are given in the literature.* (These details are *not* necessary for the application of the program.) This example shows a *pattern* which can be applied to *any* system of interest.

```
% computes matrix freq response
% set up the list of frequencies
% f = linspace(0,20,201)'; define frequency range in Hz
f = linspace(0,300,201)';   shows high-frequency phase
w = f*2*pi;                 convert to rad/sec
ii = sqrt(-1);s = w*ii;     define a new name for "i" since it
                            is used
conv = 360/(2*pi)           in the "for loop" coming up
for i = 1:201               start for loop
    A = [s(i).^2 + 5.*s(i) + 4000 -5.*s(i)-4000; Equation A.5
    -5.*s(i)-4000 10.*s(i).^2 + 55.*s(i) + 14000];   Equation A.6
    AI = inv(A);            invert the A matrix
    C1 = [1 0]';            column vector with f1 selected
    qo1 = AI*C1;            matrix multiply
    x1of1(i) = qo1(1);      calculate transfer function x1/f1
    x2of1(i) = qo1(2);      calculate transfer function x2/f1
    C2 = [0 1]';            column vector with f2 selected
    qo2 = AI*C2;            matrix multiply
    x1of2(i) = qo2(1);      calculate transfer function x1/f2
    x2of2(i) = qo2(2);      calculate transfer function x2/f2
    arx1of1 = abs(x1of1); phix1of1 = angle(x1of1)*conv;
    arx1of2 = abs(x1of2); phix1of2 = unwrap(angle(x1of2))*conv;
    arx2of1 = abs(x2of1); phix2of1 = unwrap(angle(x2of1))*conv;
    arx2of2 = abs(x2of2); phix2of2 = angle(x2of2)*conv;
end
plot(f,arx1of1);pause; plot(f,phix1of1);pause
plot(f,arx1of2);pause; plot(f,phix1of2);pause
plot(f,arx2of1);pause; plot(f,phix2of1);pause
plot(f,arx2of2);pause; plot(f,phix2of2);pause
```

---

* E.O. Doebelin, *System Dynamics: Modeling, Analysis, Simulation, Design,* Marcel Dekker, New York, 1998, pp. 640–642.

There are two alternative $f = linspace$ statements. One displays the main frequency range of interest, the other goes out to higher frequencies to confirm that the phase angles behave as they should. The A matrix is obtained directly from the set of simultaneous equations which define the system, put into the standard form of *rows and columns* of outputs on the left and inputs on the right. Our example is a 2×2 matrix. The rows and columns in the matrix correspond exactly with the coefficients of the output terms in the equation, using the s notation for derivatives. All system parameters *must* be given as numbers; you cannot use letter coefficients for mass, damping, etc. The calculation is done for *one input at a time*. The inputs are represented by a column vector C where the inputs are ordered as the left-hand sides were in the A matrix. That is, $C = [f1\ f2]'$ so that if we want to study the response to $f1$ we set $f2$ to zero and $C = C1 = [1\ 0]'$.

When defining the A matrix in the program, be sure to separate the coefficients by one or more spaces, and do not leave *any* spaces between terms in a coefficient. The A matrix will always be *square* $(2 \times 2, 3 \times 3,$ etc.) because there must be as many equations as there are unknowns. The column vector C will always have as many rows as the A matrix. As soon as we decide which of the inputs we want to study, that row of C will be nonzero, and all the other rows will be zero. In our example, the *coefficients* of inputs $f1$ and $f2$ were both equal to 1.0, so C is either $[1\ 0]'$ or $[0\ 1]'$. If the coefficients of the inputs of a system are *not* 1.0 but rather some functions of s (as the unknowns' coefficients usually are), then C will contain those s-functions. The parameter *conv* in the program simply converts angles from radians to degrees. Some of the phase angle statements use MATLAB's *unwrap* operation to make the phase angles conform to what we can see from equations such as A.9. That is, $x1/f1$ is seen to be second order over fourth order, so the high-frequency phase angle *must* be asymptotic to $-180°$, while $x1/f2$ is first order over fourth order which requires the high-frequency phase to go to $-270°$. If we had *not* earlier derived Equation A.9, but had instead gone directly to the matrix frequency-response calculation, we might have accepted "bad" phase angle curves since we would not have anything to compare with. Figures A.60 and A.61 show a sample of the results from the matrix calculation.

This completes our treatment of the frequency response of linear, time-invariant systems. MATLAB and its various toolboxes have many other capabilities that relate to frequency response, but these are usually oriented toward a specific application area, often feedback control systems, rather than the general scope of this appendix.

We next deal with the frequency content of *signals* rather than the frequency response of *systems*. That is, we show how to take a time-varying signal and find its *frequency spectrum*. This will involve the use of MATLAB's *FFT* operations.

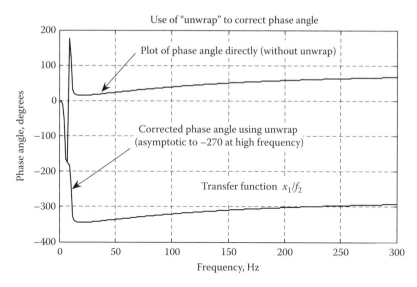

FIGURE A.60
Use of *Unwrap* to correct phase angle plot.

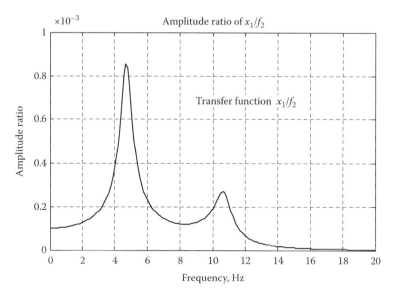

FIGURE A.61
Results of matrix frequency-response calculation.

A.2.8 Finding the Frequency Spectra of Time-Varying Signals

If we know the frequency response of any linear time-invariant system (whether by theoretical calculation or experimental measurement), we can use that information to find the response of that system to *any* time-varying input signal. To actually do this we need to convert the time history into its frequency-domain representation, that is, we need to find the *frequency spectrum* of the input signal. In fact, if we have the input, system, and output all in the frequency domain, *if we have any two of these we can always calculate the third*.[*] For example, in a *measurement* system, if we have the system's frequency response and the time response to some input, we can compute the *true* input signal even if the measurement system is not dynamically *perfect*. This is the basis of an important *dynamic correction* method[†] in measurement technology. Many other useful applications are based on knowledge of the frequency spectrum of a signal. In our brief treatment, we do not try to explain all these applications but rather concentrate on the *mechanics* of getting the frequency spectrum using MATLAB methods.

The most common way of getting frequency spectra uses FFT methods. These convert the time signal into its frequency-domain representation. MATLAB provides for the direct use of these methods and also offers various *spectrum analyzers* which have the FFT operations embedded inside them. I personally have not had much luck with the spectrum analyzers, and the direct approach is not complicated and seems to work well, so we will not address the use of the spectrum analyzers.

A.2.8.1 *Frequency Spectrum of Periodic Signals*

Perhaps, the simplest introduction to frequency spectra is through *Fourier series*. If our time signal is a *periodic function*, then Fourier series is the proper method for getting the frequency spectrum. For periodic time functions that are mathematically *simple* (like, say, a square wave), one can compute the Fourier series *analytically* using integral tables to evaluate the integrals that are part of the definition. We will not here pursue this analytical method. Whether our function is simple or not, the *numerical* methods of getting the Fourier series will get us the spectrum. It turns out that for *all* types of signals (periodic, transient, random, amplitude-modulated, etc.), the *same* FFT method is used, but for each class, we need to be careful in using the results that MATLAB directly provides. We generally need to do some routine (but not always obvious) manipulations to get what we really want. Since the MATLAB *help* does not usually explain such details (you are supposed to already know them), these notes will give *cookbook recipes* so that one can get useful results quickly and easily. The reason that *the same* FFT method is used for all kinds of signals is that the method treats all the signals *as if they were periodic*. That is, when we analyze any real signal, we *must* deal with a record of finite length. This finite-length record is treated *as if* it were repeated over and over forever. When the signal is *periodic* this is exactly true; the one-period record *is* repeated over and over. When the signal is *not* periodic, even though the record is *not* actually repeated, we treat it *as if it were*.

---

[*] *Measurement Systems*, 5th edn., pp. 149–200.
[†] *Measurement Systems*, 5th edn., pp. 202–206, 215.

Whether the periodic function is given analytically or by experimental data, to use the MATLAB FFT method to get the Fourier series, the time record must be given as a set of equally spaced *sample points*. To demonstrate the method, we will use MATLAB to make up one period of an arbitrary *function* and then use the FFT routines to get the Fourier series. A program to implement this follows.

```
% fourier series program me770newmanfourier
% let's first make up one period of our periodic function
t= 0:1:10;                          we use the spline command
f= [0 2 2 0 -2 -4 0 0.5 0 -2 0];    to make one cycle of our
tint= 0:10/511:10;                  periodic function from
fint= spline(t,f,tint);             coarsely-spaced points
plot(tint,fint); hold on; plot(t,f); graph to see the function
  pause;
fseries= fft(fint,512);             compute Fourier series
avg= fseries(1)/512                 get the average value term
mag= abs(fseries)/256;              compute the amplitudes
[mag(2:10)]'                        list a few numbers
fr= 0:0.1:1.0;                      define some frequencies
mag= [avg [mag(2:512)]]';hold off   re-define amplitude list
bar(fr(1:11),mag(1:11),0.1);pause   plot amplitude spectrum
fseries= [fseries]';                prepare for phase calc.
phi= atan2(real(fseries),imag(fseries)); compute phase, radians
phi(2:10)                           list a few phase angles
bar(fr(2:11),phi(2:11)*57.3,0.1)    plot angles in degrees
ti= 0:0.1:30;
ft= avg + mag(2)*sin(fr(2)*ti*2*pi + phi(2))
      + mag(3)*sin(fr(3)*ti*2*pi + phi(3))...
      + mag(4)*sin(fr(4)*ti*2*pi + phi(4))
      + mag(5)*sin(fr(5)*ti*2*pi + phi(5))...
      + mag(6)*sin(fr(6)*ti*2*pi + phi(6))
      + mag(7)*sin(fr(7)*ti*2*pi + phi(7));
plot(ti,ft);hold on; plot(tint,fint);pause
```

In setting up the list of sample points from our periodic function, it is best to use a number of points that is some power of 2, such as 512, 1024, 2048, etc. since all FFT routines work most efficiently with such data. (MATLAB *does* provide FFT routines which work with *any* number of points, so that is available if it were needed for some reason.) Of course the number of points should be large enough that the sampled data "captures all the wiggles" in the function. The *fseries = fft(fint,512)* statement actually computes the Fourier series as a list of complex numbers whose magnitudes are the amplitudes of the sinusoidal terms which make up the series and whose phase angles are the phase angles of these sinusoidal terms. The *frequencies* (in Hz) of the terms are integer multiples of the reciprocal of the period T of our periodic function. In our example, the period is 10 s so the first frequency is 0.1 Hz, the second is 0.2 Hz, and so forth up to the highest frequency available. The lowest frequency is called the *fundamental frequency* or, alternatively, the *first harmonic*. The second frequency is called the *second harmonic* and so forth. While the *fft* command produces in our case a list of 512 terms, only the first "half" (257) of these is needed, since the last 255 are redundant. If n is the number of time points used to compute the series ($n = 512$ in our example), then the highest computed harmonic is $(n - 2)/2$, which is 255 in our case. The first term in the computed series always relates to the *average value* of the periodic function. The relation is that the average value is always *fseries(1)/n*. The harmonic amplitudes are always *abs(fseries)/(n/2)*.

For any periodic function, the frequency spectrum is always a *discrete spectrum*; that is, the time signal has frequency content *only* at the harmonic frequencies. There is *nothing* in between these discrete points. Thus when we plot this kind of spectrum, we want *isolated "spikes,"* not a continuous curve. This kind of display is available with the MATLAB bargraph statement *bar*. The rightmost argument in the *bar* statement is the width of the plotted bars; for a narrow *spike*, a width of 0.1 seems to work well for both magnitude and angle. The last few statements in the program compute and plot the periodic function given by the average value and the first seven harmonics. The Fourier series is in general an *infinite* series, but for practical use it must always be truncated to some *finite* number of terms. Our example provides 255 harmonics, but only the first few of these are needed to get an acceptable curve fit. To see how many terms are needed we just *keep adding terms* until a graph of the original periodic function and the curve fit for some number of terms agree *well enough* for our practical purposes. In our present example, six or seven harmonics seem to do the job.

Figure A.62 shows how the desired 1-cycle sample of the periodic function was produced using the spline operation. If we were using real experimental data, we would set up our data-acquisition system to save, say, 512 points over one period of the signal. Figure A.63 shows the amplitude spectrum, including the average value (plotted at 0 frequency) and the first 10 harmonics. We see already that the higher harmonics have quite small amplitudes relative to the first few, justifying a truncation of the series to just a few terms. The average value of a periodic function may of course be either positive, negative, or zero; our example happens to have a negative average value and is plotted as such in Figure A.63. By convention, the harmonic amplitudes are always plotted as *positive numbers*, thus when we *add up* terms in the series to check the curve fit, the terms are actually *added*; the phase angle will take care of series terms that might otherwise have had a minus sign in front of them. That is: $-\sin x = +\sin(x + pi)$. The phase angles of the various harmonic terms are originally given in radians since that is how they would be entered into a sine-wave formula. When we plot them, we might prefer them in degrees, as in Figure A.64. Figure A.65 shows that the average value and the first seven harmonics seem to give a good fit for this particular periodic function.

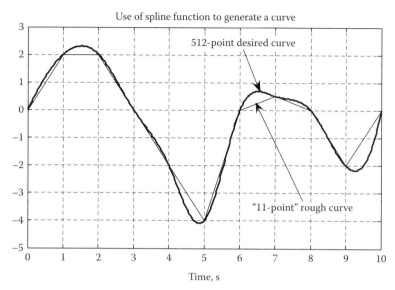

FIGURE A.62
Use of spline function to generate a desired curve.

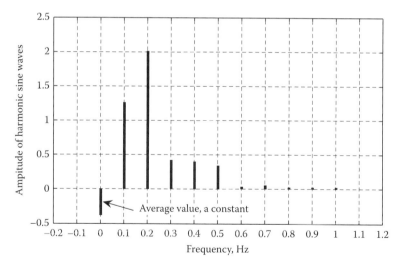

FIGURE A.63
Amplitudes of Fourier series harmonics.

A.2.8.2 *Frequency Spectrum of Transient Signals*

When we want to get the frequency spectrum of a transient analytically, the proper tool is the *Fourier transform*. This produces a *continuous* spectrum; the signal has frequency content at *every* frequency, not just at isolated harmonic frequencies as was found for periodic functions. The definition of the Fourier transform shows how it is calculated analytically:

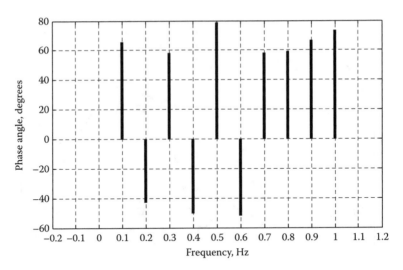

FIGURE A.64
Phase angles of Fourier series harmonics.

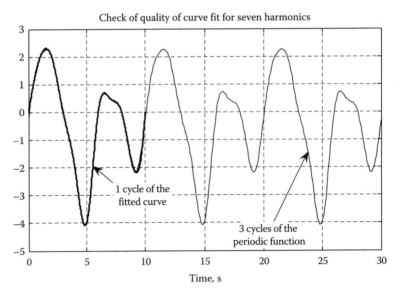

FIGURE A.65
Seven harmonics give a good curve fit.

$$F(i\omega) \triangleq \int_0^\infty f(t) \cdot \cos(\omega t)dt - i\int_0^\infty f(t) \cdot \sin(\omega t)dt \qquad (A.15)$$

Here $f(t)$ is the transient time signal and $F(i\omega)$ is its companion frequency function (frequency spectrum). If we know $f(t)$ explicitly, we can perform

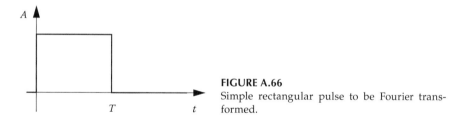

FIGURE A.66
Simple rectangular pulse to be Fourier transformed.

the integrals and thus get $F(i\omega)$. This will be a *continuous* function of frequency since ω can take on any value whatever. The *definition* of a transient is that it exists from $t = 0$ out to a time T, and is zero thereafter, thus the upper limit on the integrals can be set to T, rather than infinity. As an example, consider the rectangular pulse shown in Figure A.66. Using Equation A.15 we get

$$F(i\omega) = \frac{A\sin(\omega T)}{\omega} + i \cdot \frac{A}{\omega}(-1+\cos(\omega T)) = \frac{\sqrt{2}A}{\omega} \cdot \sqrt{1-\cos(\omega T)}\angle\alpha \qquad \alpha = -\frac{\omega T}{2}$$

(A.16)

We can only plot this function for numerical values of A and T, so let us take them both equal to 1.0 as an example. We can plot $F(i\omega)$ with two types of graphs, either the real part and the imaginary part, or the magnitude and the angle. The magnitude/angle graphs are often the most useful, so we show them in Figure A.67. If you set ω equal to zero in Equation A.15, you see that $F(i0)$ will always be the *net area* under the transient waveform, in our example 1.0. This is a good check on any software that computes transforms

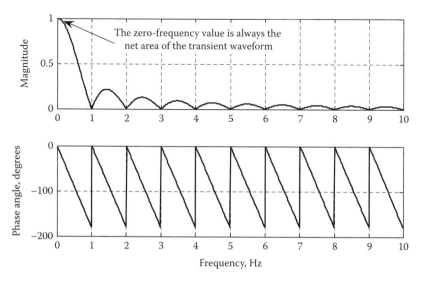

FIGURE A.67
Analytical Fourier transform of rectangular pulse.

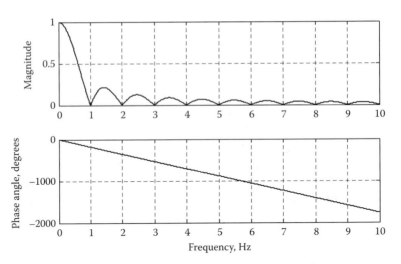

FIGURE A.68
Numerical Fourier transform of pulse, with "unwrapped" phase angle.

numerically; the zero-frequency value should always equal the pulse net area. The phase angle α varies as a straight line with frequency, getting more and more lagging without any bound. In Figure A.67, this seems to not be the case; the angle plots as a *sawtooth* curve. Actually, this curve, though visually misleading, is numerically correct since the phase angle is defined by its tangent in the calculations, and there is a trig identity stating $tan(x + pi) \equiv tan(x)$. If we use the *unwrap* command on the phase angle, we get Figure A.68.

When we use FFT methods to *numerically* (rather than analytically) compute the spectrum of a transient, the spectrum must be *discrete* even though we know that the correct spectrum is continuous. However, if the discrete frequency values are spaced closely enough, the fact that we *lose* what is going on in between them will not cause serious errors. One way to get some useful experience with this question is to do the numerical FFT for a transient that is simple enough to compute analytically. We will use the rectangular pulse of Figure A.66 for this purpose. To compute the Fourier transform numerically, we use exactly the same MATLAB statement as we used for the Fourier series, but we will change the manipulations that we perform on the raw output of this statement so that the results will compare favorably with the correct analytical values. We will see that, to get good results, we usually have to define a new version of the transient time signal by "padding" the time record with zero values out to a time somewhat longer than the *actual* transient lasts (see Figure A.69). The MATLAB FFT statement has a convenient feature: if you stipulate, say, a 1024-point time record, but you only supply, say, 200 points, MATLAB will *automatically* pad the time record with zeros to supply the extra points. A rule widely used states that if the actual transient lasts for T seconds, pad it with zeros out to about $10T$. To see why this rule

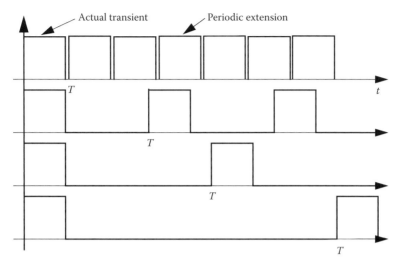

FIGURE A.69
Use of zero-padding to improve frequency resolution.

works, we will use the FFT statement on the rectangular pulse with several different amounts of padding. In the program below, we first do the FFT on the original transient (no zero-padding) and then use $T = 2.0\,s$ and finally $T = 10.0\,s$ (the recommended amount of padding).

```
% demonstrates "10T" rule for transient FFT's
% first, do FFT for original(un-padded)transient
t = linspace(0,1.00,1023);
x1 = ones(1,1023);            1023 of the 1024 points will have the
                              value 1.0
ftran1 = fft(x1,1024);        MATLAB will automatically make the
                              last (1024th) point 0.0
ftran1 = ftran1(1:512);       we always redefine the fft to use
                              only the first half of the points
delt = 1.00/1023              delt is the spacing between points in
                              the time signal
tfinal = 1.00 + delt          time value at the last point
mag= delt.*abs(ftran1);       we always multiply the "raw"
                              magnitude by the spacing between time
                              points
phi = angle(ftran1);          this is how angle is computed; I did
                              it only for this FFT and didn't plot
fr = [0:1/tfinal:511/tfinal]'; always use 1/tfinal and (n/2-1)/
                                  tfinal where n is fft "size"
% use zero pad to 2.0 seconds
t2 = linspace(0,2.0,1024);
x2 = ones(1,512);
ftran2 = fft(x2,1024);
```

```
ftran2 = ftran2(1:512);
delt2 = 2.0/1024
mag2 = delt2.*abs(ftran2);
fr2 = [0:1/2.0:511/2.0]';
% use zero pad to 10 seconds
t3 = linspace(0,10.0,1024);
x3 = ones(1,102);
ftran3 = fft(x3,1024);
ftran3 = ftran3(1:512);
delt3 = 10.0/1024;
mag3 = delt3.*abs(ftran3);
fr3 = [0:1/10.0:511/10.0]';
plot(fr,mag);axis([0 5 0 1]);pause;hold on
plot(fr2,mag2);axis([0 5 0 1]);pause;hold on
plot(fr3,mag3);axis([0 5 0 1])
% now compute and plot the exact analytical transform
A = 1; T = 1; f = 0:.0105:10; w = f.*2.*pi;
F = A.*sin(w.*T)./w + (i*A./w).*(-1 + cos(w.*T));
magt = abs(F);
plot(f,magt);
% plot(fr,phi);axis([0 5 -pi pi]);pause
```

In Figure A.70, we see that the exact analytical transform magnitude and that computed numerically with the FFT are nearly identical when we pad the transient out to 10 times its actual duration. (Figure A.71, a zoomed view, shows that there are slight differences.) We see in Figure A.70 that using the un-padded transient in this example gives magnitude

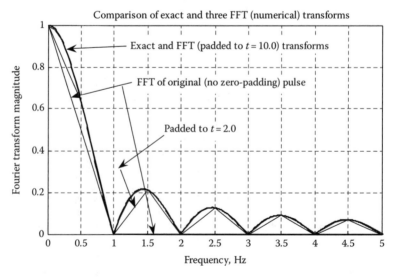

FIGURE A.70
Effect of zero-padding on Fourier transform.

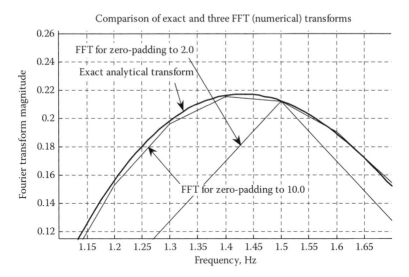

FIGURE A.71
Details of zero-padding effect.

values only at the frequencies 0, 1, 2, 3,... Hz, where unfortunately, the magnitude is zero. Padding the transient out to $t = 2.0$ s helps, but the frequency resolution is still too coarse (only every 0.5 Hz) to give a clear indication of the curve shape. Note also that the true peaks in the magnitude curve would be *missed* by this calculation. We see from these brief examples that the main problem with un-padded transient FFTs is *inadequate frequency resolution*. For transients of more complicated shape than our rectangular pulse, the effects may be more subtle, but the 10 times rule is usually adequate. If there is any question in a specific case, one can try padding farther than 10 times and compare results. If there is significant change between say, 10 and 15, we will then try 20 or more until we get no more improvement. The *rules* given in the program for modifying the raw output of MATLAB's FFT statement will work for any transient FFTs that you might need to do.

A.2.8.3 Frequency Spectrum of Random Signals

Here we are usually interested in getting the mean-square spectral density (also called PSD). Again we use the same basic tool, the MATLAB fft statement, but once we have computed the fft, the definition of PSD is that it is the square of the fft, divided by the frequency resolution. For an example, we will generate the needed time signal by using a Simulink block diagram and the band-limited white noise generator. Once we have run this diagram, the random signal is available to MATLAB where we can run an m-file to do the desired calculations and plotting.

```
% computes PSD of random signals using FFT run simulink first
% bwn is a column vector with 2561 rows, generated in Simulink
delt= 0.001                    the data generated in Simulink
                               is spaced at 0.001 seconds
tfinal= 0.255                  our time record runs from 0 to
                               0.255 seconds (256 points)
delf= 1/tfinal                 the fft frequency spacing is
                               always 1/tfinal
fr= [0:1/tfinal:127/tfinal];   define list of 128 frequencies
A= zeros(128,10);              define an empty matrix (used
                               later for averaging)
for i= 1:10                    start a for-loop to break bwn
                               into 10 segments
    p= 256.*(i-1) + 1;         p and q define start and end
                               of each segment
    q= 256.*i;
    bwna= bwn(p:q);            define the current segment
    ftr= fft(bwna,256);        perform the fft
    ftra= ftr(1:128);          we always use only the first
                               half of the fft
    mag= delt.*abs(ftra);      get magnitude of complex number
    psd= mag.*mag./delf;       compute PSD
    plot(fr,psd);pause;hold on  plot PSD versus frequency in Hz
    A(:,i)= psd;               enter the current PSD into a
                               column of A
end
avg10= (A(:,1) + A(:,2)        compute an average spectrum
+ A(:,3)+ A(:,4)+ A(:,5)+···
A(:,6)+ A(:,7)+ A(:,8)         from the 10 segment spectra
+ A(:,9)+ A(:,10) )./10
plot(fr,avg10,fr,A(:,1))       plot the average spectrum and
                               the first
                               segment spectrum for comparison
```

Figure A.72 shows the Simulink diagram used to generate the basic data for the program above. We usually set up the band-limited white noise generator to produce a *random square wave* but in this case the successive random values are connected directly with straight-line segments. Thus our frequency spectrum will *not* be the one shown in Figure A.28. (If you want the *square wave* you need to use a computing time increment which is *smaller* than the sample time of the random generator, say 10 times smaller.) The signal *bwn* runs from $t = 0$ s to $t = 2.560$ s, producing 2561 points. This total record length will be cut into 10 segments of 256 points each, and we will compute a PSD curve for each segment. As we have mentioned before, a *single* record of a random signal *no matter how long*, will not give a reliable frequency spectrum; one *must* average a number of records. Our choice of 256-point ffts has no special significance; smaller or larger values exhibit the usual trade-offs. (In Figure A.72 the Butterworth low-pass filter has nothing to do with the

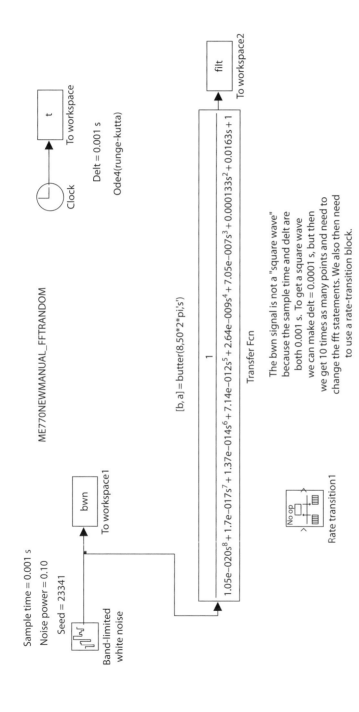

FIGURE A.72

Simulation used to generate random data for spectrum analysis.

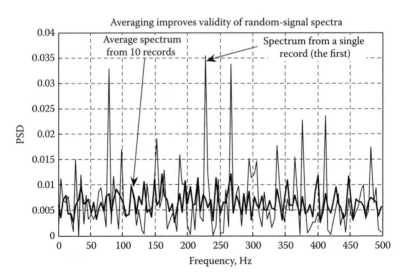

FIGURE A.73
Averaging improves quality of random-signal spectrum.

above program. We will use its output shortly for another purpose.) We did
not attach any physical dimensions to the signal called bwn in the Simulink
simulation. If it were, for example, a pressure in psi, then the dimensions of
the PSD would be psi^2/Hz.

In Figure A.73, we see that averaging 10 records does reduce the variability
of the spectrum, but it is still quite *ragged*. Ignoring this roughness, it appears
that the spectrum is more or less *flat* out to our highest frequency of 498 Hz.
For *real-world* signals which are not obvious transients or periodic functions,
but some complex mixture of random and roughly periodic components, we
usually use a *data window* on the time data before doing the fft operation.
Of the many available windows, the *Hanning* is probably the most widely
used. To use this window on the data of the above program, we can modify
the program as follows:

```
bwna = bwn(p:q);
bwnahann = bwna.*hann(256);
ftr = fft(bwnahann,256);
```

If you run the program with this modification there is not any obvious
change in the nature of the computed spectrum; it is still quite *rough* after
averaging 10 records. We will not, here, discuss the pros and cons of using
data windows; suffice it to say that it is routinely recommended except for
transients and clearly periodic signals. If you have never seen the effect of
using a window on a function, Figure A.74a shows the effect of the Hanning
window function on the first segment of our random data signal. In general, window functions (various kinds are available) multiply the original

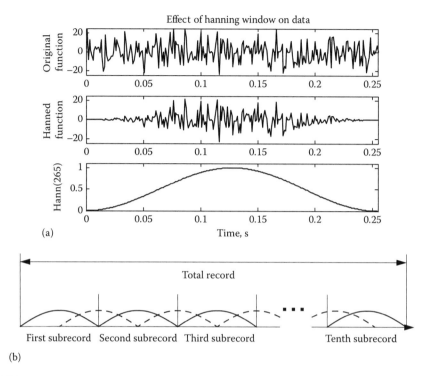

FIGURE A.74

(a) Use of Hanning data window and (b) overlap processing with 50% overlap.

data by some function that *tapers* the data at the beginning and end, so as to eliminate the sudden jumps that always occur there because the first and last sample are not the same. In the center of the data, window functions usually have the value 1.0, so that data is not modified.

A.2.9 Overlap Processing

In Figure A.74a, we see that the use of a *data window* (as routinely recommended for signals other than transients or strictly periodic functions) strongly attenuates any data near the beginning or end of the record, in a way distorting the signal to be analyzed. To compensate for this, we often use *overlap processing*. This operation is best explained graphically, as in Figure A.74b. Without overlap processing, data near the beginning and end of a subrecord do not contribute their *fair share* to the FFT calculation. If *something important* should happen in these regions, its effect would be to some extent ignored. This situation can be improved by defining another set of subrecords, each of which starts somewhere within one of the original subrecords and extends into the next subrecord. The most common amount of *overlap* is 50%, which is what is shown in Figure A.74b. Once this other set of subrecords is defined, we perform the same FFT calculation on them to generate the frequency spectra. We are doing all these operations in our MATLAB/Simulink software, but *hardware-type* analyzers are also widely used. In these, the overlap operations are implemented by providing two data buffers. The first buffer gathers samples as shown by the solid lines in Figure A.74b; the second buffer starts its data gathering when the first buffer is half full (assuming we want 50% overlap), and this sequence is continued for as many records as we desire for averaging. We can, of course, achieve the same result in a MATLAB program, which I now display.

```
% me770newmanual_randfft3_overlap.m
% computes PSD of random signlals using FFT run simulink first
% bwn is a column vector with 2561 rows, generated in
% Simulink program called ME770NEWMANUAL_FFTRANDOM.MDL
% uses Hanning window, averaging, and overlap processing
delt = 0.001
tfinal = 0.255
delf = 1/tfinal
fr = [0:1/tfinal:127/tfinal];
A = zeros(128,10);
frhann = linspace(0,255,256);
for i = 1:10
    p = 256.*(i-1)+ 1;
    q = 256.*i;
    bwna = bwn(p:q);
    bwnahann = bwna.*hann(256);
%      plot(frhann,bwna,frhann,bwnahann);pause
    plot(frhann,20.*hann(256),frhann,bwnahann);pause
    ftr = fft(bwnahann,256);
%      ftr = fft(bwna,256);
    ftra = ftr(1:128);
    mag = delt.*abs(ftra);
    psd = mag.*mag./delf;
    plot(fr,psd);pause; %hold on
    A(:,i) = psd;
end
avg10 = (A(:,1)+ A(:,2)+ A(:,3)+ A(:,4)+ A(:,5)+ ...
A(:,6)+ A(:,7)+ A(:,8)+ A(:,9)+ A(:,10) )./10;
hold off
plot(fr,avg10,fr,A(:,1) );pause
% now add the overlap processing
for k = 1:9
    r = 128.*k;
    s = r+ 255;
    bwnao = bwn(r:s);
    bwnaohann = bwnao.*hann(256);
    ftro = fft(bwnaohann,256);
```

```
      ftroa = ftro(1:128);
      mago = delt.*abs(ftroa);
      psdo = mago.*mago./delf;
      plot(fr,psdo);pause; %hold on
      B(:,k) = psdo;
end
avgB = (B(:,1) + B(:,2) + B(:,3) + B(:,4) + B(:,5) + ...
      B(:,6)+ B(:,7) + B(:,8) + B(:,9))./9;
hold off
avgtot = (avg10 + avgB)./2;
plot(fr,avgtot,fr,avgB,fr,avg10)
```

The first part of this program is just a repeat of our earlier program which dealt with the *un-overlapped* subrecords. The second part, which starts with the *% now add the overlap processing* comment, deals with the 50% overlapped subrecords. It ends with a graph comparing the averaged spectra of the un-overlapped sections, the overlapped sections, and finally a *total* average spectrum which is the average of the overlapped and un-overlapped averages, which is the *final* result that we would use. While our example is for 50% overlap, the program is easily modified for any desired percentage.

A.2.10 Experimental Modeling of Systems Using Frequency-Response Testing

To get by experiment a frequency response model of a system, we apply a suitably chosen input signal and measure both this input and the system response as time histories. The input signal could be a transient, a random signal, or some other waveform, but it must have significant frequency content out to the highest frequency for which we want an accurate model. We then get the frequency spectra of the input and response signals and divide the output spectrum by the input spectrum. This gives a complex number whose magnitude and angle are the amplitude ratio and phase angle of the system's sinusoidal transfer function. If we then need an analytical transfer function, we can try some curve-fitting tools such as that explained earlier in this appendix.

For an input which is a random signal, we can use the simulation of Figure A.72 to generate the data for a MATLAB program which finds the system transfer function.

```
% demonstrates use of freqency spectra for experimental modeling
% run simulink first to generate data; uses co-spectrum/PSD to see
% whether results differ from FFTout/FFTin
% it turns out the results are exactly the same, as we expect for
% noise-free signals and no averaging
in = bwn(1024:2048);
out = filt(1024:2048);
inft = fft(in,1024);
```

```
outft = fft(out,1024);
inft = inft(1:512);
outft = outft(1:512);
cospec = outft.*conj(inft);    to get cospec we should really divide
                               by the frequency resolution delf,
                               but psd
psd= inft.*conj(inft);         requires this same division, so delf
                               cancels out when we comput tf2 below
% now compute system transfer function as ratio of the FFT's
tf1= outft./inft;
% now compute system transfer function as ratio co-spectrum/PSD
tf2= cospec./psd;
mag1= abs(tf1);
mag2= abs(tf2);
phi1= angle(tf1);
phi2= angle(tf2);
fr= [0:1/1.023:511/1.023]';
plot(fr,mag1);axis([0 200 0 2.0]);pause;hold on
plot(fr,mag2,'k');axis([0 200 0 2.0]);pause
plot(fr,unwrap(phi1)*57.3);axis([0 200 -1000 1000]);pause;hold on
plot(fr,unwrap(phi2)*57.3,'k');axis([0 200 -1000 1000])
```

In commercial data acquisition/processing software (Labview, Dasylab, etc.) the calculation of the system transfer function from time histories of the input and output signals often provides a *choice* of two methods. Until now we have mentioned only the obvious way; take the ratio of the output and input FFTs. When the time signals are corrupted to some extent by *noise*, it can be shown that another method offers some noise rejection if *averaging* is employed.* In this method, we take the ratio of the *cospectrum* of input and output, divided by the power spectrum of the input. The cospectrum is defined as the product of the output FFT and the complex conjugate of the input FFT divided by the frequency resolution *delf*. The PSD of the input is the square of its FFT, divided by *delf*, which is the same as the product of the input FFT and its complex conjugate, divided by *delf*. Both these methods are implemented in the above program. Since our simulation produces *noise-free* signals, we find that the results are identical for the two methods. In actual practice (rather than ideal simulations), there usually *is* some noise and the second method (*together with averaging*) is routinely used. In Figure A.75, we see that the amplitude ratio, while certainly *resembling* that of a Butterworth 50 Hz low-pass filter, is very rough, showing that considerable averaging will be needed to get acceptable results. As we might expect, the phase angle curve of Figure A.76 shows similar roughness. We will not here pursue such averaging with MATLAB but will demonstrate it later using DASYLAB where the averaging is very easy to implement.

---

* E.O. Doebelin, *System Modeling and Response: Theoretical and Experimental Approaches*, Wiley, New York, 1980, pp 263–282.

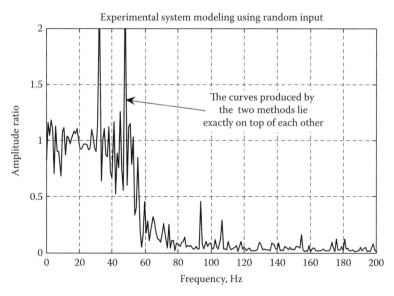

FIGURE A.75
Two alternative computing schemes agree perfectly for noise-free data.

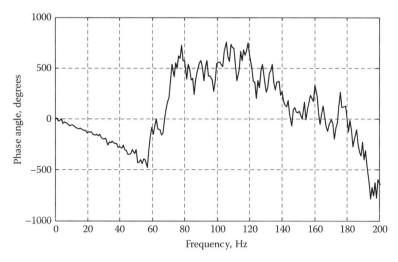

FIGURE A.76
Phase angle of tested system.

When the input signal is not random but rather deterministic, like a single pulse transient, then our noise-free simulations will usually give good results without averaging. To demonstrate this, we use the same system as in Figure A.72 but let the input signal be a simple rectangular pulse called *pulsin* (amplitude = 1.0). We saw earlier that such pulses have an FFT

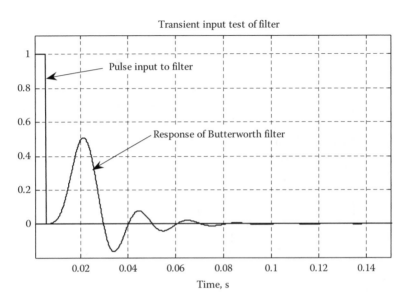

FIGURE A.77
Pulse test of dynamic system.

magnitude that starts (at 0 frequency) at the net area of the pulse and then gradually goes down to 0 magnitude at a frequency of $1/T$, where T is the pulse duration. We want the input pulse to *exercise* the tested system up to the highest frequency of interest in the model, thus we need to choose T such that the pulse FFT has "significant" magnitude out to this frequency. We know that our eighth-order Butterworth filter has nearly zero amplitude ratio above about 100 Hz, so we choose $T = 0.005$ s. (Note that in *real-world* testing, the tested system is to some extent *unknown*, so the choice of T in practice is not so easy.)

A program for simulating this pulse test is shown below. Figure A.77 shows the system input and output signals. Note that I used a shorter computing increment (0.0001 s) here since the input pulse lasts only 0.005 s, and larger FFTs (2048 time points). If you want to get some feel for the effects of changing these numbers, you can easily modify the program to try out such changes.

```
% demonstrates use of frequency spectra for experimental modeling
% run simulink first to generate data
in = pulsin(1:2048);
out = filt(1:2048);
inft = fft(in,8192);
outft = fft(out,8192);
inft = inft(1:4096);
outft = outft(1:4096);
```

```
tf = outft./inft;
mag = abs(tf);
phi = angle(tf);
fr = [0:1/.8192:4095/.8192]';
plot(fr,mag);axis([0 200 0 1.0]);pause
plot(fr,unwrap(phi)*57.3);axis([0 200 0 -1000])
```

Note that the *computed* frequencies go from 0 to 5000 Hz but we only show them to 200 Hz in Figures A.78 and A.79 since our input pulse was designed for that range. Also, the amplitude ratio at zero frequency should be exactly 1.0 but is computed as 0.9873. The phase angle plot seems to "blow up" as we approach 200 Hz (the amplitude ratio would also show this but at higher frequencies). This behavior will be observed in *every* practical test because the response of all real systems must go to zero at *sufficiently high* frequency. When this happens, the FFT computations are dealing with smaller and smaller numbers, making them more and more erratic. However, for frequencies up to about 160 Hz, the results agree closely with the filter's known behavior.

Remember that our input pulse FFT had its first *zero* at 200 Hz (recall Figure A.68), so at that frequency the input provides *no* excitation at all and the system ideal response would also have to be zero. The numerical calculations, however, are not infinitely precise, so that instead of zero values we are dealing with very small and unreliable nonzero values. These effects are not usually a practical problem since our interest is only in having a valid model for frequencies where the system *does* respond. If we were using one

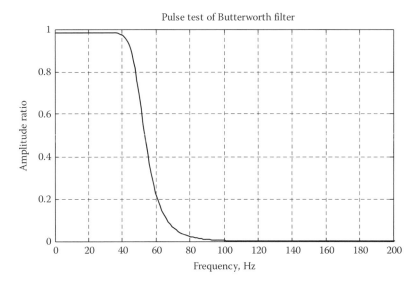

Pulse test of Butterworth filter

FIGURE A.78
Pulse test gives good result for amplitude ratio.

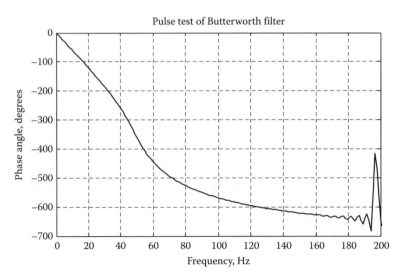

FIGURE A.79
Calculations go bad at high frequency, due to small numbers.

of the curve-fitting programs to convert the numerical data to an analytical transfer function, we would of course use only frequency points that gave *good* FFT results.

Figure A.80 shows a DASYLAB* (the data acquisition/processing software we mentioned earlier) worksheet for computing the sinusoidal transfer function of a system from measurements of its input and output signals. DASYLAB here is being used in a *simulation mode,* where instead of getting actual sensor signals from a real physical system, we are simulating the system (an eighth-order Butterworth low-pass filter set at 50 Hz) and producing the random input signal in a signal generator module. DASYLAB has a *transfer function module* which accepts as inputs, the system input and output signals. From these, it computes (using FFT methods) the amplitude ratio and phase angle curves of the system. The module gives you several choices of how to do these calculations, including the two we mentioned above. It also provides several averaging options. When I tried the two methods which were discussed earlier (and which should give *identical* results for noise-free signals) I found that the *cospectrum/psd* method gave clearly better results. This must be due to some software details which the DASYLAB programmers used and which are not apparent to the DASYLAB user. On this worksheet, I used this method in both transfer function modules, one of which was preceded by a Hanning window, which again, should not really make any difference. It turns out that using this window also improves the

---

* *Measurement Systems,* 5th edn., pp. 981–1014, www.dasylab.com

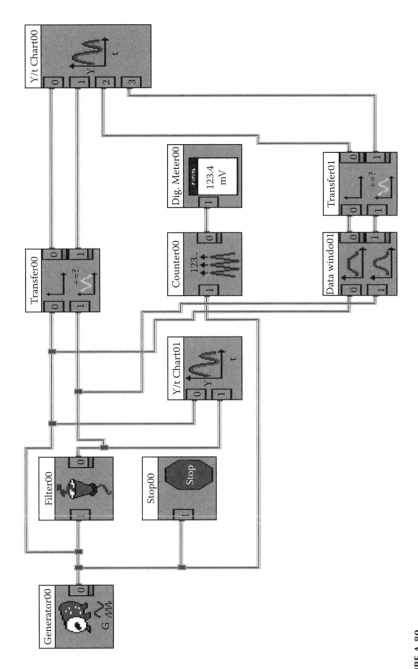

FIGURE A.80
DASYLAB worksheet for random signal dynamic testing.

performance. When using commercial software without knowing all the details of programming used, it appears that we cannot always predict how things are going to turn out. In any case, the DASYLAB methods are very easy to use and seem to work very well, so a practical person just goes ahead and uses them even though we are not able to know all the *behind-the-scenes* details.

This DASYLAB worksheet uses a sampling rate of 10,000 samples/s and an FFT block size of 8,192, so a block of date is obtained about once a second. A counter and digital meter keep track of how many blocks of data have been averaged. One chart module records the time histories of input and output, and another graphs the amplitude ratio and phase angle curves. As expected, for high frequencies the results are "garbage" so we zoom the graphs to cover only the range of interest; from 0 Hz to about 200 Hz. Figure A.81 show the results after averaging six blocks of data. We see that good results can be obtained in just a few seconds of testing for this example. More averaging would *smooth out* some of the roughness in the amplitude ratio curve below, but the results for just six averages are quite usable. As in much engineering software, DASYLAB's phase angle calculation is constrained to ±180°, so some interpretation is needed to reconcile the phase angle graph with what we know to be correct for an eighth-order Butterworth low-pass filter. Recall that this filter has a phase angle which decreases monotonically from 0 to −720°, so in this curve, the first upward *jump* of 360° should be removed so that the curve is continuous, and similarly for the second upward jump. With these changes, the curve is continuous, going from 0 to −540 at about 85 Hz,

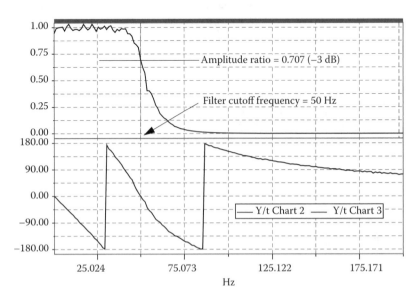

FIGURE A.81
DASYLAB test gives good results with six averages.

and then going to about −640 at 200 Hz, with a trend above that frequency which would be asymptotic to −720.

Another way to deal with this kind of phase angle problem is to have the software print and/or plot the sinusoidal transfer function as *real and imaginary parts*, rather than amplitude ratio and phase angle. The ±180° problem arises when we use the real and imaginary parts (which *are* the raw output of the FFT calculations) to compute the phase angle using an inverse tangent formula. If we focus on the graphs of the real and imaginary parts versus frequency, it is easy to manually *track* the actual angle on a sketched polar plot as we move through the four quadrants, and see the *true* phase angle variation. DASYLAB provides the real/imaginary plot as one option, and of course we can use MATLAB to do the same thing. If we use this option, we probably will also display the phase angle plot itself, since most applications want to see this graph. That is, we use the real/imaginary plot to properly interpret the phase angle plot, which we can then label in some way to give the viewer the correct interpretation.

Appendix B: Basic Statistical Calculations Using Minitab

B.1 Introduction

Engineers make use of a variety of software products to speed up and improve the calculations needed for various aspects of an engineering task. While *general-purpose* software may include some statistical capability, learning to use a *dedicated* statistics program has some advantages. Such software is written by specialists in the field and is thus usually more reliable. It will, of course, offer many features that are not regularly employed by the typical user, but may be needed only "now and then." If you get familiar with such software by using it in the more routine tasks, it will not be too difficult to extend your capability to some specialized routine when the need arises. Also, if the software is in wide industrial use, it is more likely to already be the "standard" at the company you work for. Wide use also means that, as successive versions of the program are developed, they benefit from user feedback with regard to bug removal and general improvement. While there are, of course, many statistical packages of this sort, Minitab has a long record of acceptance in academic and industrial circles.

Unfortunately, successful comprehensive software of almost any type will have gone through many versions over the years, and will be burdened by the "new and improved" features that software companies find necessary to routinely add to stay in business. Some new features are, of course, actual improvements and are welcomed, but others serve mainly to make the software more confusing and harder to learn, especially for users who are not specialists who use the package heavily. Most engineers are not statistical specialists, so they fall in this class; they are *occasional users*. While all engineering and mathematical software today offers extensive online *help*, and perhaps even printed manuals, these aids can also be self-defeating by their complexity and comprehensiveness. For these reasons, the production of short, focused, manuals addressed to specific user groups (such as engineers) can greatly shorten the learning curve for beginners. This Minitab manual is an attempt in that direction.

While the table of contents gives a summary of topics that you can use to find specific material, it might be useful to here discuss the scope of the manual since it is, of course, radically reduced from that of the total software package. The statistical tools and their practical applications are discussed in detail in *Engineering Experimentation: Planning, Execution, Reporting**; this manual concentrates on using the software to carry out the calculations and graphing. Before explaining the statistical calculations themselves, we need to show how to get data into Minitab and how to manipulate it into the forms we want or that the program requires. Once this is done, we can demonstrate the various statistical tools available. We will begin by showing how to check a set of measured data to see whether we can reasonably assume that it more or less follows some theoretical probability density function (pdf). This initial check is important because many of the statistical calculations assume that the data follows some specific distribution, usually the *normal* (also called *Gaussian*). If the data are found to clearly not follow the assumed distribution, then subsequent calculations are open to question. If we find the data sufficiently close to the assumed distribution, we can then proceed to the desired practical calculations. The simplest of these are perhaps already familiar to readers who have no experience with Minitab or other dedicated statistics packages; the calculation of the average value, and standard deviation of the data set. These simple but useful measures are well known and their calculation is included in most hand calculators and general-purpose software, so a powerful software such as Minitab is hardly needed, but this capability is, of course, provided.

While the use of the average value and standard deviation is very common, the concept of *confidence interval* is perhaps less well known. When we compute the mean (average) value and standard deviation from a data set with n items, it is intuitively clear that the reliability of the results improves with sample size n. The confidence interval puts this conjecture on a numerical basis; it gives numerically a range of values within which we are, say, 95% sure that the *true value* lies. Such information is clearly more useful than knowing only the average value itself. Often we need to compare the average values and/or standard deviations of alternative processes, for example, the tensile strength of steel from two different manufacturers. Confidence intervals can be calculated for the difference of two average values or the ratio of two standard deviations, allowing informed judgments about the relative quality of the two competing materials or processes.

Most statistical tools require use of one or more theoretical distributions. Before personal computers were common, such data was obtained from tables of numerical values, found in textbooks. Now we get them from our software and then use them as needed. While the tabular values would, of course, be the same whether obtained from a printed table or from the software, it is necessary to learn how the specific software is used; sometimes the software uses a different notation than we might be familiar with from a specific table or textbook. We also need to explain how to use Minitab's general graphing tools and also some specific statistical graphs. Finally,

---

we will explain the application of one of the most useful statistical tools: multiple regression. This is used for two main purposes: curve fitting of analytical functions to experimental data, and model building. In model building, we try to find relations among one or more independent variables and some dependent variables, based on carefully designed experiments. Model building is used both in research and development and also in manufacturing. In R&D, we might be trying to get a predictive relation between hydrocarbon emissions in an automotive engine and engine variables such as RPM, torque, air/fuel ratio, spark advance, exhaust gas recirculation, and spark plug temperature.* In manufacturing, we might want to find the most significant process factors that affect the shrinkage of plastic casing for speedometer cables. Of the 15 factors studied, 2 were found to cause 70% of the shrinkage.†

When you access Minitab you first see the screen shown in Figure B.1: a toolbar, the session window, and the worksheet window. The session window keeps track of your work, allows the entry of keyboard commands, and displays certain types of results. The worksheet window allows the entry of numerical data and may also display certain results. There is one session window but you can have several worksheets if that is desired.

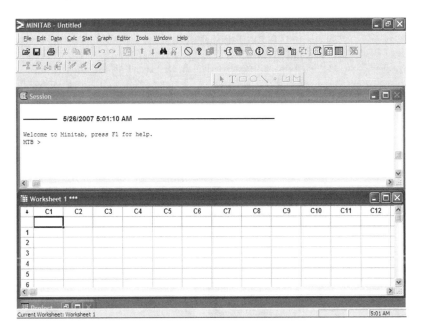

FIGURE B.1
The Minitab Toolbar, Session Window, and Worksheet Window.

---

* J.A. Tennant et al., Development and Validation of Engine Models via Automated Dynamometer Tests, Paper 790178, SAE, 1979.
† J. Quinlan, *Product Improvement by Application of Taguchi Methods*, American Supplier Institute, Dearborn, MI, 1985.

B.2 Getting Data into Minitab Using the Worksheet

Data can be imported into Minitab from other software but we here consider entering it *manually* from the keyboard. In Figure B.1, we see that the
worksheet columns are already labeled as c1, c2, etc., and the rows as 1, 2,
etc. The number of rows and columns expands automatically as needed and
without any practical upper limit. An unnumbered row just below the row
of column numbers can be used to give descriptive names to the columns of
data. This is not necessary, but highly recommended since it keeps us aware
of the physical meaning of the data. When doing calculations, one can use
either the column numbers (c1, c2, etc.) or the assigned variable names, or
both. Using column numbers gives compactness but using variable names
aids later physical understanding. To actually enter data, click the cursor in
the desired cell (row and column) and type in the numerical value, using
the *arrow* keys and/or the tab key to move among the cells. You can erase
values with the delete key. A drop-down menu under EDIT on the toolbar
allows further editing of the worksheet. You can select the numerical format
for each column from three options, using *editor > format column > numeric*.
Each column can have its own format. Sometimes a column of *data* exhibits
a known pattern, allowing the rapid entry of numerical values. Using the
sequence *calc > make patterned data > simple set of numbers*, we accessed the
menu of Figure B.2 and created column 3 of the worksheet. Figure B.3 shows
some other options available in *patterned data* but we will not explore this in
detail. Figure B.4 shows the worksheet produced by these operations. I have

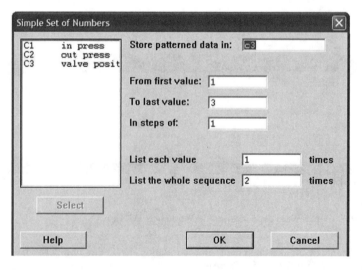

FIGURE B.2
Generating patterned data.

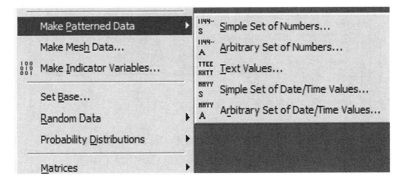

FIGURE B.3
Options for making patterned data.

↓	C1	C2	C3	C4
	in press	out press	valve position	
1	1.1000E+01	1.1000	1	
2	1.3342E+00	1.1040	2	
3	2.5517E+00	1.0050	3	
4	3.6723E+00	1.2020	1	
5	5.7781E+00	1.0023	2	
6	7.3421E+00	1.1234	3	
7				

FIGURE B.4
Sample worksheet.

named the columns *in press, out press,* and *valve position* to relate to a fluid flow process with inlet and outlet pressures, and a valve with three positions.

We will later show some useful applications that require a set of data coming from a theoretically perfect probability distribution of our choice. Minitab provides an easy way to generate data of this type. We use the sequence *calc > random data* to access a list of different distributions, from which we select the one we need, often the normal (Gaussian) distribution. Figure B.5 shows how to generate a sample of 100 random numbers with the Gaussian distribution, an average value of 3.56, and a standard deviation of 0.57. We can choose our distribution from a list of 24, which appears when we enter *calc > random data*. I put this data in column 4 but it could have been put in any column.

All random number generators need to initially be given a numerical value usually called the seed (called *base* in Minitab) to get started. If you do not

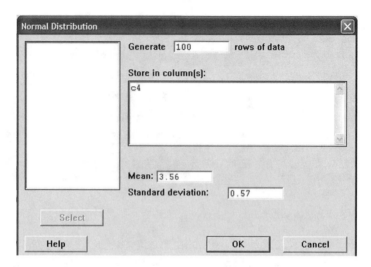

FIGURE B.5
Generating random data.

specify the base value (we did not in the above example), Minitab uses the time of day in seconds or fractions of a second, but does not display that number unless you ask for it by typing *base* in the session window. It is probably a good practice to routinely pick your own base value rather than leaving it up to Minitab; that way it will be displayed and recorded in the session window. To set the base value, use *calc > set base* and enter a number of your choice, which could be as simple as a single digit, like 5. Minitab does not give any recommendations for choosing base values, but some other statistical softwares do, such as *use a 5-digit odd integer*, which is what I personally use. In any case, be sure to choose a base value and note what it is. If you later want to *duplicate* a random sample, you must use the same base value; this will generate exactly the same set of random numbers. If you want to generate a second random sample, and you want it to be independent of the first random sample, you must use a different base value.

B.3 Manipulating Data in the Worksheet

We often need to move data around in the worksheet or perform calculations on data in the worksheet. Use the sequence *calc > calculator* to bring up the screen of Figure B.6. There, I have done a calculation involving columns 1 and 2 and stored the result in column 5. In addition to the usual arithmetic functions, this screen lists alphabetically a large number of other

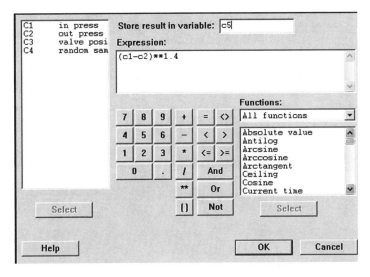

FIGURE B.6
The calculator screen.

useful operations, so that you do not have to remember them or look them up elsewhere.

We often want to put a column of data into rank order (ascending or descending); this is done using *data > sort*, which brings up the screen of Figure B.7. There, I sorted the random sample in ascending order and put

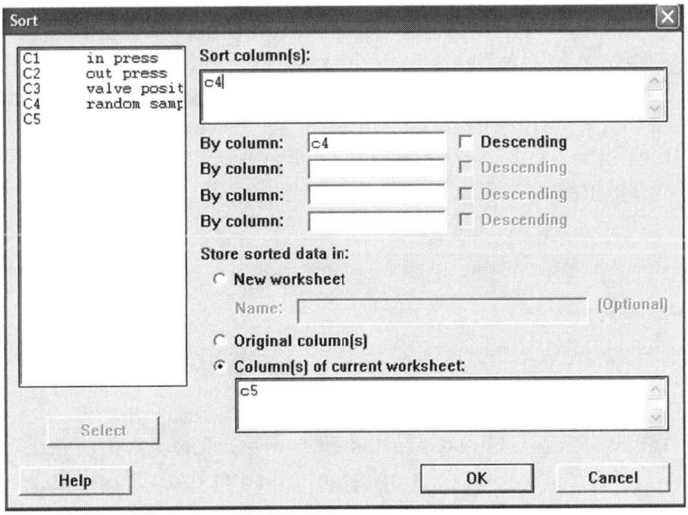

FIGURE B.7
The sort screen.

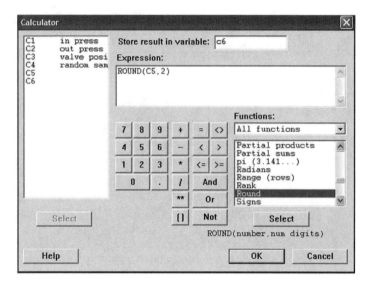

FIGURE B.8
Rounding data values.

these sorted values in column 5. Note that it is possible here to sort one column according to ascending or descending values of a *different* column. Another useful operation is *rounding* numerical values to a desired number of decimal places; this is done from the calculator *(calc > calculator)* by scrolling down the listed functions to find *round* and then entering that function in the *expression* space. You can round a single number or an entire column. In Figure B.8, I rounded all of column 5 (the sorted random values) to two digits to the right of the decimal place and put these rounded values in column 6. The *round* function is quite versatile; consult the online help for a detailed description of its other functions. A drop-down menu under *data* lists many other operations such as copying data from one column to another, subsetting and merging worksheets, and stacking and unstacking columns.

B.4 Graphing Tools for General Graphing

We need to be able to produce our own graphs for routine purposes and also use the special-purpose graphs that are offered with certain statistical operations. Here we cover only general-purpose graphing, leaving the special-purpose graphs for later when we discuss the specific statistical tools involved. Figure B.9 shows the wide variety of graphs offered; we discuss only a few that fit the *general-purpose* category.

FIGURE B.9
List of graph types.

Under *scatterplots* (Figure B.10), we would usually use *with connect line* for our *ordinary x-y* type of graphing. Clicking on OK in Figure B.10 brings up a dialog box (Figure B.11) which allows you to set up the graph. You can choose the *x* and *y* variables that you want to plot from the list of data columns shown at the upper left. You can type them in or click on them in the list and use *select* to enter them into the x or y box. If you want to produce multiple graphs, either overlaid on a single sheet, or as separate graphs, you can use the rows of *x-y* variables numbered 1, 2, 3, etc. Clicking on *scale, labels, data view, multiple graphs*, or *data options* opens dialog boxes, which allow you to customize things like tick marks; often the defaults will be acceptable, at least for a first view of the graph. Also, when the graph appears, there is a graph *editor* which allows easy changes and additions to things like axis labels, titles, symbol types, etc. I often take the defaults just mentioned and then use the graph editor to add or subtract features that I want. One nice feature of the graph editor is that, in addition to providing menus of editing items, one can also just click on a graph item (like the symbols) and a dialog box appears where you can set the type of symbol, size, color, etc. If you had, say, a graph

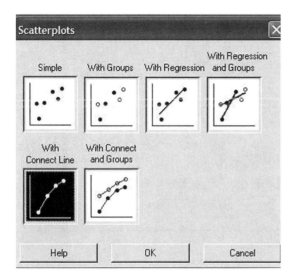

FIGURE B.10
Types of scatter plots.

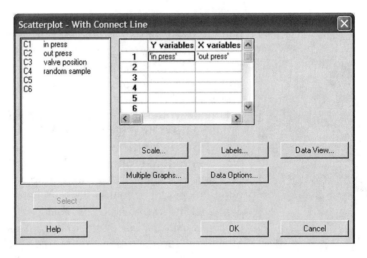

FIGURE B.11
Setting up a graph.

with *both* square and round symbols, if you click once on *any* symbol, *all* the symbols are selected for editing. If you then click again (*not* a double-click) on, say, a square symbol, *only* the square symbols are selected for editing. A *third* click selects only the *one* symbol which you have been clicking on. This is usually easier and quicker than using the drop-down menus of the editor. Figure B.12 shows the graph after I used this method to change some colors (I prefer black/white for most graphs) and changed the default *solid* circle symbols to *open* circle symbols.

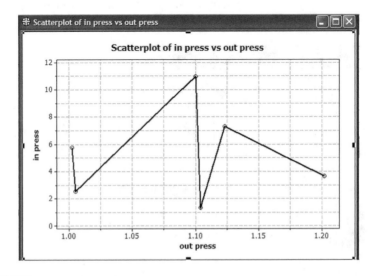

FIGURE B.12
Example scatterplot.

B.5 Checking Physical Data for Conformance to Some Theoretical Distribution

We need to first be clear that it is always *impossible* to prove that any physical data *exactly* follows some theoretical probability density function (distribution). The best that one can do is to show that the data does not deviate radically from the chosen distribution. There are various statistical tests in the literature which give a *numerical value* that indicates the quality of the fit of the data to the theoretical model; the *chi-square goodness-of-fit test** being perhaps the most common. In *Engineering Experimentation: Planning, Execution, Reporting,* I state a preference for *graphical* tests over numerical ones, so we here use that approach. The most commonly used theoretical distribution is the *normal* (also called *Gaussian*), so we will start with that. Graphical tests usually use the *cumulative* distribution function rather than the *probability density* function, and commercial *graph paper* for this test is available. Minitab provides the equivalent of this graph paper in its *probability plot*. The Gaussian cumulative distribution has an s-*shaped* curve, but by suitably distorting the plotting scale of the graph paper (or Minitab's equivalent plot), an infinite-size sample of perfect Gaussian data will plot as a *perfect straight line*, which the human eye/brain easily interprets. The problem, of course, is that we never have samples of infinite size, and *finite-size samples of perfect Gaussian data do not plot as perfect straight lines*. Since the decision as to whether we treat the data as "nearly Gaussian" depends on a visual judgment of the conformance of our data to a straight line, we need some help here, which Minitab's random number generator supplies. Since this generator is for all practical purposes *perfect*, we can generate perfect Gaussian data of the same sample size as our real-world data, and thus get some idea what the plots look like. Since there is *no hope* of ever *proving* that real-world data is Gaussian, such qualitative visual judgments are acceptable.

To get some practice with this method, let us make up some perfect data of different sample sizes and see what the plots look like. Suppose we had some *real* data that consisted of a set of five values with a mean value of 10 and standard deviation of 2. I made up four Gaussian data sets with five samples in each set; all had the same standard deviation (2) and mean value (10) and I put them in columns 7, 8, 9, and 10. I then set up a *probability graph* with four panels (Figure B.13), showing the "probability graph paper" plot for each of these data sets. We would now visually compare the probability plot of our real data with those of the four sets of perfect Gaussian data. If our real data looked no worse than any of the four ideal plots, we would from there on treat our real data as if it were Gaussian. The curved lines on either side of the

---

* E.O. Doebelin, *Engineering Experimentation: Planning, Execution, Reporting*, McGraw-Hill, New York, 1995, pp. 55–58.

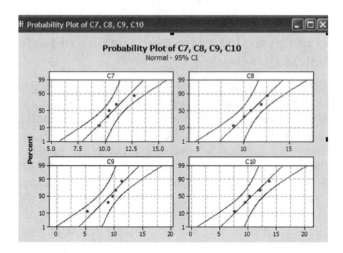

FIGURE B.13
Four sample of perfect Gaussian data.

best fit straight line represent 95% confidence intervals, which in my opinion are of little help in making our decision. They can be deleted with the editor.

Summing up, if we have a set of physical data entered into some column, we should first compute the mean and standard deviation of that data (we will show how shortly). If we want to check for conformance to the Gaussian distribution, we then generate a set of perfect Gaussian data with the same sample size, mean, and standard deviation. In Figure B.14, note that column c12 was generated as perfect Gaussian data, and we requested the same mean and standard deviation (9.524 and 4.156) as was found for the *unknown*

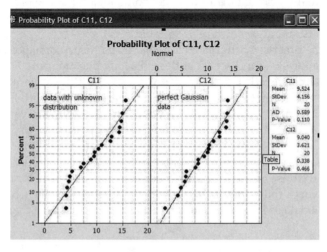

FIGURE B.14
Comparing non-Gaussian and Gaussian data.

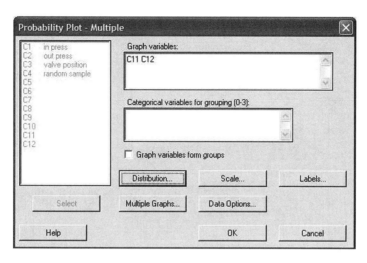

FIGURE B.15
The probability plot.

data (column c11), but the sample generated had *different* values. This is to be expected; a finite sample drawn from the perfect distribution will not have the same mean and standard deviation as the distribution itself. Note that, in Figure B.14, I have requested that the confidence interval curves not be included; I would recommend this as standard practice.

We see in the graph of the *unknown* data, a fairly obvious s-*shaped* curve, whereas the Gaussian data conforms well, though not perfectly, to the ideal straight line. Most engineers would take these displays as an indication that the unknown data is very likely not Gaussian. The sequence used to obtain Figure B.14, which can be used for all such studies is *graph > probability plot > multiple > ok*. This sequence opens a graphing window (Figure B.15) where several choices must be made. For *distribution*, choose *normal* and for *data display*, choose both *symbols* and *distribution fit* and do not check *show confidence interval*. *Scale, Data Options*, and *Labels* can be taken as the defaults. For *multiple graphs* select *in separate panels of the same graph*.

If you want to compare your data to distributions other than the Gaussian, you need to generate a perfect random sample of that distribution and in the probability plot, just request that distribution rather than the Gaussian.

B.6 Computing Mean and Standard Deviation (and Their Confidence Intervals) for a Sample of Data

This calculation is very often needed and is easily accomplished in Minitab using the sequence *stat > basic statistics > graphical summary*. Figure B.16 shows the resulting display, where I have used the 100-item perfect Gaussian sample

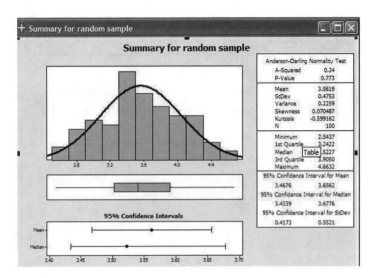

FIGURE B.16
The graphical summary display.

stored in column 4 of the worksheet. A histogram (bar chart) of the data with an overlaid perfect Gaussian density function is automatically included. The confidence intervals for mean and median are shown both numerically and graphically, while that for the standard deviation is shown only as a numerical value. Recall the *meaning* of the confidence interval. When we say that we are 95% *sure* that the true value of the mean or standard deviation lies within the confidence interval, what we mean is explained as follows. If you routinely compute confidence intervals and claim that the true value lies within the confidence interval, you will be right 19 times out of 20, and wrong once out of 20. Some additional statistical measures (which we rarely use) are also included in this display.

B.7 Transforming a Non-Gaussian Distribution to a Gaussian One

When we decide that our graphical test indicates that the distribution is not close to Gaussian, we can try other distributions to see if we get a better "fit." Another approach is to *transform* the data in some way so that the transformed version is closer to Gaussian. A very common successful transformation is routinely used with the fatigue failure of materials. For a given cyclically reversed applied stress, the lifetime in cycles to failure exhibits extreme statistical scatter and the distribution is clearly not Gaussian. It was found years ago that if we plotted the *logarithm* (either 10-base or natural) of the life, rather than the life itself, then the probability plot was close to Gaussian. (An example with actual fatigue data is found in *Engineering Experimentation: Planning, Execution, Reporting*.)* If any transformation is found successful, we can use on our transformed data all the statistical tools based on the assumption of "normality." We can then make statements about the *original* physical data by doing the *inverse* transformation.

When our data is clearly not Gaussian and we have no idea as to what kind of transformation would help, we are faced with *blindly* trying various mathematical operations (square, cube, square root, log, power law, etc.) which can be very time consuming. Fortunately a *generic* type of transformation, the *Box–Cox*, has been found to often work and it is easily tried using Minitab. The relation between the original data y and the transformed y_t is

$$y_t \triangleq y^\lambda \tag{B.1}$$

where λ is a positive or negative number automatically found by the software. Non-Gaussian distributions sometimes arise when we work, not with the raw data, but with some quantity *computed* from the raw data. For example, we might be mass producing shafts with a specified diameter and tolerance and measuring the diameters of these shafts. The distribution of the shaft diameters might be close to Gaussian, but if we really are interested in the cross-sectional *area* of the shafts (because of a stress calculation), the distribution of *area* might not be as close to Gaussian as was that of diameter. In fact, it is known that a quantity (dependent variable) which is a *linear* function of several other quantities (independent variables) will have a Gaussian distribution if all the independent variables have a Gaussian distribution. However, if the relation between the dependent and independent variables is *nonlinear*, then the dependent variable will *not* have a Gaussian distribution.

When the standard deviations of the independent variables are a small percentage of their mean values, a widely used approximation applies. This approximation relies on the well-known fact that any nonlinear function can be closely approximated, in the near neighborhood of a chosen *operating point*, by a linear function. Thus in our shaft example, if the diameter tolerance is small compared to the nominal diameter and if the diameter has a distribution close to Gaussian, the shaft area distribution may also be close to Gaussian even though the relation is nonlinear. We can use Minitab's random number generator to simulate real data and get some experience with such questions. I generated a Gaussian sample of 100 diameter values

---

* E.O. Doebelin, *Engineering Experimentation: Planning, Execution, Reporting*, McGraw-Hill, New York, 1995, pp. 58–61.

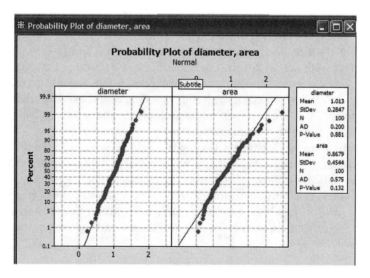

FIGURE B.17
Probability plots for shaft diameter and area.

with mean value of 1.0 in. and standard deviation of 0.3 in. (clearly an unrealistic standard deviation, but useful in demonstrating the above claims). Then I computed the 100 areas that go with these diameters. The probability plots for this data (Figure B.17) show that the diameter variable is "close to Gaussian" (actually it *is* Gaussian) but the area shows non-Gaussian behavior at the two ends of the curve. I then repeated this test but with a more realistic diameter standard deviation of 0.005 in. giving the graphs of Figure B.18. Now, even though the area/diameter relation is still nonlinear, both distributions appear close to Gaussian. Since we know that the diameter distribution *is* Gaussian, the fact that both plots look very similar in shape means that the area distribution is "as Gaussian" as the diameter distribution, which we know is Gaussian because we generated it from a nearly perfect random number generator.

For the area data of Figure B.17, if we had some physical data that looked like this, we would take it to *not* be Gaussian and might thus try the Box–Cox transformation on it. To invoke this transformation use the sequence *stat > control charts > box-cox*. Before you use this sequence, if you want Minitab to actually display and *use* the *optimum* value of lambda to transform the data, you need to invoke the sequence *tools > options > control charts and quality tools > other* and uncheck the box labeled *use rounded values for Box–Cox transformation when possible*. If you do not do this, Minitab will find the *optimum* value of lambda but may *not* use it to compute the transformed data. Instead it may use a *rounded* value for lambda. To make sure what lambda value was

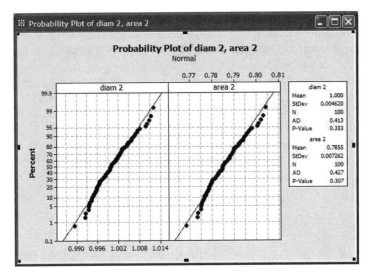

FIGURE B.18
Diameter and area probability plots for reduced standard deviation.

actually used, you can compute lambda from the columns of original and transformed data using the ratio of the logarithms.

Figure B.19 shows the dialog box that appears after you enter *stat > control charts > box-cox*. You need to enter the variable (column) that you want transformed; *area* in this example. Take the defaults for everything else,

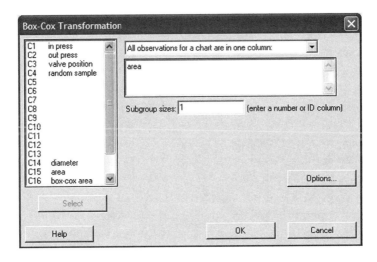

FIGURE B.19
The Box–Cox transformation display.

as shown. For *options* select *optimal or rounded lambda,* and you need to enter the column where you want the transformed data to be stored. After running this routine the plot of Figure B.20a or b appears, depending on whether you let Minitab use rounded lambda's or not, as explained above. The graph displays the *steps* that the method went through to find the *best* value of the transformation parameter λ, which in Figure B.20a came out to be the rounded value 0.5000. This is the value that is then used to compute the transformed *area* data. Note that in this case, the routine was rather clever in that a lambda of 0.50 means that it took the *square root* of the area

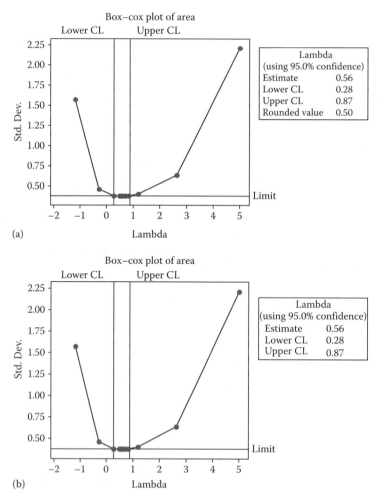

FIGURE B.20
Two different ways to specify lambda.

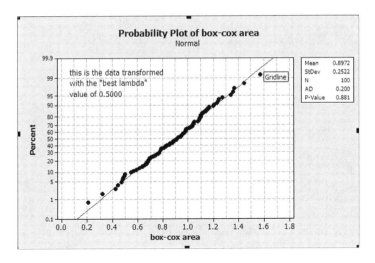

FIGURE B.21
Box–cox transformation is successful.

data, exactly what it takes to *linearize* the squaring operation that originally *made* the area data non-Gaussian. We, of course, did *not tell* the routine that it should take a square root; it was smart enough to deduce this. In Figure B.20b, we insisted that Minitab use the *optimum* value 0.56, rather than a *rounded* lambda value.

For *general* data the non-Gaussian nature often is *inherent* and not the result of some operation such as squaring, but the routine still will often work well. For the c11 data of Figure B.14, which was generated from an ideal *uniform* distribution, the Box–cox transformation does no good at all. Try it some time if you are curious. Uniform distributions have a *rectangular* distribution function (every value in the specified range is equally likely), much different from the *bell-shaped curve* of the Gaussian. The two curves are so fundamentally different that the transformation has little chance of working. For the *area* data of Figure B.17, the task is more reasonable and the transformation comes up with the good results of Figure B.21.

B.8 Comparing Means and Standard Deviations

We often need to compare two materials or processes to decide which is superior. When we compare, say, the failure stress of two materials, we want to choose the one that has the higher average value. If we are comparing two manufacturing processes with respect to variability, we want the one

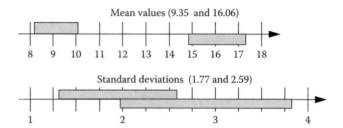

FIGURE B.22
Overlap graphs for mean and standard deviation.

which exhibits the smaller standard deviation. The obvious way to make these comparisons is to find the confidence intervals for the two means or the two standard deviations and see how much *overlap* there is between them. A convenient way to display and evaluate this overlap is by plotting a simple *overlap graph*.

I requested two Gaussian samples of 20 items each with mean values of 10 and 15 and standard deviations of 2 and 3 and used *basic statistics* to find the confidence intervals. These are plotted in Figure B.22 as *overlap graphs*, using my word processor graphics, since Minitab does not provide an "automatic" graph for this display. We see that the mean value confidence intervals show no overlap at all, so the choice of the *better* (higher) mean value material seems clear. Because of the overlap in the standard deviation plots, it is possible that the two standard deviations are actually equal. While the overlap plots are useful, particularly when making a presentation to an audience not expert in statistics, a more statistically correct method of comparing means is available in Minitab as the 2-*sample t test*. Use the sequence *stat > basic statistics > 2-sample t*, select *samples in different columns* and enter the columns where the two samples are located. Otherwise take the defaults. A 95% confidence interval will be computed and displayed in the session window. For this example, the interval went from 5.28 to 8.14, so we are 95% sure that the true value of the difference between the means lies in that range. Sometimes the two samples form a before-and-after pair. That is, the second sample of a material may be the first sample, but after heat treatment to improve the properties. In such a case, we should use the paired t test: *stat > basic statistics > paired t*. For our present example, the new 95% confidence interval is from 5.02 to 8.40.

When comparing standard deviations, the applicable test gives a confidence interval for the *ratio* of the standard deviations, whereas the above test for average values uses the *difference* of the means. Unfortunately Minitab does not provide a *readymade* routine for this test; one has to do some calculations using values from the *F* distribution, so we now need to show how to get those values. As usual, we need to choose a confidence *level* for our interval; most people use 95% (0.95), but other values can be chosen. If your chosen level is x (0.95 in the above example), then the *inverse cumulative probability* that you enter into the F distribution window is $(x + 1)/2$, which would be 0.975 for a 95% interval. The test requires that we "look up" two values from the F distribution. The two sets of data that we wish to compare can each have their own sample size, say n_1 and n_2. From these sample sizes, we compute a *numerator degree of freedom* $v_1 \triangleq (n_1 - 1)$ and a *denominator degree of freedom* $v_2 \triangleq (n_2 - 1)$. The formula for computing the confidence interval is

$$\frac{1}{F_{\left(\frac{x+1}{2}\right), v_a, v_b}} \cdot \frac{s_a^2}{s_b^2} \leq \frac{\sigma_a^2}{\sigma_b^2} \leq \frac{s_a^2}{s_b^2} \cdot F_{\left(\frac{x+1}{2}\right), v_b, v_a} \tag{B.2}$$

where
the sigmas are the *true* values of the standard deviations
the s's are the *actual* standard deviations computed from the samples in the ordinary way

Once we look up the two F values in Minitab, we can *bracket* the ratio of the two true values between a lower and an upper limit, which define the confidence interval. In the subscript of the F value, the first v value is the *numerator* degree of freedom and the second is the *denominator* degree of freedom. You can assign the "a" and "b" subscripts to either of the two samples, but you then have to be consistent throughout Equation B.2. To look up the needed F values, use the sequence *calc > probability distributions > f* to bring up the window of Figure B.23. There I selected *inverse cumulative probability* and entered the numerator and denominator degrees of freedom, which are both 19 since the two samples are both of size 20. When the two samples are the same size, the two F values in Equation B.2 are *the same*, so we need to look up only one F value. When the two samples are of *different* size, we will use the window of Figure B.23 twice, entering the proper values into the degrees of freedom locations. If you want to find the F value for only one confidence level, you could select *input constant* and enter the $(x + 1)/2$ into that box. If you want the F value stored in some empty column, enter that column number into *optional storage*. In my example, I wanted to get F values for three different confidence levels, 0.95, 0.90, and 0.85. All three F values can be computed *at once* by selecting *input column* (c25 in the example) and optional storage in c26. I entered manually into column 25 the values 0.975, 0.950, and 0.925. Running the routine produced the results

c25	c26
$(1 + x)/2$	F
0.975	2.52645
0.950	2.16825
0.925	1.96486

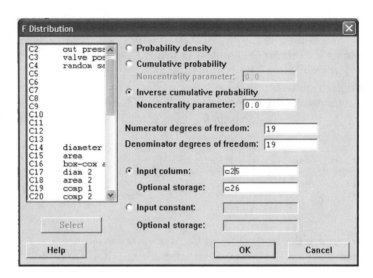

FIGURE B.23
Getting values from the *F*-distribution.

To actually find the desired confidence interval, we need the standard deviations of our two samples; these are 1.772 for the sample in column 19 and 2.592 for that in column 20. For the 95% confidence level, Equation B.2 gives the confidence interval $0.430 < \sigma_1/\sigma_2 < 1.087$; for 90% confidence the limits are 0.464 and 1.007; and for 85% confidence 0.488 and 0.952. The ratio of the standard deviations of the samples is 0.684. For sample sizes larger than 20 the confidence intervals narrow since the *F* value approaches 1.0 as the sample size approaches infinity. That is, σ_1/σ_2 approaches s_1/s_2. If one requires a very narrow confidence interval, one can easily look up the *F* values for large *n*'s to see what sample size is needed.

B.9 Multiple Regression

Probably all the readers of this book have used the simplest form of regression; the fitting of the "best" straight line to a set of experimental data points. *Best* here means that the sum of the squares of the deviations of the data points from the *best line* is the smallest possible for *any* straight line. This is the so-called *least-squares* method of curve fitting, the result of a calculus minimization process. Extensions of this familiar tool include fitting curves other than straight lines and finding the "best" models of relations between a dependent variable y and several independent variables x_1, $x_2, x_3, ..., x_k$. All these problems are easily addressed by standard regression routines found in Minitab and many other software products. The mathematical form of all these applications is

$$y = B_0 + B_1 f_1(x_1, x_2, \ldots, x_k) + B_2 f_2(x_1, x_2, \ldots, x_k) + \cdots + B_n f_n(x_1, x_2, \ldots, x_k) \quad \text{(B.3)}$$

The f functions can depend on any combination of the x's, but the B coefficients *must* appear as multiplying factors, as shown, not as exponents in the functions or in other *nonlinear* ways. This form of equation is called *linear multiple regression* and the solution for the best B values is a routine application of linear algebra, easily computerized. When the B's appear, not as in Equation B.3, but in some nonlinear way, then *nonlinear regression* is needed, a much more complex situation that does not yield to routine calculation but rather depends on iterative optimization methods which may or may not converge, and if they do converge, do not guarantee the best solution. Much statistical software includes one or more nonlinear regression routines, but Minitab does not. I have over the years suggested they add at least one such routine but so far to no avail. Fortunately, most practical problems yield to linear regression. Also, sometimes a *tranformation** can be found which converts a nonlinear problem into a linear one.

B.9.1 Curve Fitting

The simplest application of Equation B.3 is to the curve fitting of x–y data; here all the x's are the single variable x. We can demonstrate this application by *manufacturing* some data, using known f functions and then adding some random noise:

$$y = 2.66 + 3.57x - 2.23x^2 + 5.2e^{0.22x} - 1.35\sin 1.22x \quad \text{(B.4)}$$

We will vary x from 0 to 1.0 in steps of 0.05 and then add some Gaussian random noise with mean 0 and standard deviation 0.1 to each "reading." All this is easily done in Minitab using tools explained earlier and we then act as if we had just run an experiment which produced this data. At least two different situations might arise in practice:

1. We have a theoretical model that predicts the *form* of the f functions, but *not* the coefficients.
2. The x/y relation is strictly an empirical one and we have no idea whether there is some underlying mathematical relation or what it might be.

---

* E.O. Doebelin, *Engineering Experimentation: Planning, Execution, Reporting*, McGraw-Hill, New York, 1995, pp. 214–215.

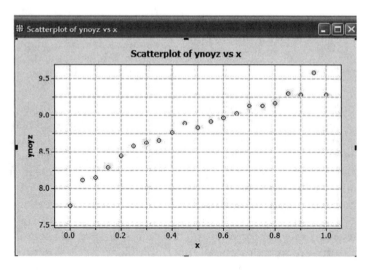

FIGURE B.24
Data made up for curve fitting practice.

The second situation is perhaps the more common, so we address it first. We need to enter the x/y data into two columns of the worksheet, giving the columns appropriate names. We should always then *plot* the data and carefully consider the *shape* of the *curve* that it displays (see Figure B.24). Based on this shape, we choose the *simplest* combination of f functions that seems suitable to fit the data and use the Minitab regression routine to compute the best set of coefficients. Then we produce a graph which displays both the experimental data points and the best-fit curve, to judge the adequacy of the fit. If the fit is inadequate, we try to interpret the nature of the misfit and use that to suggest a second trial. As most readers will know, a common approach is to try *polynomial* functions, raising the power included until a good fit is achieved, except that *too high* a power can result in *over-fitting*, where the fitted curve follows the random fluctuations rather than the underlying smooth relation. Also, recall that if the number of polynomial terms is equal to the number of data points, then we get an *exact* fitting of every point, which with noisy data gives an undesirable result; we have fitted the noise rather than the underlying relation.

Figure B.24 shows the "experimental" data plotted as a scatterplot without connecting the points. Assuming that we are not aware of any theoretical model for this phenomenon, the smooth nature of the general trend suggests a rather low-order polynomial might be a good first trial. Since a straight line is clearly too simple, we try $B_0 + B_1 x + B_2 x^2$. Whatever the form of the functions we choose (in our case, x and x^2), we must compute these functions and enter each into its own Minitab column. We than access the regression routines with *stat > regression > regression*, which brings up the window of Figure B.25. Here we need to explain a number of choices that must be made. Enter into *response* the name or column number of the dependent variable, in our case

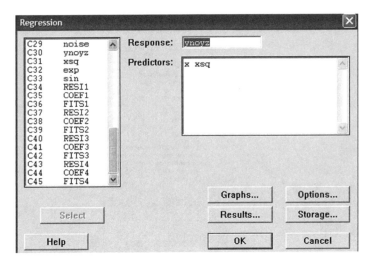

FIGURE B.25
The regression display.

ynoyz, the noisy *y* data. For *predictors*, list the columns of all the functions in your chosen formula, except for B_0, which is assumed present unless deleted by unselecting *fit intercept* under *options*. For *graphs*, take the defaults. For *options* take the defaults, except choose whether you want the constant term (intercept) or not. For *results* select the second item from the top (Regression equation...). For *storage* select *residuals, coefficients,* and *fits*.

Running this routine automatically produces some graphical results, entries in the worksheet, and statements in the session window. It does *not* produce the graph that we really want (the plot of the fitted curve "on top of" the original data points) but we can get it easily from *graph > scatterplot.* Figure B.26 shows this graph; it appears to be a reasonable fit. If we wanted to improve it, we might add the cubic term to the polynomial, but we will not pursue this detail here.

One of the "automatic" plots is that of Figure B.27, which shows several displays involving the *residuals*. In regression, the term *residual* means simply the difference between the actual data point and the fitted curve. The two upper panels are probably the most useful. In the right-hand upper panel, small residuals are, of course, desirable since that means that the fitted curve is close to the actual data, but this is obvious. What is less obvious is that the residuals in this graph should show a *random* pattern, which seems to be true for our example. If a more-or-less obvious *trend* is seen, this can be a clue suggesting a defect in the assumed functions and a need to add some suitable function. If we had tried to fit a straight line, such a trend would be obvious, suggesting the addition of the squared and/or the cubed term. The left-hand upper panel checks the residuals for *normality* (conformance to Gaussian distribution). In our made-up example data, we intentionally used

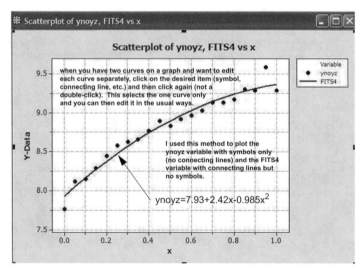

FIGURE B.26
Checking the fitted curve against the data.

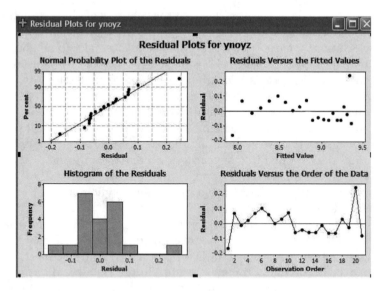

FIGURE B.27
Using the residual plots to evaluate the curve fit.

Gaussian random noise, so the *normality* of this graph should not be surprising. With real-world experimental data, we cannot assume normality, but it often is (approximately) observed in this plot.

When you run the regression routine you also get some results in the session window, the main one being a statement of the equation relating y and x, which I have added to Figure B.26. The session window display is:

Regression Analysis: ynoyz versus x, xsq

```
The regression equation is
ynoyz = 7.93 + 2.42 x - 0.985 xsq
```

Predictor	Coef	SE Coef	T	P
Constant	7.93021	0.05406	146.70	0.000
x	2.4213	0.2505	9.67	0.000
xsq	-0.9847	0.2419	-4.07	0.001

```
S = 0.0905618  R-Sq = 96.5%  R-Sq(adj) = 96.1%
```

The equation coefficients are actually computed to many digits; the equation and table show them rounded off. If you use the equation to compute the curve values (not really necessary since the quantity called FITS gives this), you might not get exactly the right values if you use the rounded coefficient numbers. Accurate coefficient values appear on the worksheet, if you earlier had asked for them to be stored.

The table also gives values for *se coef* (the *standard error* of each coefficient), and T and p, two statistical measures. These quantities are all of little practical utility in terms of judging the quality of the fit. The term R^2 (R-Sq) is called the *coefficient of determination*; the closer it is to 1.0, the better the fit. You might use it to make a decision if you were comparing two fits that *looked* about equally good and you needed to choose one. Usually, however, we judge the fit visually from graphs like Figure B.26.

Usually we are curve fitting in a situation where no underlying theoretical formula is available and we just want the simplest function which gives a good fit to the data. Sometimes, however, we *do* have some underlying theory to predict the *form* of the function but not accurate values for the coefficients of the various terms. Let us now use Equation B.4 and our earlier *experimental* data to investigate this situation. The *ideal* situation is where we have *perfect* data (no noise) and we use the regression routine to find the coefficients in Equation B.4, treating them as unknowns which the routine tries to solve for. We need to use the calculator to form a column of numerical values for each of the terms in the function except B_0; that is, for x, x^2, $\exp(x)$, and $\sin 1.22x$. Using the noise-free data for y, we get the following result in the session window:

Regression Analysis: y versus xsq, exp, sin, x

```
The regression equation is
y = 2.66 - 2.23 xsq + 5.20 exp - 1.35 sin + 3.57 x
```

Predictor	Coef	SE Coef	T	P
Constant	2.66000	0.00000	*	*
xsq	-2.23000	0.00000	*	*
exp	5.20000	0.00000	*	*
sin	-1.35000	0.00000	*	*
x	3.57000	0.00000	*	*

```
S = 0  R-Sq = 100.0%  R-Sq(adj) = 100.0%
```

We see that the routine was able to discover the correct values of all the coefficients, so a plot of the fitted curve is a perfect fit for the data, and the residuals are all essentially zero.

If we now use the *noisy* data we get quite a different result, even though the noise (as seen in Figure B.26) is not excessive.

Regression Analysis: ynoyz versus x, xsq, exp, sin

```
The regression equation is
ynoyz = 9295 + 2150 x + 215 xsq - 9287 exp - 84.1 sin

Predictor     Coef      SE Coef      T        P
Constant      9295      5497       1.69     0.110
x             2150      1265       1.70     0.108
xsq           215.3     130.1      1.65     0.117
exp           -9287     5497       -1.69    0.111
sin           -84.05    44.87      -1.87    0.079

S = 0.0752324  R-Sq = 97.9%  R-Sq(adj) = 97.3%
```

We see that the computed coefficients are *vastly* different from their *true* values, however the R^2 value is quite close to 1.0, indicating a good fit, as is verified by Figure B.28. I ran the regression with the noise reduced to 10% of the original noise and got a better fit, but the computed coefficients were still far from the correct values. We see, therefore, that for real-world data (which always has some noise) the regression method will usually *not* be able to extract accurate coefficient values for an assumed functional form, but it may produce coefficients which give a give a good curve fit.

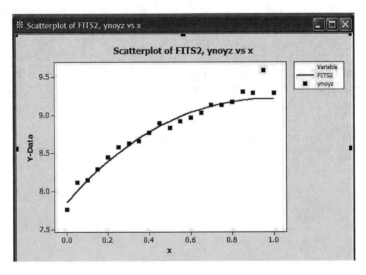

FIGURE B.28
Wrong coefficients can give a good fit.

B.9.2 Model Building

An application of regression that is perhaps of more economic importance than curve fitting is *model building*. Here, we have some machine or process that involves a *dependent variable* (called *response*), that relates to product *quality* or cost, and several *independent variables* (called *factors*) which we are able to set within a range of values. Usually there are certain combinations of factor settings that produce an *optimum* response (such as high quality or low cost) and we would like to locate this *sweet spot*. Unfortunately, many practical machines and processes are too complicated to allow accurate theoretical modeling of the *response/factor* relationships, so carefully designed *experiments* are used to develop practically useful empirical relations. Even if optimum points do not exist, the developed relations are useful in identifying which factors are most effective in changing the response, so that we do not waste time adjusting those factors which have been found to have little effect. Methods for planning, executing, and analyzing results for such experiments are discussed in *Engineering Experimentation: Planning, Execution, Reporting.*[*]

The function that we assume for our model can take many forms, but the most common is an *additive* form that involves a combination of the factors themselves (main effects) and products of the factors (interaction terms). For example, if we have picked three factors that we feel have some influence on the response, we might try a model such as

$$y = B_0 + B_1 x_1 + B_2 x_2 + B_3 x_3 + B_4 x_1 x_2 + B_5 x_1 x_3 + B_6 x_2 x_3 \qquad \text{(B.5)}$$

Interaction refers to the fact that the effect of one factor may vary depending on the value of another factor or factors. Higher order interaction terms such as $x_1 x_2 x_3$ are possible but rarely used. In designing our experiment to find the numerical values of the B's, a naive approach would say that all that is needed (since Equation B.5 has seven unknowns) is to gather seven sets of x and y values; then we can solve seven linear algebraic equations in seven unknowns using available computerized methods, such as Minitabs regression routines. This would generate a model that *exactly* fits the given set of data. If we instead had *more* than seven sets of data, then the routine must use a least-squares approach, and the model would be the best fit *among* the data points, rather than an exact fit *through* the data points. *Designing an experiment involves choosing sets of factor values that constitute a *run* and then deciding how many such runs to make. One can just use *common sense* to design the experiment. That is, in any specific application, we would know the overall *range* of each factor, thus the factor values must fall somewhere in that range. From run to run, we might want to make *large* changes

---

[*] E.O. Doebelin, *Engineering Experimentation: Planning, Execution, Reporting,* McGraw-Hill, New York, 1995, Chapter 4.

in the factor values since the effect of a factor will be more obvious if the change in the factor is large. Beyond these obvious considerations, the choice of a specific set of runs becomes less clear. Fortunately, many years of study and application have provided some concrete and well-tested guidelines for designing such experiments based on statistical principles. Software such as Minitab provides *readymade* experiment plans that are useful for the majority of applications.

Many practical model-building studies begin with so-called *screening experiments*. Here, based on our knowledge of the specific machine or process, we have decided on a tentative list of factors that we feel might be influential on the response variable. We also have established the practical upper and lower limits of each factor. The experiment plan then consists of some combination of runs in which the factors take on only their highest and lowest values. The most comprehensive such experiment is the *full factorial*, where every possible combination of the high and low values of every factor is exercised. If there are n factors, then there must be 2^n runs to cover all the combinations. For small numbers of factors, the full factorial experiment may be feasible; for large numbers it soon becomes impractical (8 factors, for example, requires 256 runs). To reduce the number of runs needed, various types of *fractional factorial* designs have been developed. The price paid for a more frugal experiment is the phenomenon of *confounding*. In a full factorial experiment, we are able to distinguish both main effects and interactions. This capability is lost, to some extent, in fractional factorial designs; the main effects and some interactions are said to be *confounded*. The degree of confounding is given by the *resolution level* of the design; common designs are designated as Resolution III, Resolution IV, or Resolution V. Resolution III designs have the smallest number of runs, but can only isolate the *main effects* of the factors; interaction terms cannot be reliably identified. Resolution IV designs require more runs, but can find both main effects and two-way interactions. Higher order (three-way, four-way, etc.) interactions are confounded and thus not identifiable. Resolution V designs can find main effects, two-way interactions, and three-way interactions. Since three-way interactions are not common, most fractional factorial designs are either Resolution III or IV. Software that offers various designs will identify the resolution level. Initially, when we have a fairly long list of potential factors, we often look for only main effects with a Resolution III design. When this first experiment narrows the field to fewer factors, we may switch to Resolution IV to look for interactions.

To set up an experiment, once you have decided on the list of factors to be studied, enter the sequence *stat > doe > factorial > create factorial design*, which brings up the screen of Figure B.29. In this screen, you need to enter the number of factors (I used six here) and usually you take the default *type of design*, 2-*level factorial (default generators)*. You then click on *Display Available Designs* to get the screen of Figure B.30. This display is very useful since it shows both full factorial and fractional factorial designs for experiments

Create Factorial Design
Type of Design
2-level factorial (default generators) (2 to 15 factors)
2-level factorial (specify generators) (2 to 15 factors)
Plackett-Burman design (2 to 47 factors)
General full factorial design (2 to 15 factors)
Number of factors: 6
Display Available Designs...
Designs...
Factors...
Options...
Results...
Help
OK
Cancel

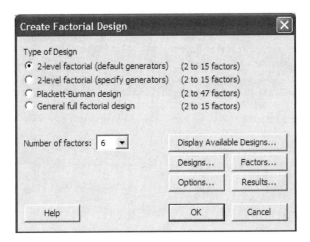

FIGURE B.29
Creating a factorial design.

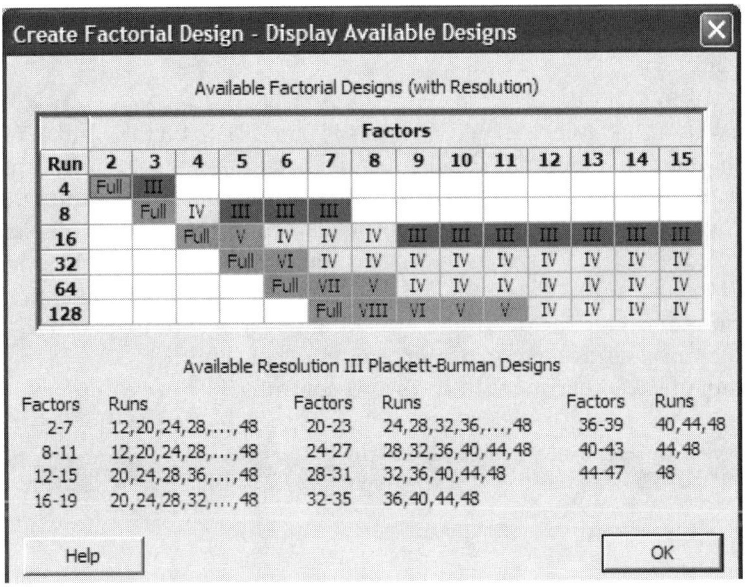

Create Factorial Design - Display Available Designs
Available Factorial Designs (with Resolution)
Factors
Run 2 3 4 5 6 7 8 9 10 11 12 13 14 15
4 Full III
8 Full IV III III III
16 Full V IV IV IV III III III III III III III
32 Full VI IV IV IV IV IV IV IV IV IV
64 Full VII V IV IV IV IV IV IV IV
128 Full VIII VI V V IV IV IV IV
Available Resolution III Plackett-Burman Designs
Factors Runs
2-7 12,20,24,28,...,48
8-11 12,20,24,28,...,48
12-15 20,24,28,36,...,48
16-19 20,24,28,32,...,48
Factors Runs
20-23 24,28,32,36,...,48
24-27 28,32,36,40,44,48
28-31 32,36,40,44,48
32-35 36,40,44,48
Factors Runs
36-39 40,44,48
40-43 44,48
44-47 48
Help
OK

FIGURE B.30
Choosing a specific design.

with a wide range of factors, and also lists their resolution (III–VIII). If we are looking only for main effects, we can use resolution III, which in this case requires 8 runs. If instead we want to also look for two-way interactions, this takes resolution IV and 16 runs. The full factorial, which allows all the interactions, takes 64 runs. Let's decide to look only for main effects; we then click

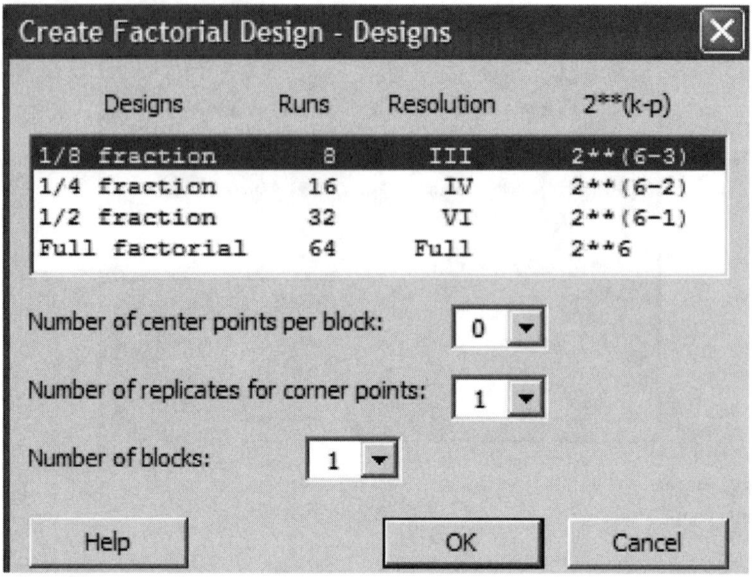

FIGURE B.31
The design screen.

on that cell in the table, which highlights it. Then click on OK, which returns you to the screen of Figure B.29. There you click on Designs, which brings up the screen of Figure B.31, where you take the defaults by clicking OK. This returns you to Figure B.29, where you click on Factors, which displays the names (A, B, C, D, E, F) of the factors and the high (1) and low (–1) coded values of the factors. A click on OK returns you to Figure B.29 where you click on Options and take the defaults by clicking OK which returns you to Figure B.29. Clicking on OK here brings up a worksheet (Figure B.32) which shows the layout of the experiment in rows and columns.

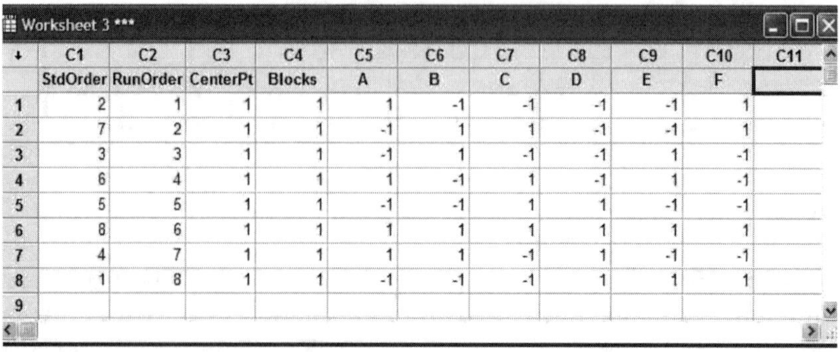

	C1	C2	C3	C4	C5	C6	C7	C8	C9	C10	C11
	StdOrder	RunOrder	CenterPt	Blocks	A	B	C	D	E	F	
1	2	1	1	1	1	-1	-1	-1	-1	1	
2	7	2	1	1	-1	1	1	-1	-1	1	
3	3	3	1	1	-1	1	-1	-1	1	-1	
4	6	4	1	1	1	-1	1	-1	1	-1	
5	5	5	1	1	-1	-1	1	1	-1	-1	
6	8	6	1	1	1	1	1	1	1	1	
7	4	7	1	1	1	1	-1	1	-1	-1	
8	1	8	1	1	-1	-1	-1	1	1	1	
9											

FIGURE B.32
Definition of the 8 runs.

Column 2 refers to the (randomized) order in which the 8 runs will actually be carried out. Column 1 gives the *standard* order as defined in statistics literature, and is displayed only for reference purposes; we will not use this column in any way. Columns 3 and 4 are also for reference only; we will not need to consider them further. The six factors have been given letter names; I prefer to replace these with *physical* names related to the actual variables in the experiment. Similarly, the *high* and *low* values of the factors are coded as 1 and −1, a standard statistical practice. I prefer, as in the example below, to replace these with the actual physical high and low values for each factor. After we complete all 8 runs, we need to enter the values of the response variable into any empty column such as c11.

To show a numerical example, I will now "manufacture" some data using a known functional relation between the response and the factors and also assume a high and low value for each factor. Using this formula, I can fill in the worksheet just as if I were running a real experiment.

$$y = 12.3 + 5.22A - 13.7B + 1.83CC - 20.6D + 2.04E + 15.8F \qquad \text{(B.6)}$$

The respective high and low values are $A(3,7)$, $B(1,12)$, $CC(6,9)$, $D(82,104)$, $E(15,25)$, $F(6,12)$. (Note that I have replaced C with CC; Minitab objected to the use of C, perhaps this notation conflicted with the standard column notation.) Using the runs defined in Figure B.32, we can calculate the y values for each of the 8 runs and enter these into column 11 of the worksheet. With the worksheet completely filled in, we can then use regression to build models of the relation. We start with perfect (noise-free) data, and as one might expect, the routine finds almost perfect values of the coefficients. Note that we have *eight equations in seven unknowns*, so that least squares must be used and we should not expect perfect results once we add some noise to the data.

B.9.2.1 Best Subsets Regression

There is a version of regression called *best subsets*, which is offered in most statistics software, that we have not mentioned so far and that can be quite useful in general model building. It uses the same worksheet data that ordinary regression uses, so we do not have to alter our worksheet to try it. What it does is that it finds the "best" models of various *sizes*. That is, when we begin model building, we usually *do not know* which of our assumed factors are really important and which might actually have little effect on the response variable. The simplest models would be those with *only one* factor, but *which* of our six factors would be the best if we insist on using only one factor? Best subsets regression answers this question for models of *all* possible sizes. That is, it finds the best 1-variable model, the best 2-variable model, the best 3-variable model, etc. Since it is quick and easy to use, we often do it *before* we run "ordinary" regression on the 6-variable model. Best subsets does not actually run the regressions; it only identifies the best models, and you can then run

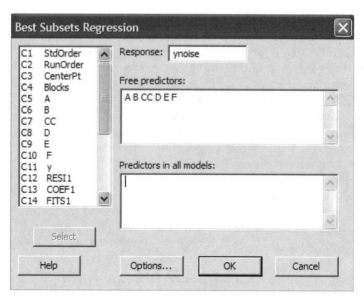

FIGURE B.33
The best subsets screen.

ordinary regressions on those models of interest. To use the *best subsets* routine, enter *stat > regression > best subsets*, which brings up the screen of Figure B.33. Here you have to enter the response variable (*ynoise* in our example) and the list of factors ("free predictors"). Leave *predictors in all models* blank. Click on *options* to get Figure B.34's screen and use the settings shown there for most applications, then click OK to return to Figure B.33, and OK there to run the routine. The sessions window then shows all the various "best" models.

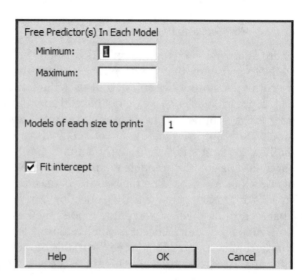

FIGURE B.34
The free predictor screen.

Best Subsets Regression: ynoise versus A, B, CC, D, E, F

```
Response is ynoise
Vars   R-Square   A   B   CC   D   E   F
 1       72.8                  X
 2       83.0      X           X
 3       95.7      X           X       X
 4       98.5   X  X           X       X
 5       99.9   X  X           X   X   X
 6      100.0   X  X   X       X   X   X
```

We see that the best 1-variable model used only factor D, while B and D were used for the 2-variable model, and B, D, and F for the 3-variable model, A, B, D, F for the 4-variable model, A, B, D, E, F, for the 5-variable, and of course all six for the 6-variable. The column of R-Sq values shows the improvement in fit as more factors are included in the model. It seems that B, D, and F would give a good fit with only three factors. To explore this with more detail, one now runs ordinary regressions using any of these models that is of interest. For the 3-factor model, we get

Regression Analysis: ynoise versus B, D, F

```
The regression equation is
ynoise = - 291 - 13.2 B - 17.2 D + 22.5 F

Predictor       Coef      SE Coef        T         P
Constant      -290.6       154.4      -1.88     0.133
B           -13.151       3.084      -4.26     0.013
D           -17.228       1.542     -11.17     0.000
F            22.463       5.653       3.97     0.016

S = 47.9694    R-Sq = 97.5%    R-Sq(adj) = 95.7%
```

Figure B.35 shows that this model does give a good fit and thus that factors A, CC, and E are not as important as B, D, and F. If we want to use all six factors in our model, we get the results shown in Figure B.36. We see that the routine does not find the *correct* values of the coefficients, but it does give a model that fits the data well.

Regression Analysis: ynoise versus A, B, CC, D, E, F

```
The regression equation is
ynoise = - 311 + 14.6 A - 13.2 B + 2.07 CC - 17.2 D - 3.42 E + 22.5 F

Predictor       Coef      SE Coef         T         P
Constant     -310.578       0.488     -636.29     0.001
A            14.5674      0.0214      680.89     0.001
B           -13.1507      0.0078    -1690.34     0.000
CC            2.07113     0.02853      72.60     0.009
D           -17.2283      0.0039    -4428.94     0.000
E            -3.41787     0.00856     -399.38     0.002
F            22.4629      0.0143     1574.89     0.000

S = 0.121027    R-Sq = 100.0%    R-Sq(adj) = 100.0%
```

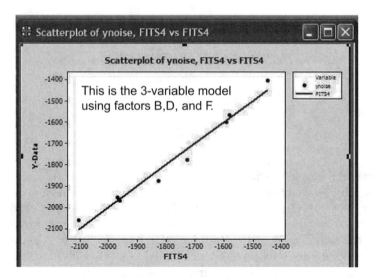

FIGURE B.35
Three-factor model gives good fit.

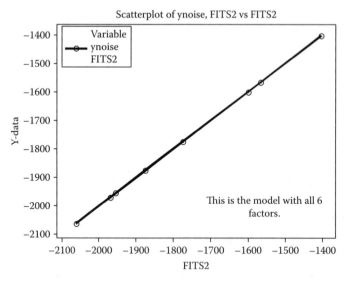

FIGURE B.36
Six-factor model gives improved fit.

B.9.2.2 Validation Experiments

When the experiment involves a fairly large number of runs, it may be prudent to apply the concept of the *validation experiment*. When we build a model from a certain combination of factor values, we really want it to predict the response variable well for combinations that were *not* used to build the model. One way of doing this is to use only about 75% of the runs to build the model and then *check* the remaining 25% to see whether the model predicts the response variable adequately.

If one decides to do this, then the question is *which* of the runs do you reserve for the validation check? Statistical software with which I am familiar (including Minitab) does not provide guidance on this choice, so we just use *common sense* here as best we can, using any knowledge we have of the process or machine we are modeling. Once the choice is made, Minitab *does* provide an easy way to do the validation calculations. On the worksheet, you need to use as many *empty* columns as there are factors and as many rows as there are validation runs to enter the data for those runs. You must use the same right-to-left sequence for these columns as you used for the "ordinary" runs on the worksheet. Using *stat > regression > regression* to get to the regression window, you then click on OPTIONS, which brings up a window that has a box labeled *prediction intervals for new observations*. You enter into this box the column numbers (say c12, c13, c14, c15 if there are four factors) where you have stored the data for the validation runs. Take the default (95%) for the confidence level. When you now run the regression, Minitab will compute the response values *predicted by the model* for all the validation runs and you can compare these with the response values actually measured in the experiment.

B.9.2.3 Analyze Factorial Design

When you enter *stat > doe > factorial design*, the drop-down menu offers ANALYZE FACTORIAL DESIGN, which uses the analysis of variance (ANOVA) method of analysis rather than regression. We have not explained this possibility because we prefer the regression method of analysis to the *ANOVA* method. Both methods are closely related but ANOVA, in my opinion, is needlessly complicated and rarely offers any advantage, especially for users who are not *statistics experts*. You thus should have no need to use this option, either for curve fitting or model building.

B.9.2.4 Nonnumerical Factors

While the response variable must *always* be given as a numerical value, some applications will involve factors that are nonnumerical. For example, we might be interested in the effect of different cutting fluids, tool materials, part materials, and machine manufacturer on the surface roughness of parts produced by milling machines. We would run experiments for various combinations of the factors and record the measured surface roughness. All four of these factors are nonnumerical; we can assign them identifying names (fluid A, fluid B, etc.) but not numerical values. Of the various ways to handle such situations, I prefer the *dummy variable* method because it allows us to continue to use the simple regression method of analysis. Actually the

ANOVA method was originally developed to deal with such experiments, but the dummy variable and regression approach is simpler to use and interpret. Minitab allows the use of nonnumerical factors but you might find the dummy variable/regression scheme easier; it is explained (with examples) in *Engineering Experimentation: Planning, Execution, Reporting.*[*]

B.9.2.5 3-Level Experiments (See Also Chapter 1)

For the initial *screening* experiments, it is conventional to use only two levels of all the factors: *high* and *low*. Such experiments are frugal but are not capable of revealing any "curvature" in the relations between factors and response. To study such effects, one needs to use at least three levels of the factors: low, medium, and high. That is, two points determine a straight line, whereas three can allow curvature. In Minitab, perhaps the easiest way to plan three-level experiments is to use the Taguchi designs. Use *stat > doe > taguchi > create taguchi design* to access these designs, which allow more than two levels, and each factor can have its own number of levels if that is wanted. The selection process is fairly easy to follow, so we will not give any detailed explanation or examples. These designs are not listed according to their resolution, but in one of the steps you can say that you want to include interaction terms and it lists the interaction terms that are available. You select the ones you want from the list and it tells you whether you have selected too many, so you can reduce the list. When the number of interactions requested is acceptable, it produces a new worksheet (similar to our Figure B.22, but with three levels for each factor) from which you can run the experiment and fill in the numerical data. Again, there is an ANALYZE TAGUCHI DESIGN option which requires some specialized knowledge to use and interpret, but we ignore this and just use our usual regression method to build the models and evaluate the results. That is, whether we use the factorial designs or the Taguchi designs, we use the Minitab routines only to *design* the experiment; we *do not* use the ANOVA or TAGUCHI analysis routines provided. Once the experiment is designed, we *always* use the same regression method from there on. That way we do not have to learn and remain *expert* in using *three* analysis methods; we need only one.

---

[*] E.O. Doebelin, *Engineering Experimentation: Planning, Execution, Reporting*, McGraw-Hill, New York, 1995, pp. 239–250.

Index